ECHINODERMS IN A CHANGING WORLD

PROCEEDINGS OF THE 13TH INTERNATIONAL ECHINODERM CONFERENCE, UNIVERSITY OF TASMANIA, HOBART TASMANIA, AUSTRALIA, 5–9 JANUARY 2009

Echinoderms in a Changing World

Editor

Craig Johnson
Institute for Marine and Antarctic Studies, University of Tasmania, Hobart, Tasmania, Australia

CRC Press
Taylor & Francis Group
Boca Raton London New York Leiden

CRC Press is an imprint of the
Taylor & Francis Group, an **informa** business

A BALKEMA BOOK

Cover photo: Test of *Tripneustes gratilla* cast as beach wrack, Western Australia
(photo credit: Jemina Stuart-Smith)

CRC Press/Balkema is an imprint of the Taylor & Francis Group, an informa business

© 2013 Taylor & Francis Group, London, UK

Typeset by V Publishing Solutions Pvt Ltd., Chennai, India

Published by: CRC Press/Balkema
P.O. Box 447, 2300 AK Leiden, The Netherlands
e-mail: Pub.NL@taylorandfrancis.com
www.crcpress.com – www.taylorandfrancis.com

ISBN: 978-1-138-00010-0 (Hbk)
ISBN: 978-0-203-63156-0 (eBook)

Echinoderms in a Changing World – Johnson (ed)
© 2013 Taylor & Francis Group, London, ISBN 978-1-138-00010-0

Table of contents

Foreword

Ever since ecologist Eugene Stoermer first coined the term, an increasing number of researchers from a diversity of scientific disciplines have acknowledged that planet Earth has entered a new geological era, the Anthropocene, as a result of Man's activities changing Earth's environment. The rate of this global environmental change is without parallel in the last 300 million years, and it affects oceans and the terrestrial environment alike. In the ocean, increasing temperatures, carbon dioxide levels and acidity receive most attention as the harbingers of this shift, but there are also many other facets. Change in the distribution and dynamics of heat energy is manifest as alterations in wind patterns and oceanic circulation, which in turn influences levels of trace metals and other nutrients, while levels of chemical contaminants, rates of marine habitat degradation or outright destruction and, for some marine species, rates of direct harvesting are generally unprecedented.

A key question is how the changes in ocean physics, chemistry and habitat characteristics, sometimes combined with effects of harvesting, will influence marine organisms and the ecologies in which they are embedded. Are echinoderms, with their calcareous structures in larval and adult form, and their largely sedentary habit as adults, particularly at risk? Will they show large changes in abundance and distribution? Will there be extinctions or can they adapt to change in the physical and chemical environment, or respond through phenotypic plasticity? What can we learn about environmentally induced changes to echinoderms from the rich fossil record of this group? What are the most useful and parsimonious means to measure and monitor changes in fossil and extant echinoderm populations? Are there means to mitigate deleterious effects of environmental change? How will changes in the distribution and abundance of echinoderms directly and indirectly affect the dynamics of the ecological communities in which they live?

In one way or another, the majority of talks and posters presented at the 13th International Echinoderm Conference in Hobart, January 2009, touched on these important questions, and the papers presented in this volume comprise an interesting sample of the topics covered. With a little over 100 delegates, it was a relatively small meeting, which may explain it being a particularly lively one. The community of echinoderm scholars is highly engaged and active, and the IEC meetings are remarkable in the way they bring together researches from a great diversity of disciplines, united in their focus on this fascinating group of animals. The 13th IEC was no exception.

The organizers are grateful for the assistance of the anonymous reviewers who provided advice to correct and improve papers (note that works presented in this volume as abstracts only were not reviewed externally). Dr. Jemina Stuart-Smith provided wonderful support to every aspect of the conference organization, and Dr. Jessica Melbourne-Thomas played a crucial role in supporting the editorial process. The conference was managed superbly by Conference Design Pty Ltd, and could not have happened without the assistance of a large team of volunteers (too numerous to name individually). Finally, the organizers wish to thank the sponsors whose contributions enabled the event to occur at all: The principal sponsor was the Tasmanian Aquaculture & Fisheries Institute, University of Tasmania, while supporting sponsors were CSIRO, the Fisheries Research and Development Corporation (FRDC), Myriax Software, and the BueNet Project.

Craig Johnson
Editor

Plenary papers

Echinoderms in a Changing World – Johnson (ed)
© 2013 Taylor & Francis Group, London, ISBN 978-1-138-00010-0

Ocean acidification and echinoderms: How bad will it be for our favourite phylum?

M.A. Sewell
School of Biological Sciences, University of Auckland, Auckland, New Zealand

G.E. Hofmann
Department of Ecology, Evolution and Marine Biology, University of California, Santa Barbara, California, USA

ABSTRACT: Marine ecosystems world-wide will be affected by Ocean Acidification (OA) and the accompanying changes to carbonate concentrations. Here we consider how OA might impact echinoderms by considering two contrasting reproductive strategies in sea urchins: planktotrophic larvae, and the brooded young of deep-sea Antarctic echinoids. In planktotrophic sea urchins previous research has shown that larvae developing in elevated CO_2 are smaller, with skeletal abnormalities at high $[CO_2]$. Recent research with *Lytechinus pictus* using gene expression profiling further shows that numerous genes central to energy metabolism and biomineralization are down-regulated in high CO_2 conditions. Skeletal formation in adult Antarctic sea urchins, and their developing broods, is also likely to be affected by shallowing of the high Mg-calcite saturation horizon, potentially impacting growth, bathymetric distribution and local population persistence. Thus, OA will affect sea urchins with both brooding and larval developmental modes, suggesting broad-scale impacts on other calcified echinoderm classes.

1 INTRODUCTION

Ocean Acidification (OA) has been described by Richard Feely as "global warming's evil twin" (quoted in Parks 2008) and is considered to be one of the greatest environmental challenges to marine organisms in the 21st century (reviewed in Raven et al. 2005, Kleypas et al. 2006, Fabry et al. 2008, Guinotte & Fabry 2008, Doney et al. 2009, Hofmann et al. 2010). The average ocean pH has decreased by 0.1 units since the beginning of the industrial age (from 8.21 to 8.10, Raven et al. 2005), and is expected to decrease a further 0.3 to 0.4 units by 2100, resulting in seawater that contains 150% more H^+ than present (Orr et al. 2005).

There has been considerable research in recent years on the effects of OA on a variety of marine organisms with calcium carbonate skeletons, including coccolithophores, coralline algae, corals, and molluscs (reviewed in Raven et al. 2005, Kleypas et al. 2006, Fabry et al. 2008, Doney et al. 2009, Hofmann et al. 2010, Kroeker et al. 2010). A growing body of research is now being conducted on the affects of OA on the Phylum Echinodermata (reviewed in Doney et al. 2009, Dupont et al. 2010a, b, Kroeker et al. 2010), which because of their high magnesium calcite endoskeleton are particularly vulnerable to calcium carbonate dissolution under OA conditions (Orr et al. 2005).

In this paper we briefly introduce the important aspects of carbonate chemistry that influence the acidity of sea water, then review the effects of OA in echinoderms with two contrasting reproductive strategies: the effect of lower pH conditions on the larvae of planktotrophic sea urchins during early development; and the potential for OA to affect both growth and reproduction in deep-sea Antarctic sea urchins. Along with other papers presented at the Echinoderm Conference, and reviews in the recent literature (Kurihara 2008, Dupont et al. 2010a, b, Hendricks et al. 2010, Hendricks & Duarte 2010, Kroeker et al. 2010), we conclude that echinoderms may be negatively affected by OA, with impacts likely to include growth, reproduction and local population persistence.

2 CARBONATE CHEMISTRY 101

The carbonate ion (CO_3^{2-}) is part of a series of complex chemical reactions, known as the carbonic acid system:

$$CO_{2(atmos)} \leftrightarrow CO_{2(aq)} + H_2O \leftrightarrow H_2CO_3 \leftrightarrow H^+ + HCO_3^- \leftrightarrow 2H^+ + CO_3^{2-} \quad (1)$$

Dramatic changes in atmospheric CO_2 content over the last 250 years (pre-industrial = ca 280 ppm to ca 384 ppm in 2007; IPCC 2007) have therefore resulted in increased CO_2, carbonic acid (H_2CO_3), bicarbonate (HCO_3^-) and hydrogen ion (H^+) concentrations in seawater. As pH = $-\log_{10}[H^+]$, it follows that increased atmospheric CO_2 results in a lower, and more acid pH, or ocean acidification. Further, the increased availability of H^+ ions in seawater promotes the formation of bicarbonate from carbonate, resulting in lower concentrations of CO_3^{2-} in a high-CO_2 ocean.

The impact of ocean acidification on organisms with calcium carbonate skeletons ($CaCO_3$) is through the formation/dissolution reaction:

$$CaCO_3 \leftrightarrow Ca^{2+} + CO_3^2 \qquad (2)$$

The $CaCO_3$ saturation state (Ω) is defined as the ion product of the calcium and carbonate concentrations, divided by the stoichiometric solubility product $K_{sp}*$:

$$\Omega = [Ca^{2+}] [CO_3^2]/K_{sp}* \qquad (3)$$

Values of $\Omega > 1$ favour formation of $CaCO_3$ skeletons, whereas when $\Omega < 1$ seawater is corrosive to skeletons and dissolution will occur. An important contribution to the saturation state of seawater is made by $K*_{sp}$, which depends on local conditions, such as temperature, pressure and salinity, and also the mineral phase of calcite. High magnesium calcite, characteristically found in the skeletal material of echinoderms, can have a solubility in seawater that exceeds aragonite and calcite (Morse et al. 2006), making this phylum particularly vulnerable to the effects of ocean acidification.

In the interests of space, we have here provided only an explanation of the "essentials" of carbonate chemistry. For a more detailed explanation the reader is referred to resources such as Broecker & Peng (1982), Morse & McKenzie (1990), Raven et al. (2005), Kleypas et al. (2006), Morse et al. (2006), Andersson et al. (2008), Fabry et al. (2008) and Doney et al. (2009).

3 CASE STUDY 1: PLANKTOTROPHIC SEA URCHINS

Sea urchin larvae have long been a model system for developmental biology and details of skeletal formation are well known at both morphological (e.g. McEdward 1984, McEdward & Herrera 1999) and biomineralization levels, with the $CaCO_3$ skeleton forming after gastrulation (Wilt 2005, Ettensohn 2009). Additionally, because methods of spawning and culturing sea urchin larvae were developed early (Hinegardner 1969) they have long been used in studies of ecotoxicology (Hoffman et al. 2003).

In combination, the large knowledge base, and the ease of experimentally manipulating the culture environment provides an ideal model system for examining the effects of OA on a marine calcifier.

Two important studies on OA using planktotrophic sea urchin larvae were conducted on *Hemicentrotus pulcherrimus* and *Echinometra mathaei* in Japan by Kurihara et al. (2004) and Kurihara & Shirayama (2004a, b). These studies tested two high-CO_2 scenarios: the effects of increased atmospheric [CO_2] up to 2000 ppm, and the impacts of direct injection of CO_2 into the deep ocean (=carbon sequestration: 5000, 10,000 ppm). In this discussion we will focus on just the results from the first scenario.

Although high-CO_2/lower pH conditions affected fertilization and early development of both *H. pulcherrimus* and *E. mathaei* (Kurihara et al. 2004, Kurihara & Shirayama 2004a, b) the impact of [CO_2] on fertilization differed between females, perhaps indicating a genetic difference in CO_2 tolerance between populations (Kurihara 2008). Later larval development, however, showed more dramatic effects of elevated [CO_2], with larval and arm sizes significantly smaller at higher [CO_2] and evidence of abnormal development particularly of the larval skeleton (Kurihara et al. 2004, Kurihara & Shirayama 2004a, b).

A critical development in the use of sea urchins as a model system for OA experiments has been the availability of the genome resources provided by the purple sea urchin *Strongylocentrotus purpuratus* (Sea Urchin Genome Sequencing Consortium 2006). Although the sub-lethal effects of elevated [CO_2] on organismal, developmental and physiological features of marine calcifiers has been identified in echinoderms (Kurihara et al. 2004, Kurihara & Shirayama 2004a, b) and other marine calcifiers (Fabry et al. 2008, Guinotte & Fabry 2008, Doney et al. 2009), the cellular level mechanisms that are responsible for the observed effects are unknown (Hofmann et al. 2008).

As a first step to understanding how OA and elevated [CO_2] affect biomineralization and skeletogenesis in both larval and adult sea urchins, a suite of genes has been identified and used to prepare an oligonucleotide microarray (Hofmann et al. 2008, Todgham & Hofmann 2009). Three classes of genes were included: coding for proteins in the organic matrix, for membrane transporters, and for an enzyme, carbonic anhydrase, which is involved in the elimination of CO_2 from cells (Hofmann et al. 2008, Todgham & Hofmann 2009). The potential for this approach to provide information on the physiological response to elevated [CO_2] was shown in recent experiments with *S. purpuratus* larvae. At 540 ppm CO_2 there was a broad-scale down-regulation of genes for skeletogenesis, the cellular stress response and energy metabolism compared to controls (Todgham & Hofmann 2009).

We combined these approaches—morphological and mechanistic—in an experiment on *Lytechinus pictus* (O'Donnell et al. 2010) where we asked the question: can sea urchin larvae compensate for the impact of CO_2 on skeletal formation by changes in the expression of the biomineralization genes? (Hofmann et al. 2008). The first challenge in this approach was whether the microarray developed by Todgham & Hofmann (2009) for *Strongylocentrotus purpuratus* would show cross-species hybridization with mRNA samples extracted from *Lytechinus pictus*. Our assumption in this case was that many of the genes on the oligonucleotide microarray would be conserved and that, as phylogenetic analysis had shown these urchin genera are in the same clade (Smith et al. 2006), there would be good hybridization; an assumption confirmed during the course of the experiments.

The combined morphological and mechanistic approach taken in the OA experiments with *Lytechinus pictus* was successful in beginning to tease apart how sea urchin larvae are affected by growth in OA conditions (O'Donnell et al. 2010). Morphologically, as in previous studies (Kurihara et al. 2004, Kurihara & Shirayama 2004a, b), sea urchin larvae raised in high-CO_2/lower pH conditions were smaller and with shorter skeletal rods than the control larvae. A multivariate analysis using the length of six major arm and body rods revealed that, as well as being smaller in overall size, there were also changes to larval shape. High-CO_2/lower pH larvae were "shorter and stumpier", with the shape of the main body compartment (defined as the area within the boundaries of the ventral transverse rods and body rods, see McEdward & Herrera 1999, Fig. 2), more triangular than box-like (O'Donnell et al. 2010).

The results from the oligonucleotide microarray showed a distinct down-regulation of gene expression under high-CO_2/lower pH conditions, with the exception of a few genes involved in ion regulation and acid-base balance pathways that were induced (O'Donnell et al. 2010). Down-regulation was seen in genes involved in metabolic processes, including the electron-transport chain and Tricarboxylic Acid (TCA) cycle, the cellular defensome (e.g. ubiquitin) and genes involved in protein synthesis. Of particular interest to our question regarding biomineralization, we observed down-regulation in eight genes involved in producing the organic matrix of the endoskeleton, including two spicule matrix proteins, SM-29 and SM30-like (O'Donnell et al. 2010). This suggests that sea urchin larvae are unable to compensate for the impact of CO_2 on skeletal formation by changes

in the expression of the biomineralization genes (O'Donnell et al. 2010).

Taken together these results provide a physiological fingerprint of larvae that are developing under OA conditions. The morphological measurements are a visual reflection of the physiological challenges to skeletal growth under high-CO_2/lower pH conditions, and by examining the transcriptome we can begin to understand the ability of sea urchin larvae to cope with these conditions. The overall trend for down-regulation of genes on the microarray suggests that sea urchin larvae have undergone a suppression of metabolism and retreated into a hypometabolic state (Storey & Storey 2007). Metabolic rate depression is a well-documented survival strategy of organisms when environmental conditions are poor and involves the coordinated down-regulation of ATP-consuming and ATP-generating cellular functions (Storey & Storey 2007). However, at least in the sea urchin larvae studied to date, there is not the expected up-regulation of genes associated with cellular defense mechanisms, such as molecular chaperones, or antioxidants, that promote the stability of macromolecules and contribute to the viability of the cell when the animal resumes regular metabolism (Todgham & Hofmann 2009, O'Donnell et al. 2010). As the majority of maternal yolk lipids are utilized by sea urchin larvae in the formation of the 4-arm pluteus (Sewell 2005), it may be that there are insufficient energy reserves to regulate recovery mechanisms and resume routine metabolism (O'Donnell et al. 2010). This is an area in which we will be focusing our experimental work in the future.

The OA experiments to date (reviewed by Kurihara 2008), presented at the 2009 Echinoderm Conference and recently published (Clark et al. 2009, O'Donnell et al. 2010, Sheppard Brennand et al. 2010) suggest that decreased skeletal formation and changes to larval morphology when cultured in elevated [CO_2] is a pattern seen in planktotrophic sea urchins from a variety of locations. OA experiments conducted on the ophioplutei of *Ophiothrix fragilis* similarly show abnormalities in skeletogenesis, including asymmetry and morphometric changes, and higher mortality (Dupont et al. 2008).

It would be of interest to conduct experiments more broadly in echinoderms with planktotrophic larvae, initially with a focus on the pluteal larval forms (echinoplutei, ophioplutei) where there is a greater degree of skeletal formation. However, insights into the effects of OA on echinoderms in general will also require experiments to be conducted in the bipinnaria and auricularia forms, as transcriptome studies (Todgham & Hofmann 2009, O'Donnell et al. 2010) have shown that some of the effects of OA may be more subtle than morphological changes to the skeletal structures alone.

Finally it is important that in the design and implementation of OA experiments that we are cognizant of the need to study the ecological consequences of changes to larval morphology and gene expression. For example, Dupont et al. (2008) showed marked increases in mortality in high-CO_2/lower pH conditions after days 4–6 which they suggested might be a result of compromised larval feeding performance. Direct measurements of larval feeding in larvae raised in OA conditions are therefore needed to address this hypothesis. Further, most OA experiments on echinoderm larvae have been relatively short-term and have not yet considered the second major period of calcification, when the rudiment is formed. Although perhaps technically challenging, it would be of interest to conduct longer term OA experiments to determine the impacts on developmental times and size of the resultant rudiment. The early echinopluteus is only the very beginning of the sea urchin life cycle and we need to remember that later echinoplutei and settled juveniles will need to grow to sexually reproductive adults in OA conditions as well.

4 CASE STUDY 2: BROODING ANTARCTIC SEA URCHINS

CO_2 gas is more soluble in seawater as the temperature decreases and the pressure increases, therefore the impacts of OA are predicted to be especially negative in the deep waters of the polar regions (see Equation 1). At high-latitudes the depth at which the $CaCO_3$ saturation state shifts to values of $\Omega < 1$ will shallow for all forms of calcite (Feely et al. 2002, 2004, Gehlen et al. 2007, Gangsto et al. 2008, Andersson et al. 2008), impacting calcifying organisms in both pelagic and benthic environments (Raven et al. 2005, Kleypas et al. 2006, Guinotte et al. 2006, Fabry et al. 2008, Hofmann et al. 2010). Distributional and experimental studies have shown that in Antarctic and Southern ocean habitats, pelagic pteropods (Fabry et al. 2008), deep sea corals (Guinotte et al. 2006), and foraminifera (Moy et al. 2009) will have problems with growth and maintenance of their $CaCO_3$ skeletons as a result of changes to the aragonite saturation horizon, and that both vertical and horizontal distributions will be affected.

Changes to the depths of the saturation horizons have been modeled for aragonite and calcite (e.g. Feely et al. 2002, 2004, Gehlen et al. 2007, Gangsto et al. 2008) where the stoichiometric solubility products (K_{sp}*, see Equation 3) are known (Andersson et al. 2008). In contrast, the K_{sp}* for Mg-calcite minerals are the subject of considerable debate, with two different experimental solubility curves for biogenic Mg-calcites (see Morse et al. 2006, Andersson et al. 2008 and Sewell & Hofmann 2011 for a detailed discussion of this problem). Models using both these curves have shown that high latitude surface seawater will become under-saturated with respect to Mg-calcite phases containing 12 mol% and higher $MgCO_3$ during the 21st century, and that this environment could be in metastable equilibrium with a 4–5 mol% Mg-calcite by 2100 (Andersson et al. 2008).

Echinoids are one of the most calcified of the echinoderm classes, typically contain ratios of 4–5% Mg-calcite (Dubois & Chen 1989), and as a dominant member of the Antarctic macrofauna are likely to be particularly vulnerable to changes in Mg-calcite saturation at high latitudes (Sewell & Hofmann 2011). The Antarctic echinoids also show an extremely high incidence of brooding (Poulin & Feral 1996, Pearse & Lockhart 2004) where the young are retained externally on the female often in elaborate brooding structures (David et al. 2005). Calcification of the skeletal plates in brooded embryos and juveniles therefore occur in the same $\Omega_{Mg\text{-}calcite}$ environment as the mother.

If there is shallowing of the Mg-calcite saturation horizon brooding Antarctic echinoids (52% of the 81 species are known brooders; Poulin & Feral 1996, David et al. 2005) may be particularly impacted for two reasons. Firstly, as the amorphous precursor phase of high-Mg calcite is 30 times more soluble than calcite without magnesium (Beniash et al. 1997, Politi et al. 2004) both the growth of the mother and the calcification of the developing young may be impacted by unfavourable values of $\Omega_{Mg\text{-}calcite}$ (Sewell & Hofmann 2011). This could result in greater costs to producing skeletons, changes to growth rates, and/or thinner and weaker skeletons being formed at depths at or near the point where $\Omega_{Mg\text{-}calcite} < 1$ (Sewell & Hofmann 2011). Secondly, lower reproductive rates at depths where $\Omega_{Mg\text{-}calcite} < 1$, coupled with low dispersal of brooded young, except for relatively rare drifting or rafting events (Pearse & Lockhart 2004), could reduce the densities of deeper populations to low levels or even to local extinction (Sewell & Hofmann 2011). Thus, there may be contractions in the bathymetric ranges of many Antarctic echinoid species, and changes to the distribution patterns of species within Antarctica, as changes to the calcium carbonate saturation horizons are not spatially uniform in the Southern Ocean (Feely et al. 2004, Gehlen et al. 2007, Gangsto et al. 2008, McNeil & Matear 2008). As >80% of the global $CaCO_3$ production from echinoderms is at shallow depths (0–800 m), changes to bathymetric limits may also have consequences for global $CaCO_3$ cycling (Lebrato et al. 2010).

5 META-ANALYSES OF OA EFFECTS IN ECHINODERMS

Meta-analyses of the effects of OA on marine organisms (including 44 echinoderm studies, Hendriks et al. 2010; 24 echinoderm studies, Kroeker et al. 2010) and echinoderms alone (19 studies, Dupont et al. 2010a) have recently been published. There is considerable overlap in the source data being used, and varying approaches to the statistical analyses used for the meta-analyses which has itself prompted a lively debate (see Hendriks & Duarte 2010 responding to Dupont et al. 2010b, Kroeker et al. 2010).

When considering just pH changes that are predicted for the next 100 years (<0.5 units), Hendriks et al. (2010) found overall no "general consistent effect" of OA (effect size = 1), although there were significant differences between taxa and biological processes. In sea urchins, the only echinoderms used in their meta-analysis, there were negative impacts on fertility, growth and survival (Hendriks et al. 2010).

Dupont et al. (2010a, b), using the same measure of effect size as Hendriks et al. (2010), but a smaller data set for echinoderms only, found generally negative effects of OA on larvae and juveniles in survival, calcification and growth, but positive effects on calcification and growth in adults. As noted by Hendriks & Duarte (2010), however, this meta-analysis did not use a statistical approach, which makes interpretation more difficult.

Finally, Kroeker et al. (2010), using a different measure of effect size, found for echinoderms a non-significant positive effect on calcification, and a non-significant negative effect on growth. When separating into developmental stages, there was a significantly greater negative effect on growth in juveniles than larvae, with a non-significant difference in survival between larvae and adults. In contrast to Hendriks et al. (2010) this meta-analysis showed a strong overall negative effect of OA on marine organisms (Kroeker et al. 2010).

6 FUTURE DIRECTIONS

Understanding the impact of OA on marine organisms is a rapidly growing research field, with 6 presentations and 4 posters devoted to OA and echinoderms at this conference alone. One point that seems to be emerging from the recent literature,

however, is that echinoderms show a high degree of inter- and intra-specific response to experimental OA conditions (Dupont et al. 2010a). It is not yet clear if differences between echinoderm studies are related to differences in experimental procedures, or reflect real differences related to local adaptation to natural variability in pH conditions at the species or population level.

As the field develops, there is an increasing move to standardize methodologies, with publications on the procedures for measuring seawater chemistry (Dickson et al. 2007), and general aspects of the design and analysis of OA experiments (Riebesell et al. 2010). Laboratories conducting OA research are also publishing detailed descriptions of equipment designed for the production of seawaters of varying pH and for long-term maintenance of OA conditions within tight chemical parameters (e.g. Findlay et al. 2008, Fangue et al. 2010). It is our hope that as these methodologies converge we will have a greater ability to differentiate between experimental and biological differences at both the inter- and intra-specific level.

Finally, in this plenary we have focused on OA, although there will be concomitant changes in seawater temperature that may also cause thermal stress (Pörtner et al. 2005). In a recent experiment, O'Donnell et al. (2009) used quantitative PCR (qPCR) to measure the levels of the molecular chaperone hsp70 in 4-arm pluteus of *Strongylocentrotus purpuratus*. Larvae raised at high $[CO_2]$ showed lower levels of expression of hsp70 and an upward shift in the temperature (+2°C) at which the hsp70 response was maximal. Development in OA conditions, therefore, has the potential to modify an organism's response to other environmental stresses. In future OA experiments it is, therefore, recommended that we consider a multistressor approach to reveal the presence of any synergistic impacts of temperature and OA climate changes; an approach that has recently been undertaken in echinoderms (e.g. Byrne et al. 2009, 2010a, 2010b, Sheppard Brennand et al. 2010).

Teasing apart the complexities of the effects of OA on echinoderms will, however, require the integration of different research approaches. Collaborations are suggested between echinoderm biologists interested in OA as an ecological problem, and scientists with research skills in other areas (e.g. microarray, proteomics, metabolomics, oceanography, chemistry, biomaterials). The approach suggested by Greg Wray in his plenary presentation on Gene Regulatory Networks (GRN, see review in Ettensohn 2009) may also prove useful in future investigations.

As echinoderm biologists we share an interest and, in some cases, an inordinate fondness for the phylum that we study. Unfortunately, the high-Mg-calcite skeleton of echinoderms makes them particularly vulnerable to the effects of OA. Although research on OA has focused primarily on reproductive and developmental effects, we encourage all echinoderm biologists, regardless of your research focus, to give some thought to the effects of an acid-ocean on the biology of your chosen species. While the focus to date has been on skeletal formation, it may be that the greatest impacts of increased CO_2 and OA may be in yet-to-be-discovered areas of echinoderm physiology.

REFERENCES

Andersson, A.J., Mackenzie, F.T. & Bates, N.R. 2008. Life on the margin: implications of ocean acidification on Mg-calcite, high latitude and cold-water marine calcifiers. *Mar. Ecol. Prog. Ser.* 373: 265–273.

Beniash, E., Aizenberg J., Addadi L. & Weiner S. 1997. Amorphous calcium carbonate transforms into calcite during sea urchin larval spicule growth. *Proc. R. Soc. Lond. B* 264: 461–465.

Broecker, W.S. & Peng, T-H. 1982. *Tracers in the Sea.* New York: Lamont Doherty Geological Observatory, Columbia University.

Byrne, M., Ho, M.A., Selvakumaraswamy, P., Nguyen, H.D., Dworjanyn, S.A. & Davis, A.R. 2009. Temperature, but not pH, compromises sea urchin fertilization and early development under near-future climate change scenarios. *Proc. Royal Soc. B, Biol. Sci.* 276: 1883–1888.

Byrne, M., Soars, N.A., Ho, M.A., Wong, E., McElroy, D., Selvakumaraswamy, P., Dworjanyn, S.A. & Davis, A.R. 2010a. Fertilization in a suite of coastal marine invertebrates from SE Australia is robust to near-future ocean warming and acidification. *Mar. Biol.* 157: 2061–2069.

Byrne, M., Soars, N.A., Selvakumaraswamy, P., Dworjanyn, S.A. & Davis, A.R. 2010b. Sea urchin fertilization in a warm, acidified and high pCO₂ ocean across a range of sperm densities. *Mar. Environ. Res.* 69: 234–239.

Clark, D., Lamare, M. & Barker, M. 2009. Response of sea urchin pluteus larvae (Echinodermata: Echinoidea) to reduced seawater pH: a comparison among a tropical, temperate, and a polar species. *Mar. Biol.* 156: 1125–1137.

David, B., Choné, T., Mooi, R. & De Ridder, C. 2005. *Antarctic Echinoidea.* In: Wägele, J.W. & Sieg, J. (eds). Synopses of the Antarctic benthos, Vol. 10. Königstein: Koeltz Scientific Books.

Dickson, A.G., Sabine, C.L. & Christian, J.R. 2007. Guide to best practice for ocean CO_2 measurements, PICES Special Publication 3.

Doney, S.C., Fabry, V.J., Feely, R.A. & Kleypas, J.A. 2009. Ocean acidification: The other CO_2 problem. *Annu. Rev. Mar. Sci.* 1: 169–192.

Dubois, P. & Chen, C.P. 1989. Calcification in echinoderms. In: Jangoux, M., Lawrence, J.M. (eds.), *Echinoderm Studies 3:* 109–178. Rotterdam: A.A. Balkema.

Dupont, S., Havenhand, J., Thorndyke, W., Peck, L. & Thorndyke, M. 2008. Near-future level of CO_2-driven ocean acidification radically affects larval survival and development in the brittlestar *Ophiothrix fragilis. Mar. Ecol. Prog. Ser.* 373: 285–294.

Dupont, S., Ortega-Martínez, O. & Thorndyke, M. 2010a. Impact of near-future ocean acidification on echinoderms. *Ecotoxicology* 19: 449–462.

Dupont, S., Dorey, N. & Thorndyke, M. 2010b. What meta-analysis can tell us about vulnerability of marine biodiversity to ocean acidification? *Estuar. Coastal Shelf Sci.* 89: 182–185.

Ettensohn, C.A. 2009. Lessons from a gene regulatory network: echinoderm skeletogenesis provides insights into evolution, plasticity and morphogenesis. *Development* 136: 11–21.

Fabry, V.J., Seibel, B.A., Feely, R.A. & Orr, J.C. 2008. Impacts of ocean acidification on marine fauna and ecosystem processes. *ICES J. Mar. Sci.* 65: 414–432.

Fangue, N.A., O'Donnell, M.J., Sewell, M.A., Matson, P.G., MacPherson, A.C. & Hofmann, G.E. 2010. A laboratory-based experimental system for the study of ocean acidification effects on marine invertebrate larvae. *Limnol. Oceanogr.: Methods* 8: 441–452.

Feely, R.A., Sabine, C.L., Lee, K., Millero, F.J., Lamb, M.F., Greeley, D., Bullister, J.L., Key, R.M., Peng, T.-H., Kozyr, A., Ono, T. & Wong, C.S. 2002. In situ calcium carbonate dissolution in the Pacific Ocean. *Global Biogeochem. Cycles* 16: 1144, doi:10.1029/2002GB001866.

Feely, R.A., Sabine, C.L., Lee, K., Berelson, W., Kleypas, J., Fabry, V.J. & Millero, F.J. 2004. Impact of anthropogenic CO₂ on the CaCO₃ system in the oceans. *Science* 305: 362–366.

Findlay, H.S., Kendall, M.A., Spicer, J.I., Turley, C. & Widdicombe, S. 2008. Novel microcosm system for investigating the effects of elevated carbon dioxide and temperature on intertidal organisms. *Aquatic Biology* 3: 51–62.

Gangsto, R., Gehlen, M., Schneider, B., Bopp, L., Aumont, O. & Joos, F. 2008. Modeling the marine aragonite cycle: changes under rising carbon dioxide and its role in shallow water CaCO₃ dissolution. *Biogeosciences* 5: 1057–1072.

Gehlen, M., Gangsto, R., Schneider, B., Bopp, L., Aumont, O. & Ethe, C. 2007. The fate of pelagic CaCO₃ production in a high CO_2 ocean: A model study. *Biogeosciences* 4: 505–519.

Guinotte, J.M., Orr, J., Cairns, S., Freiwald, A., Morgan, L. & George, R. 2006. Will human-induced changes in seawater chemistry alter the distribution of deep-sea scleractinian corals? *Front. Ecol. Environ.* 4: 141–146.

Guinotte, J.M. & Fabry V.J. 2008. Ocean acidification and its potential effects on marine ecosystems. *Ann. N.Y. Acad. Sci.* 1134: 320–342.

Hendriks, I.E. & Duarte, C.M. 2010. Ocean acidification: Separating evidence from judgment—A reply to Dupont et al. *Estuar. Coastal Shelf Sci.* 89: 186–190.

Hendriks, I.E., Duarte, C.M. & Álvarez, M. 2010. Vulnerability of marine biodiversity to ocean acidification: a meta-analysis. *Estuar. Coastal Shelf Sci.* 86: 157–164.

Hinegardner, R.T. 1969. Growth and development of the laboratory cultured sea urchin. *Biol. Bull.* 137: 465–475.

Hoffman, D.J., Rattner, B.A. & Burton, G.A. Jr (eds). 2003. *Handbook of Ecotoxicology*. 2nd Edition. Boca Raton: CRC Press.

Hofmann, G.E., Barry, J.P., Edmunds, P.J., Gates, R.D., Hutchins, D.A., Klinger, T. & Sewell, M.A. 2010. The effect of ocean acidification on calcifying organisms in marine ecosystems: An organism-to-ecosystem perspective. *Annu. Rev. Ecol. Evol. Syst.* 41: 127–147.

Hofmann, G.E., O'Donnell, M.J. & Todgham, A.E. 2008. Using functional genomics to explore the effects of ocean acidification on calcifying marine organisms. *Mar. Ecol. Prog. Ser.* 373: 219–225.

IPCC (2007). Climate Change 2007: The Physical Science Basis. Contribution of Working Group I to the Fourth Assessment Report of the Intergovernmental Panel on Climate Change [Solomon, S., Qin, D., Manning, M., Chen, Z., Marquis, M., Averyt, K.B., Tignor, M. & Miller, H.L. (eds)]. Cambridge: Cambridge University Press.

Kleypas, J.A., Feely, R.A., Fabry, V.J., Langdon, C., Sabine, C.L. & Robbins, L.L. 2006. Impacts of ocean acidification on coral reefs and other marine calcifiers: A guide for future research. Report of a workshop sponsored by NSF, NOAA and the US Geological Survey, 88 pp.

Kroeker, K.J., Kordas, R.L., Crim, R.N. & Singh, G.G. 2010. Meta-analysis reveals negative yet variable effects of ocean acidification on marine organisms. *Ecology Letters* 13: 1419–1434.

Kurihara, H. 2008. Effects of CO_2-driven ocean acidification on the early developmental stages of invertebrates. *Mar. Ecol. Prog. Ser.* 373: 275–284.

Kurihara, H., Shimode, S. & Shirayama, Y. 2004. Sub-lethal effects of elevated concentration of CO_2 on planktonic copepods and sea urchins. *J. Oceanogr.* 60: 743–750.

Kurihara, H. & Shirayama, Y. 2004a. Effects of increased atmospheric CO_2 on sea urchin early development. *Mar. Ecol. Prog. Ser.* 274: 161–169.

Kurihara, H. & Shirayama, Y. 2004b. Effects of increased atmospheric CO_2 and decreased pH on sea urchin embryos and gametes. In: Heinzeller, T. & Nebelsick J.H. (eds) *Echinoderms: München*: 31–36. Leiden: A.A. Balkema.

Lebrato, M., Iglesias-Rodríguez, D., Feely, R.A., Greeley, D., Jones, D.O.B., Suarez-Bosche, N., Lampitt, J.E., Cartes, J.E., Green, D.R.H. & Alker, B. 2010. Global contribution of echinoderms to the marine carbon cycle: $CaCO_3$ budget and benthic compartments. *Ecol. Monogr.* 80: 441–467.

McEdward, L.R. 1984. Morphometric and metabolic analysis of the growth and form of echinopluteus. *J. Exp. Mar. Biol. Ecol.* 82: 259–287.

McEdward, L.R. & Herrera, J.C. 1999. Body form and skeletal morphometrics during larval development of the sea urchin *Lytechinus variegatus* Lamark. *J. Exp. Mar. Biol. Ecol.* 232: 151–176.

McNeil, B.I. & Matear, R.J. 2008. Southern Ocean acidification: a tipping point at 450-ppm atmospheric CO_2. *Proc. Natl. Acad. Sci. USA* 105: 18860–18864.

Morse, J.W. & Mackenzie, F.T. 1990. *Geochemistry of sedimentary carbonates*. Netherlands: Elsevier Science.

Morse, J.W., Andersson, A.J. & Mackenzie, F.T. 2006. Initial responses of carbonate-rich shelf sediments to rising atmospheric pCO2 and "ocean acidification": Role of high Mg-calcites. *Geochimica et Cosmochimica Acta* 70: 5814–5830.

Moy, A.D, Howard, W.R., Bray, S.G. & Trull, T.W. 2009. Reduced calcification in modern Southern Ocean planktonic foraminifera. *Nature Geoscience* 2: 276–280.

O'Donnell, M.J., Hammond, L.M. & Hofmann, G.E. 2009. Predicted impact of ocean acidification on a marine invertebrate: elevated CO_2 alters response to thermal stress in sea urchin larvae. *Mar. Biol.* 156: 439–446.

O'Donnell, M.J., Todgham, A.E., Sewell, M.A., Hammond, L.M., Ruggiero, K., Fangue, N.A., Zippay, M.L. & Hofmann, G.E. 2010. Ocean acidification alters skeleton formation of larvae of the sea urchin *Lytechinus pictus*: evidence from morphometrics and microarray data. *Mar. Ecol. Prog. Ser.* 398: 157–171.

Orr, J.C., Fabry, V.J., Aumont, O., Bopp, L., Doney, S.C., Feely, R.A., et al. 2005. Anthropogenic ocean acidification over the twenty-first century and its impact on calcifying organisms. *Nature* 437: 681–686.

Parks, N. 2008. Sea change underway with ocean acidification. *Front. Ecol. Environ.* 6: 460.

Pearse, J.S. & Lockhart, S.J. 2004. Reproduction in cold water: paradigm changes in the 20th century and a role for cidaroid sea urchins. *Deep Sea Res. II* 51: 1533–1549.

Politi, Y., Arad, T., Klein, E., Weiner, S. & Addadi, L. 2004. Sea urchin spine calcite forms via a transient amorphous calcium carbonate phase. *Science* 306: 1161–1164.

Pörtner, H.O., Langenbuch, M. & Michaelidis, B. 2005. Synergistic effects of temperature extremes, hypoxia, and increases in CO_2 on marine animals: from earth history to global change. *J. Geophys. Res.* 110: C09S10.

Poulin, E. & Feral, J-P. 1996. Why are there so many species of brooding Antarctic echinoids? *Evolution* 50: 820–830.

Raven, J., Caldeira, K., Elderfield, H., Hoegh-Guldberg, O., Liss, P., Riebesell, U., Shepherd, J., Turley, C. & Watson, A. 2005. Ocean acidification due to increasing atmospheric carbon dioxide, Policy Document 12/05, Royal Society, London.

Riebesell, U., Fabry, V.J., Hansson, L. & Gattuso, J.P. [eds.]. 2010. Guide to best practices for ocean acidification research and data reporting. Publications Office of the European Union.

Sea Urchin Genome Sequencing Consortium. 2006. The genome of the sea urchin *Strongylocentrotus purpuratus*. *Science* 314: 941–952.

Sewell, M.A. 2005. Utilization of lipids during early development of the echinometrid sea urchin *Evechinus chloroticus*. *Mar. Ecol. Prog. Ser.* 304: 133–142.

Sewell, M.A. & Hofmann, G.E. 2011. Antarctic echinoids and climate change: a major impact on the brooding forms. *Global Change Biology* 17: 734–744.

Sheppard Brennand, H., Soars, N., Dworjanyn, S.A., Davis, A.R. & Byrne, M. 2010. Impact of ocean warming and ocean acidification on larval development and calcification in the sea urchin *Tripneustes gratilla*. *PLoS ONE* 5(6): e11372. doi:10.1371/journal.pone.0011372.

Smith, A.B., Pisani, D., Mackenzie-Dodds, J.A., Stockley, B., Webster, B.L. & Littlewood, D.T.J. 2006. Testing the molecular clock: molecular and paleontological estimates of divergence times in the Echinoidea (Echinodermata). *Mol. Biol. Evol.* 23: 1832–1851.

Storey, K.B. & Storey, J.M. 2007. Tribute to P.L. Lutz: putting life on 'pause'—molecular regulation of hypometabolism. *J. Exp. Biol.* 210: 1700–1714.

Todgham, A.E. & Hofmann, G.E. 2009. Transcriptomic response of sea urchin *Strongylocentrotus purpuratus* to CO_2-driven seawater acidification. *J. Exp. Biol.* 212: 2579–2594.

Wilt, F.H. 2005. Developmental biology meets materials science: Morphogenesis of biomineralized structures. *Develop. Biol.* 280: 15–25.

Echinoderms in a Changing World – Johnson (ed)
© *2013 Taylor & Francis Group, London, ISBN 978-1-138-00010-0*

The legacy of ocean climate and chemistry change in the echinoderm fossil record: A review

J.H. Nebelsick
Institute for Geosciences, University of Tübingen, Tübingen, Germany

A. Kroh
Natural History Museum Vienna, Austria

A. Roth-Nebelsick
State Museum of Natural History, Stuttgart, Germany

ABSTRACT: The geological history of chemistry of the oceans is complex and results from the interactions between different Earth systems. The broad historical development of ocean chemistry is reviewed with respect to different proxies used to acquire data for temperature, CO_2 and Ca values as well as the presence of calcite and aragonite dominated seas in the past. The role of echinoderms in determining ocean chemistry of the past is reviewed with respect to taphonomy, diagenesis and the possibilities for measuring palaeotemperatures, and ancient seawater composition including isotopic ratios. Finally, the possible effects of changing ocean chemistry on the evolution and diversity of echinoderms are reviewed with respect to both long-term secular changes and short term drastic events associated with mass extinctions.

1 GENERAL INTRODUCTION

Ocean acidification has rapidly become a field of primary interest among ocean chemists and biologists due to the rapid changes within the historical time frame as well as potential implications for the calcification and survival of organisms (e.g. Kleypas et al. 1999, 2001, 2006, Riebesell et al. 2000, Crowley & Berner 2001, Caldeira & Wickett 2003, Raven 2005, De´ath et al. 2009, Doney et al. 2009). A number of basic questions can be asked of the rock record as well as echinoderm fossils with respect to the current debate of climate change and ocean acidification:

1. What changes can be seen in the fossil and sedimentary record?
2. How does the magnitude and rates of oceanic change in the past compare to that we are experiencing at the present?
3. Can echinoderms be used to determine ocean geochemistry? In what respect does the mineralogy of the echinoderm skeleton reflect the chemical composition of the ambient sea water?
4. Has ocean geochemistry influenced echinoderm evolution and diversity with respect to long-term developments and abrupt short term changes?

Rates of changes invoked for both Recent and fossil environments range from millions of years in the fossil record to decades in historical records. Crowley & Berner (2001), for example, state that CO_2 concentrations and extremes developed over the geological time scale (millions of years) which dictate tectonic change and biological evolution. The sensitivity of change of ocean pH appears to be time dependent: ocean pH is relatively sensitive to added CO_2 when CO_2 change occurs over a short time interval (less than 10,000 yrs.) whereas ocean chemistry is buffered when change occurs over a longer time interval (100,000 yrs.) (Caldeira & Wickett 2003). Very quick changes, in contrast, have been shown to occur within historical time periods. For example, by studying calcification, extension and density of Great Barrier Reef corals, De´ath et al. (2009) showed how since industrialisation, global average atmospheric CO_2 and the concentration of hydrogen ions in ocean surface waters have increased while the aragonite saturation state has decreased.

2 GEOLOGICAL AND BIOLOGICAL CON-STRAINTS ON OCEAN GEOCHEMISTRY

There are numerous factors affecting the chemical composition of the Earth's oceans in the present and past. It has also become evident that the earth's hydrosphere (with oceans as its main component) has undergone dramatic changes in the past. The hydrosphere has evolved in close association with other earth systems such as the lithosphere, atmosphere, cryosphere and biosphere, showing both directional developmental tendencies as well as secular variations in the 4.5 billion years since the Earth's existence. Historical data locked within the earth's crust including the sedimentary and fossil record have shown extremes of both atmospheric and ocean climates in part comparable, and in part, very different from the world we know today. The study of proxy data from the rock and fossil record has led to the realizations that long interacting systematic changes and secular trends exist which continue to this day. These changes and trends include aspects of: (1) plate tectonics, (2) climate change, and sea level variations, (3) fluctuations in CO_2 as well as alternating (4) magnesium/calcium, and (5) aragonite/calcite seas.

2.1 Plate tectonics

Plate tectonics affects ocean geochemistry and evolution in numerous ways. Most obvious is the changing position of the continents and ocean on the globe. Higher spreading activity at rapidly expanding mid oceanic ridges not only causes higher sea levels by water displacement. Further influences are the placement of larger land masses in polar regions providing the base for continental ice production, altering marine current patterns and profoundly altering the biogeography of both terrestrial and marine realms. There is a very broad pattern in the Phanerozoic showing disparate continents and oceans in Cambrian (550 MA), the consolidation of the super-continent Pangaea and the all encompassing Panthalassa Ocean in the Permian (250 MA), and the breaking up of Pangaea through the Mesozoic and Cenozoic to the present situation of widely distributed continents and ocean.

The most important effects for sea water and atmospheric chemistry is that different spreading rates affect CO_2 levels and Mg/Ca levels through the circulation of seawater through hot hydrothermal activity at the spreading centres. Higher spreading rates imply high CO_2 production, low Mg/Ca levels and calcite production, while low spreading rates lead to low CO_2 production and high Mg/Ca levels, both of which favour aragonite production.

2.2 Greenhouse/icehouse climate and sea level change

Long-term secular changes in climatic conditions during the Phanerozoic (Fischer 1981, 1982) are designated as "Greenhouse", corresponding to generally warmer periods or "Icehouse" conditions corresponding to cooler periods. These phases more or less correlate with Ice ages during the late Pre-Cambrian, the Permo-Carboniferous and the current ongoing Ice age starting at the Eocene/Oligocene boundary. A prominent exception is the late Ordovician Ice Age which falls within the Lower Palaeozoic Greenhouse period.

Sea level changes are influenced by various factors including changes in spreading rates of mid-oceanic ridges as well as the amount of ice locked in continental ice sheets. Changes of sea level can vary over hundreds of meters and lead to transgressions and regressions (e.g. Vail et al. 1977, Haq et al. 1987, Hallam 1992). Changes in sea level not only have a profound effect on the distribution of land and sea but also influence biogeochemical cycles for example by dictating erosion and sedimentation rates of organic matter and CO_2 with respect to sedimentary rocks.

2.3 CO_2 production and the carbon cycle

CO_2 is well-known as an important greenhouse gas and has received much attention due to its anthropogenic increase in the atmosphere. Besides its impact on global temperature and on marine pH, it has additional biological significance as the substrate of photosynthesis (assimilation) and the as the product of respiration (dissimilation). These biological processes are part of the global terrestrial carbon cycle which comprises Long-Term and Short-Term processes as well as biological and geological processes (see Fig. 1). The Carbon content in rocks is of several magnitudes higher than for the atmosphere. Small changes in the geological turnover rates would thus have an excessively high impact on the atmospheric CO_2 content (Berner 1998). That the atmospheric CO_2 content appears to show no fluctuations of this magnitude suggests that the geologic processes are quite well-balanced.

The atmospheric CO_2 content is, however, far from being constant over Earth's history. During the Cambrian until the Early Devonian it was much higher than today (see Fig. 2). For the Early Devonian, for example, atmospheric CO_2 contents of more than ten times the pre-industrial value are indicated by various methods (Berner 1998, Konrad et al. 2000, Berner and Kothavala 2001, Royer et al. 2004, Royer 2006). It then declined rapidly during the Middle Devonian until the Permo-Carboniferous. After the Permian, atmospheric CO_2 rose again and showed then various fluctuations towards the pre-industrial level of about 280 ppm.

There are various approaches for reconstructing fossil CO_2 content of the atmosphere in more

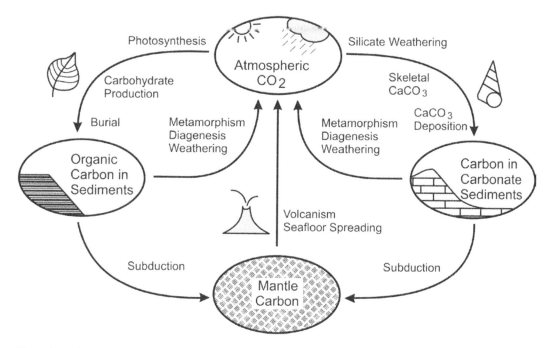

Figure 1. Principle mechanism of the carbon cycle comprising biotic and abiotic (geological) processes, redrawn from http://www.magazine.noaa.gov.

detail. Geochemical modelling is based on integrating all processes contributing to the carbon cycle and to obtain mass balances over time, such as GEOCARB or other models (Berner 2006, Tajika 1998, Wallmann 2001). Besides the difficult task of parameterisation, geochemical models do not provide a high temporal resolution. Carbon isotopic composition, $\delta13C$, of autotrophic phytoplankton is based on the fractionation of carbon isotopes during photosynthesis. Several processes and parameters then contribute to the final isotopic value, including the concentration of carbon dioxide dissolved in seawater. It is then attempted to recover the atmospheric CO_2 (Freeman and Hayes 1992, Pagani et al. 2005).

Another isotopic method utilizes the $\delta13C$ of pedogenic carbonates. These are authigenic carbonates that develop under conditions of low precipitation rates. Soil CO_2 represents a mixture of soil-respired CO_2 (by biological processes) and atmospheric CO_2. Under the assumption that soil-respired CO_2 has the same isotopic composition as soil organic matter, the $\delta13C$ of pedogenic carbonates can then be used to calculate atmospheric CO_2 content (Cerling 1999).

A further marine proxy is provided by Boron isotopic composition of calcareous shells. The uptake of Boron reflects the isotopic composition

of Boron dissolved in seawater and this parameter depends on pH. The data of pH of seawater can then be used to calculate atmospheric CO_2 (Pearson and Palmer 1999, 2000).

Fossils of land plants are also utilized for reconstructing CO_2. The leaves of land plants possess micropores (stomata) which allow for gas exchange with the atmosphere. It was observed that the density of these pores per leaf area changes reversely with atmospheric CO_2 content (Woodward 1987). This method is particularly difficult for reconstructing CO_2 in deep time since the response of stomata to changes in atmospheric CO_2 is species specific (Beerling and Royer 2002, Roth-Nebelsick 2005). It is attempted to utilize so-called "living fossils" for epochs earlier than the Miocene (Beerling and Royer 2002).

The role of CO_2 fluctuations for Earth's climate is much debated. Since CO_2 represents an important greenhouse gas, changes in atmospheric CO_2 concentrations are largely credited for the development of global temperatures during the Palaeogene and the Cenozoic (Raymo 1991, Berner & Kothvala 2001, Crowley & Berner 2001, Zachos et al. 2001). Tight coupling between Ca and temperature is indeed documented for the past 740,000 years by data obtained from ice cores (Petit et al. 1999, EPICA community members 2004). Whether

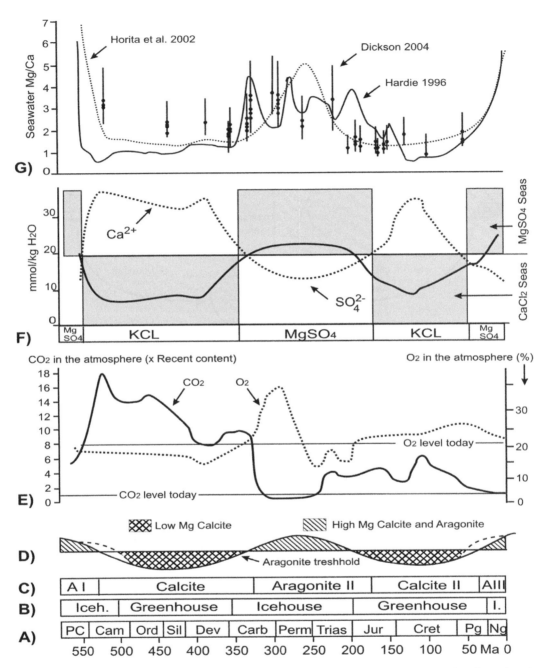

Figure 2. Development of ocean geochemistry in the Phanerozoic, comparison of various approaches. Curves have in part been readjusted to the latest stratigraphic dates. A) Stratigraphy and dates following www.stratigraphy.org. PC = Precambrian, Cam = Cambrian, Ord = Ordovician, Sil = Silurian, Dev = Devonian, Carb = Carboniferous, Perm = Permian, Trias = Triassic, Jur = Jurassic, Cret = Cretaceous, PG = Palaeogene, NG = Neogene. B) Greenhouse/icehouse distribution following Fischer (1982). Iceh.; I = Icehouse. C) Calcite/aragonite seas following Stanley & Hardie (1998); A I = Aragonite I, AIII = Aragonite III. D) Distribution of aragonite/calcite seas redrawn after Sandberg (1983). E) Atmospheric CO_2 content from the Precambrium onwards. The Y-axis shows the atmospheric CO_2 content as ratio of the recent value (y = fossil content/Recent content); after Berner and Kothvala (2001). F) Secular variations in Ca and Mg concentrations based on fluid inclusions in evaporites (based on Lowenstein et al. 2003). G) Mg/Ca ratio curves following Hardie (1996) and Horita et al. (2002). Values of echinoderm measurements redrawn after Dickson (2004).

this tight coupling between Ca and climate was also valid throughout other time periods or not is insecure. For example, a decoupling of climate and Ca is suggested for the Miocene by Pagani et al. (1999). Kürschner et al. (2008), however, found evidence for a coupling of Ca and temperature during the Miocene. A tight coupling between CO_2 and global climate is also suggested for the Palaeozoic (Montañez et al. 2007).

Significant change of atmospheric CO_2 content (and other greenhouse gases) and climate is therefore a common event in Earth's history. Usually, however, these natural processes occurred on much larger time scales than the current anthropogenic environmental change whose sheer velocity poses a particular problem to organisms.

2.4 *Magnesium/calcite seas*

Changes in ocean geochemistry and the determination of secular variations in Ca and Mg concentrations and thus Mg/Ca ratios for the past (Fig. 2) have been reconstructed by using both fluid inclusions in evaporites (Lowenstein et al. 2001, 2003) as well as the primary mineralogy of ooids which are non-skeletal carbonate components commonly found in tropical shelf environments. Theses methods are thus independent of the vital affects occurring when organisms draw different mineral phases from sea water to construct their skeletons. The ultimate isotopic composition of the skeletons may thus differ from that of the ambient sea-water through selective inclusion within the crystal lattices of the skeletal frameworks. Further studies on ancient variations of calcium isotopes and other biochemical controls through time are presented by Steuber & Veizer (2002), Holland (2005, 2006), Farkaš et al. (2007), James & Austin (2008), Prokoph et al. (2008) and Ries (2008).

2.5 *Aragonite/calcite seas*

Variation in the calcite/aragonite seas follows fluctuations of the magnesium/calcite ratios and general climate conditions. High-Mg calcite and aragonite dominate during Icehouse conditions; low-Mg calcite dominates during Greenhouse conditions. These two conditions are separated by the so-called "aragonite threshold". These are long ranging phases and using the terminologies of Stanley & Hardie (1998) can be differentiated into several distinct periods characterized by prevailing aragonite or calcite precipitation: Aragonite I leading into the Phanerozoic, Calcite I from the Upper Cambrian to the early Carboniferous, Aragonite II lasting into the Jurassic, Calcite II up to the Paleogene/Neogene boundary and then Aragonite III phase till the present day (see Fig. 2). Palaeoecological evidence of aragonite dissolution in calcite seas was studied by Palmer et al. (1988).

3 ECHINODERMS AS A PROXY FOR ANCIENT OCEAN GEOCHEMISTRY

Can echinoderms be used as climate proxies for past ocean condition? Their long fossil record, starting in the Cambrian, and prevalence in various marine environments certainly lets them appear well suited for such a purpose.

3.1 *The echinoderm skeleton and its mineralogical properties*

The skeleton of echinoderms is remarkably consistent among all representatives and throughout the long history of the phylum. Echinoderms posses a mesodermal, and thus internal skeleton, consisting of numerous distinct calcitic elements. The number of elements involved varies strongly, but can exceed 250,000 ossicles per individual in some taxa (e.g. comatulid crinoids, Meyer & Meyer 1986).

The skeletal elements consist of a mesh-like, three-dimensional network of rods and pillars called stereom. This stereom is very variable in shape and structure ranging from loosely interconnected struts to almost solid elements with few pores (Macurda et al. 1978, Smith 1980). The geometry of the stereom varies widely according to the position and its function within the skeleton as a whole or within a single skeletal element. The production of the skeleton is intracellular and the porous stereom is penetrated by the soft tissue of the stroma consisting of various cells (for example sclerocytes and phagocytes) and liquids.

The skeleton generally consists of so called "high Mg calcium carbonate" meaning that a significant portion (3 to 18.5 weight%; Weber 1969) of the calcite ions in the crystal lattice are replaced by magnesium ions. The amount of Mg varies highly according to taxa, type of plate involved and ambient environmental conditions such as temperature and sea water chemistry (Weber 1969, 1973). Extremely high values of Mg (up to 40 mol%) have been measured in such elements subject to high pressures including the teeth of regular echinoids (Schroeder 1969, Schroeder et al. 1969, Märkel et al. 1971). Additionally, minor amounts (less than 1 weight%) of iron and strontium may be incorporated. Manganese, aluminium and silicon occur as trace elements.

3.2 *Echinoderm taphonomy*

Echinoderm taphonomy has been extensively reviewed (Lewis 1980, Donovan 1991, Brett et al. 1997, Ausich 2001, Nebelsick 2004). The multi-element skeleton of echinoderms can rapidly fall apart after death (Kier 1977, Allison 1990). Depending on decay rates, biostratinomic processes and sedimentation events, preservation may range from almost complete animals to disarticulated, isolated ossicles making the reconstruction and identification of these animals difficult. The type of tissue connecting the individual elements further influences the preservation potential of the animals, in part causing the poor fossil record of some of the echinoderm groups (e.g. holothurians, asteroids and ophiuroids)—except in cases of exceptional preservation (i.e. rapid burial or dysoxic conditions). Articulations between individual elements can also be reinforced by interlocking skeletal structures increasing the fossilisation potential of some groups.

3.3 Echinoderm diagenesis

The high-magnesium-calcite of the echinoderm skeleton is thermodynamically metastable under earth surface conditions. Recrystallization of echinoderm calcite during early diagenesis, however, is usually inhibited by organic or inorganic linings and interaction of Mg ions in the sea water with the skeletal calcite (Berner 1966, Weber 1969).

Nevertheless, in most cases, fossil echinoderms are transformed to low-magnesium-calcite (Weber & Raup 1968, Neugebauer 1979), especially during vadose meteoric diagenesis (Richter 1974). Similar to the magnesium loss during diagenesis, changes in the isotopic signature of the echinoderm skeletal remains may occur. Data, however, are scare and equivocal. While Manze & Richter (1979) found a correlation between Mg loss and that of ^{13}C under meteoric-vadose conditions, Weber & Raup (1968) did not find such evidence in a large survey of fossil echinoids. Despite the mineralogical alteration, the macro- and microstructure of the skeleton is largely unaffected by the process.

Early in echinoderm diagenesis the growth of cement crystals starts, usually as soon as the soft tissue is decomposed. Cement growth depends on Ca ion availability in the ambient pore water/sea water, but is intense in most cases, usually completely filling the pores of the stereom during the process.

3.4 Determining and using mineralogies as geological proxies

Considering the diagenetic processes mentioned above—can echinoderm material then be used for geochemical studies at all? The answer is a cautious "yes". Echinoderm skeletal material can be altered

along different diagenetic pathways (Dickson 2004). In some instances, high-magnesium-calcite has been shown to be preserved, and examples dating back to the Silurian are known (Dickson 1995, 2004). Here cement coating clearly plays an important role, as so called "crystal caskets" (Dickson 2001) may be formed, preventing ion-exchange with surrounding pore water.

Echinoderms thus represent a potential tool for the investigation of palaeo-temperatures, geochemical cycles and ancient seawater composition. Yet, only a small number of geochemical studies have actually employed echinoderm material (e.g. Bill et al. 1995). Echinoderms, nevertheless, represent an unexploited source of original data, as emphasized by Dickson (2004) considering their durable skeleton and their common occurrence in many marine deposits.

3.4.1 Echinoderms and palaeotemperature

Initial studies on the chemistry of the echinoderm skeleton (e.g. Clarke & Wheeler 1914) suggested a good correlation between magnesium content and temperature in all extant echinoderm orders, suggesting the potential use of the apparent magnesium-temperature correlation for (palaeo-) climatological studies.

Numerous detailed investigations with wider taxon sampling and multiple samples per taxon, however, have shown that things are much more complex that initially thought. Numerous vital effects exist, influencing magnesium uptake, including (1) differing fractionation within single skeletal elements (in echinoid spines; Weber 1969), (2) between different functional elements of the skeleton (in echinoids; Clarke & Wheeler 1915, Weber 1969), (3) between different echinoderm groups, (4) effects of salinity (Pilkey & Hower 1960), (5) effects of regeneration (Davies et al. 1972, Weber 1973), (6) effects of magnesium availability in the sea water, (7) effects of metabolic rate and energy allocated to biomineralization (Roux et al. 1995), and (8) effects of growth rates (Ebert 2001).

Most of the data available relate to echinoids, whereas data on crinoids and even more so on holothurians is almost absent. Surprisingly, asteroids and ophiuroids, in contrast to echinoids, showed only small differences between different species collected from the same locality (Weber 1969) and genetic control of Mg-uptake seems to be less pronounced in these groups (Weber 1973).

Clearly, the use of echinoderm $MgCO_3$ content as a palaeo-temperature proxy is not as straightforward as once hoped for and direct comparison between data from different taxa may be highly problematic, without intense background research. The availability of suitable samples in high-Mg calcite preservation, however, will likely be the largest obstacle.

Other applications utilizing Mg-content of fossil echinoderms may be more feasible. An innovative approach, for example, was employed by Richter (1974), who facilitated Mg-loss under subaerial diagenesis for the relative dating of Pleistocene carbonate terraces.

3.4.2 *Echinoderms and ancient seawater compositions*

The chemistry of seawater is known to have varied considerably during the Phanerozoic based on data derived from sediment mineralogy (Sandberg 1983), geochemical modelling (Wilkinson & Algeo 1989, Hardie 1996) and fluid inclusion studies (Lowenstein et al. 2001, Horita et al. 2002).

Recently, Dickson (1995, 2002, 2004) and Ries (2004) employed fossil echinoderm ossicles as an independent source for Mg/Ca oscillation of ancient seawater. The resulting curves Dickson (2002, 2004) generally follow the first-order Mg/Ca seawater oscillations documented in earlier studies (Hardie 1996, Horita et al. 2002) especially following the post Triassic. Considerable differences exist, however, at a finer scale (<100 Myr) and additional data is required. Nevertheless, despite the uncertainties involved, echinoderms seem to have a high potential as a "seawater archive" given that enough material in suitable preservation can be located.

3.4.3 *Echinoderms and isotopes*

Stable isotopes analysis of carbonate shells is one of the most powerful tools in palaeoenvironmental reconstruction. Echinoderms, however, have rarely been facilitated for such studies (see compilation in Kroh & Nebelsick 2010). Existing data indicate strong vital effects and high interspecific, intraspecific and even intraindividual variation (Weber & Raup 1966a,b, 1968, Weber 1968). Both the $\delta^{13}C$- and the $\delta^{18}O$-values appear to be largely genetically controlled. Where present, correlation of carbon isotopes with temperature is positive, while in oxygen isotopes the correlation is negative. In asteroids, crinoids and echinoids $\delta^{13}C$ is negatively correlated with depth, while a positive correlation was found for $\delta^{18}O$ (Weber & Raup 1966b, Weber 1968, Roux et al. 1995, Baumiller 2001). According to Roux et al. (1995) the $\delta^{13}C$-depth correlation may be related to the metabolic rate in crinoids and thus an effect of temperature, food supply and ambient $\delta^{13}C$-level.

Stable isotope analysis in fossil echinoderms is, furthermore, complicated by diagenesis. According to Weber & Raup (1968) and Manze & Richter (1979) isotope values may decline during meteoric diagenesis and samples need to be checked for the degree of diagenetic alteration. Isotope analysis

in echinoderms, however, may be applied to a wide range of other (palaeo-)biological issues. Oji (1989), for example, used $\delta^{18}O$-data in extant isocrinoids to reconstruct seasonal temperature variation in bottom seawater and annual stem growth rates. Baumiller (2001), on the other hand, could show that regenerated parts of the skeleton and the type of soft-tissue attached can be identified by isotope signature in crinoid ossicles.

4 INFLUENCE OF GEOCHEMISTRY ON ECHINODERM EVOLUTION AND DIVERSITY

4.1 *Correlation of aragonite/calcite seas and the origination of major taxa*

Porter (2008) investigated the timing of the origin of major taxa and their relationship to changing ocean chemistries. The time of interest is in the basal Cambrian with the suggested change from an aragonite to a calcite sea. For example mollusc along with a number of lesser known taxa have predominately aragonite skeletons and originated in the aragonite sea of the most basal Cambrian while in the late Early Cambrian echinoderms, brachiopods and later on in the Ordovician other major taxa such as bryozoans and corals originated, which have "calcite" skeletons. The problems here are (1) identifying and especially dating the earliest first appearance of major taxonomic groups and (2) the designation of echinoderms as calcite taxa. Echinoderms with their high-Mg skeleton have in some cases they have been lumped together with aragonite skeletons (due to higher Mg content), in other cases lumped together with the (non-aragonitic) calcite skeletons.

4.2 *Diversity changes of echinoderms and ocean chemistry with respect to mass extinction events*

Large scale geodynamic events can be followed over hundreds of millions of years and represent repeating long-term cycles. Short-term events are also present in Earth History, most spectacularly with respect to mass extinctions representing the abrupt widespread extinction of species across taxonomic and ecological boundaries. Mass extinction events lead, not only to the demise of dominating taxonomic groups, but also to the restructuring of ecosystems opening the way for new taxa to develop. Five major mass extinction events dominate with respect to the amount of taxa at the Cambrian/Ordovician, Frasnian/Fammenian (one stage boundary before the Devonian/Carboniferous boundary), the Permian/

Triassic, the Triassic/Jurassic and the Cretaceous/Palaeogene boundaries.

The role of changes in ocean geochemistry on these boundaries can be exemplarily shown for the Permian/Triassic boundary. This boundary is apparently the most intense mass extinction event with up to 95% of the all marine and terrestrial taxa having gone extinct. Extinction patterns for echinoderms show for all taxa not only a dramatic reduction of taxonomic diversity at the Permian/Triassic boundary (for example echinoids), but also the extinction of major groups (for example the blastoids and most crinoid and asteroid lineages; Kier, 1971, 1973, Broadhead & Waters 1980, Smith 1985, Hess et al. 1999). Twitchett & Oji (2005) in their study of the affects of the Permian/Triassic extinction event show not only the echinoids, crinoids and probably the asteroids were severely affected by the event, but also that the recovery of echinoderms was apparently slow. Lower Triassic echinoderm taxa were small-sized (the "Lilliput affect" Twitchett 2005, 2007a) and restricted to low latitude environments characterized by well oxygenated, shallow water conditions (Twitchett & Oji 2005).

The Permian/Triassic boundary has long been a subject of intense research and several scenarios have been suggested for the mass extinction, most involving complex interaction of environmental factors including major changes of ocean chemistry (e.g. Wignall & Hallam 1992, Erwin 1993, 2001, 2006, Twitchett, 1999, 2007b, 2007c). Recent studies have suggested that global warming compounded by extensive degassing related to the vast Siberian basaltic volcanisms at the Permian/Triassic boundary may have triggered the dramatic changes in ocean chemistry. Geodynamic events led to a dramatic transgression (rise in sea-level) and methane release from shelf sediments thus resulting in ocean anoxia, higher concentrations of calcium and CO_2 and resulting acidic oceans (e.g. Wignall 2001, Berner 2002, Racki & Wignall 2005, Twitchett 2006, Knoll et al. 2007, Heydari et al. 2008). These combined processes lead to drastic environmental instability and eventually to the worldwide extinction of faunas in both marine and terrestrial environments.

5 CONCLUSIONS

1. The palaeoclimate and geochemistry of the oceans is the result of complex interactions between different Earth systems. These have shown, on the one hand, a distinct development, for example in the evolution of organisms

and the development of ecosystems. Secular changes, on the other hand, are postulated for climate change and geochemical cycles.
2. Geochemical cycles have shown strong fluctuations in the past. Long term variations have been demonstrated for concentrations of calcium, CO_2, O_2 as well as aragonite or low magnesium dominated oceans.
3. Well preserved echinoderms have been used as a proxy for changes in ocean geochemistry. This can only be done if exceptionally well preserved material is present protected from diagenetic alterations thus preserving original geochemical signals.
4. Ocean geochemistry may have influenced echinoderm evolution with respect to origination of the high magnesium calcitic shell as well as the fate of echinoderms during mass extinction events.
5. Rates of change at the present are much higher than the long-term secular variations of ocean geochemistry in the past. Ocean acidification is pushing us back into a calcite sea at a rate not yet experienced in long term developments of geochemical cycles and can only be compared with postulated short-term changes experienced during mass extinction events.

REFERENCES

Allison, P.A. 1990. Variation in rates of decay and disarticulation of Echinodermata: implications for the application of actualistic data. *Palaios* 5: 432–440.

Ausich, W.I. 2001. Echinoderm taphonomy. In M. Jangoux & J.M. Lawrence (eds) *Echinoderm studies* 6: 171–227. Lisse: Balkema.

Baumiller, T.K. 2001. Light stable isotope geochemistry of the crinoid skeleton and its use in biology and paleobiology. In M.F. Barker (ed.) *Echinoderms 2000.— Proceedings of the 10th International Echinoderm Conference Dunedin, New Zealand, 31 Jan.–4 Feb. 2000*: 107–112. Lisse: A.A. Balkema Publishers (Swets & Zeitlinger B.V.).

Beerling, D.J. & Royer, D.L. 2002. Reading a CO2 signal from fossil stomata. *New Phytologist* 153: 387–397.

Berner, R.A. 1966. Diagenesis of carbonate sediments: interaction of magnesium in sea water with mineral grains. *Science* 153: 188–191.

Berner, R.A. 1998. The carbon cycle and CO2 over Phanerozoic time: the role of land plants. *Philosophical Transactions of the Royal Society of London* B 353: 75–82.

Berner, R.A. 2002. Examination of hypotheses for the Permo-Triassic boundary extinction by carbon cycle modelling. *Proceedings of the National Academy of Science* 99: 4172–4177.

Berner, R.A. 2006. GEOCARBSULF: A combined model for Phanerozoic atmospheric O_2 and CO_2. *Geochimica et Cosmochimica Acta* 70: 5653–5664.

Berner R.A. & Kothavala, Z. 2001. GEOCARB III: a revised model of atmospheric CO2 over Phanerozoic time. *American Journal of Science* 301: 182–204.

Bill, M., Baumgartner, P.O. & Hunziker, J.C. 1995. Carbon isotope stratigraphy of the Liesberg Beds Member (Oxfordian, Swiss Jura) using echinoids and crinoids. *Eclogae Geologicae Helvetiae* 88: 135–155.

Brett, C.E., Moffat, H.A. & Taylor, W.L. 1997. Echinoderm taphonomy, taphofacies, and Lagerstätten. In J.A. Waters & C.G. Maples, (eds), *Geobiology of echinoderms: Paleontological Society Papers* 3: 147–190.

Broadhead, W. & Waters, J.A. (eds) 1980. *Echinoderms, Notes for a Short Course.* University of Tennessee: Department of Geologica Sciences.

Caldeira, K & Wickett, M.E. 2003. Anthropogenic carbon and ocean pH. *Nature* 425: 365.

Cerling T.E. 1999. Stable carbon isotopes in palaeosol carbonates. *Special Publications of the International Association of Sedimentologists* 27: 43–60.

Clarke, F.W. & Wheeler, W.C. 1914. The composition of crinoid skeletons. *United States Geological Survey Professional Papers* 90D: 33–37.

Clarke, F.W. & Wheeler, W.C. 1915. The inorganic constituents of echinoderms. *United States Geological Survey Professional Papers* 90L: 191–196.

Crowley, T.J. & Berner, R.A. 2001. CO$_2$ and climate change. *Science* 292: 870–872.

Davies, T.T., Crenshaw, A. & Heatfield, B.M. 1972. The effect of temperature on the chemistry and structure of echinoid spine regeneration. *Journal of Paleontology* 46: 874–883.

De'ath, G., Lough, J.M. & Fabricius, K.E. 2009. Declining Coral Calcification on the Great Barrier Reef. *Science* 323: 116–119.

Dickson, J.A.D. 1995. Paleozoic Mg calcite preserved: Implications for the Carboniferous ocean *Geology* 23: 535–538.

Dickson, J.A.D. 2001. Diagenesis and crystal caskets: Echinoderm Mg calcite transformation, Dry Canyon, New Mexico, U.S.A. *Journal of Sedimentary Research* 71: 764–777.

Dickson, J.A.D. 2002. Fossil Echinoderms as Monitor of the Mg/Ca Ratio of Phanerozoic Oceans. *Science* 298: 1222–1224.

Dickson, J.A.D. 2004. Echinoderm skeletal preservation: Calcite-aragonite seas and the Mg/Ca ratio of Phanerozoic oceans. *Journal of Sedimentary Research* 74: 355–365.

Doney, S.C., Fabry, V.J., Feely, R.A. & Kleypas, J.A. 2009. Ocean acidification: the other CO2 problem. *Annual Review of Marine Science* 1, 169–192.

Donovan, S.K. 1991. The taphonomy of echinoderms: calcareous multi-element skeletons in the marine environment. In S.K. Donovan (ed.) *The Processes of Fossilisation:* 241–269. London: Belhaven Press.

Ebert, T.A. 2001. Growth and survival of post-settlement sea urchins. In J.M. Lawrence (ed.) *Edible Sea Urchins: Biology and Ecology. Developments in Aquaculture and Fisheries Science:* 79–102. Amsterdam: Elsevier.

EPICA Community Members (2004) Eight glacial cycles from an Antarctic ice core. *Nature* 429: 623–628.

Erwin, D.H., 1993. *The great Paleozoic Crisis, life and death in the Permian.* New York: Columbia University Press.

Erwin, D.H. 2001. Lessons from the past: Biotic recoveries from mass extinctions. *Proceedings of the National Association of Science* 98: 5399–5403.

Erwin, D.H. 2006. *Extinction: How Life Nearly Died 250 Million Years Ago.* Princeton: Princeton University Press.

Farkaš, J., Böhm, F., Wallmann, K., Blenkinsop, J., Eisenhauer, A., van Geldern, R., Munnecke, A., Voigt, S. & Veizer, J. 2007. Calcium isotope record of Phanerozoic oceans: Implications for chemical evolution of seawater and its causative mechanisms. *Geochimica et Cosmochimica Acta* 71: 5117–5134.

Fischer, A.G. 1981. Climatic oscillations in the biosphere. In M. Nitecki, (ed.) *Biotic Crises in Ecological and Evolutionary Time:* 103–131. New York. Academic Press.

Fischer, A.G. 1982. Long-term climatic oscillations recorded in stratigraphy. In W. Berger (ed.) *Climate in Earth History:* 97–104. Washington, National Academy of Sciences.

Freeman K.H. & Hayes, J.M. 1992. Fractionation of carbon isotopes by phytoplankton and estimates of ancient CO2 levels. *Global Biogeochemical Cycles* 6: 185–198.

Hallam, A. 1992. *Phanerozic Sea-Level Changes.* New York: Columbia University Press.

Haq, B.U., Hardenbol, J. & Vail, P.R. 1987. Chronology of fluctuating sea levels since the Triassic. *Science* 235: 1156–1167.

Hardie, L.A. 1996. Secular variation in seawater chemistry: An explanation for the coupled secular variation in the mineralogies of marine limestones and potash evaporates over the past 600 my. *Geology* 24: 279–283.

Hess, H., Ausich, W.I., Brett, C.E. & Simms, M.S. 1999. *Fossil Crinoids.* Cambridge: Cambridge University Press.

Heydari, E., Arzani, N. & Hassanzadeh, J. 2008. Mantle plume: The invisible serial killer—Application to the Permian-Triassic boundary mass extinction. *Palaeogeography, Palaeoclimatology, Palaeoecology* 264: 147–162.

Holland, H.D. 2005. Sea level, sediments and the composition of seawater. *American Journal of Science.* 305: 220–239.

Holland, H.D. 2006. The geologic history of seawater. In H. Elderfield (ed.) *Treatise on Geochemistry* 6: 583–625. Amsterdam: Elsevier.

Horita J., Zimmermann H. & Holland H.D. 2002. Chemical evolution of seawater during the Phanerozoic: implications from the record of marine evaporites. *Geochimica et Cosmochimica Acta* 66: 3733–3756.

James, R.H. & Austin, W.E.N. 2008. Biogeochemical controls on palaeoceanographic environmental proxies: a review. *Geological Society of London, Special Publications* 303: 3–32.

Kier, P.M. 1971. The echinoderms and Permian-Triassic time *Bulletin of Canadian Petroleum Geology* 19: 331–332.

Kier, P.M. 1973. The echinoderms and Permian-Triassic time. *Memoir, Canadian Society of Petroleum Geologists* 2: 622–629.

Kier, P.M. 1977. The poor fossil record of the regular echinoids *Paleobiology* 3: 168–174.

Kleypas, J.A., Buddemeier, R.W. & Gattuso, J.-P. 2001. The future of coral reefs in an age of global change. *International Journal of Earth Sciences* 90: 426–437.

Kleypas, J.A., Buddemeier, R.W. Archer, D., Gattuso, J.-P., Langdon C. & Opdyke, B.N. 1999. Geochemical consequences of increased atmospheric carbon dioxide on coral reefs. *Science* 284: 118–120.

Kleypas, J.A., Feely, R.A., Fabry, V.J., Langdon, C., Sabine, C.L. & Robbins, L.L. 2006. Impacts of ocean acidification on coral reefs and other marine calcifiers: A guide for future research, report of a workshop held 18–20 April 2005, St. Petersburg, FL, sponsored by NSF, NOAA, and the U.S. Geological Survey.

Knoll, A.H., Bambach, R.K., Payne, J.L., Pruss, S. & Fischer, W.W. 2007. Paleophysiology and end-Permian mass extinction. *Earth and Planetary Science Letters* 256: 295–313.

Konrad, W., Roth-Nebelsick, A., Kerp, H. & Hass. H. 2000. Transpiration and assmiliation of Early Devonian land plants with axially symmetric telomeres—simulations on the tissue level. *Journal of Theoretical Biology* 206: 91–107.

Kroh A. & Nebelsick, J.H. 2010. Echinoderms and Oligo-Miocene Carbonate Systems: Potential applications in sedimentology and environmental reconstruction. *International Association of Sedimentologist, Special. Publications* 42, 201–228.

Kürschner, W.M., Kvacek, Z. & Dilcher, D.L. 2008. The impact of Miocene atmospheric carbon dioxide fluctuations on climate and the evolution of terrestrial ecosystems. *Proceedings of the National Association of Science* 105: 449–453.

Lewis, R. 1980. Taphonomy. In T.W. Broadhead & J.A. Waters (eds) Echinoderms, notes for a short course: University of Tennessee, Studies in Geology 3: 27–39.

Lowenstein, T.K., Timofeeff, M.N., Brennan, S.T., Hardie, L.A. & Demicco, R.V. 2001. Fluid inclusions in marine evaporite deposits provide the most definitive constraints on the Mg2+/Ca2+ ratio of ancient seawater. *Science* 294: 1086.

Lowenstein T.K., Hardie M.N., Timofeeff R.V. & Demicco R.V. 2003. Secular variations in seawater chemistry and the origin of calcium chloride basinal brines. *Geology* 31: 857–860.

Macurda, D.B. Jr., Meyer, D.L. & Roux, M. 1978. The crinoid stereom. In R.C. Moore (ed.) *Treatise on Invertebrate Paleontology, Part T. Echinodermata, 2 (1)*: T217-T228. Boulder, Colorado & Lawrence, Kansas: Geological Society of America & University Kansas Press.

Manze, U. & Richter, D.K. 1979. Die Veränderung des C^{13}/C^{12}—Verhältnisses in Seeigelcoronen bei der Umwandlung von Mg-Calcit in Calcit unter meteorisch-vadosen Bedingungen. *Neues Jahrbuch für Geologie und Paläontologie, Abhandlungen* 158: 334–345.

Märkel, K., Kubanek, F. & Willgallis, A. 1971. Polykristalliner Calcit bei Seeigeln (Echinodermata, Echinoidea). *Zeitschrift für Zellforschung und mikroskopische Anatomie* 119(3): 355–377.

Meyer, D.L. & Meyer, K.B. 1986. Biostratinomy of Recent crinoids (Echinodermata) at Lizard Island, Great Barrier Reef, Australia. *Palaios* 1: 294–302.

Montañez, I.P., Tabor, N.J., Niemeier, D., DiMichele, W.A., Frank, T.D., Fielding, C.R., Isbell, J.L., Birgenheier, L.P. & Rygel, M.C. 2007. CO_2-forced climate and vegetation instability during Late Paleozoic deglaciation. *Science* 315: 87–91.

Nebelsick, J.H. 2004: Taphonomy of Echinoderms: A Review. In Th. Heinzeller & J.H. Nebelsick (eds) Echinoderms München. Proceedings of the 11th International Echinoderm Meeting: 471–478. Rotterdam: Taylor & Francis.

Neugebauer, J. 1979. Drei Probleme der Echinodermendiagenese: Innere Zementation, Mikroporenbildung und der Übergang von Magnesiumcalcit zu Calcit *Geologische Rundschau* 68: 856–875.

Oji, T. 1989. Growth rate of stalk of *Metacrinus rotundus* (Echinodermata: Crinoidea) and its functional significance *Journal of the Faculty of Science University of Tokyo, Section II* 22: 39–51.

Pagani, M., Arthur, M.A. & Freeman, K.H. 1999. Miocene evolution of atmospheric carbon dioxide. *Palaeoceanography* 14: 273–293.

Pagani, M., Zachos, J.C., Freeman, K.H., Tipple, B. & Bohaty, S. 2005. Marked Decline in Atmospheric Carbon Dioxide Concentrations During the Paleogene. *Science* 309: 600–603.

Palmer, T.M., Hudson, J.D. & Wilson, M.A. 1988. Palaeoecological evidence for early aragonite dissolution in ancient calcite seas. *Nature* 335: 809–810.

Pearson, P.N. & Palmer, M.R. 1999. Middle Eocene seawater pH and atmospheric carbon dioxide concentrations. *Science* 284: 1824–1826.

Pearson, P.N. & Palmer, M.R. 2000. Atmospheric carbon dioxide concentrations over the past 60 million years, *Nature* 406: 695–699.

Petit, J.R., Jouzel, J., Raynaud, D., Barkov, N.I., Barnola, J.-M., Basile, I., Bender, M., Chappellaz, J., Davis, M., Delaygue, G., Delmotte, M., Kotlyakov, V.M., Legrand, M., Lipenkov, V.Y., Lorius, C., Pepin, L., Ritz, C., Saltzman, E. & Stievenard, M. 1999. Climate and history of the past 420,000 years from the Vostok ice core, Antarctica. *Nature* 399: 429–436.

Pilkey, O.H. & Hower, J. 1960. The effect of environment on the concentration of skeletal Magnesium and Strontium in *Dendraster. Journal of Geology* 68: 203–216.

Porter, S.M. 2008. Seawater chemistry and early carbonate biomineralization. *Science* 316: 1302.

Prokoph, A., Shields, G.A. & Veizer, J. 2008. Compilation and time-series analysis of a marine carbonate $\delta 18O$, $\delta 13C$, 87Sr/86Sr and $\delta 34S$ database through Earth history. *Earth-Science Reviews* 87: 113–133.

Racki, G. & Wignall, P.B., 2005. Late Permian double-phased mass extinction and volcanism: an oceanographic perspective. In D.J. Over, J.R. Morrow & P.B. Wignall (eds) Understanding Late Devonian and Permian-Triassic biotic and climatic events: Towards an integrated approach. *Developments in Palaeontology and Stratigraphy* 20: 263–297.

Raven, J. 2005. Ocean acidification due to increasing atmospheric carbon dioxide. *The Royal Society Policy Document* 12/05: 1–60.

Raymo, M.E. 1991. Geochemical evidence supporting T.C. Chamberlin´s theory of glaciation. *Geology* 19: 344–347.

Richter, D.K. 1974. Zur subaerischen Diagenese von Echinidenskeletten und das relative Alter pleistozäner Karbonatterrassen bei Korinth (Griechenland) *Neues Jahrbuch für Geologie und Paläontologie, Abhandlungen.* 146: 51–77.

Riebesell, U., Zondervan, I., Rost, B., Tortell, P.D., Zeebe, R.E. & Morel, F.M.M. 2000. Reduced calcification of marine plankton in response to increased atmospheric CO2. *Nature* 407: 364–367.

Ries, J.B. 2004. Effect of ambient Mg/Ca ratio on Mg fractionation in calcareous marine invertebrates: A record of the oceanic Mg/Ca ratio over the Phanerozoic. *Geology* 32: 981–984.

Ries, J.B. 2008. Palaeoceanography—Seeing changes in a changing sea. *Nature Geoscience* 1: 497–498.

Roth-Nebelsick, A. 2005. Reconstructing atmospheric carbon dioxide with stomata: possibilities and limitiations of a botanical pCO2-sensor. *Trees—Structure and Function* 19: 251–265.

Roux, M., Renard, M., Améziane, N. & Emmanuel, L. 1995. Zoobathymetrie et composition chimique de la calcite des ossicules du pedoncule des crinoides. *Comptes Rendus de l'Académie des Sciences Paris, Série IIa* 321: 675–680.

Royer, D.L. 2006. CO2-forced climate thresholds during the Phanerozoic. *Geochimica et Cosmochimica Acta* 70: 5665–5675.

Royer, D.L., Berner, R.A., Montanez, I.P., Tabor, N.J., Beerling, D.J. 2004. CO2 as a primary driver of Phanerozoic climate. *GSA Today* 3: 4–10.

Sandberg, P.A. 1983. An oscillating trend in Phanerozoic nonskeletal carbonate mineralogy. *Nature* 305: 19–22.

Schroeder, J.H. 1969. Experimental dissolution of calcium, magnesium, and strontium in Recent skeletal carbonates: A model of diagenesis. *Journal of Sedimentary Petrology* 39: 1057–1073.

Schroeder, J.H., Dwornik, E.J. & Papike, J.J. 1969. Primary protodolomite in echinoid skeletons. *Geological Society of America, Bulletin* 80: 1613–1618.

Smith, A.B. 1980. Stereom microstructure of the echinoid test. *Special Papers in Palaeontology* 25: 1–81.

Smith, A.B. 1985. Crinoidea. In Echinodermata, Chapter 7. In J.W. Murray (ed.) *Atlas of invertebrate macrofossils:* 153–181. London: The Palaeontological Association.

Stanley, S.M. & Hardie, L.A. 1998. Secular oscillations in the carbonate mineralogy of reef-building and sediment-producing organisms driven by tectonically forced shifts in seawater chemistry. *Palaeogeography, Palaeoclimatology, Palaeoecology* 144: 3–19.

Steuber, T. & Veizer J. 2002. Phanerozoic record of plate tectonic control of seawater chemistry and carbonate sedimentation. *Geology* 30: 1123–1126.

Tajika, E. 1998. Climate change during the last 150 million years: reconstruction from a carbon cycle model. *Earth and Planetary Science Letters* 160: 695–707.

Twitchett, R.J. 1999. Palaeoenvironments and faunal recovery after the end-Permian mass extinction. *Palaeogeography, Palaeoclimatology, Palaeoecology* 154: 27–37.

Twitchett, R.J. 2005. The Lilliput Effect in the aftermath of the end-Permian extinction event. *Albertiana* 33: 79–81.

Twitchett, R.J. 2006. The palaeoclimatology, palaeoecology and palaeoenvironmental analysis of mass extinction events. *Palaeogeography, Palaeclimatology, Palaeoecology* 232: 190–213.

Twitchett, R.J. 2007a. The Lilliput affect in the aftermath of the end–Permian extinction event. *Palaeogeography, Palaeclimatology, Palaeoecology* 252: 132–144.

Twitchett, R.J. 2007b. The Late Permian mass extinction event and recovery: biological catastrophe in a greenhouse world. In: P.R. Sammonds & J.M.T. Thompson (eds) *Advances in Earth Science—from earthquakes to global warming, Royal Society Series on Advances in Science—Vol. 2*: 69–90. London: Imperial College Press.

Twitchett, R.J. 2007c. Climate change across the Permian/Triassic boundary. In: M. Williams, A.M. Haywood, F.J. Gregory & D.N. Schmidt (eds), *Deep-Time Perspectives on Climate Change: Marrying the Signal from Computer Models and Biological Proxies. The Micropalaeontological Society, Special Publications*: 191–200. London: The Geological Society.

Twitchett, R.J. & Oji, T. 2005. Early Triassic recovery of echinoderms. *Comptes Rendus Palevol* 4: 531–542.

Vail, P.R., Mitchum, R.M. Jr. & Thompson, S. III, 1977. Global cycles of relative changes of sea level. *American Association of Petroleum Geologists Memoir* 26: 83–97.

Wallmann K. 2001. Controls on the Cretaceous and Cenozoic evolution of seawater composition, atmospheric CO2 and climate. *Geochimica et Cosmochimica Acta* 65: 3005–3025.

Weber, J.N. 1968. Fractionation of stable isotopes of carbon and oxygen in calcareous marine invertebrates—the Asteroidea, Ophiuroidea, and Crinoidea. *Geochimica et Cosmochimica Acta* 32: 33–70.

Weber, J.N. 1969. The incorporation of magnesium into the skeletal calcite of echinoderms. *American Journal of Science* 267: 537–566.

Weber, J.N. 1973. Temperature dependence of Magnesium in echinoid and asteroid skeletal calcite: a reinterpretation of its significance. *Journal of Geology* 81: 543–556.

Weber, J.N. & Raup, D.M. 1966a. Fractionation of stable isotopes of carbon and oxygen in marine calcareous organisms—the Echinoidea. Part I. Variation of C^{13} and O^{18} content within individuals. *Geochimica et Cosmochimica Acta* 30: 681–703.

Weber, J.N. & Raup, D.M. 1966b. Fractionation of stable isotopes of carbon and oxygen in marine calcareous organisms—the Echinoidea. Part II. Environmental and genetic factors. *Geochimica et Cosmochimica Acta* 30: 705–736.

Weber, J.N. & Raup, D.M. 1968. Comparison of C^{13}/C^{12} and O^{18}/O^{16} in skeletal calcite of recent and fossil echinoids. *Journal of Paleontology* 42: 37–50.

Wignall P.B. 2001. Large igneous provinces and mass extinctions. *Earth-Science Reviews* 53: 1–33.

Wignall, P.B. & Hallam, A. 1992. Anoxia as a cause of the Permian/Triassic extinction: facies evidence from northern Italy and the western United States. *Palaeogeography, Palaeoclimatology, Palaeoecology* 93: 21–46.

Wilkinson, P.A. & Alego, T.J. 1989. Sedimentary carbonate record of calcium-magnesium cycling. *American Journal of Science* 289: 1158–1194.

Woodward, F.I. 1987. Stomatal numbers are sensitive to increases in CO_2 concentration from pre-industrial levels. *Nature* 327:617–618.

Zachos, J., Pagani, M., Sloan, L., Thomas, E. & Billups, K. 2001. Trends, rhythms, and aberrations in global climate 65 Ma to present. *Science* 282: 686–693.

Echinoderms in a Changing World – Johnson (ed)
© *2013 Taylor & Francis Group, London, ISBN 978-1-138-00010-0*

How many species of fossil holothurians are there?

M. Reich

Geoscience Centre of the Georg-August University, Göttingen, Germany

ABSTRACT: The species-level diversity of fossil Holothuroidea is summarised for each geological period since the Ordovician. In chronological order they are: Ordovician (9 published fossil paraspecies/species names), Silurian (1), Devonian (41), Carboniferous (65), Permian (26), Triassic (252), Jurassic (264), Cretaceous (103), Paleogene (103), Neogene (39), Quaternary (17). Based on a compilation of more than 1,300 primary literature references on fossil sea cucumbers, about 500 to 600 valid species can be recognised from 948 published names. Triassic and Jurassic holothurian taxa thus appear to dominate the fossil record, compared to their Phanerozoic diversity. By contrast, Early and Late Paleozoic as well as Neogene holothurians are noticeably under-represented as fossils, for various reasons. A detailed summary is provided of all important localities (fossil lagerstätten), yielding holothurian body fossils.

1 INTRODUCTION

Smiley (1994) and Kerr (2003) reported a total of more than 1,400 (1,427/1,430) valid species of Recent Holothuroidea. But how many species of fossil sea cucumbers are there? Older figures of ~115 valid fossil holothurian species names can be culled from Durham (1954) and, some years later, Frizzell & Exline (1956, 1958) have catalogued 128 and respectively 153 fossil species. In the *Treatise on Invertebrate Paleontology* Frizzell & Exline (1966: U658–U662) listed 295 holothurian paraspecies and species. In the next decades the total number of fossil sea cucumber species increased. Pawson (1980) reported 454 species based on sclerites and body fossils; Gilliland (1993a) mentioned 559, Reich (2004c) mentioned 794 paraspecies and species of fossil holothurians.

Here I offer a modern summary of crude fossil holothurian diversity (Figs. 1–3) and the most important localities (fossil lagerstätten; Table 1) of holothurian body fossils as a baseline for future research. Around

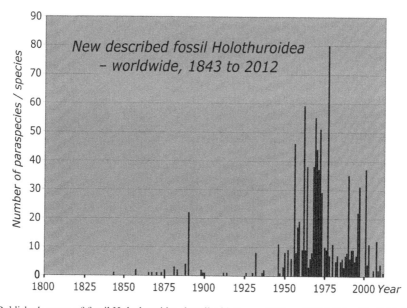

Figure 1. Published names of fossil Holothuroidea described between 1843 and 2012 (March), divided into decades. In comparison with a bibliography of all published fossil holothurian references, several specific time intervals of 'active holothurian researchers' were visible.

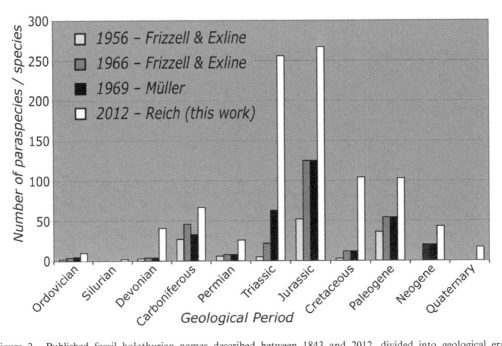

Figure 2. Published fossil holothurian names described between 1843 and 2012, divided into geological eras. Comparison of datasets from Frizzell & Exline (1956, 1966), Müller (1969) and Reich (2012, this work). Data includes newly described subfossil (Quaternary) holothurian species, probably synonymous with Recent species.

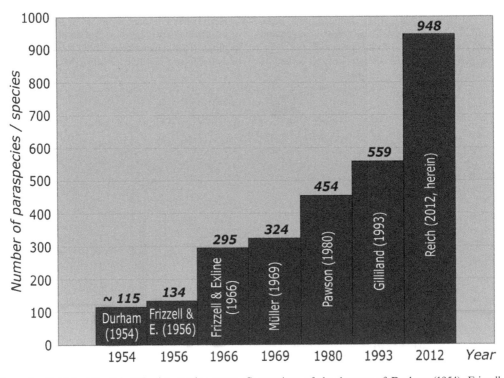

Figure 3. Published fossil holothurian species names. Comparison of the datasets of Durham (1954), Frizzell & Exline (1956, 1966), Müller (1969), Pawson (1980), Gilliland (1993a), and Reich (2012, this work).

Table 1. Significant localities yielding holothurian body fossils, including details of their stratigraphic position and approximate age in millions of years (Ma). Ages based primarily on the 'International stratigraphic chart' http://www.stratigraphy.org/ics%20chart/09_2010/StratChart2010.pdf. Systematic designations were taken from the primary publication.

Locality	Country	Period	Epoch	Ma	Taxa and notes
'Nawashiroda Formation': Ibaraki Prefecture, Kanto	Japan	Middle Miocene	Langhian	15	Less than 10 specimens of two species; *Cucumaria igoi* (Dendrochirotida: Cucumariidae) and '*Ypsilothuria bitentaculata*' (Dactylochirotida: Ypsilothuriidae)
'Upper Chalk': England	U.K.	Late Cretaceous	Santonian	84	One undescribed specimen (Aspido- chirotida: Holothuriidae)
'Fish Shale': Hjoŭla	Lebanon	Late Cretaceous	Early Late Cenomanian	95	Several undescribed specimens (Aspido- chirotida: Holothuriidae; more than 20 specimens)
Tlayúa near Tepexi de Rodriguez	Mexico	Early Cretaceous	Middle to Late Albian	105	Two described dendrochirote species (*Paleopentacta alencasterae*, *Parapsolus tlayuensis*) and several yet undescribed specimens (less than 10 specimens)
'Solnhofen limestone': Solnhofen, Eichstätt, Bavaria	Germany	Late Jurassic	Early Tithonian	149	*Protholothuria armata* and a few undescribed specimens (Aspidochirotida: Holothuriidae) (total of less than 10 specimens)
'Hauptrogenstein': Schinznach and Auenstein, Aargau	Switzerland	Middle Jurassic	Bathonian	166	*Holothuriopsis pawsoni* (Apodida) and one specimen with a preliminary description (Aspidochirotida)
Collbató, Catalonia	Spain	Middle Triassic	(Late) Ladinian	230	*Collbatothuria danieli* (Aspidochirotida), *Momilipsolus mirabile* (Dendrochirotida: Psolidae), *Strobilothyone rogenti* (Dendrochirotida: Heterothyonidae) (in total less than 30 specimens)
Between Montral and Alcover, Tarragona	Spain	Middle Triassic	Late Ladinian	230	*Bathysynactites viai* (Aspidochirotida: Synallactidae), *Oneirophantites tarragonemis* (Elasipoda: Deimatina) (2 specimens)
Nitzenhausen near Künzelsau, Baden- Württemberg	Germany	Middle Triassic	Earliest Ladinian	235	Numerous preliminary described specimens (Aspidochirotida: Holothuri-idae) on one broken slab
'Mazon Creek biota': near Essex, Illinois	U.S.A.	Late Carboniferous	Pennsylvanian, Late Moscovian	308	More than seven thousand specimens of (probably one) unnamed apodid species with *Achistrum*-like body-wall ossicles
Near Drolshagen, North Rhine-Westphalia	Germany	Middle Devonian	Early Givetian	389	Less than 5 specimens of *Nudicorona seilacheri* (Apodida) and one specimen of *Podolepithuria walliseri*
'Hunsrück Slate': Bundenbach, Gemünden etc., Rhineland-Palatinate	Germany	Early Devonian	Emsian	405	*Palaeocucumaria hunsrueckiana* ('Arthro- chirotida'; ~50 specimen) and at least one more undescribed species (Elasipoda; less than 5 specimens)
SW of Jachal, Argentine Precordillera	Argentina	Early Devonian	?	414-400	Less than 15 specimens of *Prokrustia tabulifera* and *Andenothyone gond- wanensis* (Dendrochirotida: ?Hetero- thyonidae)
Yass Basin, New South Wales	Australia	Late Silurian	Upper Ludlow and Přídolian	419 and 417	One recently new described species: *Porosolhyone picketti* (Dendrochirotida: Heterothyonidae; less than 5 specimens)
Llandrindod, Wales	U.K.	Middle Ordovician	Late Darriwilian	461	Around 8 poorly preserved specimens, recently described as the new genus and species: *Oesolecumaria eostre*

500 to 600 valid species can be probably recognised here from 948 published names (see appendix).

2 METHODS

No single modern catalogue of fossil sea cucumbers and their relatives exists, thus data had to be compiled from the primary literature (>1,300 references; Reich 2011). Here I restrict counts of species to published names only, i.e., excluding thesis work or papers in progress. The total number of described names (948) incorporates synonyms, nomina dubia, nomina nuda, and incorrectly assigned material (28 taxa; see appendix). The detailed list includes species, original genus, author(s), type stratum and locality.

3 RESULTS

3.1 Ordovician

Only a handful of Ordovician paraspecies, often based on simple sieve plates are known today, mostly from Upper Ordovician rocks of the U.S.A. and Europe (overview in Reich 1999a).

New well-preserved material (synallactid crosses, stem-group apodid wheels) comes from partly silicified limestones of Sweden (Sandbian, Katian, and Hirnantian in age; Reich 2001, 2010a). Reich (1999b) and Reich & Lehnert (subm.) reported the stratigraphically oldest unequivocal record of Holothuroidea, based on calcareous ring elements and associated ossicles from the late Middle Ordovician (early Darriwilian) of Sweden.

Recently, the oldest (late Darriwilian) articulated holothurians yet known were mentioned (Botting et al. 2004, Muir & Botting 2005) and later described by Botting & Muir (2012), but the specimens are doubtful and too poorly preserved for a specific assignment to be made. Moreover, the new erected genus and species does not provide further insights into early holothurian evolution or the origin of the calcareous ring.

3.2 Silurian

Holothurian ossicles from Silurian strata were mostly overlooked by all palaeontologists, since the first mentions and figures by Mostler (1968b: Fig. 10), Franzén (1979), and Wang & Li (1986: Pl. 1, Fig. 10). During the last fifty years most holothurian workers thought that the first unequivocal holothurians appeared in the Early Devonian, because of the missing Silurian record and the often poorly preserved simple Ordovician sieve plates.

Over decades, there were no published holothurian taxa from Silurian strata worldwide. Therefore Silurian sea cucumbers have definitely been understudied in comparison to their other Palaeozoic and Mesozoic counterparts. Well-preserved calcareous ring material comes from the Middle and Upper Silurian of Sweden (e.g., Reich & Kutscher 2001, Reich 2010b, Reich et al. 2011, and work in progress by the author). There is one mentioned holothurian body fossil from the Late Silurian (Přídolian) of Australia, unpublished since the mid 1980s (Gilliland 1993a, Haude 1995a), recently described by Jell (2010) in detail.

3.3 Devonian

More than 40 paraspecies and species have been reported from the Devonian, mostly from Europe (Germany and Poland; Beckmann 1965, Boczarowski 2001). The latter author recently monographed a diverse holothurian assemblage, based on ossicles and calcareous ring elements.

Lehmann (1958) was the first to describe a Devonian holothurian body fossil (*Palaeocucumaria*) from the Hunsrück Slate fossil lagerstätte; a new study showed a more diverse holothurian body fossil assemblage (Smith & Reich 2009, Reich & Smith 2010, work in progress) with 2–3 taxa. A few other holothurian body fossil taxa from Germany and Argentina were described by Haude (1995b, 1997, 2002). Other Gondwanan body fossil records (e.g., Devonian of South Africa) are yet unstudied.

3.4 Carboniferous

Around 65 holothurian paraspecies from Carboniferous strata have been reported since 1881 (Etheridge), mostly from the U.S.A. and P.R. China (e.g., Gutschick et al. 1967, Ding 1985). Only a few taxa were described from Europe (U.K., Germany, Poland) and Afghanistan (e.g., Alexandrowicz 1971, Mostler 1971, Weber 1997). No modern work on Carboniferous sea cucumbers is available.

Late Carboniferous holothurian body fossils from the Mazon Creek fossil lagerstätte were only preliminarily studied by Sroka (1988) and Sroka & Blake (1997). More than 7,000 specimens were known (Reich pers. observ.), originally deposited in a brackish estuarine environment, and showing well-preserved *in situ* ossicles and the entire calcareous ring (Reich & Stegemann 2010).

3.5 Permian

Less then 30 paraspecies of Holothuroidea from the Permian are known worldwide, mostly reported from the U.S.A., Iran, and Italy (e.g., Mostler & Rahimi-Yazd 1976; Kozur & Mostler 1989, 2008).

There are no reported holothurian body fossils from Permian sediments. Permian sea cucumbers have been relatively understudied in comparison to their other Palaeozoic and Mesozoic counterparts.

3.6 Triassic

More than 250 paraspecies and species are recorded from the Triassic, belonging to all modern orders and various environments. An extensive holothurian fauna has been described from Europe, mostly from Austria, Germany, Italy, Slovakia, Hungary, and Spain (Kristan-Tollmann 1963ff., Mostler 1968aff., Kozur 1969ff.); a few reports are from Turkey, Jordan, India, and the P.R. China. No modern catalogue of Triassic Holothuroidea exists at the moment, and future work has to be done to effect a better understanding of early Mesozoic diversification.

Well-preserved holothurian body fossils were reported from north-eastern Spain and southern Germany (Cherbonnier 1978, Smith & Gallemí 1991, Reich 2004a).

3.7 Jurassic

The holothurian fauna of the Jurassic has received a considerable amount of study since the first report of fossil holothurian ossicles in 1843 (synaptid anchor; Late Jurassic of Germany) by von Münster. 264 paraspecies and species are known today, mostly from Europe (Austria, France, Germany, Poland, Russia, Switzerland, U.K.; e.g., Deflandre-Rigaud 1962, Mostler 1972), and others from Egypt, India, and Jordan. No modern catalogue of Jurassic sea cucumbers exists as yet.

Jurassic body fossils of sea cucumbers were reported from the Middle Jurassic of Switzerland (Hess 1973, Hess & Holenweg 1998), as well as from the Late Jurassic Solnhofen fossil lagerstätte in southern Germany (e.g., Giebel 1857). Frickhinger (1999) reported new undescribed findings from the Solnhofen area. Not all previously published taxa belong to the Holothuroidea (e.g., Ziegler 1991).

3.8 Cretaceous

For nearly 150 years, only a very sparse holothurian fauna was recorded from Cretaceous sediments. About 100 paraspecies and species are known at present, mostly from Europe (Germany, Poland, Austria, Hungary), as well as from the U.S.A. and Jordan, and these were described by Reich (1995ff.), Sadeddin (1996ff.), Reich & Wiese (2010), and others.

Cretaceous holothurian body fossils were recorded from Albian lithographic limestones near Tepexí de Rodríguez (Mexico; e.g., Buitrón-Sánchez 1991, Applegate et al. 2009), and the Cenomanian 'Fish Shale' of Lebanon (Reich 2004b), but are yet undescribed or only incompletely investigated.

3.9 Paleogene

More than 100 paraspecies of sea cucumbers were reported from the Paleogene, most from Europe (France, Germany, Poland; Schlumberger 1888f.,

Reich 2002), India (Tandon & Saxena 1977), the U.S.A, and Brazil. No modern catalogue of these forms exists.

No holothurian body fossils have been reported from Paleogene sediments.

3.10 Neogene

It is surprising that only nearly 40 Neogene holothurian paraspecies and species are recorded worldwide, mostly described or reported from Europe (Austria, France, Poland; Kristan-Tollmann 1964, Rioult 1965, Walkiewicz 1977, Górka 1997, Reich & Kroh 2010), as well as from the U.S.A. and Japan (Ishijima & Hatai 1973).

Well-preserved holothurian body fossils were described from Miocene sediments of the Kantō region, Japan, and partly designated as belonging to modern species (Kikuchi & Nikaido 1996). Neogene holothurians have definitely been understudied in comparison to their other Mesozoic and Cenozoic counterparts.

3.11 Quaternary

For the most part, Pleistocene and Holocene (subfossil) records of sea cucumbers have been referred to Recent species by older palaeontologists (e.g., Bell 1897, Madsen 1904). Unfortunately nearly 20 'new' paraspecies were described by Soodan (1977a, 1977b) from Quaternary sediments of the Arabian Sea and the Bay of Bengal; these forms can very probably be assigned to modern Indo–Pacific species.

There are no reported Quaternary body fossils of Holothuroidea.

4 SHORT DISCUSSION AND PROBLEMS

Comparing levels of fossil holothurian biodiversity between all the Phanerozoic geological eras (Fig. 2), my species counts imply that the holothurian fossil record is biased for various reasons: (1) low fossilisation potential of sea cucumbers, (2) holothurian body fossils are very rare, with a few exceptions (fossil lagerstätten, ~15 localities), (3) problems concerning the taxonomy and systematics of fossil Holothuroidea (paraspecies/ formspecies vs. species), (4) large areas worldwide are unstudied or understudied with regard to fossil sea cucumbers, and (5) general lack of interest or lack of potential holothurian researchers.

Furthermore, we have to accept the reality (and need) of separate systems in describing fossil holothurians, i.e. (i): for body fossils, 'natural assemblages/associations', isolated 'diagnostic' calcareous ring elements, isolated 'diagnostic' ossicles like wheels, crosses, tables etc. (= 'species'), and (ii): for isolated 'undiagnostic' ossicles, like sieve plates, scales and rods (= 'form-/paraspecies'). For the latter one, there is vague hope that enough 'natural assemblages/associations' will be found in the near future, to resolve the systematic/taxonomic status of 'form-/paraspecies'. Hereby the strict use of the oldest available species ('form-/paraspecies') name when transferring already known 'form-/paraspecies' from 'formgenera'/'paragenera' to 'biological genera' is important.

All current available biodiversity diagrams (e.g., Sprinkle 1980) or stratigraphic distribution charts (e.g., Fletcher et al. 1967, Soodan 1993, Gilliland 1993b, Sepkoski 2002) of fossil Holothuroidea are not significant, due to the application of a mixture of genera/'paragenera' and/or 'parafamilies'.

Therefore, the holothurian fossil record needs a careful updating and evaluation, including a revision of all available holothurian taxa. This will help to provide a more comprehensive picture of diversity change, origination and extinction rates.

5 FUTURE PROSPECTS

Future work is urgently needed to extend our knowledge of modern holothurian anatomies. Only a modern catalogue of all the different calcareous hardparts, such as ossicles and calcareous ring elements in 3D-view (including the variability represented within individual specimens), will allow a better understanding (and prospecting) of fossil holothurians and the relationships within the group as well as within other eleutherozoan echinoderms.

It is likely to improve the dating of holothurian phylogenies by providing evidence of robust dates. Therefore, several catalogues of fossil Holothuroidea are in preparation by the author and will be published in the near future. A short annotated bibliography of fossil sea cucumbers is also available (Reich 2011).

ACKNOWLEDGEMENTS

Most of this work has been done during my stay at Innsbruck University, Austria, for my Ph.D. thesis during the years 1999–2002. I thank Professor Dr H. Mostler for fruitful discussions, and two anonymous referees for comments.

This study was supported in part by Synthesys grants, a programme financed by European Community Research Infrastructure Action (GB-TAF-2446, SE-TAF-799, AT-TAF-787, FR-TAF-1618). Funding (in part) of this project by the German Science Foundation (DFG RE2599/4-1, RE2599/6-1) is also gratefully acknowledge.

I wish to thank the following people for the supply of literature, for numerous comments, for donation or loan of material, and for any other help I received, in alphabetical order: M. Al-Tamimi (Irbid), J. Ansorge (Horst/Greifswald), C. Bartels (Bochum), A. Bartolini (Paris), A. Bebiolka (Berlin), D.B. Blake (Urbana, Ill.), W. Blind (Gießen), A. Boczarowski (Sosnowiec), E. Brand (†, Kassel), B.E. Buitrón-Sanchez (México), J.-B. Caron (Toronto), W. Etter (Basel), C. Franzén (Stockholm), P. Frenzel (Weimar/Jena), E. Frey (Karlsruhe), H. Furrer (Zurich), A.S. Gale (Portsmouth), J. Gallemí (Barcelona), U. Gebhardt (Freiberg/Sa.), P.M. Gilliland (formerly Exeter), H. Hagdorn (Ingelfingen), K. Haldimann (Basel), R. Haude (Göttingen), A. Heinke (Berlin), K. Henne (Stuttgart), E. Herrig (Hinrichshagen/Greifswald), H. Hess (Basel/Binningen), G. Heumann (Bonn), M. Hiß (Krefeld), D. Holloway (Melbourne), M. Jäger (Dotternhausen), A. Jamnik (Ljubljana), U. Jansen (Frankfurt/M.), A.M. Kerr (Mangilao, Guam), C. Klein (Bochum), H. Kozur (Budapest), E. Kristan-Tollmann (†, Vienna), R. Kohring (†, Berlin), A. Kroh (Vienna), M. Kutscher (Sassnitz), W. Langer (Bonn); P. Lebrun (Paris), F. Lehmann (Krefeld), J. Lehmann (Bremen), W. Lindert (Berlin), A.R. Lord (London/Frankfurt/M.), F.W. Luppold (Hannover), H. Lutz (Mainz), H. Mäschker (Lohme), D. Merle (Paris), W. Mette (Innsbruck), W. Munk (Karlsruhe), H. Nestler (Greifswald), C. Neumann (Berlin), M. Nose (Munich), P.M. O'Loughlin (Melbourne), R. Panchaud (Basel), D.L. Pawson (Washington), E. Pietzeniuk (Berlin), A. Pisera (Warszawa), R. Prokop (Prague), A. Ramovš (Ljubljana), W. Riegraf (Münster), D. Rudkin (Toronto), W. Sadeddin (Irbid), E. Schindler (Frankfurt/M.), O. Schmidt (Basel), S. Schneider (Berlin), G. Schweigert (Stuttgart), E. Seibertz (Wolfsburg), A.B. Smith (London), T.R. Stegemann (Göttingen), S. Stöhr (Stockholm), J. Sztejn (Warszawa), M. Urlichs (Stuttgart), M.-T. Vénec-Peyré (Paris), L. Villier (Marseille), E. Voigt (†, Hamburg), J. Waddington (Toronto), H. Wagner (Hamburg), H.M. Weber (formerly Cologne), F. Wiese (Berlin/Göttingen), and T. Wiese (Hannover).

I also thank the organisers of the 13th International Echinoderm Conference (IEC), University of Tasmania, as well as Professor C. Johnson for the invitation to give a plenary talk on fossil sea cucumbers at this meeting. This work is a part of it.

REFERENCES

Alexandrowicz, Z. 1971. Carboniferous Holothuroidea sclerites in the Upper Silesia Coal Basin (Southern Poland). *Rocznik Polskiego Towarzystwa Geologicznego [= Annales de la Société Géologique de Pologne]* 41(2): 281–291.

Applegate, S., Buitrón-Sánchez, B.E., Solís-Marín, F.A. & Laguarda-Figueras, A. 2009. Two Lower Cretaceous (Albian) fossil holothurians (Echinodermata) from Tepexi de Rodriguez, Puebla, Mexico. *Proceedings of the Biological Society of Washington* 122(1): 91–102.

Beckmann, H. 1965. Holothuriensklerite aus dem Givet der Paffrather Mulde (Rheinisches Schiefergebirge). *Fortschritte in der Geologie von Rheinland und Westfalen* 9: 195–208.

Bell, A. 1897. A Synopsis of the Crustacea and Echinodermata of the Upper Tertiairies. *Annual Report of the Yorkshire Philosophical Society* [1896]: 1–12.

Boczarowski, A. 2001. Isolated sclerites of Devonian non-pelmatozoan echinoderms. *Palaeontologia Polonica* 59: 3–220.

Botting, J.P. & Muir, L.A. 2012. Fauna and ecology of the holothurian bed, Llandrindod, Wales, UK (Darriwilian, Middle Ordovician), and the oldest articulated holothuria. *Palaeontologia Electronica* 15(1)[9A]: 1–28. palaeo-electronica.org/content/2012-issue-1-articles/ 191-welsh-holothurian-bed.

Botting, J.P., Muir, L.A. & Barnie, T. 2004. A Welsh Ordovician Hunsrück. In The Palaeontological Association, 48th Annual Meeting, 17th–20th December 2004, University of Lille. Abstracts. *The Palaeontological Association Newsletter* 57: p. 143.

Branson, C.C. 1964. Sclerites? From Bromide Formation. *Oklahoma Geology* 24(5): 106–107.

Buitrón-Sánchez, B.E. 1991. Los Equinodermos del Cretácico Temprano. *Spectra. Información científica y tecnólogica* 13(179–180)[Agosto-Septiembre]: 15–18.

Cherbonnier, G. 1978. Note sur deux empreintes d'Holothuries fossiles du Trias moyen de la region de Tarragone (Espagne). In Proceedings of the 2nd Echinoderm Conference. Rovinj, September 26–October 1, 1975. *Thalassia Jugoslavica* 12[for 1976](1): 75–79.

Clark, A.H. 1912. Restoration of the genus *Eldonia*, a genus of free swimming Holothurians from the Middle Cambrian. *Zoologischer Anzeiger* 39: 723–725.

Clark, A.H. 1913. Cambrian holothurians. *The American Naturalist* 47(560): 488–507.

Clark, H.L. 1912. Fossil holothurians. *Science* 35(894): 274–278.

Conway Morris, S. 1978. *Laggania cambria* Walcott: a composite fossil. *Journal of Paleontology* 52(1): 126–131.

Croneis, C. & McCormack, J. 1932. Fossil Holothuroidea. *Journal of Paleontology* 6(2): 111–148.

Deflandre-Rigaud, M. 1962. Contribution à la connaissance des sclérites d'Holothurides fossiles. *Mémoires du Muséum National d'Histoire Naturelle, (N. S., Série C: Sciences de la Terre)* 11(1): 1–123.

Ding Hui 1985. [Discovery of holothurian sclerites from the Taiyuan Formation (Upper Carboniferous), Henan, China.] *Acta Micropalaeontologica Sinica [= Weiti-gushengwu-xuebao]* 2(4): 339–350.

Durham, J.W. 1954. Status of Invertebrate Paleontology, 1953. IV. Echinodermata: Eleutherozoa. *Bulletin of the Museum of Comparative Zoology at Harvard College* 112(3): 151–160.

Etheridge, R. [jr.] 1881. On the presence of the scattered skeletal remains of Holothuroidea in the Carboniferous limestone series of Scotland. *Proceedings of the Royal Physical Society of Edinburgh* 6: 183–198.

Fletcher, B.N., Philip, G.M. & Wright, C.W. 1967. Echinodermata: Eleutherozoa. In V. Jaanusson, S. Laufeld & R. Skoglund (eds), W.B. Harland, C.H. Holland, M.R. House, N.F. Hughes, A.B. Reynolds, M.J.S. Rudwick, G.E. Satterthwaite, L.B.H. Tarlo & E.C. Willey (eds), *The Fossil Record*: 583–599. London: Geological Society.

Franzén, C. 1979. Echinoderms. In V. Jaanusson, S. Laufeld & R. Skoglund (eds), Lower Wenlock faunal and floral dynamics—Vattenfallet section, Gotland. *Sveriges Geologiska Undersökning (C: Avhandlingar och uppsatser)* 762: 216–224.

Frickhinger, K.A. 1999. The Fossils of Solnhofen. A documentation of fauna and flora of the Solnhofen Formation 2: New specimens, new details, new results: 190 pp. Korb: Goldschneck.

Frizzell, D.L. & Exline, H. 1956. Monograph of Fossil Holothurian Sclerites. *Bulletin of School of Mines and Metallurgy (Technical Series)* 89[for 1955](1): 1–204.

Frizzell, D.L. & Exline, H. 1958. *The Frizzell-Exline Card Catalogue of Holothurian Sclerites*. 216 cards. Alexandria, Vir.: The McLean Paleontological Laboratory.

Frizzell, D.L. & Exline, H. 1966. Holothuroidea—Fossil Record. In R.C. Moore (eds), *Treatise on Invertebrate Paleontology, U, Echinodermata 3 [Asterozoa-Echinozoa] (2)*: U646-U672. Lawrence, Kan.: University of Kansas Press & Boulder, Colo.: Geological Society of America.

Gilliland, P.M. 1993a. The skeletal morphology, systematics and evolutionary history of holothurians. *Special Papers in Palaeontology* 47: 1–147.

Gilliland, P.M. 1993b. Class Holothuroidea de Blainville, 1834. In M.J. Simms, A.S. Gale, P.M. Gilliland, E.P.F. Rose & G.D. Sevastopulo: Echinodermata. In M.J. Benton (ed), *The Fossil Record 2*: 509–513. London: Chapman & Hall.

Giebel, C.G.A. 1857. Zur Fauna des lithographischen Schiefers von Solnhofen (neue Insekten, Krebs, Holothurien). *Zeitschrift für die Gesammten Naturwissenschaften* 9: 373–388.

Górka, H. 1997. Holothurian Sclerites from the Miocene Deposits of the Gliwice Area (Upper Silesia, Poland). *Bulletin of the Polish Academy of Sciences (Earth Sciences)* 45(2–4): 107–118.

Gutschick, R.C., Canis, W.F. & Brill, K.G. [jr.] 1967. Kinderhook (Mississippian) holothurian sclerites from Montana and Missouri. *Journal of Paleontology* 41(6): 1461–1480.

Haude, R. 1995a. Die Holothurien-Konstruktion: Evolutionsmodell und ältester Fossilbericht. In W.-E. Reif (ed), Festschrift, Herrn Professor Adolf Seilacher zur Vollendung des 70. Lebensjahres am 24. Februar 1995 gewidmet. *Neues Jahrbuch für Geologie und Paläontologie, Abhandlungen* 195(1–3): 181–198.

Haude, R. 1995b. Echinodermen aus dem Unter-Devon der argentinischen Präkordillere. *Neues Jahrbuch für Geologie und Paläontologie, Abhandlungen* 197(1): 37–86.

Haude, R. 1997. *Nudicorona*, eine devonische Holothurie. *Fossilien* 14(1): 50–57.

Haude, R. 2002. Origin of the holothurians (Echinodermata) derived by constructional morphology. In M. Aberhan, W.-D. Heinrich & S. Schultka (eds), Festband Hans-Peter Schultze zum 65. Geburtstag. *Mitteilungen aus dem Museum für Naturkunde Berlin. Geowissenschaftliche Reihe* 5: 141–153.

Hess, H. 1973. Neue Echinodermenfunde aus dem mittleren Dogger des Aargauer Juras. *Eclogae geologicae Helvetiae* 66(3): 625–656.

Hess, H. & Holenweg, H. 1998. Ein neuer Holothurienfund aus dem Schweizer Jura. *Fossilien* 15(5): 306–309.

Ishijima, W. & Hatai, K. 1973. An Interesting Microfossil from the Mizunami Group, Toki-Mizunami District, Gifu Prefecture, Japan. *St. Paul's Review of Science [= Rikkyō-daigaku-kenkyū-hōkūku]* 3(3): 69–70.

Jell, P.A. 2010. Late Silurian echinoderms from the Yass Basin, New South Wales—the earliest holothurian body fossil and two diploporitan cystoids (Sphaeronitidae and Holocystitidae). *Memoirs of the Association of Australasian Palaeontologists* 39: 27–41.

Kerr, A.M. 2003. Holothuroidea (Sea Cucumbers). In D. Thoney (ed), *Grzimek's Animal Life Encyclopedia. Second Edition. Volume I. Lower Metazoans and Lesser Deuterostomes:* 417–431. New York: Gale Group Publishers.

Kikuchi, Y. & Nikaido, A. 1996. Miocene Holothurian fossils from the Middle Miocene Nawashiroda Formation in Ibaraki Prefecture, northern Kanto, Japan. In Anonymous (ed), *Professor H. Igo Commemorative Volume:* 155–162.

Kozur, H. 1969. Holothuriensklerite aus der germanischen Trias. *Monatsberichte der Deutschen Akademie der Wissenschaften zu Berlin* 11(2): 146–154.

Kozur, H. & Mostler, H. 1989. Echinoderm Remains from the Middle Permian (Wordian) from Sosio Valley (Western Sicily). *Jahrbuch der Geologischen Bundesanstalt* 132(4): 677–685.

Kozur, H. & Mostler, H. 2008. Neue Holothurien-Sklerite aus dem unteren Wordian vom Nordwestrand des Delaware Beckens am Südosthang der Guadalupe Mountains (West-Texas, USA). *Geo.Alp* 5: 107–119.

Kristan-Tollmann, E. 1963. Holothurien-Sklerite aus der Trias der Ostalpen. *Sitzungsberichte der Österreichischen Akademie der Wissenschaften, Mathematisch-naturwissenschaftliche Klasse (Abteilung I: Biologie, Mineralogie, Erdkunde und verwandte Wissenschaften)* 172: 350–380.

Kristan-Tollmann, E. 1964. Holothurien-Sklerite aus dem Torton des Burgenlandes, Österreich. *Sitzungsberichte der Österreichischen Akademie der Wissenschaften, Mathematisch-naturwissenschaftliche Klasse (Abteilung I: Biologie, Mineralogie, Erdkunde und verwandte Wissenschaften)* 173(1–2): 75–100.

Lane, N.G. 1976. A crinoid tegmen composed of holothurian sclerites. *Journal of Paleontology* 50(2): 240–244.

Lehmann, W.M. 1958. Eine Holothurie zusammen mit *Palaenectria devonica* und einem Brachiopoden in den unterdevonischen Dachschiefern des Hunsrücks durch Röntgenstrahlen entdeckt. *Notizblatt des Hessischen Landesamtes für Bodenforschung zu Wiesbaden* 86: 81–86.

Madsen, V. 1904. Nils Olof Holst: Kvartär-studier i Danmark och norra Tyskland. En kritisk Anmeldelse. *Geologiska Föreningens i Stockholm förhandlingar* 26(7): 529–534.

Madsen, F.J. 1956. *Eldonia* a Cambrian siphonophore—formerly interpreted as a holothurian. *Videnskabelige Meddelelser fra Dansk Naturhistorisk Forening* 118: 7–14.

Madsen, F.J. 1957. On Walcott's supposed Cambrian holothurians. *Journal of Paleontology* 31(1): 281–283.

Mostler, H. 1968a. Conodonten und Holothuriensklerite aus den norischen Hallstätter Kalken von Hernstein (Niederösterreich). *Verhandlungen der Geologischen Bundesanstalt* [1967](1/2): 177–188.

Mostler, H. 1968b. Das Silur im Westabschnitt der Nördlichen Grauwackenzone (Tirol und Salzburg). *Mitteilungen der Gesellschaft der Geologie- und Bergbaustudenten in Wien* 18[1967]: 89–150.

Mostler, H. 1971. Mikrofaunen aus dem Unter-Karbon vom Hindukusch. *Geologisch-Paläontologische Mitteilungen Innsbruck* 1(12): 1–19.

Mostler, H. 1972. Holothuriensklerite aus dem Jura der Nördlichen Kalkalpen und Südtiroler Dolomiten. *Geologisch-Paläontologische Mitteilungen Innsbruck* 2(6): 1–29.

Mostler, H. & Rahimi-Yazd, A. 1976. Neue Holothuriensklerite aus dem Oberperm von Julfa in Nordiran. *Geologisch-Paläontologische Mitteilungen Innsbruck* 5(7): 1–35.

Müller, A.H. 1969. Reste seltener Holothurien (Echinodermata) aus dem Mesozoikum Europas. *Monatsberichte der Deutschen Akademie der Wissenschaften zu Berlin* 11(8/9): 662–671.

Münster, G.G. von 1843. Ueber einige Theile fossiler Holothurien im Jura-Kalk von Streitberg. *Beiträge zur Petrefacten-Kunde* 6: 92–93.

Muir, L.A. & Botting, J.P. 2005. The oldest known complete holothurian (early Llandeilian, Ordovician), and the origin of the calcareous ring. In The Palaeontological Association, 49th Annual Meeting, 18th–21st December 2005, University of Oxford. Abstracts. *The Palaeontological Association Newsletter* 60: p. 26.

Pawson, D.L. 1980. Holothuroidea. In T.W. Broadhead & J.A. Waters (eds), Echinoderms. Notes for a Short Course. *University of Tennessee, Department of Geological Sciences, Studies in Geology* 3: 175–189, 215–235 [references].

Reich, M. 1995. Erster sicherer Nachweis der Elasipoda (Holothuroidea, Echinodermata) aus der Kreide, sowie Bemerkungen zu den Holothurienresten der Oberkreide. *Archiv für Geschiebekunde* 1(11): 681–688.

Reich, M. 1999a. Ordovizische und silurische Holothurien (Echinodermata). In M. Reich (ed), Festschrift zum 65. Geburtstag von Ekkehard Herrig. *Greifswalder Geowissenschaftliche Beiträge* 6: 479–488.

Reich, M. 1999b. Die ältesten fossilen Holothurien (Echinodermata). In Anonymous (ed), *2. Jahrestagung der Gesellschaft für Biologische Systematik, GfBS—1999, Friedrich-Schiller-Universität Jena, 2.–4. September 1999:* p. 47. Jena.

Reich, M. 2001. Ordovician holothurians from the Baltic Sea area. In M. Barker (ed), *Echinoderms 2000:* 93–96. Lisse etc.: A. A. Balkema Publishers.

Reich, M. 2002. Holothurienreste (Echinodermata) aus dem Oligozän (Rupelium) Norddeutschlands. *Neues Jahrbuch für Geologie und Paläontologie, Abhandlungen* 224(1): 97–113.

Reich, M. 2004a. Aspidochirote holothurians (Echinodermata) from the Middle Triassic of southern Germany. In T. Heinzeller & J. Nebelsick (eds), *Echinoderms: München:* 485–486. Leiden etc.: A. A. Balkema Publishers.

Reich, M. 2004b. Holothurians from the Late Cretaceous 'Fish shales' of Lebanon. In T. Heinzeller & J. Nebelsick (eds), *Echinoderms: München*: 487–488. Leiden etc.: A. A. Balkema Publishers.

Reich, M. 2004c. Fossil Holothuroidea (Echinodermata): An overview. In T. Heinzeller & J. Nebelsick (eds), *Echinoderms: München*: p. 602. Leiden etc.: A. A. Balkema Publishers.

Reich, M. 2006. Cambrian holothurians? In B. Lefebvre, B. David, E. Nardin & E. Poty (eds), *Journées G. Ubaghs. 30–31 janvier 2006. Université de Bourgogne, Dijon (France)*: 36–37. Dijon.

Reich, M. 2010a. The oldest synallactid sea cucumber (Echinodermata: Holothuroidea: Aspidochirotida). *Paläontologische Zeitschrift* 84(4): 541–546.

Reich, M. 2010b. The early evolution and diversification of holothurians (Echinozoa). In L.G. Harris, S.A. Böttger, C.W. Walker & M.P. Lesser (eds), *Echinoderms: Durham*: 55–59. London etc.: Taylor & Francis.

Reich, M. 2011. Annotated bibliography of fossil Holothuroidea (Echinodermata: Echinozoa)—1829–2010. Unpublished manuscript. Göttingen.

Reich, M. & Kroh, A. 2010. Middle Miocene holothurians (Echinodermata) from the Vienna Basin (Austria). In M. Harzhauser & W.E. Piller (eds), *4. Jahrestagung von NOBIS Austria, 2.–3. Dezember 2010. Berichte des Institutes für Erdwissenschaften, Karl-Franzens-Universität Graz 15: p. 40*.

Reich, M. & Kutscher, M. 2001. Ophiocistioids and holothurians from the Silurian of Gotland (Sweden). In M. Barker (ed), *Echinoderms 2000*: 97–101. Lisse etc.: A. A. Balkema Publishers.

Reich, M., Kutscher, M. & Stegemann, T.R. 2011. Sea cucumbers from the Silurian of the Isle of Gotland, Sweden. In *6th North American Echinoderm Conference, 14–19 August 2011, Rosario Beach Marine Laboratory, Anacortes, Washington, USA*: p. 36. College Place, WA: Walla Walla University.

Reich, M. & Lehnert, O. subm. The oldest unequivocal record of fossil Holothuroidea (Echinodermata: Echinozoa).

Reich, M. & Smith, A.B. 2010. A new deep-sea holothurians representative from the Palaeozoic Hunsrück Slate Fossillagerstätte (Early Devonian, Germany). In M. Reich, J. Reitner, V. Roden & B. Thuy (eds), *Echinoderm Research 2010. 7th European Conference on Echinoderms, Göttingen, October 2–9, 2010. Abstract Volume and Field Guide to Excursions*: 89–90. Göttingen: Universitätsverlag.

Reich, M. & Stegemann, T. 2010. Holothurians (Apodida) from the Late Palaeozoic Mazon Creek Fossillagerstätte (Illinois, USA). In M. Reich, J. Reitner, V. Roden & B. Thuy (eds), *Echinoderm Research 2010. 7th European Conference on Echinoderms, Göttingen, October 2–9, 2010. Abstract Volume and Field Guide to Excursions*: 90–91. Göttingen: Universitätsverlag.

Reich, M. & Wiese, F. 2010. Apodid sea cucumbers (Echinodermata: Holothuroidea) from the Upper Turonian of the Isle of Wolin, NW Poland. *Cretaceous Research* 31(4): 350–363.

Rioult, M. 1965. Sclérites d'Holothuries tertiaires (Éocène du Bassin de Paris et Miocène du Bassin d'Aquitaine). *Revue de Micropaléontologie* 8(3): 165–174.

Sadeddin, W. 1996. Holothurian sclerites from the Albian (Early Cretaceous) of the Fort Worth area, Texas, USA. *Neues Jahrbuch für Geologie und Paläontologie, Abhandlungen* 200(3): 325–351.

Schlumberger, M. 1888. Note sur les Holothuridées du Calcaire grossier. *Bulletin de la Société géologique de France (sér. 3)* 16: 437–441.

Sepkoski, J.J. [jr.] (eds D. Jablonski & M. Foote) 2002. A Compendium of Fossil Marine Animal Genera. *Bulletins of American Paleontology* 363: 1–560.

Smiley, S. 1994. Holothuroidea. In F.W. Harrison & Chia Fu-Shiang (eds), *Microscopic Anatomy of Invertebrates. Volume 14*: 401–471. New York etc.: Wiley-Liss.

Smith, A.B. & Gallemí, J. 1991. Middle Triassic holothurians from northern Spain. *Palaeontology* 34(1): 49–76.

Smith, A.B. & Reich, M. 2009. Getting beneath the skin of fossil holothurians. In *The Palaeontological Association, 53rd Annual Meeting, 13th–16th December 2009, University of Birmingham. Abstracts*: p. 32.

Soodan, K.S. 1977a. Holothurian sclerites from the Recent sediments. *Proceedings of the Indian Colloquium on Micropalaeontology and Stratigraphy* 4: 104–109.

Soodan, K.S. 1977b. Fossil Holothuroidea from the subsurface Pleistocene rocks of the Bay of Bengal. *Proceedings of the Indian Colloquium on Micropalaeontology and Stratigraphy* 6: 279–282.

Soodan, K.S. 1993. Palaeozoic—Cainozoic fossil Holothuroidea zonation scheme. *Geoscience Journal* 14(1–2): 127–144.

Sprinkle, J. 1980. An Overview of the Fossil Record. In T.W. Broadhead & J.A. Waters (eds), *Echinoderms, notes for a short course. Studies in Geology* 3: 15–26, 215–235 [references].

Sroka, S.D. 1988. Preliminary studies on a complete fossil holothurian from the Middle Pennsylvanian Francis Creek Shale of Illinois. In R.D. Burke, P.V. Mladenov, P. Lambert & R.L. Parsley (eds), *Echinoderm biology. Proceedings of the Sixth International Echinoderm Conference*: 159–160.

Sroka, S.D. & Blake, D.B. 1997. Echinodermata. In C.W. Shabica & A.A. Hay (eds), *Richardson's Guide to The Fossil Fauna of Mazon Creek*: 223–225. Chicago, Ill.: Northeastern Illinois University.

Tandon, K.K. & Saxena, R.K. 1977. Fossil holothuroids from Middle Eocene rocks of Kutch, India. *Geophytology* 7(2): 229–259.

Walkiewicz, A. 1977. Holothurian sclerites from the Korytnica Clays (Middle Miocene; Holy Cross Mountains, Poland). *Acta Geologica Polonica* 27(2): 177–192.

Wang Cheng-yuan & Li Dong-jin 1986. Conodonts from the Ertaogou Formation in Central Jilin Province. *Acta Micropalaeontologica Sinica [= Weiti-Gushengwu-Xuebao]* 3(4): 421–428.

Weber, H.M. 1997. Holothurien- und Ophiocistioiden-Reste (Echinodermata) aus dem Unterkarbon des Velberter Sattels (Rheinisches Schiefergebirge). In R.H. Weiß & H.G. Herbig (eds), *Festschrift Eugen Karl Kempf. Sonderveröffentlichungen des Geologischen Instituts der Universität zu Köln* 114: 485–497.

Ziegler, B. 1991. Was ist *Laffonia helvetica* Heer? *Stuttgarter Beiträge zur Naturkunde (B: Geologie und Paläontologie)* 172: 1–10.

APPENDIX—LIST OF ALL PUBLISHED FOSSIL HOLOTHURIAN NAMES

The list excludes all published records from Cambrian strata (*Eldonia, Laggania, Louisella, Mackenzia, Portalia, Redoubtia* etc.; cf. H.L. Clark 1912, A.H. Clark 1912, 1913, Croneis & McCormack 1932, Madsen 1956, 1957, Conway Morris 1978, Reich 2006) as well as Recent species reported from Neogene or Quaternary sediments worldwide. The total number of described names incorporates synonyms, nomina dubia, nomina nuda, and incorrectly assigned material. There are more than two dozens (28) of named species within the Holothuroidea which can be assigned to other invertebrate groups (cf. in part Branson 1964, Lane 1976, Pawson 1980), e.g., echinoids (*Eothuria beggi, Kaliobullites umbo, Pachopsites annulatus*), ophiuroids (*Frizzellus* spp., *Mortensenites* spp., *Prosynapta* [= *Calclyra*] *eiseliana*), crinoids (*Etheridgella* spp.), Echinodermata inc. sed. (*Alexandrites alexandri*), foraminifers (*Hampsancora brownendensis*), trilobites (*Thuroholia* spp. [p.p.]) or to trace fossils (*Palaeosynapta flaccida*). There are also a few holothurian species originally described within other invertebrate groups e.g., asteroids (*Astrophyton permianum*), foraminifers (*Calcarina rotula, Gromia liasica*), ophiocistioids (*Microantyx sosioensis, M. pseudobotoni, Pararotasaccus permicus*) or even within diatoms (*Actinoclava Frankei*).

The detailed list includes the species, original genus, author(s), type stratum and locality. All cited authors can be found in Reich (2011). A few fossil species are designated in part as follows:

* sclerite assemblage
** body fossil
*** non-Holothuroidea or very probably non-Holothuroidea

1. *acanthica* Mostler, 1972 [1972f]; *Eocaudina* [Triassic: Lower Norian; Germany]
2. *acanthica* Mostler, 1972 [1972f]; *Uncinulina* [Triassic: Middle Norian; Germany]
3. *acanthicus* Mostler, 1968 [1968b]; *Priscopedatus* [Triassic: Upper Anisian; Austria]
4. *acanthocaudinoides* Mostler, 1970; *Eocaudina* [Triassic: Norian; Austria]
5. *acanthotheelioides* Mostler, 1968 [1968b]; *Spinites* [Triassic: Upper Anisian; Turkey]
6. *accipitris* Boczarowski, 2001*; *Devonothyonites* [Devonian: Early Eifelian; Poland]
7. *acmaea* Matyja, 1972; *Protocaudina* [Jurassic: late Middle Oxfordian; Poland]
8. *acuta* Soodan, 1975; *Fletcherina* [Jurassic: Bathonian; India]
9. *acutus* Mostler, 1972 [1972e]; *Priscopedatus* [Jurassic: ?Hettangian–?Toarcian; Austria]
10. *adamsi* Soodan & Whatley, 1990; *Yorkshirella* [Jurassic: Hettangian; U.K.]
11. *aegyptiacus* Said & Barakat, 1958; *Priscopedatus* [Jurassic: Bajocian; Egypt]
12. *aequiperforatus* Mostler, 1972 [1972f]; *Kuehnites* [Triassic: Upper Norian; Austria]
13. *aequiperforatus* Mostler & Rahimi-Yazd, 1976; *Punctatites* [Permian: Lopingian: Wuchiapingian; Iran]
14. *affinis* Deflandre-Rigaud, 1946 [1946b]; *Priscopedatus* [Jurassic: Oxfordian; France]
15. *agariciformis* Kristan-Tollmann, 1964 [1964b]; *Theelia* [Triassic: Rhaetian; Austria]
16. *ageri* Soodan & Whatley, 1990; *Yorkshirella* [Jurassic: Hettangian; U.K.]
17. *agighiolensis* Gheorghian, 1995; *Acanthotheelia* [Triassic: Anisian; Romania]
18. *agrawali* Tandon & Saxena, 1977; *Calclamnella* [Paleogene: Eocene: Lutetian; India]
19. *ahmadi* Tandon & Saxena, 1977; *Calcancora* [Paleogene: Eocene: Lutetian; India]
20. *albiansis* Sadeddin, 1996 [1996b]; *Spinopriscopedatus* [Cretaceous: Albian; U.S.A.]
21. *albiensis* Mutterlose, 1982; *Rigaudites*? [Cretaceous: early Lower Albian; Germany]
22. *alencasterae* Applegate, Buitrón-Sánchez, Solís-Marín & Laguarda-Figueras, 2009**; *Paleopentacta* [Cretaceous: Albian; Mexico]
23. *alexandri* Kristan-Tollmann, 1964 [1964c]***; *Alexandrites* [Neogene: Miocene: Middle Tortonian; Austria]
24. *alta* Speckmann, 1968; *Theelia* [Triassic: Anisian: Pelsonian; Germany]
25. *alternata* Mostler, 1972 [1972f]; *Uncinulinoides* [Triassic: Upper Norian; Austria]
26. *alveata* Mostler & Rahimi-Yazd, 1976; *Theelia* [Permian: Lopingian: Wuchiapingian; Iran]
27. *ambiguus* Deflandre-Rigaud, 1962 [1962b]; *Cucumarites* [Jurassic: Oxfordian; France]
28. *anceps* Deflandre-Rigaud, 1962 [1962b]; *Paracucumarites*? [Jurassic: Oxfordian; France]
29. *anceps* Schlumberger, 1890; *Priscopedatus* [Paleogene: Eocene: Upper Lutetian; France]
30. *ancile* Boczarowski, 2001*; *Palaeocucumaria* [Devonian: Late Eifelian; Poland]
31. *ancora* Boczarowski, 2001; *Bracchiothuria* [Devonian: Late Givetian; Poland]
32. *andrusovi* Kozur & Mock, 1972; *Kuehnites* [Triassic: Norian: earliest Sevatian; Slovak Republic]
33. *andrusovi* Kozur & Mostler, 1970; *Theelia* [Triassic: Anisian: Bithynian; Germany]
34. *anguinea* Mostler, 1971 [1971b]; *Theelia* [Triassic: Carnian: Cordevolian; Austria]

35. *angulata* Soodan, 1977 [1977b]; *Eocaudina* [Quaternary: Holocene; Arabian Sea]

36. *angulata* Frizzell & Exline, 1956; *Uncinulina* [Carboniferous: Mississippian: Serpukhovian; U.S.A.]

37. *angulatus* Deflandre-Rigaud, 1950; *Chiridotites* [Jurassic: Bathonian; France]

38. *angulatus* Deflandre-Rigaud, 1946 [1946b]; *Priscopedatus* [Jurassic: Oxfordian; France]

39. *angulatus* Krainer & Mostler, 1997; *Punctatites* [Jurassic: Sinemurian/Pliensbachian; Austria]

40. *anguliferus* Zankl, 1966; *Priscopedatus* [Triassic: Norian/Rhaetian; Germany]

41. *angusta* Reich in Herrig et al., 1997; *Stueria* [Cretaceous: late Upper Campanian; Germany]

42. *angustiperforata* Mostler, 1971 [1971b]; *Acanthotheelia* [Triassic: Norian; Austria]

43. *anisica* Mostler, 1968 [1968b]; *Acanthotheelia* [Triassic: Upper Anisian; Austria]

44. *anisica* Mostler, 1970; *Calclamnella* [Triassic: Norian; Austria]

45. *annulata* Giebel, 1857**/***; *Protholothuria* [Jurassic: Lower Tithonian; Germany]

46. *annulatus* Kristan-Tollmann, 1964 [1964c]***; *Pachopsites* [Neogene: Miocene: Middle Tortonian; Austria]

47. *ansorgei* Reich, 2003 [2003d]; *Calcligula* [Cretaceous: Lower Maastrichtian; Germany]

48. *antiquus* Boczarowski, 1997 [1997a]*; *Achistrum* [Permian: Lopingian: Wuchiapingian; Poland]

49. *antyx* Kristan-Tollmann, 1964 [1964b]; *Protocaudina* [Triassic: Rhaetian; Austria]

50. *apertus* Deflandre-Rigaud, 1962 [1962b]; *Priscopedatus* [Jurassic: Oxfordian; France]

51. *apertus* Mostler, 1971 [1971b]; *Priscopedatus* [Triassic: Anisian: Illyrian; Montenegro]

52. *appensa* Mostler, 1968 [1968e]; *Calclamnella* [Triassic: Norian; Austria]

53. *aequibrachiata* Krainer & Mostler, 1997; *Biacumina* [Jurassic: Pliensbachian; Austria]

54. *aequiperforatus* Mostler & Rahimi-Yazd, 1976; *Punctatites* [Permian: Lopingian: Wuchiapingian; Iran]

55. *aquitanica* Rioult, 1961; *Calclamnella* [Neogene: Miocene: Upper Burdigalian; France]

56. *arcuatus* Deflandre-Rigaud, 1950; *Auricularites* [Jurassic: Oxfordian; France]

57. *arcuatus* Deflandre-Rigaud, 1962 [1962b]; *Spandelites* [Jurassic: Oxfordian]

58. *arcuatus* Deflandre-Rigaud, 1952; *Stichopites* [Carboniferous: Mississippian: Serpukhovian; U.S.A.]

59. *arduohamata* Kristan-Tollmann, 1964 [1964c]; *Calcancora* [Neogene: Miocene: Middle Tortonian; Austria]

60. *arkelli* Soodan & Whatley, 1990; *Yorkshirella* [Jurassic: Hettangian; U.K.]

61. *armata* Giebel, 1857**/***; *Protholothuria* [Jurassic: Lower Tithonian; Germany]

62. *artus* Mostler, 1971 [1971b]; *Priscopedatus* [Triassic: Norian; Austria]

63. *aspergillum* Schlumberger, 1890; *Priscopedatus* [Paleogene: Eocene: Upper Lutetian; France]

64. *aspis* Kristan-Tollmann, 1964 [1964c]; *Synaptites* [Neogene: Miocene: Middle Tortonian; Austria]

65. *assymmetrica* Mostler, 1969; *Theelia* [Triassic: Norian; Austria]

66. *astralis* Boczarowski, 2001; *Gagesiniotrochus* [Devonian: Late Givetian; Poland]

67. *asymmetricus* Zhang Jin-jian, 1983; *Paracucumarites* [Carboniferous: Mississippian: Lower Tournaisian; P.R. China]

68. *asymmetricus* Deflandre-Rigaud, 1962 [1962b]; *Priscopedatus* [Jurassic: Oxfordian; France]

69. *atava* Waagen, 1867; *Chirodota* [Jurassic: Bajocian; Germany]

70. *austini* Soodan & Whatley, 1990; *Calclamnoidea* [Jurassic: Hettangian; U.K.]

71. *austriacus* Deflandre-Rigaud, 1962 [1962b]; *Synaptellus* [Neogene: Miocene: Middle Tortonian; Austria]

72. *aviformis* Kubiatowicz & Matyja, 1977; *Priscopedatus* [Cretaceous: Upper Valanginian; Poland]

73. *avis* Boczarowski, 2001*; *Devonothyonites* [Devonian: Late Givetian; Poland]

74. *bajocica* Kaptarenko-Černousova, 1954; *Chirodota (Hemisphaeranthos)* [Jurassic: Bajocian; Russia]

75. *barkeyi* Kozur & Simón, 1972; *Theelia* [Triassic: Lower Carnian; Spain]

76. *barnardi* Soodan & Whatley, 1990; *Binoculites* [Jurassic: Hettangian; U.K.]

77. *bartensteini* Frizzell & Exline, 1956; *Achistrum* [Jurassic: Sinemurian/Pliensbachian; Germany]

78. *bartensteini* Frentzen, 1964; *Crux* [Jurassic: Upper Pliensbachian; Germany]

79. *bartensteini* Deflandre-Rigaud, 1952; *Staurocumites* [Jurassic: Upper Pliensbachian; Germany]

80. *bastropanus* Frizzell & Exline, 1957; *Rigaudites* [Paleogene: Eocene: Lutetian; U.S.A.]

81. *bathonianum* Frizzell & Exline, 1956; *Achistrum* [Jurassic: Bathonian; France]

82. *batoniensis* Górka & Łuszczewska, 1969; *Priscopedatus* [Jurassic: Upper Bathonian; Poland]

83. *beckmanni* Kozur & Mostler, 1971; *Achistrum* [Triassic: Ladinian: Langobardian; Hungary]

84. *beggi* MacBride & Spencer, 1938; *Eothuria* [Ordovician: Katian; U.K.]
85. *bengalensis* Soodan, 1977 [1977c]; *Calcancora* [Quaternary: Pleistocene; Bay of Bengal]
86. *bengalensis* Soodan, 1977 [1977c]; *Eocaudina* [Quaternary: Pleistocene; Bay of Bengal]
87. *beurleni* Tinoco, 1963; *Calcancora* [Paleogene: Paleocene: Danian; Brazil]
88. *bhatiai* Tandon & Saxena, 1977; *Eocaudina* [Paleogene: Eocene: Lutetian; India]
89. *bhujensis* Tandon & Saxena, 1977; *Eocaudina* [Paleogene: Eocene: Lutetian; India]
90. *bichordata* Fletcher, 1962; *Achistrum (Cancellrum)* [Jurassic: Upper Oxfordian; U.K.]
91. *biconvexa* Summerson & Campbell, 1958***; *Etheridgella* [Carboniferous: Pennsylvanian: Moscovian; U.S.A.]
92. *bifidus* Hodson, Harris & Lawson, 1956; *Rhabdotites* [Jurassic: Oxfordian; U.K.]
93. *bifurcata* Soodan, 1975; *Fletcherina* [Jurassic: Bathonian; India]
94. *bifurcata* Kozur & Sadeddin, 1990; *Pseudoexlinella* [Jurassic: Bathonian; Jordan]
95. *bifurcata* Soodan & Whatley, 1990; *Yorkshirella* [Jurassic: Hettangian; U.K.]
96. *bikanerensis* Soodan, 1980; *Calclamna* [Paleogene: Eocene: Ypresian; India]
97. *bilamellaris* Górka, 1997***; *Mortensenites* [Neogene: Miocene; Poland]
98. *billetti* Boczarowski, 2001*; *Gagesiniotrochus* [Devonian: Early Givetian; Poland]
99. *bimembris* Boczarowski, 2001; *Ballistocucumis* [Devonian: Early Eifelian; Poland]
100. *binocularis* Frentzen, 1964; *Cucumariopsis* [Jurassic: Upper Sinemurian; Germany]
101. *binoculata* Boczarowski, 2001; *Ocellothuria* [Devonian: Late Givetian; Poland]
102. *biperforata* Kozur & Sadeddin, 1990; *Jordanocaudina* [Jurassic: Bathonian; Jordan]
103. *blakei* Soodan & Whatley, 1988; *Stueria* [Jurassic: Upper Oxfordian; U.K.]
104. *bogschi* Kozur & Mock, 1972; *Priscopedatus* [Triassic: Lower–Middle Norian; Slovak Republic]
105. *bohemicum* Prantl, 1947; *Achistrum* [Devonian: Emsian; Czech Republic]
106. *bolkoviensis* Górka & Łuszczewska, 1969; *Priscopedatus* [Jurassic: Kimmeridgian; Poland]
107. *bombayensis* Soodan, 1977 [1977b]; *Eocaudina* [Quaternary: Holocene; Arabian Sea]
108. *bonhourei* Deflandre-Rigaud, 1962 [1962b]; *Ornaticannula* [Jurassic: Oxfordian; France]
109. *botoni* Gutschick, 1959; *Microantyx* [Carboniferous: Mississippian: late Tournaisian; U.S.A.]
110. *brachyura* Broili, 1926**/***; *Pseudocaudina* [Jurassic: Lower Tithonian; Germany]
111. *brevis* Gutschick, 1959; *Achistrum* [Carboniferous: Mississippian: late Tournaisian; U.S.A.]
112. *brevis* Mostler, 1971 [1971b]; *Achistrum* [Triassic: Norian; Austria]
113. *britannica* Soodan & Whatley, 1987; *Achistrum (Cancellrum)* [Jurassic: Upper Oxfordian; U.K.]
114. *britannica* Soodan & Whatley, 1990; *Calclamna* [Jurassic: Hettangian; U.K.]
115. *brownendensis* Morris, 1970***; *Hampsancora* [Carboniferous: Mississippian: Viséan; U.K.]
116. *brownwoodensis* Croneis in Croneis & McCormack, 1932 [1932c]; *Ancistrum* [Carboniferous: Pennsylvanian: Kasimovian; U.S.A.]
117. *burdigalensis* Rioult, 1961; *Cucumarites* [Neogene: Miocene: Upper Burdigalian; France]
118. *burgioensis* Kozur & Mostler, 1989; *Bicornuticaudina* [Permian: Guadalupian: Wordian; Italy]
119. *bystrickyi* Kozur & Mock, 1972; *Uniramosa* [Triassic: Lower–Middle Norian; Slovak Republic]
120. *callosus* Krainer & Mostler, 1997; *Centripedatus* [Jurassic: Pliensbachian; Austria]
121. *campbelli* Gutschick, 1959; *Rota* [Carboniferous: Mississippian: late Tournaisian; U.S.A.]
122. *canalifera* Kristan-Tollmann, 1963; *Calclamnoidea* [Triassic: Carnian: Cordevolian; Italy]
123. *canina* Boczarowski, 2001; *Staurocaudina* [Devonian: Late Frasnian; Poland]
124. *carpenteri* Moore, 1873; *Chirodota* [Jurassic: Bajocian/Lower Bathonian; U.K.]
125. *cassianensis* Frizzell & Exline, 1956; *Eocaudina* [Triassic: Carnian; Italy]
126. *chandrai* Tandon & Saxena, 1977; *Koteshwaria* [Paleogene: Eocene: Lutetian; India]
127. *chatterjii* Tandon & Saxena, 1977; *Rigaudites* [Paleogene: Eocene: Lutetian; India]
128. *chaussiensis* Frizzell & Exline, 1956; *Calcancora* [Paleogene: Eocene: Lutetian]
129. *chaussiensis* Rioult, 1961; *Exlinella* [Paleogene: Eocene: Upper Lutetian; France]
130. *chiplonkari* Tandon & Saxena, 1977; *Rigaudites* [Paleogene: Eocene: Lutetian; India]
131. *chrysanthemum* Frentzen, 1964; *Myriotrochus* [Jurassic: Lower Oxfordian; Germany]
132. *circularis* Frizzell & Exline, 1956***; *Mortensenites* [Jurassic: Upper Sinemurian; Germany]

133. *circularis* Zhang Jin-jian, 1993; *Paraeocaudina* [Carboniferous: Mississippian: late Viséan; P.R. China]

134. *circularis* Schlumberger, 1890; *Synapta* [Paleogene: Eocene: Upper Lutetian; France]

135. *circumvallata* Kristan-Tollmann, 1964 [1964b]; *Eocaudina?* [Triassic: Rhaetian; Austria]

136. *clavata* Walkiewicz, 1977; *Calclamnella* [Neogene: Middle Miocene; Poland]

137. *clavatus* Deflandre-Rigaud, 1950; *Chiridotites* [Jurassic: Sinemurian; France]

138. *clavatus* Deflandre-Rigaud, 1946 [1946b]; *Priscopedatus* [Jurassic: Oxfordian; France]

139. *clypeus* Deflandre-Rigaud, 1962 [1962b]; *Sclerothurites* [Jurassic: Oxfordian; France]

140. *cognata* Boczarowski, 2001*; *Palaeohemioedema* [Devonian: Late Givetian; Poland]

141. *collaris* Deflandre-Rigaud, 1946 [1946b]; *Priscopedatus* [Jurassic: Oxfordian; France]

142. *coloculum* Gutschick, Canis & Brill, 1967; *Achistrum* [Carboniferous: Mississippian: Lower Tournaisian; U.S.A.]

143. *columcanthus* Gutschick, Canis & Brill, 1967; *Eocaudina* [Carboniferous: Mississippian: Lower Tournaisian; U.S.A.]

144. *compactus* Deflandre-Rigaud, 1962 [1962b]; *Cucumarites* [Jurassic: Oxfordian; France]

145. *complexus* Kozur & Mock, 1972; *Praeeuphronides* [Triassic: Lower–Middle Norian; Slovak Republic]

146. *compressa* Mostler, 1972 [1972f]; *Uncinulina* [Triassic: Norian; Germany]

147. *concameratus* Mostler, 1980; *Priscopedatus* [Triassic: Anisian: Pelsonian; Italy]

148. *concavus* Mostler, 1971 [1971b]; *Praeeuphronides* [Triassic: Norian; Austria]

149. *concentrica* Langer, 1991; *Eocaudina* [Devonian: Lower Givetian; Germany]

150. *concinnus* Sadeddin, 1996 [1996b]; *Priscopedatus* [Cretaceous: Albian; U.S.A.]

151. *conglobata* Mostler, 1971 [1971b]; *Theelia* [Triassic: Norian; Austria]

152. *consona* Mostler & Parwin, 1973; *Calclamnella* [Triassic: Carnian; Austria]

153. *consonus* Carini, 1962; *Thalattocanthus* [Carboniferous: Pennsylvanian: Moscovian; U.S.A.]

154. *conspicuus* Deflandre-Rigaud, 1959; *Priscopedatus* [Paleogene: Oligocene: Rupelian; Germany]

155. *convexa* Whidborne, 1883; *Chirodota* [Jurassic: Bathonian; U.K.]

156. *corbisema* Deflandre-Rigaud, 1959; *Dictyothurites* [Paleogene: Oligocene: Rupelian; Germany]

157. *corbula* Zankl, 1966; *Theelia* [Triassic: Norian/Rhaetian; Germany]

158. *cordatum* Hampton, 1957; *Achistrum* [Jurassic: Upper Bathonian; U.K.]

159. *corolla* Schlumberger, 1890; *Priscopedatus* [Paleogene: Eocene: Upper Lutetian; France]

160. *costata* Reich, 2003 [2003d]; *Calcligula* [Cretaceous: Lower Maastrichtian; Germany]

161. *costatus* Sztejn, 1992; *Tomasovites* [Cretaceous: Lower Valanginian; Poland]

162. *costifera* Terquem & Berthelin, 1875; *Hemisphaeranthos* [Jurassic: Upper Pliensbachian; France]

163. *crassa* Mostler, 1972 [1972f]; *Eocaudina* [Triassic: Norian: Alaunian; Austria]

164. *crassa* Zhang Jin-jian, 1983; *Thuroholia* [Carboniferous: Mississippian: Lower Tournaisian; P.R. China]

165. *crassidentatus* Deflandre-Rigaud, 1950; *Chiridotites* [Jurassic: Upper Pliensbachian; Germany]

166. *crassirimosus* Mostler, 1972 [1972e]; *Praeeuphronides* [Jurassic: ?Pliensbachian; Austria]

167. *crassomarginata* Kristan-Tollmann, 1973; *Calclamnoidea* [Triassic: Ladinian: Fassanian; France]

168. *crassus* Mostler, 1971 [1971b]; *Priscopedatus* [Triassic: Norian; Austria]

169. *crassus* Schlumberger, 1890; *Priscopedatus* [Paleogene: Eocene: Upper Lutetian; France]

170. *cretacea* Reich, 2003 [2003d]; *Calcligula* [Cretaceous: upper Lower Maastrichtian; Germany]

171. *cretacea* Reich, 1997 [1997b]***; *Tricalclamnella* [Cretaceous: lower Upper Maastrichtian; Germany/southern Baltic Sea]

172. *cribellum* Frizzell & Exline, 1956; *Eocaudina* [Carboniferous: Mississippian: Serpukhovian; U.S.A.]

173. *cribellum* Schlumberger, 1890; *Priscopedatus* [Paleogene: Eocene: Lutetian; France]

174. *cribriformis* Gutschick, 1954; *Thuroholia* [Ordovician: Sandbian; U.S.A.]

175. *cribrum* Frizzell & Exline, 1956; *Eocaudina* [Carboniferous: Mississippian: Serpukhovian; U.S.A.]

176. *crinerensis* Reso & Wegner, 1964; *Thuroholia* [Ordovician: Sandbian; U.S.A.]

177. *croneisi* Gutschick, 1954; *Thuroholia* [Ordovician: Sandbian; U.S.A.]

178. *crucensis* Boczarowski, 2001; *Priscocaudina* [Devonian: Mid/?Late Frasnian; Poland]

179. *cruciformis* Mostler, 1968 [1968e]; *Calclamnella* [Triassic: Norian; Austria]

180. *cruciformis* Mostler, 1969; *Stichopitella* [Triassic: Norian; Austria]
181. *cruciformis* Sadeddin & Saqqa, 1997; *Spinopriscopedatus* [Cretaceous: Lower Turonian; Jordan]
182. *cruciformis* Walkiewicz, 1977; *Calclamna* [Neogene: Middle Miocene; Poland]
183. *crux* Deflandre-Rigaud, 1962 [1962b]; *Priscopedatus* [Jurassic: Oxfordian; France]
184. *cumunis* Mostler, 1969; *Triradites* [Triassic: Norian; Austria]
185. *cuneus* Frizzell & Exline, 1956***; *Mortensenites* [Jurassic: Upper Sinemurian; Germany]
186. *Cunninghami* Deflandre-Rigaud, 1949 [1949b] *; *Synaptites* [Paleogene: Eocene/Oligocene: Upper Priabonian/Lower Rupelian; U.S.A.]
187. *curriculum* Schlumberger, 1890; *Chirodota* [Paleogene: Eocene: Upper Lutetian; France]
188. *curta* Frizzell & Exline, 1956; *Tetravirga* [Carboniferous: Mississippian: Serpukhovian; U.S.A.]
189. *curvus* Sadeddin & Saqqa, 1997; *Spinopriscopedatus* [Cretaceous: Lower Turonian; Jordan]
190. *Cuvillieri* Deflandre-Rigaud, 1949 [1949b]; *Synaptites* [Paleogene: Eocene: Ypresian; France]
191. *danieli* Smith & Gallemí, 1991**; *Collbatothuria* [Triassic: Ladinian; Spain]
192. *dattai* Soodan, 1977 [1977b]; *Rigaudites* [Quaternary: Holocene; Arabian Sea]
193. *deeckei* Reich, 2001; *Pravuscucumis* [Cretaceous: Upper Turonian; Poland]
194. *deflandreae* Walkiewicz, 1977; *Calclamnella* [Neogene: Middle Miocene; Poland]
195. *deflandreae* Frizzell & Exline, 1956; *Priscopedatus* [Jurassic: Oxfordian; France]
196. *deflandreae* Frizzell & Exline, 1956; *Theelia* [Paleogene: Eocene: Upper Lutetian; France]
197. *delicata* Boczarowski, 2001*; *Palaeocucumaria* [Devonian: Late Givetian; Poland]
198. *densusmaculis* Sadeddin & Al-Tamimi, 2006; *Octopedatus* [Cretaceous: Late Cenomanian–Early Turonian; Jordan]
199. *dentata* Sadeddin, 1996 [1996b]; *Eocaudina* [Cretaceous: Albian; U.S.A.]
200. *dentata* Górka & Łuszczewska, 1969; *Theelia* [Jurassic: Bajocian; Poland]
201. *dentatus* Deflandre-Rigaud, 1962 [1962b]; *Cucumarites* [Jurassic: Oxfordian; France]
202. *denticulatus* Górka & Łuszczewska, 1969; *Priscopedatus* [Jurassic: Bajocian; Poland]
203. *dentigerus* Deflandre-Rigaud, 1962 [1962b]; *Cucumarites* [Jurassic: Oxfordian; France]

204. *diffissa* Mostler, 1972 [1972f]; *Acanthotheelia* [Triassic: Ladinian: Langobardian; Austria]
205. *diffissus* Mostler, 1971 [1971b]; *Uncinulinoides* [Triassic: Norian; Austria]
206. *diplococcus* Deflandre-Rigaud, 1962 [1962b]; *Cucumarites* [Jurassic: Oxfordian; France]
207. *divergens* Hodson, Harris & Lawson, 1956; *Rhabdotites* [Jurassic: Oxfordian; U.K.]
208. *diversimeatus* Deflandre-Rigaud, 1962 [1962b]; *Cucumarites* [Jurassic: Oxfordian; France]
209. *dobrogensis* Gheorghian, 1995; *Eocaudina* [Triassic: Anisian; Romania]
210. *dombrowiana* Boczarowski, 2001; *Staurocaudina* [Devonian: Early Eifelian; Poland]
211. *donofrioi* Kozur & Sadeddin, 1990***; *Frizzellus* [Jurassic: Bathonian; Jordan]
212. *doreckae* Kozur & Mostler, 1970; *Theelia* [Triassic: Anisian: Pelsonian; Austria]
213. *dorsetensis* Soodan & Whatley, 1988; *Protocaudina* [Jurassic: Callovian; U.K.]
214. *dorsetensis* Hodson, Harris & Lawson, 1956; *Rhabdotites* [Jurassic: Oxfordian; U.K.]
215. *dracoformis* Mostler, 1968 [1968e]; *Calclamnella* [Triassic: Norian; Austria]
216. *dubius* Deflandre-Rigaud, 1946 [1946b]; *Priscopedatus* [Jurassic: Oxfordian; France]
217. *dumosus* Mostler, 1972 [1972f]; *Kuehnites* [Triassic: Norian; Austria]
218. *dumosus* Mostler, 1971 [1971b]; *Priscopedatus* [Triassic: Norian; Austria]
219. *duncani* Tandon & Saxena, 1977; *Eocaudina* [Paleogene: Eocene: Lutetian; India]
220. *dushanensis* Zhang Jin-jian, 1993; *Thuroholia* [Carboniferous: Mississippian: late Viséan; P.R. China]
221. *duvergieri* Rioult, 1961; *Cucumarites* [Neogene: Miocene: Upper Burdigalian; France]
222. *dzhulfaensis* Mostler & Rahimi-Yazd, 1976; *Theelia* [Permian: Lopingian: Wuchiapingian; Iran]
223. *echinatus* Schlumberger, 1890; *Priscopedatus* [Paleogene: Eocene: Upper Lutetian; France]
224. *echinocucumioides* Mostler, 1968 [1968b]; *Tetravirga* [Triassic: Anisian: Illyrian; Austria]
225. *Eiffeli* Schlumberger, 1890; *Priscopedatus* [Paleogene: Eocene: Upper Lutetian; France]
226. *Eiseliana* Spandel, 1898; *Prosynapta* [Permian: Lopingian: Wuchiapingian; Germany]
227. *eisenstadtensis* Kristan-Tollmann, 1964 [1964c]; *Theelia* [Neogene: Miocene: Middle Tortonian; Austria]
228. *elegans* Malagoli, 1888; *Chiridota* [Neogene: Pliocene; Italy]

229. *elegans* Mostler, 1970; *Rhabdotites* [Triassic: Norian; Austria]
230. *elegans* Mostler, 1972 [1972f]; *Stichopitella* [Triassic: Norian: Sevatian; Austria]
231. *elegans* Schlumberger, 1888; *Stueria* [Paleogene: Eocene: Upper Lutetian; France]
232. *elenae* Gheorghian, 1995; *Acanthotheelia* [Triassic: Anisian; Romania]
233. *elgeri* Deflandre-Rigaud, 1959; *Calcligula* [Paleogene: Oligocene: Rupelian; Germany]
234. *elgeri* Deflandre-Rigaud, 1962 [1962b]*; *Molpadioites* [Paleogene: Oligocene: Rupelian; Germany]
235. *elliptica* Soodan, 1980; *Calclamna* [Paleogene: Eocene: Ypresian; India]
236. *elliptica* Soodan, 1977 [1977d]; *Protocaudina* [Jurassic: Bajocian; India]
237. *ellipticus* Deflandre-Rigaud, 1946 [1946b]; *Priscopedatus* [Jurassic: Oxfordian; France]
238. *ellipticus* Soodan, 1973 [1973b]; *Stueria* [Jurassic: Bathonian; India]
239. *elliptiferus* Zawidzka, 1971; *Priscopedatus* [Triassic: Anisian: ?Illyrian; Poland]
240. *elongata* Al-Tamimi & Saqqa, 2008; *Calclamnella* [Cretaceous: Upper Cenomanian; Jordan]
241. *elongata* Frizzell & Exline, 1956; *Eocaudina* [Carboniferous: Mississippian: Serpukhovian; U.S.A.]
242. *elongatus* Deflandre-Rigaud, 1962 [1962b]***; *Mortensenites*? [Jurassic: Oxfordian; France]
243. *elongatus* Mostler, 1972 [1972f]; *Priscopedatus* [Triassic: Norian; Austria]
244. *empeldensis* Kristan-Tollmann, 1986 [1986a]; *Priscopedatus* [Jurassic: Upper Pliensbachian; Germany]
245. *eocoena* Schlumberger, 1888; *Synapta* [Paleogene: Eocene: Upper Lutetian; France]
246. *eocenica* Soodan, 1980; *Calclamna* [Paleogene: Eocene: Ypresian; India]
247. *eostre* Botting & Muir, 2012**; *Oesolcucumaria* [Ordovician: Darriwilian; U.K.]
248. *ernsti* Reich & Wiese, 2010; *Rigaudites* [Cretaceous: Upper Turonian; Poland]
249. *erratica* Reich in Herrig et al. 1997; *Stueria* [Cretaceous: upper Upper Campanian; Germany]
250. *etheridgei* Frizzell & Exline, 1956; *Tetravirga* [Carboniferous: Mississippian: Viséan: Brigantian; U.K.]
251. *etheridgei* Soodan & Whatley, 1990; *Yorkshirella* [Jurassic: Hettangian; U.K.]
252. *eurymarginata* Kristan-Tollmann, 1963; *Eocaudina* [Triassic: Carnian: Cordevolian; Italy]
253. *exlineae* Mostler, 1970; *Acanthocaudina* [Triassic: Norian: Sevatian; Austria]
254. *exlineae* Said & Barakat, 1958; *Priscopedatus* [Jurassic: Bajocian; Egypt]
255. *exporrigus* Boczarowski, 2001*; *Devonothyonites* [Devonian: Early Givetian; Poland]
256. *extensus* Mostler, 1968 [1968e]; *Binoculites* [Triassic: Norian; Austria]
257. *fastigata* Mostler, 1972 [1972f]; *Theelia* [Triassic: Ladinian: Lower Fassanian; Germany]
258. *feddeni* Soodan, 1973 [1973b]; *Stueria* [Jurassic: Bathonian; India]
259. *Feifeli* Mortensen, 1937; *Cucumaria* [Jurassic: Upper Toarcian; Germany]
260. *ficta* Deflandre-Rigaud, 1962 [1962b]; *Calcligula*(?) [Jurassic: Oxfordian; France]
261. *fissa* Mostler, 1971 [1971b]; *Theelia* [Triassic: Norian; Austria]
262. *fissus* Mostler, 1970; *Rhabdotites* [Triassic: Norian; Austria]
263. *flaccida* Weiss, 1954***; *Palaeosynapta* [Ordovician: Katian; U.S.A.]
264. *fletcheri* Soodan & Whatley, 1990; *Calclamna* [Jurassic: Hettangian; U.K.]
265. *florealis* Frentzen, 1964; *Chirodota* [Jurassic: Lower Bajocian; Germany]
266. *florida* Terquem & Berthelin, 1875; *Hemisphaeranthos* [Jurassic: Upper Pliensbachian; France]
267. *floydensis* Summerson & Campbell, 1958; *Thuroholia* [Carboniferous: Pennsylvanian: Moscovian; U.S.A.]
268. *foliosa* Mostler, 1971 [1971b]; *Calclamnella* [Triassic: Norian: Sevatian; Austria]
269. *foliosus* Mostler, 1971 [1971b]; *Priscopedatus* [Triassic: Norian; Austria]
270. *follicula* Mostler, 1968 [1968e]; *Calclamnella* [Triassic: Norian; Austria]
271. *fordelensis* Frizzell & Exline, 1956; *Tetravirga* [Carboniferous: Mississippian: Viséan: Brigantian; U.K.]
272. *formosa* Mostler, 1972 [1972f]; *Kozurella* [Triassic: Norian: Sevatian; Germany]
273. *fortworthi* Sadeddin, 1996 [1996b]; *Eocaudina* [Cretaceous: Albian; U.S.A.]
274. *fragosa* Deflandre-Rigaud, 1962 [1962b]; *Calclamnella* [Jurassic: Oxfordian; France]
275. *Frankei* O. Müller, 1912, *Actinoclava* [Cretaceous: Cenomanian(Turonian); Germany]
276. *frenzeli* Reich, 2003 [2003d]; *Calcligula* [Cretaceous: upper Lower Maastrichtian; Germany]
277. *frizzelli* Langenheim [jr.] & Epis, 1957; *Achistrum* [Carboniferous: Mississippian: Tournaisian; U.S.A.]
278. *frizzelli* Deflandre-Rigaud, 1962 [1962b]; *Exlinella* [Jurassic: Oxfordian; France]
279. *frizzelli* Said & Barakat, 1958; *Priscopedatus*(?) [Jurassic: Bajocian; Egypt]

280. *frontis* Sadeddin in Al-Tamimi, Saqqa & Sadeddin, 2001; *Paracucumarites* [Cretaceous: Cenomanian; Jordan]

281. *fusiformis* Deflandre-Rigaud, 1959; *Calclamnella* [Paleogene: Oligocene: Rupelian; Germany]

282. *gallica* Frizzell & Exline, 1956; *Calcancora* [Paleogene: Eocene: Upper Lutetian; France]

283. *gamma* Hodson, Harris & Lawson, 1956; *Achistrum* [Jurassic: Oxfordian; U.K.]

284. *Geinitziana* Spandel, 1898; *Chirodota* [Permian: Lopingian: Wuchiapingian; Germany]

285. *germanica* Frizzell & Exline, 1956; *Calclamna* [Jurassic: Lower Hettangian; Germany]

286. *germanica* Kozur, 1969; *Theelia* [Triassic: Anisian: Bithynian; Germany]

287. *ghoshi* Soodan, 1977 [1977b]; *Rigaudites* [Quaternary: Holocene; Arabian Sea]

288. *giganta* Soodan, 1975; *Fletcherina* [Jurassic: Bathonian; India]

289. *gigantea* Sadeddin, 1996 [1996b]; *Priscoligula* [Cretaceous: Albian; U.S.A.]

290. *girondensis* Rioult, 1961; *Cucumarites* [Neogene: Miocene: Upper Burdigalian; France]

291. *glaber* Mostler, 1972 [1972f]; *Frangerites* [Triassic: Norian: Alaunian; Germany]

292. *glivicensis* Górka, 1997; *Calcancora* [Neogene: Miocene; Poland]

293. *globosus* Sadeddin in Al-Tamimi, Saqqa & Sadeddin, 2001; *Rigaudites* [Cretaceous: Upper Cenomanian; Jordan]

294. *godpurensis* Soodan, 1977 [1977a]; *Achistrum* [Jurassic: ?Oxfordian/Kimmeridgian; India]

295. *godpurensis* Soodan, 1977 [1977a]; *Fletcherina* [Jurassic: ?Oxfordian/Kimmeridgian; India]

296. *gondwanensis* Haude, 1995 [1995b]**; *Andenothyone* [Devonian: Lochkovian–Emsian; Argentina]

297. *goniaia* Kristan-Tollmann, 1964 [1964c]; *Calclamnoidea* [Neogene: Miocene: Middle Tortonian; Austria]

298. *goodlandi* Sadeddin, 1996 [1996b]; *Eocaudina* [Cretaceous: Albian; U.S.A.]

299. *gordoni* Soodan & Whatley, 1990; *Uncinulina* [Jurassic: Hettangian; U.K.]

300. *gornensis* B.A. Matyja, H. Matyja & Szulczewski, 1973; *Eocaudina* [Devonian: early Frasnian; Poland]

301. *gothathadensis* Tandon & Saxena, 1977; *Koteshwaria* [Paleogene: Eocene: Lutetian; India]

302. *gowdai* Soodan, 1977 [1977b]; *Eocaudina* [Quaternary: Holocene; Arabian Sea]

303. *gowdai* Tandon & Saxena, 1977; *Koteshwaria* [Paleogene: Eocene: Lutetian; India]

304. *gracilis* Reich, 2003 [2003d]; *Calcligula* [Cretaceous: upper Lower Maastrichtian; Germany]

305. *gracilis* Mostler, 1968 [1968e]; *Tetravirga* [Triassic: Norian; Austria]

306. *gracilis* Mostler, 1972 [1972e]; *Uncinulina* [Jurassic: ?Hettangian; Austria]

307. *gracillima* Whidborne, 1883; *Chirodota* [Jurassic: Bathonian; U.K.]

308. *grandis* Kristan-Tollmann, 1963; *Eocaudina* [Triassic: Rhaetian; Austria]

309. *granulosa* Mostler, 1970; *Theelia* [Triassic: Norian; Austria]

310. *guembeli* Frizzell & Exline, 1956; *Eocaudina* [Triassic: Carnian; Italy]

311. *guembeli* Kristan-Tollmann, 1963; *Theelia* [Triassic: Carnian: Cordevolian; Italy]

312. *guizhouensis* Zhang Jin-jian, 1993; *Eocaudina* [Carboniferous: Mississippian: late Viséan; P.R. China]

313. *gujaraticus* Soodan, 1977 [1977b]; *Rigaudites* [Quaternary: Holocene; Arabian Sea]

314. *gujeratica* Soodan, 1973 [1973c]; *Jumaraina* [Jurassic: Bathonian; India]

315. *gujeratica* Soodan, 1973 [1973b]; *Stueria* [Jurassic: Bathonian; India]

316. *gutschicki* Tandon & Saxena, 1977; *Croneisites* [Paleogene: Eocene: Lutetian; India]

317. *gutschicki* Frizzell & Exline, 1956; *Eocaudina* [Carboniferous: Pennsylvanian: Moscovian; U.S.A.]

318. *guyaderi* Rioult, 1959 [1959b]; *Priscopedatus* [Jurassic: Oxfordian; France]

319. *hagdorni* Mostler & Reich, 2001; *Achistrum* [Triassic: Norian; Austria]

320. *hallstattensis* Mostler, 1971 [1971b]; *Kuehnites* [Triassic: Norian; Austria]

321. *hamptoni* Deflandre-Rigaud, 1962 [1962b]; *Paracucumarites* [Jurassic: Oxfordian; France]

322. *hannae* Tandon & Saxena, 1977; *Koteshwaria* [Paleogene: Eocene: Lutetian; India]

323. *hannai* Croneis in Croneis & McCormack, 1932 [1932c]; *Protocaudina* [Carboniferous: Mississippian: Serpukhovian; U.S.A.]

324. *harpago* Kornicker & Imbrie, 1958; *Parvispina* [Permian: Cisuralian: Asselian; U.S.A.]

325. *harudiensis* Tandon & Saxena, 1977; *Calclamna* [Paleogene: Eocene: Lutetian; India]

326. *hastata* Mostler, 1972 [1972f]; *Uncinulinoides* [Triassic: Norian: Sevatian; Austria]

327. *haynesi* Soodan & Whatley, 1988; *Jumaraina* [Jurassic: Oxfordian; U.K.]

328. *heisseli* Mostler, 1968 [1968b]; *Priscopedatus* [Triassic: Anisian: Illyrian; Austria]

329. *helios* Kozur & Mock, 1972; *Acanthotheelia* [Triassic: Norian: Lower Sevatian; Slovak Republic]

330. *helvetica* Kübler & Zwingli, 1870; *Chirodota* [Jurassic: Lower Tithonian; Switzerland]

331. *hemisphaericus* Kristan-Tollmann, 1964 [1964c]***; *Mortensenites* [Neogene: Miocene: Middle Tortonian; Austria]

332. *hendersoni* Soodan & Whatley, 1988; *Stueria* [Jurassic: Oxfordian; U.K.]

333. *heptalampra* Bartenstein, 1936; *Chiridota* [Jurassic: Upper Pliensbachian; Germany]

334. *hernsteini* Mostler, 1969; *Stichopitella* [Triassic: Norian; Austria]

335. *herrigi* Reich, 1995; *Protocaudina* [Cretaceous: upper Upper Maastrichtian; Germany/southern Baltic Sea]

336. *heteroporus* Deflandre-Rigaud, 1962 [1962b]; *Cucumarites* [Jurassic: Oxfordian; France]

337. *heteroporus* Deflandre-Rigaud, 1962 [1962b]; *Priscopedatus* [Jurassic: Oxfordian; France]

338. *hexacneme* Summerson & Campbell, 1958; *Theelia* [Carboniferous: Pennsylvanian: Moscovian; U.S.A.]

339. *hexagona* Kristan-Tollmann, 1963; *Eocaudina* [Triassic: Rhaetian; Austria]

340. *hexagona* Soodan, 1977 [1977b]; *Eocaudina* [Quaternary: Holocene; Arabian Sea]

341. *hexagona* Mostler, 1970; *Praecaudina* [Triassic: Norian; Austria]

342. *hexagonaria* Martin, 1952; *Protocaudina* [Devonian: Givetian; U.S.A.]

343. *holsaticus* Deflandre-Rigaud, 1959; *Cucumarites* [Paleogene: Oligocene: Rupelian; Germany]

344. *horrida* Matyja, 1972; *Stueria* [Jurassic: middle Upper Oxfordian; Poland]

345. *horridus* Mostler, 1968 [1968b]; *Staurocumites* [Triassic: Anisian: Illyrian; Austria]

346. *hothamensis* Soodan & Whatley, 1990; *Harisina* [Jurassic: Hettangian; U.K.]

347. *huckei* Frizzell & Exline, 1956; *Calcligula*? [Cretaceous: Aptian/Albian; Poland]

348. *hummari* Sadeddin in Al-Tamimi, Saqqa & Sadeddin, 2001; *Spinopriscopedatus* [Cretaceous: Upper Cenomanian; Jordan]

349. *hunsrueckiana* Lehmann, 1958**; *Palaeocucumaria* [Devonian: Lower Emsian; Germany]

350. *huntingdonshirensis* Soodan & Whatley, 1988; *Lawsonina* [Jurassic: Upper Oxfordian; U.K.]

351. *hystrix* Deflandre-Rigaud, 1962 [1962b]; *Priscopedatus* [Jurassic: Oxfordian; France]

352. *igoi* Kikuchi & Nikaido, 1996**; *Cucumaria* [Neogene: Middle Miocene: Langhian; Japan]

353. *illyricus* Mostler, 1971 [1971b]; *Priscopedatus* [Triassic: Anisian: Illyrian; Italy]

354. *immissorbicula* Mostler, 1968 [1968b, 1968d]; *Theelia* [Triassic: Anisian: Illyrian; Austria]

355. *imperfecta* Soodan, 1975; *Sastriella* [Jurassic: Bathonian; India]

356. *imperforata* Gutschick, Canis & Brill, 1967; *Rotoides* [Carboniferous: Mississippian: Lower Tournaisian; U.S.A.]

357. *imperforata* Frizzell & Exline, 1956; *Tetravirga* [Carboniferous: Pennsylvanian: Moscovian; U.S.A.]

358. *inaequalis* Mostler, 1969; *Kuehnites* [Triassic: Norian; Austria]

359. *inaequalis* Mutterlose, 1982; *Paracucumarites* [Jurassic: lower Lower Albian; Germany]

360. *inæqualis* Schlumberger, 1890; *Priscopedatus* [Paleogene: Eocene: Upper Lutetian; France]

361. *inaequiperforatus* Krainer & Mostler, 1997; *Centripedatus* [Jurassic: Pliensbachian; Austria]

362. *inaequiporus* A. H. Müller, 1964; *Cucumarites* [Cretaceous: Lower Maastrichtian; Germany]

363. *incertus* Deflandre-Rigaud, 1950; *Micradites* [Jurassic: Oxfordian; France]

364. *incisus* Mostler, 1972 [1972f]; *Priscopedatus* [Triassic: Norian; Germany]

365. *inclinatus* Mostler, 1972 [1972f]; *Binoculites* [Triassic: Norian: Sevatian; Austria]

366. *inclinatus* Mostler, 1972 [1972e]; *Rhabdotites* [Jurassic: Hettangian/Sinemurian; Austria]

367. *inconstans* Mostler, 1970; *Biacumina* [Triassic: Norian; Austria]

368. *inconstans* Mostler, 1971 [1971b]; *Ramusites* [Triassic: Norian; Austria]

369. *incrassatus* Kristan-Tollmann, 1964 [1964c]; *Croneisites* [Neogene: Miocene: Middle Tortonian; Austria]

370. *incurvatus* Krainer & Mostler, 1997; *Punctatites* [Jurassic: Sinemurian; Austria]

371. *incurvatus* Mostler, 1970; *Rhabdotites* [Triassic: Norian; Austria]

372. *indica* Soodan, 1973 [1973c]; *Jumaraina* [Jurassic: Bathonian; India]

373. *indica* Soodan, 1980; *Calclamna* [Paleogene: Eocene: Ypresian; India]

374. *indica* Soodan, 1977 [1977b]; *Eocaudina* [Quaternary: Holocene; Arabian Sea]

375. *indicus* Soodan, 1977 [1977b]; *Rigaudites* [Quaternary: Holocene; Arabian Sea]

376. *indicus* Soodan, 1973 [1973b]; *Stueria* [Jurassic: Bathonian; India]

377. *inflatus* Deflandre-Rigaud, 1962 [1962b]; *Cucumarites* [Jurassic: Oxfordian; France]

378. *inflexus* Mostler, 1971 [1971b]; *Priscopedatus* [Triassic: Norian; Austria]

379. *ingens* Joshua, 1914; *Chiridota* [Paleogene: ?Eocene; Australia]

380. *ingridae* Mostler in Krainer et al., 1994; *Neomicroantyx* [Jurassic: Upper Hettangian–Upper Pliensbachian; Austria]

381. *innienensis* Deflandre-Rigaud, 1959; *Elgerius* [Paleogene: Oligocene: Rupelian; Germany]

382. *innsbrucki* Gilliland, 1993; *Priscopedatus* [Triassic: Anisian: Illyrian; Montenegro]

383. *insignis* Deflandre-Rigaud, 1962 [1962b]; *Amphitriodites* [Jurassic: Oxfordian; France]

384. *insignis* Kristan-Tollmann, 1964 [1964c]; *Croneisites* [Neogene: Miocene: Middle Tortonian; Austria]

385. *insolica* Al-Tamimi & Saqqa, 2008; *Eocaudina* [Cretaceous: Upper Cenomanian; Jordan]

386. *insolitus* Kristan-Tollmann, 1963***; *Mortensenites* [Triassic: Carnian: Cordevolian; Italy]

387. *intercessus* Deflandre-Rigaud, 1959; *Myriotrochites* [Paleogene: Oligocene: Rupelian; Germany]

388. *intermedia* Soodan & Whatley, 1990; *Calclamna* [Jurassic: Hettangian; U.K.]

389. *intermedia* Soodan & Whatley, 1990; *Hodsonina* [Jurassic: Hettangian; U.K.]

390. *inusitata* Kozur & Mostler, 1971; *Fissobractites* [Triassic: Carnian: Cordevolian; Hungary]

391. *iranica* Mostler & Rahimi-Yazd, 1976; *Jolfacaudina* [Permian: Lopingian: Wuchiapingian; Iran]

392. *irregularis* Frizzell & Exline, 1956; *Binoculites* [Jurassic: Lower Pliensbachian; Germany]

393. *irregularis* Frizzell & Exline, 1956; *Calclamnoidea* [Jurassic: Upper Pliensbachian; Germany]

394. *irregularis* Hampton, 1958 [1958d]***; *Frizzellus* [Jurassic: Upper Bathonian; U.K.]

395. *irregularis* Mostler, 1968 [1968b]; *Multivirga* [Triassic: Upper Anisian; Austria]

396. *irregularis* Sadeddin & Saqqa, 1997; *Pentapriscopedatus* [Cretaceous: Lower Turonian; Jordan]

397. *irregularis* Schlumberger, 1890; *Priscopedatus* [Paleogene: Eocene: Upper Lutetian; France]

398. *irregularis* Sadeddin, 1996 [1996b]; *Prisculatrites* [Cretaceous: Albian; U.S.A.]

399. *irregularis* Mostler, 1969; *Pseudostaurocumites* [Triassic: Norian; Austria]

400. *irregularis* Hodson, Harris & Lawson, 1956; *Rhabdotites* [Jurassic: Oxfordian; U.K.]

401. *irregularis* Deflandre-Rigaud, 1962 [1962b]; *Spandelites* [Jurassic: Oxfordian; France]

402. *irregularis* Deflandre-Rigaud, 1949 [1949b]; *Synaptites*? [Jurassic: Oxfordian; France]

403. *irregularis* Summerson & Campbell, 1958; *Thuroholia* [Carboniferous: Pennsylvanian: Moscovian; U.S.A.]

404. *issleri* Croneis in Croneis & McCormack, 1932 [1932c]; *Ancistrum* [Jurassic: Hettangian and Sinemurian; Germany]

405. *issleri* Deflandre-Rigaud, 1952; *Binoculites* [Jurassic: Upper Sinemurian; Germany]

406. *jaffari* Soodan, 1977 [1977b]; *Eocaudina* [Quaternary: Holocene; Arabian Sea]

407. *jagti* Reich, 2003 [2003b]; *Palaeotrochodota* [Cretaceous: upper Upper Maastrichtian; Germany/southern Baltic Sea]

408. *jaini* Tandon & Saxena, 1977; *Calclamnella* [Paleogene: Eocene: Lutetian; India]

409. *janetscheki* Mostler, 1983; *Microantyx* [Permian: Guadalupian: ?Wordian; Iran]

410. *janinae* Sadeddin, 1996 [1996b]; *Prisculatrites* [Cretaceous: Albian; U.S.A.]

411. *jaworznicensis* Górka & Łuszczewska, 1969; *Priscopedatus* [Jurassic: Upper Bathonian; Poland]

412. *jhadwaensis* Tandon & Saxena, 1977; *Eocaudina* [Paleogene: Eocene: Lutetian; India]

413. *jhingrani* Tandon & Saxena, 1977; *Kutchia* [Paleogene: Eocene: Lutetian; India]

414. *jhurioensis* Soodan, 1973 [1973c]; *Jumaraina* [Jurassic: Bathonian; India]

415. *jordanica* Sadeddin, 1991; *Acanthotheelia* [Triassic: Anisian: Pelsonian; Jordan]

416. *jordanica* Kozur & Sadeddin, 1990; *Huniella* [Jurassic: Bathonian; Jordan]

417. *jordanica* Kozur & Sadeddin, 1992; *Schizotheelia* [Triassic: Ladinian: Fassanian; Jordan]

418. *jordanicus* Sadeddin in Al-Tamimi, Saqqa & Sadeddin, 2001; *Spinopriscopedatus* [Cretaceous: Cenomanian; Jordan]

419. *jumaraensis* Soodan, 1975; *Feddenella* [Jurassic: Bathonian; India]

420. *jumaraensis* Soodan, 1975; *Sastriella* [Jurassic: Bathonian; India]

421. *jumaraensis* Soodan, 1973 [1973b]; *Stueria* [Jurassic: Bathonian; India]

422. *jurassica* Frizzell & Exline, 1956; *Calclamnella* [Jurassic: Upper Sinemurian; Germany]

423. *jurassica* Frizzell & Exline, 1956; *Calcligula*? [Jurassic: Upper Pliensbachian; France]

424. *jurassica* Said & Barakat, 1958; *Calcligula* [Jurassic: Bajocian; Egypt]

425. *jurassicus* Mostler, 1972 [1972e]; *Syneuphronides* [Jurassic: ?Pliensbachian; Austria]

426. *jurensis* Frentzen, 1964; *Prostichopus* [Jurassic: Upper Toarcian; Germany]

427. *kampschuuri* Kozur in Kozur et al., 1980; *Zawidzkella* [Triassic: Carnian: Cordevolian; Spain]

428. *kansasensis* Hanna, 1930; *Laetmophasma*(?) [Carboniferous: Pennsylvanian: Kasimovian/Gzhelian; U.S.A.]
429. *kashimi* Soodan, 1973 [1973b]; *Stueria* [Jurassic: Bathonian; India]
430. *khadirensis* Soodan, 1977 [1977d]; *Protocaudina* [Jurassic: Bajocian; India]
431. *khariensis* Tandon & Saxena, 1977; *Calclamnella* [Paleogene: Eocene: Lutetian; India]
432. *kielcensis* Walkiewicz, 1977***; *Mortensenites* [Neogene: Miocene: Langhian; Poland]
433. *kistnai* Jafar, 1970; *Calcancora* [Quaternary: Pleistocene; Arabian Sea]
434. *koeveskalensis* Kozur & Mostler, 1971; *Theelia* [Triassic: Carnian: Cordevolian; Hungary]
435. *kolayatensis* Soodan, 1980; *Calclamna* [Paleogene: Eocene: Ypresian; India]
436. *kolayatensis* Soodan, 1980; *Exlinella* [Paleogene: Eocene: Ypresian; India]
437. *korytnicensis* Walkiewicz, 1977; *Calclamnella* [Neogene: Miocene: Langhian; Poland]
438. *kotlickii* Kozur & Mostler, 1970; *Priscopedatus* [Triassic: Anisian: Pelsonian; Poland]
439. *kozuri* Sadeddin, 1996 [1996b]; *Calcligula* [Cretaceous: Albian; U.S.A.]
440. *kozuri* Mostler, 1970; *Priscopedatus* [Triassic: Norian; Austria]
441. *kristani* Mostler, 1969; *Theelia* [Triassic: Norian; Austria]
442. *krystyni* Kozur & Simón, 1972; *Theelia* [Triassic: uppermost Ladinian/Lower Carnian; Spain]
443. *kuepperi* Mostler, 1969; *Acanthotheelia* [Triassic: Norian; Austria]
444. *kukaviensis* Sztejn, 1992; *Priscopedatus* [Cretaceous: Lower Valanginian; Poland]
445. *kupperi* Deflandre-Rigaud, 1962 [1962b]; *Cucumarites* [Neogene: Miocene: Middle Tortonian; Austria]
446. *kutchensis* Soodan, 1977 [1977a]; *Achistrum* [Jurassic: ?Oxfordian/Kimmeridgian; India]
447. *kutchensis* Tandon & Saxena, 1977; *Eocaudina* [Paleogene: Eocene: Lutetian; India]
448. *kutchensis* Soodan, 1977 [1977a]; *Fletcherina* [Jurassic: ?Oxfordian/Kimmeridgian; India]
449. *kutchensis* Soodan, 1973 [1973c]; *Jumaraina* [Jurassic: Bathonian; India]
450. *kutchensis* Soodan, 1973 [1973b]; *Stueria* [Jurassic: Bathonian; India]
451. *kutscheri* Reich, 2003 [2003b]; *Theelia* [Cretaceous: upper Upper Maastrichtian; Germany/southern Baltic Sea]
452. *lacrimaeformis* Walkiewicz, 1977; *Eocaudina* [Neogene: Miocene: Langhian; Poland]
453. *ladinica* Kozur & Mostler, 1971; *Acanthotheelia* [Triassic: Ladinian: Langobardian; Hungary]

454. *lævigata* Schlumberger, 1890; *Synapta* [Paleogene: Eocene: Upper Lutetian; France]
455. *lanceolata* Schlumberger, 1890; *Chirodota* [Paleogene: Eocene: Upper Lutetian; France]
456. *langeri* Boczarowski, 2001; *Mercedescaudina* [Devonian: Late Givetian; Poland]
457. *lansulata* Al-Tamimi & Saqqa, 2008; *Prisculatrites* [Cretaceous: Lower Turonian; Jordan]
458. *lata* Kozur & Mostler, 1971; *Theelia* [Triassic: Carnian: Cordevolian; Hungary]
459. *latiareata* Krainer & Mostler, 1997; *Biacumina* [Jurassic: Pliensbachian; Austria]
460. *latidentata* Krainer & Mostler, 1997; *Kristanella* [Jurassic: Pliensbachian; Austria]
461. *latimarginata* Mostler, 1971 [1971b]; *Theelia* [Triassic: Carnian: Cordevolian; Austria]
462. *latus* Mostler, 1972 [1972f]; *Praeeuphronides* [Triassic: Norian: Alaunian; Germany]
463. *lehmani* Deflandre-Rigaud, 1962 [1962b]; *Spandelites* [Jurassic: Oxfordian; France]
464. *levis* Kozur & Mostler, 1970; *Tetravirga* [Triassic: Anisian: Pelsonian; Austria]
465. *liasica* Terquem, 1866; *Gromia* [Jurassic: Pliensbachian; France]
466. *liassica* Mostler, 1972 [1972e]; *Eocaudina* [Jurassic: ?Hettangian–?Toarcian; Austria]
467. *liassica* Soodan & Whatley, 1990; *Hamptonina* [Jurassic: Hettangian; U.K.]
468. *liassica* Krainer & Mostler, 1997; *Kristanella* [Jurassic: (Sinemurian)/Pliensbachian; Austria]
469. *liassica* Mostler in Krainer et al., 1994; *Theelia* [Jurassic: Upper Hettangian–Upper Pliensbachian; Austria]
470. *liassicus* Gilliland, 1992 [1992a]; *Palaeoypsilus* [Jurassic: Sinemurian; U.K.]
471. *liguliformis* Kristan-Tollmann, 1973; *Calcligula* [Cretaceous: upper Lower Hauterivian; Germany]
472. *liptovskaensis* Kozur & Mock in Gaździcki et al., 1978; *Eocaudina* [Triassic: Carnian: Cordevolian; Slovak Republic]
473. *liptovskaensis* Kozur & Mock in Gaździcki et al., 1978; *Theelia* [Triassic: Carnian: Cordevolian; Slovak Republic]
474. *lobatus* A. H. Müller, 1964; *Cucumarites* [Cretaceous: Lower Maastrichtian; Germany]
475. *lobatus* Al-Tamimi & Saqqa, 2008; *Prisculatrites* [Cretaceous: Upper Cenomanian; Jordan]
476. *loferensis* Mostler in Krainer et al., 1994; *Theelia* [Jurassic: Upper Hettangian–Upper Pliensbachian; Austria]
477. *longa* Kozur & Mock, 1972; *Eocaudina* [Triassic: Norian: Alaunian; Slovak Republic]

41

478. *longipontinum* Frentzen, 1964; *Cibrum* [Jurassic: Upper Hettangian; Germany]

479. *longirameus* Mostler, 1968 [1968e]; *Punctatites* [Triassic: Norian; Austria]

480. *longiramosus* Kozur & Mock, 1972; *Semperites* [Triassic: Norian: Alaunian; Slovak Republic]

481. *longirostrum* Mostler, 1971 [1971b]; *Achistrum* [Triassic: Norian; Austria]

482. *longistriata* Beckmann, 1965 [1965b]; *Uncinulina* [Devonian: Givetian; Germany]

483. *longitubus* Kozur & Mostler, 1989; *Tubocaudina* [Permian: Guadalupian: Wordian; Italy]

484. *lordi* Soodan & Whatley, 1990; *Yorkshirella* [Jurassic: Hettangian; U.K.]

485. *ludwigi* Croneis in Croneis & McCormack, 1932 [1932c]; *Ancistrum* [Carboniferous: Pennsylvanian: Kasimovian; U.S.A.]

486. *lunata* Kornicker & Imbrie, 1958; *Uncinulina* [Permian: Cisuralian: Asselian; U.S.A.]

487. *magnidentata* Kozur & Simón, 1972; *Theelia* [Triassic: uppermost Ladinian/Lower Carnian; Spain]

488. *magnispinosus* Kozur & Sadeddin, 1990; *Spinopriscopedatus* [Jurassic: Bathonian; Jordan]

489. *magnispinosus* Sadeddin, 1996 [1996b]; *Rigaudites* [Cretaceous: Albian; U.S.A.]

490. *malmensis* Frizzell & Exline, 1956; *Hemisphaeranthos* [Jurassic: Lower Oxfordian; Germany]

491. *malmensis* Mostler, 1972 [1972e]; *Ramusites* [Jurassic: ?Oxfordian–?Tithonian; Austria]

492. *maniaraensis* Tandon & Saxena, 1977; *Eocaudina* [Paleogene: Eocene: Lutetian; India]

493. *margaritatus* Schlumberger, 1890; *Priscopedatus* [Paleogene: Eocene: Upper Lutetian; France]

494. *marginata* Langenheim [jr.] & Epis, 1957; *Thuroholia* [Carboniferous: Mississippian: Tournaisian; U.S.A.]

495. *margostapedus* Sadeddin in Al-Tamimi, Saqqa & Sadeddin, 2001; *Prisculatrites* [Cretaceous: Upper Cenomanian; Jordan]

496. *marhensis* Tandon & Saxena, 1977; *Eocaudina* [Paleogene: Eocene: Lutetian; India]

497. *martini* Langenheim [jr.] & Epis, 1957; *Rota* [Carboniferous: Mississippian: Tournaisian; U.S.A.]

498. *mazoviensis* Kubiatowicz & Matyja, 1977; *Priscopedatus* [Cretaceous: Upper Valanginian; Poland]

499. *mccormacki* Frizzell & Exline, 1956; *Eocaudina* [Carboniferous: Pennsylvanian: Moscovian; U.S.A.]

500. *medioangusta* Kristan-Tollmann, 1964 [1964c]; *Calclamnoidea* [Neogene: Miocene: Middle Tortonian; Austria]

501. *meltonensis* Soodan & Whatley, 1987; *Achistrum (Cancellrum)* [Jurassic: Upper Oxfordian; U.K.]

502. *merhi* Tandon & Saxena, 1977; *Koteshwaria* [Paleogene: Eocene: Lutetian; India]

503. *mesojurassica* Kozur & Sadeddin, 1990; *Helfriedella* [Jurassic: Bathonian; Jordan]

504. *mesoliassica* Frentzen, 1964; *Chirodota* [Jurassic: Upper Pliensbachian; Germany]

505. *mesopermiana* Kozur & Mostler, 1989; *Theelia* [Permian: Guadalupian: Wordian; Italy]

506. *michaeli* Mutterlose, 1982; *Calcancora* [Cretaceous: lower Lower Albian; Germany]

507. *micralcyonarites* Deflandre-Rigaud, 1962 [1962b]; *Ornaticannula* [Jurassic: Oxfordian; France]

508. *microporus* Deflandre-Rigaud, 1962 [1962b]; *Cucumarites* [Jurassic: Oxfordian; France]

509. *minima* Kozur & Mock in Birkenmajer et al., 1990; *Acanthotheelia* [Triassic: Norian: Upper Sevatian; Poland]

510. *minularis* Zhang Jin-jian, 1983; *Paraeocaudina* [Carboniferous: Mississippian: Lower Tournaisian; P.R. China]

511. *mirabilis* Smith & Gallemí, 1991**; *Monilipsolus* [Triassic: Ladinian; Spain]

512. *mirabilis* Deflandre-Rigaud, 1959; *Myriotrochites* [Paleogene: Oligocene: Rupelian; Germany]

513. *misiki* Kozur & Mock, 1972; *Calclamna* [Triassic: Norian: earliest Sevatian; Slovak Republic]

514. *mississippiensis* Frizzell & Exline, 1956; *Calcancora* [Paleogene: Eocene/Oligocene: Upper Priabonian/Lower Rupelian; U.S.A.]

515. *mittali* Tandon & Saxena, 1977; *Eocaudina* [Paleogene: Eocene: Lutetian; India]

516. *mizunamiensis* Ishijima & Hatai, 1973; *Hemisphaeranthos* [Neogene: Miocene: Aquitanian/Burdigalian; Japan]

517. *mocki* Kozur in Kozur et al., 1985; *Acanthotheelia* [Triassic: Carnian: Cordevolian; Spain]

518. *monicae* Mostler & Rahimi-Yazd, 1976; *Theelia* [Permian: Lopingian: Wuchiapingian; Iran]

519. *monochordata* Hodson, Harris & Lawson, 1956; *Achistrum (Cancellrum)* [Jurassic: Oxfordian; U.K.]

520. *moorei* Soodan & Whatley, 1988; *Stueria* [Jurassic: Oxfordian; U.K.]

521. *Mortenseni* Deflandre-Rigaud, 1946 [1946b]; *Protocaudina* [Jurassic: Oxfordian; France]

522. *Mortenseni* Deflandre-Rigaud, 1950; *Chiridotites* [Jurassic: Upper Sinemurian; Germany]

523. *mortenseni* Frentzen, 1964; *Chirobaculus* [Jurassic: Bathonian; Germany]

524. *mortenseni* Frizzell & Exline, 1956; *Eocaudina* [Jurassic: Upper Sinemurian; Germany]

525. *mortenseni* Deflandre-Rigaud, 1952; *Rhabdotites* [Jurassic: Lower Oxfordian; Germany]

526. *mortenseni* Deflandre-Rigaud, 1952; *Stichopites* [Jurassic: Upper Toarcian; Germany]

527. *mostleri brouweri* Kozur & Simón, 1972; *Acanthotheelia* [Triassic: Carnian: Cordevolian; Spain]

528. *mostleri mostleri* Kozur & Simón, 1972; *Acanthotheelia* [Triassic: uppermost Ladinian/Lower Carnian; Spain]

529. *mostleri* Kozur & Mock in Birkenmajer et al., 1990; *Canisia* [Triassic: Norian: Upper Sevatian; Poland]

530. *mostleri* Kozur & Mock, 1972; *Eocaudina* [Triassic: Norian: Alaunian; Slovak Republic]

531. *mostleri* Kozur & Mock in Gaździcki et al., 1978; *Praecaudina* [Triassic: Carnian: Cordevolian; Slovak Republic]

532. *mostleri* Kozur, 1969; *Theelia* [Triassic: latest Olenekian; Germany]

533. *mostleri* Schallreuter, 1975; *Mercedescaudina* [Ordovician: Katian/Hirnantian; Sweden]

534. *mostleri* Stefanov, 1970; *Priscopedatus* [Triassic: Anisian: Illyrian; Bulgaria]

535. *mostleri* Zawidzka, 1971; *Priscopedatus* [Triassic: Anisian: ?Illyrian; Poland]

536. *mudgei* Gutschick, Canis & Brill, 1967; *Microantyx* [Carboniferous: Mississippian: Lower Tournaisian; U.S.A.]

537. *muelleri* Kozur & Mostler, 1970; *Theelia* [Triassic: Anisian: Pelsonian; Germany]

538. *muellendorfensis* Kristan-Tollmann, 1964 [1964c]; *Theelia* [Neogene: Miocene: Middle Tortonian; Austria]

539. *multiangulatus* Mostler, 1971 [1971b]; *Priscopedatus* [Triassic: Norian; Austria]

540. *multiforaminis* Sadeddin in Al-Tamimi, Saqqa & Sadeddin, 2001; *Priscopedatus* [Cretaceous: Upper Cenomanian; Jordan]

541. *multiforis* Schlumberger, 1890; *Priscopedatus* [Paleogene: Eocene: Upper Lutetian; France]

542. *multilaminaris* Walkiewicz, 1977***; *Mortensenites* [Neogene: Miocene: Langhian; Poland]

543. *multipartitus* Mostler, 1970; *Theniusites* [Triassic: Upper Norian; Austria]

544. *multiperforata* Kozur & Mock in Birkenmajer et al., 1990; *Biacumina* [Triassic: Norian: Upper Sevatian; Poland]

545. *multiperforata* Soodan & Whatley, 1990; *Binoculites* [Jurassic: Hettangian; U.K.]

546. *multiperforata* Mostler, 1968 [1968b]; *Praeuphronides* [Triassic: Anisian: Illyrian; Austria]

547. *multiperforata* Mostler, 1968 [1968b]; *Priscopedatus* [Triassic: Anisian: Illyrian; Austria]

548. *multiperforatum* Beckmann, 1965 [1965b]; *Achistrum (Porachistrum)* [Devonian: Givetian; Germany]

549. *multiplex* Speckmann, 1968; *Theelia* [Triassic: Ladinian: Langobardian; Italy]

550. *multiporata* Kozur & Sadeddin, 1992; *Schizotheelia* [Triassic: Ladinian; Jordan]

551. *multiporus* A. H. Müller, 1964; *Cucumarites* [Cretaceous: Lower Maastrichtian; Germany]

552. *multiradiata* Soodan & Whatley, 1988; *Lawsonina* [Jurassic: Oxfordian; U.K.]

553. *multiradiata* Mostler, 1971 [1971b]; *Stueria*? [Triassic: Norian; Austria]

554. *multiradiata* Kozur, 1969; *Theelia* [Triassic: Anisian: Bithynian; Germany]

555. *multiradiatus* Haude & Thomas, 1994**; *Rothamus* [Carboniferous: Mississippian: Viséan: Upper Asbian; Germany]

556. *multiundulata* Mostler in Krainer et al., 1994; *Theelia* [Jurassic: Upper Hettangian–Upper Pliensbachian; Austria]

557. *narainsarovarensis* Tandon & Saxena, 1977; *Rigaudites* [Paleogene: Eocene: Lutetian; India]

558. *naredaensis* Tandon & Saxena, 1977; *Kutchia* [Paleogene: Eocene: Lutetian; India]

559. *nasiformis* Krainer & Mostler, 1997; *Biacumina* [Jurassic: Sinemurian/Pliensbachian; Austria]

560. *nawarensis* Mostler, 1971 [1971d]; *Priscopedatus* [Carboniferous: Mississippian: Viséan; Afghanistan]

561. *Nicholsoni* Etheridge, 1881; *Achistrum* [Carboniferous: Mississippian: Viséan: Brigantian; U.K.]

562. *nigrivaccae* Deflandre-Rigaud, 1962 [1962b]; *Cucumarites* [Jurassic: Oxfordian; France]

563. *norica* Kozur & Mock, 1972; *Calclamna* [Triassic: Norian: Alaunian; Slovak Republic]

564. *norica* Mostler, 1968 [1968e]; *Palelpidia* [Triassic: Norian; Austria]

565. *norica* Mostler, 1969; *Theelia* [Triassic: Norian; Austria]

566. *Normani* Schlumberger, 1890; *Priscopedatus* [Paleogene: Eocene: Upper Lutetian; France]

567. *normannus* Deflandre-Rigaud, 1962 [1962b]; *Priscopedatus* [Jurassic: Oxfordian; France]

568. *novosandgarica* Kaptarenko-Černousova, 1954; *Chirodota (Hemisphaeranthos)* [Jurassic: Lower Oxfordian; Russia]

569. *nuda* Mostler, 1971 [1971b]; *Calclamnella* [Triassic: Norian; Austria]

570. *nudus* Reich, 2003 [2003c]; *Rigaudites* [Cretaceous: upper Lower Maastrichtian; Germany]

571. *oberalmiensis* Mostler, 1996; *Palactinopyga* [Jurassic: Kimmeridge/Lower Tithonian; Austria]

572. *obliquobrachiatus* Górka & Łuszczewska, 1969; *Priscolongatus* [Paleogene: Oligocene: Rupelian; Poland]

573. *oblonga* Ding Hui, 1985; *Eocaudina* [Carboniferous: Pennsylvanian: Kasimovian/Gzhelian; P.R. China]

574. *ocellata* Kristan-Tollmann, 1964 [1964c]; *Calclamnoidea* [Neogene: Miocene: Middle Tortonian; Austria]

575. *octoperforatus* Górka & Łuszczewska, 1969; *Priscopedatus* [Jurassic: Lower Bathonian; Poland]

576. *octoperforatus* Sadeddin, 1996 [1996b]; *Prisculatrites* [Cretaceous: Albian; U.S.A.]

577. *oertlii* Kozur & Simón, 1972; *Acanthotheelia* [Triassic: uppermost Ladinian/Lower Carnian; Spain]

578. *ogrodzieniecensis* Górka & Łuszczewska, 1969; *Hemisphaeranthos* [Jurassic: Middle Bathonian; Poland]

579. *oligocaenica* Spandel, 1900; *Synapta* [Paleogene: Oligocene: Chattian; Germany]

580. *operculum* Schlumberger, 1890; *Myriotrochus* [Paleogene: Eocene: Upper Lutetian; France]

581. *orbiculatus* Mostler, 1972 [1972f]; *Priscopedatus* [Triassic: Norian: Alaunian; Germany]

582. *ordovicicus* Reich, 2010 [2010c]; *Tribrachiodemas* [Ordovician: Katian/Hirnantian; Sweden]

583. *oreli* Kaptarenko-Černousova, 1954; *Chirodota (Hemisphaeranthos)* [Jurassic: Oxfordian; Russia]

584. *ostrea* Deflandre-Rigaud, 1959; *Elgerius* [Paleogene: Oligocene: Rupelian; Germany]

585. *ovalis* B. A. Matyja, H. Matyja & Szulczewski, 1973; *Eocaudina* [Devonian: early Frasnian; Poland]

586. *ovalis* Mostler, 1970; *Hamptonites* [Triassic: Norian; Austria]

587. *ovalis* Mostler, 1968 [1968b, 1968d]; *Priscopedatus* [Triassic: Anisian: Illyrian; Austria]

588. *overbrookensis* Reso & Wegner, 1964; *Thuroholia* [Ordovician: Sandbian; U.S.A.]

589. *oweni* Soodan & Whatley, 1990; *Calclamnella* [Jurassic: Hettangian; U.K.]

590. *palestiniensis* Sadeddin & Al-Tamimi, 2006; *Septapedatus* [Cretaceous: Late Cenomanian–Early Turonian; Jordan]

591. *pappi* Deflandre-Rigaud, 1962 [1962b]; *Synaptellus* [Neogene: Miocene: Middle Tortonian; Austria]

592. *parviperforatus* Mutterlose, 1982; *Paracucumarites* [Jurassic: lower Lower Albian; Germany]

593. *parviradiatus* Deflandre-Rigaud, 1950; *Auricularites* [Jurassic: Oxfordian; France]

594. *parvispinosa* Mostler, 1972 [1972e]; *Uncinulina* [Jurassic: ?Hettangian–?Toarcian; Austria]

595. *parvulus* Mostler, 1971 [1971d]; *Priscopedatus* [Carboniferous: Mississippian: Viséan; Afghanistan]

596. *parvus* Kozur & Mostler, 1970***; *Mortensenites*? [Triassic: Anisian: Bithynian; Germany]

597. *parvus* Sadeddin in Al-Tamimi, Saqqa & Sadeddin, 2001; *Rigaudites* [Cretaceous: Upper Cenomanian; Jordan]

598. *parvus* Mostler, 1971 [1971b]; *Solopedatus* [Triassic: Norian; Austria]

599. *patella* Boczarowski, 2001; *Eocaudina* [Devonian: Late Givetian; Poland]

600. *patinaformis* Mostler, 1970; *Theelia* [Triassic: Norian; Austria]

601. *pauciperforatus* Deflandre-Rigaud, 1962 [1962b]; *Cucumarites* [Jurassic: Oxfordian; France]

602. *pauciperforatus* Mostler, 1969; *Priscopedatus* [Triassic: Norian; Austria]

603. *pauciperforatus* Mostler, 1972 [1972f]; *Cucumarites* [Triassic: Norian: Alaunian; Germany]

604. *paucispinosa* Deflandre-Rigaud, 1962 [1962b]; *Protocaudina* [Jurassic: Oxfordian, France]

605. *pawsoni* Hess, 1973**; *Holothuriopsis* [Jurassic: Bathonian; Switzerland]

606. *pentagona* Sadeddin, 1996 [1996b]; *Eocaudina* [Cretaceous: Albian; U.S.A.]

607. *pentagonia* Kristan-Tollmann, 1963***; *Etheridgella* [Triassic: Carnian: Cordevolian; Italy]

608. *pentaradiata* Soodan, 1977 [1977a]; *Fletcherina* [Jurassic: ?Oxfordian/Kimmeridgian; India]

609. *pentaradiatus* Górka & Łuszczewska, 1969; *Priscopedatus* [Jurassic: Upper Bathonian; Poland]

610. *pentaramus* Sadeddin, 1996 [1996b]; *Ramusites* [Cretaceous: Albian; U.S.A.]

611. *perforata* Frizzell & Exline, 1956; *Calcligula* [Jurassic: Upper Pliensbachian; Germany]

612. *perforata* Frentzen, 1964; *Palaeocucumaria* [Jurassic: Upper Pliensbachian; Germany]

613. *perforata* Sadeddin, 1996 [1996b]; *Priscoligula* [Cretaceous: Albian; U.S.A.]

614. *perforata* Mostler, 1968 [1968b, 1968d]; *Tetravirga* [Triassic: Anisian: Lower Illyrian; Austria]

615. *perforatus* Frizzell & Exline, 1956; *Binoculites* [Jurassic: Upper Pliensbachian; Germany]

616. *permiana* Kozur & Mostler, 2008; *Acanthorota* [Permian: Guadalupian: Wordian; U.S.A.]

617. *permiana* Kornicker & Imbrie, 1958; *Microantyx* [Permian: Cisuralian: Asselian; U.S.A.]

618. *permianum* Spandel, 1898; *Astrophyton*(?) [Permian: Lopingian: Wuchiapingian; Germany]

619. *permicus* Kozur & Mostler, 1989***; *Pararotasaccus* [Permian: Guadalupian: Wordian; Italy]

620. *permotriassica* Mostler & Rahimi-Yazd, 1976; *Calclamnella* [Permian: Lopingian: Wuchiapingian; Iran]

621. *perpusillus* Mostler 1972 [1972f]; *Priscopedatus* [Triassic: Norian; Germany]

622. *persanensis* Gheorghian, 1995; *Fissobractites* [Triassic: Anisian; Romania]

623. *petasiformis* Kristan-Tollmann, 1964 [1964b]; *Theelia* [Triassic: Rhaetian; Austria]

624. *picketti* Jell, 2010**; *Porosothyone* [Silurian: Přídolian; Australia]

625. *pilgrimi* Fletcher, 1962; *Achistrum (Cancellrum)* [Jurassic: Upper Oxfordian; U.K.]

626. *pilicensis* Kubiatowicz & Matyja, 1977; *Priscopedatus* [Cretaceous: Upper Valanginian; Poland]

627. *pinguis* Deflandre-Rigaud, 1946 [1946b]; *Priscopedatus* [Jurassic: Oxfordian; France]

628. *piparensis* Tandon & Saxena, 1977; *Eocaudina* [Paleogene: Eocene: Lutetian; India]

629. *piveteaui* Deflandre-Rigaud, 1962 [1962b]; *Costigerites* [Jurassic: Oxfordian; France]

630. *plaga* Boczarowski, 2001*; *Eocaudina* [Devonian: Early Eifelian; Poland]

631. *plagiacanthus* Gilliland, 1992 [1992a]; *Acutisclerus* [Jurassic: Upper Hettangian; U.K.]

632. *planata* Mostler, 1968 [1968b]; *Theelia* [Triassic: Anisian: Illyrian; Turkey]

633. *planorbicula* Mostler, 1968 [1968b, 1968d]; *Theelia* [Triassic: Anisian: Illyrian; Austria]

634. *plenus* Deflandre-Rigaud, 1962 [1962b]; *Priscopedatus* [Jurassic: Oxfordian; France]

635. *ploechingeri* Mostler, 1969; *Priscopedatus* [Triassic: Norian; Austria]

636. *plummerae* Croneis in Croneis & McCormack, 1932 [1932c]; *Paleochiridota* [Carboniferous: Pennsylvanian: Moscovian; U.S.A.]

637. *plummerae* Frizzell & Exline, 1957; *Rigaudites* [Cretaceous: Albian; U.S.A.]

638. *polandica* Soodan, 1975; *Hannaina* [Paleogene: Oligocene: Lower Rupelian; Poland]

639. *polonica* Sztejn, 1993; *Huniella* [Jurassic: Lower Kimmeridgian; Poland]

640. *polonica* Matyja, 1972; *Theelia* [Jurassic: middle Upper Oxfordian; Poland]

641. *polonicus* Górka & Łuszczewska, 1969; *Croneisites* [Neogene: Miocene: Serravallian: Lower Sarmatian; Poland]

642. *polydenticulata* Mostler & Parwin, 1973; *Theelia* [Triassic: Ladinian/Carnian; Austria]

643. *polygona* Weber, 1997; *Kempfia* [Carboniferous: Mississippian: Upper Tournaisian; Germany]

644. *polymorpha* Terquem, 1862; *Uncinulina* [Jurassic: Liassic; France]

645. *polymorphus* Boczarowski, 2001; *Devonothyonites* [Devonian: Late Eifelian; Poland]

646. *polymorphus* Krainer & Mostler, 1997; *Punctatites* [Jurassic: Sinemurian/Pliensbachian; Austria]

647. *polypora* Frentzen, 1964; *Cucumariopsis* [Jurassic: Upper Pliensbachian; Germany]

648. *pomerania* Reich, 2003 [2003c]; *Calcancora* [Cretaceous: upper Lower Maastrichtian; Germany]

649. *pompatus* Matyja, 1972; *Priscopedatus* [Jurassic: late Middle Oxfordian; Poland]

650. *porosa* Croneis in Croneis & McCormack, 1932 [1932c]***; *Etheridgella* [Carboniferous: Pennsylvanian: Gzhelian; U.S.A.]

651. *porosa* Mostler, 1971 [1971d]; *Gutschickia* [Carboniferous: Mississippian: Viséan; Afghanistan]

652. *porosa* Mostler, 1972 [1972f]; *Uncinulinoides* [Triassic: Middle Norian; Austria]

653. *porosus* Deflandre-Rigaud, 1962 [1962b]; *Paracucumarites* [Jurassic: Oxfordian; France]

654. *porosus* Mostler, 1970; *Rhabdotites* [Triassic: Norian; Austria]

655. *praeacuta* Mostler & Rahimi-Yazd, 1976; *Theelia* [Permian: Lopingian: Wuchiapingian; Iran]

656. *praenorica* Kozur & Mock, 1972; *Theelia* [Triassic: Norian: Alaunian; Slovak Republic]

657. *praeseniradiata* Kozur & Mock, 1972; *Theelia* [Triassic: Norian: Alaunian; Slovak Republic]

658. *pralongiae* Kristan-Tollmann, 1963; *Theelia* [Triassic: Carnian: Cordevolian; Italy]

659. *prebritannica* Soodan & Whatley, 1990; *Calclamna* [Jurassic: Hettangian; U.K.]

660. *prima* Mostler, 1969; *Stichopitella* [Triassic: Norian; Austria]

661. *primæva* Etheridge, 1881; *Cheirodota*(?) [Carboniferous: Mississippian: Viséan: Brigantian; U.K.]

662. *procerus* Mostler, 1971 [1971b]; *Priscopedatus* [Triassic: Anisian: Pelsonian; Austria]
663. *propinquus* Schlumberger, 1890; *Priscopedatus* [Paleogene: Eocene: Upper Lutetian; France]
664. *proteus* Mortensen, 1937; *Cucumaria* [Jurassic: Upper Toarcian; Germany]
665. *protrusus* Kozur & Mostler, 1970; *Priscopedatoides* [Triassic: Anisian: Pelsonian; Poland]
666. *pseudaffinis* Deflandre-Rigaud, 1962 [1962b]; *Priscopedatus* [Jurassic: Oxfordian; France]
667. *pseudobotoni* Kozur & Mostler, 1989; *Microantyx* [Permian: Guadalupian: Wordian; Italy]
668. *pseudonormani* Sadeddin, 1996 [1996b]; *Priscopedatus* [Cretaceous: Albian; U.S.A.]
669. *pseudoplanata* Kozur & Mock, 1972; *Theelia* [Triassic: Norian; Slovak Republic]
670. *pseudospinosa* Kozur & Mock, 1972; *Acanthotheelia* [Triassic: Norian: Alaunian; Slovak Republic]
671. *pulcher* Gellai, 1973; *Cucumarites* [Cretaceous: Albian; Hungary]
672. *pulchra* Kozur & Mock, 1972; *Acanthotheelia* [Triassic: Norian; Slovak Republic]
673. *pulchrum* Kozur, 1969; *Achistrum* [Triassic: Anisian: Bithynian; Germany]
674. *punctatus* Mutterlose, 1982; *Rigaudites* [Cretaceous: lower Lower Albian; Germany]
675. *punctiferus* Deflandre-Rigaud, 1962 [1962b]; *Cucumarites* [Jurassic: Oxfordian; France]
676. *pyramidalis* Schlumberger, 1890; *Priscopedatus* [Paleogene: Eocene: Upper Lutetian; France]
677. *quadratus* Kozur & Mostler, 1970; *Priscopedatus* [Triassic: Anisian: Pelsonian; Poland]
678. *quadratus* Kozur & Sadeddin, 1990; *Spinopriscopedatus* [Jurassic: Bathonian; Jordan]
679. *quadriforamina* Sadeddin, 1996 [1996b]; *Priscoligula* [Cretaceous: Albian; U.S.A.]
680. *quadriperforatus* Górka & Łuszczewska, 1969; *Priscolongatus* [Paleogene: Oligocene: Rupelian; Poland]
681. *quadriramosus* Sadeddin, 1996 [1996b]; *Calcancora* [Cretaceous: Albian; U.S.A.]
682. *quadrispinosa* Mostler, 1969; *Ludwigia* [Triassic: Norian; Austria]
683. *quinqueloba* Terquem, 1866; *Annulina* [Jurassic: Liassic; France]
684. *quinquelobata* Mostler, 1970; *Multivirga* [Triassic: Anisian; Austria]
685. *quinqueordinata* B. A. Matyja, H. Matyja & Szulczewski, 1973; *Eocaudina* [Devonian: Eifelian; Poland]
686. *quinquespinosus* Mostler & Rahimi-Yazd, 1976; *Priscopedatus* [Permian: Lopingian: Wuchiapingian; Iran]

687. *radiata* Croneis in Croneis & McCormack, 1932 [1932c]; *Paleochiridota* [Carboniferous: Pennsylvanian: Moscovian; U.S.A.]
688. *radiata* Summerson & Campbell, 1958; *Petropegia* [Carboniferous: Pennsylvanian: Moscovian; U.S.A.]
689. *radiatus* Deflandre-Rigaud, 1962 [1962b]; *Cucumarites* [Jurassic: Oxfordian; France]
690. *radiatus* Mostler, 1971 [1971b]; *Semperites* [Triassic: Norian; Austria]
691. *rajasthanensis* Soodan, 1980; *Calclamna* [Paleogene: Eocene: Ypresian; India]
692. *rajnathi* Tandon & Saxena, 1977; *Rigaudites* [Paleogene: Eocene: Lutetian; India]
693. *rajui* Soodan, 1977 [1977b]; *Rigaudites* [Quaternary: Holocene; Arabian Sea]
694. *ramosa* Mostler, 1971 [1971d]; *Calclamnella* [Carboniferous: Mississippian: Viséan; Afghanistan]
695. *ramosa* Kozur & Mostler, 1971; *Eocaudina* [Triassic: Ladinian: Langobardian; Hungary]
696. *ramosus* Sadeddin, 1996 [1996b]; *Prisculatrites* [Cretaceous: Albian; U.S.A.]
697. *rampurensis* Tandon & Saxena, 1977; *Elgerius* [Paleogene: Eocene: Lutetian; India]
698. *ramwaraensis* Tandon & Saxena, 1977; *Calclamna* [Paleogene: Eocene: Lutetian; India]
699. *raoi* Soodan, 1973 [1973b]; *Stueria* [Jurassic: Bathonian; India]
700. *rara* Mostler, 1972 [1972e]; *Biacumina* [Jurassic: ?Hettangian; Austria]
701. *rariperforata* Kozur & Mock, 1972; *Biacumina* [Triassic: Norian; Slovak Republic]
702. *rariperforata* Zankl, 1966; *Calclamnella* [Triassic: Norian/Rhaetian; Germany]
703. *rarus* A. H. Müller, 1964; *Chiridotites* [Cretaceous: Lower Maastrichtian; Germany]
704. *rarus* Kubiatowicz & Matyja, 1977; *Auriculites* [Cretaceous: Upper Valanginian; Poland]
705. *ratcheloensis* Tandon & Saxena, 1977; *Elgerius* [Paleogene: Eocene: Lutetian; India]
706. *rectangularia* Soodan & Whatley, 1990; *Yorkshirella* [Jurassic: Hettangian; U.K.]
707. *rectangularis* Górka, 1997; *Calclamnella* [Neogene: Miocene; Poland]
708. *rectus* Frizzell & Exline, 1956; *Rhabdotites* [Triassic: Carnian; Italy]
709. *regia* Boczarowski, 2001; *Palaeocaudina* [Devonian: Late Eifelian; Poland]
710. *regularis* Ding Hui, 1985***; *Mortensenites* [Carboniferous: Pennsylvanian: Kasimovian/Gzhelian; P.R. China]
711. *regularis* Mostler, 1968 [1968e]; *Unculinoides* [Triassic: Norian; Austria]

712. *regularis* Stefanov, 1970; *Calclamnella* [Triassic: Anisian: Illyrian; Bulgaria]
713. *regularis* Zhang Jin-jian, 1983; *Thuroholia* [Carboniferous: Mississippian: Lower Tournaisian; P.R. China]
714. *reichi* Al-Tamimi & Saqqa, 2008; *Hexapriscopedatus* [Cretaceous: Upper Cenomanian; Jordan]
715. *remigia* Boczarowski, 2001*; *Propinquoohshimella* [Devonian: Early Givetian; Poland]
716. *renifera* Schlumberger, 1890; *Synapta* [Paleogene: Eocene: Upper Lutetian; France]
717. *reschi* Sadeddin in Al-Tamimi, Saqqa & Sadeddin, 2001; *Rigaudites* [Cretaceous: Upper Cenomanian; Jordan]
718. *reticulatus* Kristan-Tollmann, 1964 [1964c]***; *Mortensenites* [Neogene: Miocene: Middle Tortonian; Austria]
719. *rhaetica* Kristan-Tollmann, 1963; *Acanthotheelia* [Triassic: Rhaetian; Austria]
720. *riasanensis* Sztejn, 1992; *Spinopriscopedatus* [Cretaceous: Lower Valanginian; Poland]
721. *rigaudae* Mostler, 1970; *Protocaudina* [Triassic: Carnian; Austria]
722. *rigaudae* Rioult, 1961; *Chiridotites* [Jurassic: Lower Toarcian; France]
723. *rimosa* Boczarowski, 2001; *Eocaudina* [Devonian: Early Givetian; Poland]
724. *rinconensis* Huddleston, Finger & Kirwan, 1986; *Jumaraina* [Neogene: Miocene: ?Aquitanian; U.S.A.]
725. *Robertsoni* Etheridge, 1881; *Cheirodota*(?) [Carboniferous: Mississippian: Viséan: Brigantian; U.K.]
726. *robusta* Deflandre-Rigaud, 1962 [1962b]; *Calclamnella* [Jurassic: Oxfordian; France]
727. *robusta* Soodan, 1973 [1973b]; *Stueria* [Jurassic: Bathonian; India]
728. *robustus* Deflandre-Rigaud, 1962 [1962b]; *Cucumarites* [Jurassic: Oxfordian; France]
729. *robustus* Mostler, 1970; *Praeeuphronides* [Triassic: Norian; Austria]
730. *rogenti* Smith & Gallemí, 1991**; *Strobilothyone* [Triassic: Ladinian; Spain]
731. *rosetta* Kristan-Tollmann, 1963; *Theelia* [Triassic: Rhaetian; Austria]
732. *rotheri* Beckmann, 1965 [1965b]; *Tetravirga*? [Devonian: Givetian; Germany]
733. *rotula* Egger, 1899, *Calcarina* [Cretaceous: Cenomanian; Germany]
734. *salhoubensis* Al-Tamimi & Saqqa, 2008; *Spinopriscopedatus* [Cretaceous: Lower Turonian; Jordan]
735. *samarica* Kaptarenko-Černousova, 1954; *Chirodota (Hemisphaeranthos)* [Jurassic: Bajocian; Russia]
736. *sanctacrucensis* Walkiewicz, 1977; *Calclamnella* [Neogene: Miocene: Langhian; Poland]

737. *sandlingensis* Kristan-Tollmann, 1973; *Cosmatites* [Jurassic: ?Hettangian–?Pliensbachian; Austria]
738. *sandlingi* Mostler, 1969; *Priscopedatus* [Triassic: Norian; Austria]
739. *sastrii* Soodan, 1977 [1977b]; *Rigaudites* [Quaternary: Holocene; Arabian Sea]
740. *sastrii* Soodan, 1973 [1973b]; *Stueria* [Jurassic: Bathonian; India]
741. *satyendrai* Tandon & Saxena, 1977; *Calclamnella* [Paleogene: Eocene: Lutetian; India]
742. *saucatensis* Rioult, 1961; *Cucumarites* [Neogene: Miocene: Upper Burdigalian; France]
743. *scaber* Deflandre-Rigaud, 1959; *Cucumarites* [Paleogene: Oligocene: Rupelian; Germany]
744. *schallreuteri* Kozur & Simón, 1972; *Kuehnites* [Triassic: uppermost Ladinian/Lower Carnian; Spain]
745. *scheibelbergensis* Krainer & Mostler, 1997; *Centripedatus* [Jurassic: Sinemurian/Pliensbachian; Austria]
746. *schizotoma* Kristan-Tollmann, 1973; *Schizotheelia* [Triassic: Carnian; Italy]
747. *Schlumbergeri* Deflandre-Rigaud, 1946 [1946b]; *Priscopedatus* [Jurassic: Oxfordian; France]
748. *scotica* Frizzell & Exline, 1956; *Eocaudina* [Carboniferous: Mississippian: Viséan: Brigantian; U.K.]
749. *seilacheri* Haude, 1997**; *Nudicorona* [Devonian: Early Givetian; Germany]
750. *seniradiata* Zankl, 1966; *Theelia* [Triassic: Norian/Rhaetian; Germany]
751. *septaforaminalis* Martin, 1952; *Eocaudina* [Devonian: Givetian; U.S.A.]
752. *serratus* Mostler, 1972 [1972f]; *Kuehnites* [Triassic: Middle Norian; Austria]
753. *serta* Speckmann, 1968; *Theelia* [Triassic: Ladinian: Langobardian; Italy]
754. *siciliensis* Kozur & Mostler, 1989; *Pediculicaudina* [Permian: Guadalupian: Wordian; Italy]
755. *Sieboldi* Schwager, 1865; *Chirodota* [Jurassic: Oxfordian; Germany]
756. *Sieboldii* Münster, 1843; *Synapta* [Jurassic: Lower Kimmeridgian; Germany]
757. *sievertsae* Deflandre-Rigaud, 1962 [1962b]; *Schlumbergerites* [Jurassic: Oxfordian; France]
758. *sievertsi* Deflandre-Rigaud, 1952***; *Mortensenites* [Jurassic: Upper Pliensbachian; Germany]
759. *similis* Górka & Łuszczewska, 1969; *Paracucumarites* [Jurassic: Middle Bathonian; Poland]
760. *simoni* Kozur & Mock, 1972; *Theelia* [Triassic: Norian: Middle–Upper Sevatian; Slovak Republic]

761. *simplex* A. H. Müller, 1964; *Hemisphaeranthos* [Cretaceous: Lower Maastrichtian; Germany]

762. *simplex* Mostler, 1969; *Praeeuphronides* [Triassic: Norian; Austria]

763. *sinaiensis* Said & Barakat, 1958; *Theelia* [Jurassic: Bajocian; Egypt]

764. *sinensis* Zhang Jin-jian, 1986; *Parastaurocumites* [Carboniferous: Mississippian: Tournaisian; P.R. China]

765. *sinensis* Zhang Jin-jian, 1993; *Theelia* [Carboniferous: Mississippian: late Viséan; P.R. China]

766. *singhi* Soodan, 1977 [1977b]; *Rigaudites* [Quaternary: Holocene; Arabian Sea]

767. *sinhai* Tandon & Saxena, 1977; *Costigerites* [Paleogene: Eocene: Lutetian; India]

768. *sinuatus* Reich, 2003 [2003d]; *Prisculatrites* [Cretaceous: upper Lower Maastrichtian; Germany]

769. *sinuosa* Rioult, 1961; *Calclamnella* [Neogene: Miocene: Upper Burdigalian; France]

770. *sinuosus* Kozur & Mostler, 1970; *Priscopedatus* [Triassic: Anisian: Pelsonian; Italy]

771. *slovakensis* Kozur & Mock, 1972; *Priscopedatus* [Triassic: Anisian: Upper Illyrian; Slovak Republic]

772. *slovakensis* Kozur & Mock in Gaździcki et al., 1978; *Kuehnites* [Triassic: Carnian: Cordevolian; Slovak Republic]

773. *smirnovi* Reich, 2002 [2002b]; *Trematrochus* [Cretaceous: upper Upper Maastrichtian; Germany/southern Baltic Sea]

774. *solidus* Deflandre-Rigaud, 1946 [1946b]; *Priscopedatus* [Jurassic: Oxfordian; France]

775. *solveigae* Reich, 2003 [2003a]; *Tripuscucumis* [Cretaceous: upper Lower Maastrichtian; Germany]

776. *soodani* Tandon & Saxena, 1977; *Eocaudina* [Paleogene: Eocene: Lutetian; India]

777. *soodanii* Kalita, Kulshreshtha & Sahni, 2002; *Cucumarites* [Jurassic: Callovian–Oxfordian; India]

778. *sosioensis* Kozur & Mostler, 1989; *Microantyx* [Permian: Guadalupian: Wordian; Italy]

779. *spandeli* Frizzell & Exline, 1956; *Calcancoroidea* [Paleogene: Oligocene: Chattian; Germany]

780. *spania* Kristan-Tollmann, 1964 [1964c]; *Calclamnoidea* [Neogene: Miocene: Middle Tortonian; Austria]

781. *sparsispinosus* Deflandre-Rigaud, 1962 [1962b]; *Cucumarites* [Jurassic: Oxfordian; France]

782. *spathi* Soodan & Whatley, 1990; *Hodsonina* [Jurassic: Hettangian; U.K.]

783. *spatuligerus* Deflandre-Rigaud, 1959; *Dictyothurites* [Paleogene: Oligocene: Rupelian; Germany]

784. *speciosa* Deflandre-Rigaud, 1946 [1946b]; *Chiridota* [Jurassic: Oxfordian; France]

785. *speciosus* Deflandre-Rigaud, 1959; *Cucumarites* [Paleogene: Oligocene: Rupelian; Germany]

786. *spectabilis* Deflandre-Rigaud, 1962 [1962b]; *Priscopedatus* [Jurassic: Oxfordian; France]

787. *spectabilis* Deflandre-Rigaud, 1959; *Synaptites (Calcancora)* [Paleogene: Oligocene: Rupelian; Germany]

788. *spicata* Mostler, 1970; *Uncinulina* [Triassic: Norian; Austria]

789. *spicatus* Gutschick, 1959; *Thuroholia* [Carboniferous: Mississippian: late Tournaisian; U.S.A.]

790. *spicaudina* Gutschick, Canis & Brill, 1967; *Priscopedatus* [Carboniferous: Mississippian: Lower Tournaisian; U.S.A.]

791. *spiniferus* Deflandre-Rigaud, 1962 [1962b]; *Priscopedatus* [Jurassic: Oxfordian; France]

792. *spiniperforata* Zawidzka, 1971; *Acanthotheelia* [Triassic: Anisian: ?Illyrian; Poland]

793. *spinosa* Frizzell & Exline, 1956; *Acanthotheelia* [Triassic: Carnian; Italy]

794. *spinosa* Kozur & Mock, 1972; *Biacumina* [Triassic: Norian: earliest Sevatian; Slovak Republic]

795. *spinosa* Mostler, 1968 [1968b]; *Eocaudina* [Triassic: Anisian: Illyrian; Turkey]

796. *spinosa* Mostler, 1970; *Crucivirga* [Triassic: Norian: Sevatian; Austria]

797. *spinosa* Mostler, 1971 [1971b]; *Curvatella* [Triassic: Norian; Austria]

798. *spinosa* Summerson & Campbell, 1958; *Petropegia* [Carboniferous: Pennsylvanian: Moscovian; U.S.A.]

799. *spinosus* Deflandre-Rigaud, 1962 [1962b]; *Parvioctoidus* [Jurassic: Oxfordian; France]

800. *spinosus* Sadeddin & Saqqa, 1997; *Prisculatrites* [Cretaceous: Lower Turonian; Jordan]

801. *spinosus* Frizzell & Exline, 1957; *Rigaudites* [Cretaceous: Albian; U.S.A.]

802. *spinosus* Sztejn, 1993; *Semicucumarites* [Cretaceous: Lower Valanginian; Poland]

803. *spinosus* Frizzell & Exline, 1956; *Stichopites* [Carboniferous: Mississippian: Serpukhovian; U.S.A.]

804. *spiritus* Boczarowski, 2001; *Devonothyonites* [Devonian: Mid Frasnian; Poland]

805. *spitiensis* Soodan, 1986; *Cucumarites* [Triassic: Ladinian–Norian; India]

806. *spitiensis* Soodan, 1986; *Multivirga* [Triassic: Ladinian–Norian; India]

807. *squamma* Deflandre-Rigaud, 1962 [1962b]; *Cucumarites* [Jurassic: Oxfordian; France]
808. *staurocumitoides* Mostler, 1968 [1968b]; *Priscopedatus* [Triassic: Anisian: Illyrian; Austria]
809. *staurolithensis* Kristan-Tollmann, 1973; *Theelia* [Cretaceous: Aptian/Albian; Austria]
810. *stellatiformis* Gheorghian, 1995; *Theelia* [Triassic: Carnian; Romania]
811. *stellifera* Zankl, 1966; *Theelia* [Triassic: Norian/Rhaetian; Germany]
812. *stellifera bistellata* Kozur & Mock, 1972; *Theelia* [Triassic: Norian: lower Upper Sevatian; Slovak Republic]
813. *stelliformis* Mostler, 1970; *Priscopedatus* [Triassic: Norian; Austria]
814. *Stueri* Schlumberger, 1890; *Synapta* [Paleogene: Eocene: Upper Lutetian; France]
815. *subcircularis* Al-Tamimi & Saqqa, 2008; *Hexapriscopedatus* [Cretaceous: Upper Turonian; Jordan]
816. *subcircularis* Soodan, 1977 [1977b]; *Rigaudites* [Quaternary: Holocene; Arabian Sea]
817. *subcirculata* Mostler, 1968 [1968b]; *Theelia* [Triassic: Upper Anisian; Austria]
818. *subhexagona* Gutschick, Canis & Brill, 1967; *Eocaudina* [Carboniferous: Mississippian: Lower Tournaisian; U.S.A.]
819. *subhexagonus* Ding Hui, 1985; *Priscopedatus* [Carboniferous: Pennsylvanian: Kasimovian/Gzhelian; P.R. China]
820. *subovalis* Sadeddin in Al-Tamimi, Saqqa & Sadeddin, 2001; *Rigaudites* [Cretaceous: Upper Cenomanian; Jordan]
821. *subquadrata* Mostler, 1971 [1971b]; *Eocaudina* [Triassic: Anisian: Illyrian; Bosnia]
822. *subquadratus* Al-Tamimi & Saqqa, 2008; *Hexapriscopedatus* [Cretaceous: Upper Turonian; Jordan]
823. *subquadratus* Sadeddin & Saqqa, 1997; *Madwarites* [Cretaceous: Lower Turonian; Jordan]
824. *subquadratus* Sadeddin & Saqqa, 1997; *Prisculatrites* [Cretaceous: Lower Turonian; Jordan]
825. *subrecta* Frizzell & Exline, 1956; *Uncinulina* [Jurassic: ?Pliensbachian; France]
826. *subrotunda* Krainer & Mostler, 1997; *Eocaudina* [Jurassic: Sinemurian/Pliensbachian; Austria]
827. *subsymmetrica* Kristan-Tollmann, 1963; *Fissobractites* [Triassic: Rhaetian; Austria]
828. *subsymmetricus* Kristan-Tollmann, 1964 [1964c]; *Stichopites* [Neogene: Miocene: Middle Tortonian; Austria]
829. *subtilis* Kozur & Mostler, 1970; *Semperites* [Triassic: Anisian: Pelsonian; Austria]

830. *subtriangularis* Al-Tamimi & Saqqa, 2008; *Hexapriscopedatus* [Cretaceous: Upper Cenomanian; Jordan]
831. *subtriangularis* Al-Tamimi & Saqqa, 2008; *Prisculatrites* [Cretaceous: Lower Turonian; Jordan]
832. *subtriangularis* Kozur & Sadeddin, 1990; *Spinopriscopedatus* [Jurassic: Bathonian; Jordan]
833. *subtrigonalis* Kristan-Tollmann, 1964 [1964c]; *Eocaudina* [Neogene: Miocene: Middle Tortonian; Austria]
834. *sureshi* Tandon & Saxena, 1977; *Koteshwaria* [Paleogene: Eocene: Lutetian; India]
835. *symmetrica* Mostler, 1968 [1968e]; *Calclamnella* [Triassic: Norian; Austria]
836. *symmetrica* Mostler, 1969; *Ludwigia* [Triassic: Norian; Austria]
837. *symmetrica* Zhang Jin-jian, 1983; *Protocaudina* [Carboniferous: Mississippian: Lower Tournaisian; P.R. China]
838. *symmetrica* Zhang Jin-jian, 1993; *Thuroholia* [Carboniferous: Mississippian: late Viséan; P.R. China]
839. *synapta* Gilliland, 1992 [1992a]; *Theelia* [Jurassic: Middle Hettangian; U.K.]
840. *szoerenyiae* Gellai, 1973; *Cucumarites* [Cretaceous: Albian; Hungary]
841. *tabulifera* Haude, 2002**; *Prokrustia* [Devonian: Lochkovian–Emsian; Argentina]
842. *tallali* Frizzell & Exline, 1957; *Rigaudites* [Cretaceous: Upper Campanian; U.S.A.]
843. *tandoni* Soodan, 1980; *Calclamna* [Paleogene: Eocene: Ypresian; India]
844. *tarazi* Mostler & Rahimi-Yazd, 1976; *Microantyx* [Permian: Lopingian: Wuchiapingian; Iran]
845. *tarragonensis* Cherbonnier, 1978**; *Oneirophantites* [Triassic: Ladinian; Spain]
846. *teat* Zhang Jin-jian, 1983***; *Mortensenites* [Carboniferous: Mississippian: Lower Tournaisian; P.R. China]
847. *teneromarginata* Mostler, 1971 [1971b]; *Theelia* [Triassic: Anisian: Illyrian; Austria]
848. *terquemi* Croneis in Croneis & McCormack, 1932 [1932c]; *Paleochiridota* [Carboniferous: Pennsylvanian: Moscovian; U.S.A.]
849. *Terquemi* Deflandre-Rigaud, 1950; *Chiridotites* [Jurassic: Upper Pliensbachian; France]
850. *terquemi* Deflandre-Rigaud, 1952; *Binoculites* [Jurassic: Upper Sinemurian; Germany]
851. *terquemi* Frizzell & Exline, 1956; *Uncinulina* [Jurassic: ?Pliensbachian; France]
852. *tesseyrei* Deflandre-Rigaud, 1962 [1962b]; *Ornaticannula* [Jurassic: Oxfordian; France]
853. *tetrabrachiatus* Krainer & Mostler, 1997; *Punctatites* [Jurassic: Sinemurian/Pliensbachian; Austria]

854. *tetraporatus* Kozur & Mostler, 1989; *Staurocumites* [Permian: Guadalupian: Wordian; Italy]

855. *tewarii* Tandon & Saxena, 1977; *Kutchia* [Paleogene: Eocene: Lutetian; India]

856. *thalattocanthoides* Mostler, 1968 [1968b]; *Theelia* [Triassic: Anisian: Illyrian; Austria]

857. *thayuensis* Applegate, Buitrón-Sánchez, Solís-Marín & Laguarda-Figueras, 2009**; *Parapsolus* [Cretaceous: Albian; Mexico]

858. *thornicus* Soodan, 1975; *Sastriella* [Jurassic: Bathonian; India]

859. *thuringensis* Kozur & Mostler, 1970; *Priscopedatus* [Triassic: Anisian: Pelsonian; Germany]

860. *tokarniensis* Matyja, 1972; *Cucumarites* [Jurassic: late Middle Oxfordian; Poland]

861. *tollmannae* Walkiewicz, 1977; *Calclamnoidea* [Neogene: Miocene: Langhian; Poland]

862. *tomasoviensis* Sztejn, 1993; *Paracucumarites* [Cretaceous: Lower Valanginian; Poland]

863. *tornatus* Mostler, 1972 [1972f]; *Binoculites* [Triassic: Norian; Austria]

864. *tortoniensis* Deflandre-Rigaud, 1962 [1962b]; *Cucumarites* [Neogene: Miocene: Middle Tortonian; Austria]

865. *trammeri* Kozur & Mock in Gaździcki et al., 1978; *Theelia* [Triassic: Carnian: Cordevolian; Slovak Republic]

866. *transitus* Kozur & Mock, 1972; *Triradites* [Triassic: Norian: Alaunian; Slovak Republic]

867. *transversa* Deflandre-Rigaud, 1962 [1962b]; *Calclamnella* [Jurassic: Oxfordian; France]

868. *trapezoides* Sadeddin, 1996 [1996b]; *Prisculatrites* [Cretaceous: Albian; U.S.A.]

869. *Traquairii* Etheridge, 1881; *Cheirodota*(?) [Carboniferous: Mississippian: Viséan: Brigantian; U.K.]

870. *trema* Kristan-Tollmann, 1963; *Eocaudina* [Triassic: Rhaetian; Austria]

871. *tretomesota* Gutschick, Canis & Brill, 1967; *Rotoides* [Carboniferous: Mississippian: Lower Tournaisian; U.S.A.]

872. *trettoensis* Mostler, 1980; *Calclamna* [Triassic: Anisian: Pelsonian; Italy]

873. *triangularis* Soodan & Whatley, 1990; *Calclamna* [Jurassic: Hettangian; U.K.]

874. *triangularis* Górka & Łuszczewska, 1969; *Priscopedatus* [Jurassic: Middle Bathonian; Poland]

875. *triangularis* Kozur & Mock, 1972; *Priscopedatus* [Triassic: Anisian: Upper Illyrian; Slovak Republic]

876. *triangularis* Mostler, 1968 [1968e]; *Calclamnella* [Triassic: Norian; Austria]

877. *triangularis* Langer, 1991; *Devonothyonites* [Devonian: Lower Givetian; Germany]

878. *triangulatus* Mostler, 1971 [1971b]; *Priscopedatus* [Triassic: Norian: Sevatian; Austria]

879. *triassica* Speckmann, 1968; *Acanthotheelia* [Triassic: Anisian: Illyrian; Germany]

880. *triassica* Soodan, 1986; *Jagdipina* [Triassic: Ladinian–Norian; India]

881. *triassicum* Frizzell & Exline, 1956; *Achistrum* [Triassic: Carnian; Italy]

882. *triassicus* Mostler, 1968 [1968b]; *Priscopedatus* [Triassic: Anisian: Illyrian; Austria]

883. *triceratium* Deflandre-Rigaud, 1962 [1962b]; *Prisculatrites* [Jurassic: Oxfordian; France]

884. *trichordata* Fletcher, 1962; *Achistrum (Cancellrum)* [Jurassic: Upper Oxfordian; U.K.]

885. *tricorniculata* Walkiewicz, 1977; *Calclamna* [Neogene: Miocene: Langhian; Poland]

886. *tricostatus* Deflandre-Rigaud, 1962 [1962b]; *Prisculatrites* [Jurassic: Oxfordian; France]

887. *tridens* Hodson, Harris & Lawson, 1956; *Rhabdotites* [Jurassic: Oxfordian; U.K.]

888. *trifida* Frizzell & Exline, 1956; *Calcancoroidea* [Paleogene: Oligocene: Chattian; Germany]

889. *trigonalis* Walkiewicz, 1977; *Calclamnella* [Neogene: Miocene: Langhian; Poland]

890. *triperforata* Mostler, 1971 [1971b]; *Calclamnella* [Triassic: Norian; Austria]

891. *triperforata* Schallreuter, 1968; *Protocaudina* [Ordovician: Sandbian; Germany/Baltic Sea]

892. *triperforatus* Sadeddin, 1996 [1996b]; *Prisculatrites* [Cretaceous: Albian; U.S.A.]

893. *triplex* Mostler, 1972 [1972e]; *Punctatites* [Jurassic: ?Hettangian–?Toarcian; Austria]

894. *triradiatus* Mostler, 1968 [1968e]; *Cucumarites* [Triassic: Norian; Austria]

895. *trisulcus* Mostler, 1970; *Theniusites* [Triassic: Norian; Austria]

896. *truncata* Schlumberger, 1890; *Synapta* [Paleogene: Eocene: Upper Lutetian; France]

897. *tubercula* Kristan-Tollmann, 1963; *Theelia* [Triassic: Carnian: Cordevolian; Italy]

898. *tubercula parvituberculata* Kozur & Simón, 1972; *Theelia* [Triassic: Carnian: Cordevolian; Spain]

899. *tudorowiensis* Boczarowski, 2001*; *Devonothyonites* [Devonian: Mid/?Late Frasnian; Poland]

900. *turgidus* Mostler, 1972 [1972f]; *Kuehnites* [Triassic: Norian; Germany]

901. *turoniansis* Sadeddin & Saqqa, 1997; *Calclamna* [Cretaceous: Lower Turonian; Jordan]

902. *tuto* Boczarowski, 2001*; *Achistrum* [Devonian: Early Eifelian; Poland]

903. *tyrolensis* Mostler, 1968 [1968b]; *Priscopedatus* [Triassic: Anisian: Illyrian; Austria]

904. *umbo* Kristan-Tollmann, 1963***; *Kaliobullites* [Triassic: Rhaetian; Austria]

905. *undata* Mostler, 1968 [1968b]; *Theelia* [Triassic: Anisian: Illyrian; Turkey]

906. *undatus* Deflandre-Rigaud, 1962 [1962b]; *Cucumarites* [Jurassic: Oxfordian; France]

907. *undosus* Deflandre-Rigaud, 1959; *Myriotrochites* [Paleogene: Oligocene: Rupelian; Germany]

908. *undulata* Schlumberger, 1888; *Chirodota* [Paleogene: Eocene: Upper Lutetian; France]

909. *undulatus* Krainer & Mostler, 1997; *Centripedatus* [Jurassic: Pliensbachian; Austria]

910. *ungersteinensis* Mostler, 1970; *Semperites* [Triassic: Norian; Austria]

911. *unkenensis* Krainer & Mostler, 1997; *Centripedatus* [Jurassic: Sinemurian/Pliensbachian; Austria]

912. *unispinosus* Kozur & Sadeddin, 1990; *Spinopriscopedatus* [Jurassic: Bathonian; Jordan]

913. *urkutica* Gellai, 1973; *Calclamnella* [Cretaceous: Albian; Hungary]

914. *valanginensis* Kubiatowicz & Matyja, 1977; *Priscopedatus* [Cretaceous: Upper Valanginian; Poland]

915. *valdiyai* Tandon & Saxena, 1977; *Rigaudites* [Paleogene: Eocene: Lutetian; India]

916. *variabilis* Kozur & Mostler, 2008; *Acanthoschizorota* [Permian: Guadalupian: Wordian; U.S.A.]

917. *variabilis* Rioult, 1961; *Cucumarites* [Neogene: Miocene: Upper Burdigalian; France]

918. *variabilis* Mostler, 1972 [1972e]; *Priscopedatus* [Jurassic: ?Hettangian–?Toarcian; Austria]

919. *variabilis* Zankl, 1966; *Theelia* [Triassic: Rhaetian; Germany]

920. *variabilis slovakensis* Kozur & Mock, 1972; *Theelia* [Triassic: Norian: Alaunian; Slovak Republic]

921. *varicum* Boczarowski, 2001*; *Achistrum* [Devonian: Late Givetian; Poland]

922. *veghae* Kozur & Mostler, 1971; *Acanthotheelia* [Triassic: Ladinian: Langobardian; Hungary]

923. *venusta* Reich & Wiese, 2010; *Calcancora* [Cretaceous: Upper Turonian; Poland]

924. *venusta* Mostler, 1972 [1972f]; *Calclamnella* [Triassic: Norian: Alaunian; Germany]

925. *venustus* A. H. Müller, 1964; *Chiridotites* [Cretaceous: Lower Maastrichtian; Germany]

926. *vermai* Tandon & Saxena, 1977; *Eocaudina* [Paleogene: Eocene: Lutetian; India]

927. *vetusta* Schwager, 1866; *Chirodota* [Jurassic: Oxfordian; Germany]

928. *viai* Cherbonnier, 1978**; *Bathysynactites* [Triassic: Ladinian; Spain]

929. *virgiliae* Kozur in Kozur et al., 1985; *Acanthotheelia* [Triassic: Carnian: Cordevolian; Spain]

930. *vitoldi* Sztejn, 1992; *Spinopriscopedatus* [Cretaceous: Lower Valanginian; Poland]

931. *vonvalensis* Kubiatowicz & Matyja, 1977; *Priscopedatus* [Cretaceous: Upper Valanginian; Poland]

932. *vonvalensis* Kubiatowicz & Matyja, 1977; *Theelia* [Cretaceous: Upper Valanginian; Poland]

933. *waghopadarensis* Tandon & Saxena, 1977; *Rigaudites* [Paleogene: Eocene: Lutetian; India]

934. *waiorensis* Tandon & Saxena, 1977; *Eocaudina* [Paleogene: Eocene: Lutetian; India]

935. *walliseri* Haude, 2002**; *Podolepithuria* [Devonian: Lower Givetian; Germany]

936. *wanlessi* Summerson & Campbell, 1958; *Thuroholia* [Carboniferous: Pennsylvanian: Moscovian; U.S.A.]

937. *warboysensis* Soodan & Whatley, 1987; *Achistrum (Cancellrum)* [Jurassic: Upper Oxfordian; U.K.]

938. *wartensis* Garbowska & Wierzbowski, 1967; *Theelia* [Jurassic: Lower Kimmeridgian; Poland]

939. *waylandensis* Croneis in Croneis & McCormack, 1932 [1932c]; *Paleochiridota* [Carboniferous: Pennsylvanian: Gzhelian; U.S.A.]

940. *wessexensis* Hodson, Harris & Lawson, 1956; *Theelia* [Jurassic: Oxfordian; U.K.]

941. *wrighti* Soodan & Whatley, 1990; *Calclamna* [Jurassic: Hettangian; U.K.]

942. *wynnei* Soodan, 1975; *Sastriella* [Jurassic: Bathonian; India]

943. *wynnei* Tandon & Saxena, 1977; *Eocaudina* [Paleogene: Eocene: Lutetian; India]

944. *yarmoukensis* Sadeddin in Al-Tamimi, Saqqa & Sadeddin, 2001; *Rigaudites* [Cretaceous: Upper Cenomanian; Jordan]

945. *zankli* Mostler, 1969; *Ludwigia* [Triassic: Norian; Austria]

946. *zankli* Kozur & Simón, 1972; *Theelia* [Triassic: Carnian: Cordevolian; Spain]

947. *zapfei* Kozur & Mostler, 1970; *Theelia* [Triassic: Anisian: Pelsonian; Poland]

948. *zawidzkae* Kozur & Mock, 1972; *Theelia* [Triassic: Norian: Alaunian; Slovak Republic]

Echinoderms in a Changing World – Johnson (ed)
© 2013 Taylor & Francis Group, London, ISBN 978-1-138-00010-0

Arm loss and regeneration in stellate echinoderms: An organismal view[1]

J.M. Lawrence

University of South Florida, Tampa, Florida, USA

ABSTRACT: Arm loss and regeneration are characteristic of stellate echinoderms. They are integrally related and must be considered together. Arm loss and regeneration of crinoids and ophiuroids differ in some ways from asteroids because of structural and functional differences in their arms. Arm loss is primarily from predation and stellate echinoderms have mechanisms to prevent it. Arm loss has a cost that must be less than the benefit of regeneration. The degree of arm loss and the ability to recover from it affect the basic life-history strategies of stellate echinoderms. Understanding arm loss and re-generation at the organismal level in stellate echinoderms is important to understand their population biology and ecology and to have guidelines for investigating the molecular mechanisms involved.

Cuénot (1948) and Hyman (1955) seem to be the sources of the generalization that stellate echinoderms often lose their arms and regenerate them quickly, although Cuénot did note that it is rare in asteroids with a pentagonal body or those with arms with a broad base. Although fundamentally alike, crinoids and ophiuroids differ from asteroids as a result of structural and functional differences in the arms. This generalization that arm loss and regeneration are common is often repeated. Fleming et al. (2007) recently stated in a review of autotomy in invertebrates that "Starfish will regularly lose their arms" and "... in fact it is uncommon or rare to find individuals (brittlestars) without regeneration scars".

The generalization is not completely true. No arm loss has been documented in some species of stellate echinoderms and the frequency varies with populations of species of ophiuroids (Emson and Wilkie 1980) and asteroids (Lawrence 1992, Lawrence et al. 1999). But it is important to recognize that absence of evidence is not evidence of absence. In addition, the ability to autotomize arms may be present although the ability to regenerate them because of body structure is not. Although Scheibling (pers. comm.) did not observe any arm regeneration of an estimated several thousand *Oreaster reticulatus* in the field, he induced autotomy by threading a line through the proximal part of arms and observed the wound could not be closed and the individuals died.

It is important to know the characteristics of arm loss and regeneration in stellate echinoderms at the organismal level. Allee et al. (1949) pointed out the pivotal role of the organism, stating it is the lowest unit considered by the ecologist and the upper limit considered by the physiologist. The characteristics of the organism provide the bases for those of populations and provide guidelines for investigations of their molecular mechanisms. This review characterizes arm loss and arm regeneration in stellate echinoderms and then considers the phenomena in terms of life-history strategies.

1 ARM LOSS

1.1 *Cause of arm loss*

Arm loss is generally considered to result from sublethal predation (Emson and Wilkie 1980). This is primarily based on analysis of gut contents of fish. Physical factors may be responsible in some high energy shallow-water habitats or those with pebbles or stones (e.g. Woodley et al. 1981, Alva and Jangoux 1990, Makra and Keegan 1999). Retreat and withdrawal of arms by crinoids (Meyer 1973) suggest potential damage by physical factors. Ameziane and Roux (1997) implied tropical storms and hurricanes cause arm loss in crinoids. Mladenov (1983) discounted possible arm loss by the crinoid *Florometra serratissima* from water movement because of its depth of occurrence.

1.2 *Position of arm loss*

Crinoids, ophiuroids and asteroids have differences in the position at which arm loss by autotomy occurs and the evolution of this is of interest. Autotomy of arms of crinoids occurs at specialized ligamentary articulations, syzygies and cryptosyzygies (Oji 2001). Oji (2001) suggested Paleozoic crinoids could not autotomize their arms because they lacked localized articulations. He said Mesozoic and Cenozoic

[1]Dedicated to the memory of Valery Levin.

crinoids, including modern representatives, are characterized by dominance of mucular articulations and few ligamentary articulations. Although only one syzygy is found in each brachitaxes (one branching series) except for the primibrachials, syzygies are generally spaced at regular intervals in distal arms of comatulids, millercrinids and a few isocrinids (Hess, Oji, pers. comms.). Oji and Okamoto (1994) concluded the particular distribution of these articulations determines the amount of arm loss. Because autotomy occurs proximally to the damage site, the shorter spacing between the syzygies of the comatulid *Oreaster japonicas* would result in less arm loss than in the isocrinid *Metacrinus rotundus*. Despite the presumed occurrence of syzygies along of the arm of the comatulid *F. serratissima*, arms that are grasped autotomize near the base (Mladenov 1983). Baumiller (2008) pointed out the rigid syzygies that provide for autotomy decrease flexibility and mobility provided by muscular articulations.

The possibility of autotomy at every joint in ophiuroid arms (Emson and Wilkie 1980) results in the greatest possible decrease in amount of arm loss from distal damage. Because crinoid and ophiuroid arms are used for locomotion and particulate feeding, a mechanism for decreasing the amount of arm loss would be adaptive. Less loss of reproductive output could result from location of genital pinnules at the base of crinoid arms (Meyer 1985, Vail 1987, Holland 1991, Nichols 1994) and gonads within the body of ophiuroids or proximal part of the arms in some genera of Euryalidae (Stewart 1996), Asteroschematidae and Ophiocanopidae.

In contrast to crinoids and ophiuroids, arm loss in most asteroids is near the disc (Wilkie 2001). Why should autotomy be near the disc instead of distally? One would think as little of the arm should be lost as possible as in crinoids and ophiuroids. This may be related to the structure of the body wall of asteroids (Blake 1989) that does not allow for breakage zones in an autotomy plane as described by Wilkie (2001) except in unusual paxillosids such as the luidiids. Although *Luidia clathrata* can autotomize the arm at any position, it still has a primary autotomy plane near the disc (Lawrence, unpub.). Also, in contrast to crinoids and ophiuroids, the arms of asteroids are not only involved in locomotion and maintenance of position, but contain pyloric caeca and reproductive organs. The pyloric caeca and gonads have ducts at their proximal end. King (1898) said the gonads of *Asterias rubens* are attached to the last plate remaining in the stump. Wound healing and regeneration should be simpler if these ducts are severed by autotomy instead of rupture of the pyloric caeca and gonads. Anderson (1962, 1965) showed *Henricia leviuscula* regenerated pyloric caeca from their ducts.

Damage to asteroid arms does not necessarily lead to autotomy at the disc. Marrs et al. (2000) reported distal arm loss in *A. rubens* This may be related to Hancock's (1955) observation that arm tips of *A. rubens* damaged by the spider crab *Hyas araneus* wither. Bingham et al. (2000) similarly reported crushed distal ends of arms of *Leptasterias hexactis* sloughed off after several days. This kind of arm loss differs greatly from the sharp break usually associated with autotomy.

Although distal arm loss is considered a mechanism to minimize amount of arm loss, it also prevents structural damage to the calyx in crinoids and to the body wall of the disc of ophiuroids and asteroids.

1.3 *Arm loss and predation*

Meyer (1985) used frequency of arm regeneration as evidence of the intensity of predation on comatulids, and suggested it could be used to evaluate predation intensity in fossil species. Gahn and Baumiller (2005) noted the lack of data for frequency of arm loss in fossil crinoids and emphasized its importance for the hypothesis of the Mesozoic marine revolution. They found the frequency of arm loss in Mississippian crinoids ranged up to 27% for one species and suggested this indicates predation may have been great in the Paleozoic. This relation between frequency of sublethal arm loss and predation intensity has been used since for ophiuroids (e.g. Bowmer and Keegan 1983, Aronson 1987, 1991, Munday 1993, Sköld and Rosenberg 1996).

Schneider (1988) used frequency of arm loss to conclude large crinoids are preyed on preferentially. Gahn and Baumiller (2005) noted only *Rhodocrinites kirbyi* with a calyx height greater than 6 mm and a crown height greater than 20 mm had arm regeneration. Viviani (1978) documented an increase in frequency of arm loss with an increase in body size for *Stichaster striatus* and *Meyenaster gelatinosus*. He suggested this was because predation of small individuals is lethal for *S. striatus* and *M. gelatinosus,* i.e. these species have an escape from lethal predation in size. A similar lower frequency of arm loss in small individuals has been reported for *Acanthaster planci* (McCallum et al. 1989). Hopkins et al. (1994) found a conspicuous increase in frequency of arm loss with size for one population of *Astropecten articulatus* but not for another. Neither of these authors suggested lethal predation of small individuals. Sköld and Rosenberg (1996) also suggested lethal predation could be higher on small ophiuroids. Stewart (1996) found small *Astrobrachion constrictum* had significantly fewer regenerating arms than large individuals. In contrast to these studies, Marrs et al. (2000) reported a decrease in frequency of arm loss with increase in body size in six of seven populations of *Asterias rubens*.

1.4 *Arm loss and autotomy*

Arm autotomy as an escape response from predators has been suggested since the late 19th century (Cuénot 1891), Emson and Wilkie (1980) and Wilkie (2001) have maintained the restricted definition of autotomy in echinoderms: "the adaptive detachment of animal body parts where this serves a defensive function, is achieved by an intrinsic mechanism, and is nervously mediated". This restricted definition was used in the review of autotomy in invertebrates by Fleming et al. (2007). Wilkie (2001) called arm loss by the mechanisms listed but with non-defensive function exaptations, derived functions. Maginnis (2006) listed three current usages of the term: the reflexive loss of an appendage at a preformed breakage plane, the (non-reflexive) loss of an appendage at a preformed breakage plane, and the general loss of an appendage with no preformed breakage plane. She defined it broadly as any appendage loss with no implication of mechanism.

To be effective as an escape response to predation, autotomy should be an immediate response to being grasped by a predator. Forbes (1841) noted some ophiuroids fragment quickly when held while others did not. Wilkie (1978) induced immediate (within seconds) autotomy in ophiuroids by grasping the arm with forceps. Dupont and Thorndyke (2006) induced autotomy in ophiuroids by gently scraping a blade between vertebrae of an arm.

Because Paleozoic crinoids lacked the necessary structures or are not specialized for autotomy, Gahn and Baumiller (2005) suggested their arm loss implies mechanical failure due to a concentrated force that would be expected from the bite of fish or invertebrate jaws. Clark (1915) reported severe pinching of crinoid arms did not induce autotomy and had "the impression that comatulids are surprisingly callous to mechanical stimuli". Oji and Okamoto (1994) induced autotomy in crinoids by incising the arm.

King (1898) reported she was unable to induce autotomy of the arm by *Asterias vulgaris* (= *Asterias rubens*) by cutting it in several places and succeeded only by cutting off nearly all the tube feet. Similarly, Hotchkiss et al. (1990) were unable to induce arm autotomy in *A. vulgaris* and *A. forbesi* by pinching the aboral surface with a hemostat, a blow with a hammer, or cutting the arm. They did induce autotomy within several minutes by vigorously scraping out most of the tube feet and radial nerve from an ambulacral groove and within seconds by injection with KCl. Marrs et al. (2000)

and Ramsay et al. (2001) induced arm autotomy by *A. rubens* by crushing them with pliers. Marrs et al. (2000) reported autotomy in less than 30 seconds for individuals with a radius of <ca. 30 mm and ca. 40–75 seconds for those with a radius of ca. 40 mm. *Luidia clathrata* autotomized only 80% of the arms cut from the center of the disc to the arm tip (Lawrence, unpub.). Time to autotomy of grasped arms in *Coscinasterias muricata* was short (<three minutes) (Mazzone et al. 2003).

Arm autotomy could not be defensive as an escape mechanism for attached stalked crinoids and unlikely for cryptic comatulids or buried or cryptic ophiuroids. Rosenberg and Lundberg (2004) suggested autotomy might prevent the entire individual being pulled from the sediment by a predator. But the disc of *Amphiura filiformis* is 6–10 cm below the sediment surface (Solan and Kennedy 2002) and has two arms extended into the water column (Loo et al. 1996). Buchanan (1964) noted the deeply buried disc of *A. filiformis* should provide protection for it although the exposed arms were regularly cropped by predation. Turner (1974) and Muus (1981) attributed differential growth of arms of small *Ophiophragmus filograneus* and *A. filiformis*, respectively, to facilitate burrowing to escape from predators while having feeding arms exposed. Stancyk et al. (1994) knew of no reports of the behavior of amphiurids during attack by predators. Sköld (1998) observed fleeing but no autotomy by four epibenthic ophiuroid species in the laboratory in response to simulated predation. He referred to Emson and Wilkie (1980) for stating arm autotomy is an antipredatory characteristic of the species studied, but they did not document this. I know of no published account of autotomy by crinoids or ophiuroids as an escape response in the field.

In contrast to this, arm autotomy at the disc by asteroids can be quick in response to contact with arms of predatory asteroid species (*Asterias rubens,* Hancock 1955; *Heliaster helianthus, Meyenaster gelatinosus, Luidia magellanica, Stichaster striatus,* Viviani 1978; *Luidia clathrata,* Pomory, pers. comm.). The mode of predation is different between asteroids and that of fish and crabs. Fish and crabs are mechanical in their attacks and capable of severing of the arm while asteroids are not.

Autotomy in the restricted sense occurs in some asteroids. Hancock (1955) reported *A. rubens* in the laboratory autotomized an arm after *Crossaster papposus* attached to it and that the detached arm was ingested. Witman et al. (2003) provided a photograph of *A. rubens* with an autotomized arm of a conspecific in its mouth. Observations suggesting sublethal predation of asteroids in the field were made by Mauzey et al. (1968). They found *Solaster dawsoni* consuming presumably autotomized arms of *Solaster stimpsoni*. In the laboratory, *Evasterias troscheli* and *Pycnopodia helianthoides* autotomize arms in the presence of *S. dawsoni*. *S. dawsoni* ingested *S. stimpsoni, Mediaster aequalis* and *Crossaster papposus* whole without their autotomizing arms.

The most complete field observations on arm autotomy as an escape response by asteroids are those of Viviani (1978). *Heliaster helianthus* autotomizes its arms from attacks by *Meyenaster gelatinosus; Stichaster striatus*, from attacks by *Luidia magellanica; M. gelatinosus*, from attacks by *M. gelatinosus* and *L, magellanica;* and *L. magellanica*, from attacks by *L. magellanica* and *M. gelatinosus*. He reported *S. striatus* stops fleeing and autotomizes an arm in about 90 seconds after *L. magellanica* had ingested it. On numerous occasions he observed *H. helianthus* autotomizing 3 to 9 arms in attacks by *M. gelatinosus*. *Luidia clathrata* autotomizes an arm in response to the mere touch by the arm of *Luidia altaernata* (Pomory, pers. comm.). It is possible that chemicals from predators induce autotomy more easily than physical stimuli in these asteroids.

Voluntary arm loss, even if it follows predation but does not function in escape, requires another explanation. An alternative explanation is that autotomy follows predation in buried ophiuroids because it retains an ancestral trait that is adaptive because it facilitates wound healing and regeneration (Emson and Wilkie 1980, Wilkie 2001).

1.5 *Defense against arm loss*

1.5.1 *Structural defense*
Requirements for flexibility and movement of arms of crinoids and ophiuroids mean it would be difficult for them to have substantial structural defense. The spines that characterize them may be effective against small predators but their development is limited by the arm's functional requirements. Consequently their defense against large fish and crustacean predators seems limited. The asteroid arm has greater potential for defensive development. Blake (1989) pointed out they show great variety in shape and sturdiness, e.g. Scheibling and Metaxas (2008) suggested the large spines of *Protoreaster nodosus* are effective preventing fish predation because they observed no arm loss of an estimated one thousand individuals (Scheibling, pers. comm.).

1.5.2 *Alarm and escape responses*
Perceiving predators and avoiding them is one of the best anti-predator defenses for mobile animals. An *escape response* is flight when a predator is detected. *Florometra seratissima* responds to the presence

of predatory asteroids by swimming (Mladenov 1983, Shaw and Fontaine 1990). Predatory asteroid species are detected by other asteroids by a chemical stimulus that leads to escape (Hancock 1955, 1974, Mauzey et al. 1968, Castilla and Crisp 1970, Mayo and Mackie 1976, Van Veldhuizen and Oakes 1981, Birkeland et al. 1982, Gaymer and Himmelmann 2008, McClintock et al. 2008,). The epibenthic bathyal *Ophiura sursi* flees active *Leptychaster arcticus* (Fujita and Ohta 1989). Yee et al. (1987) stimulated escape by *Ophiopterus papillosa* by touching specimens with tube feet of *Pycnopodia helianthoides*. Drolet et al. (2004) observed *Ophiopholus aculeate* does not move with contact of the arm tip or ambulacral groove of *Asterias vulgaris* and moves only a few centimeters with contact of the stomach. Arm autotomy did not occur in either of these contact responses. Solan and Kennedy (2002) and Solan and Battle (2003) reported retraction of feeding arms of *Amphiura filiformis* to tactile stimulation.

An *alarm response* is flight when predation on a conspecific is detected. Arm loss should result in loss of tissue and coelomic fluid. An alarm response to conspecific tissue and/or body fluid in the laboratory has been reported for *Crossaster papposus* (Sloan and Northway 1982), *Pycnopodia helianthoides* (Lawrence 1991a), *Coscinasterias tenuispina* (Swenson and McClintock 1998), *Asterias amurensis* and *Distolasterias nipon* (Levin et al. 1984). In contrast, *Asterias rubens* shows no alarm response to coelomic fluid or gonad and pyloric caeca extracts (Campbell et al. 2001). Although *Luidia clathrata* is normally not cannibalistic, it shows situational cannibalism. Intact individuals attack conspecifics after arm amputation in the laboratory and in the field (Lawrence et al., in press). Gaymer and Himmelmann (2008) observed a tactile alarm response by *Heliaster helianthus* in the field.

Rosenberg and Selander (2000) reported an alarm response in *Amphiura filiformis* to arm loss. Individuals responded to arm loss in others by bending down arms stretched up in the filter feeding posture to the surface and partly into the substrate. I know of no reports of alarm responses by crinoids.

1.5.3 *Bioluminesence*
Morin (1983) proposed most bioluminescent signals by benthic invertebrates are predator deterrents. Herring (1995) suggested bioluminescence is primarily an anti-predator behavior in echinoderms. Bioluminescence is widespread in ophiuroids (Mallefet et al. 2008, Mallefet and O'Hara 2009). Gotto (1963) interpreted bioluminescence of the autotomized arm of *Acrocnida brachiata* as a behavior to "distract and occupy a predator

56

long enough for the animal to escape." Because *A. brachiata* lives in burrows 10–20 cm deep with only the tips of the arms exposed (Bourgoin and Guillou 1994) it seems improbable the bioluminescence functions as an escape response. Buchanan (1964) noted *Amphiura chiajei* is a surface feeder while *A. filiformis* is a suspension feeder. He suggested *A. filiformis* with its arms extended for feeding would be more vulnerable to predation and pointed out it is the species with bioluminescent arms. Because amputation of an arm causes a flash from all arms and then rapid withdrawal below the surface, he thought this would have survival value. It should decrease sublethal predation. Predator-deterrent bioluminescence was demonstrated for *Ophiopsila californica* (Basch 1988) and *Ophiopsila riissei* (Grober 1988a). Deheyne et al. (2000) observed bioluminescence by the epifaunal *Amphipholis squamata* in response to contact with a crab (*Cancer maenas*). Arm autotomy in the laboratory occurred only after prolonged interaction.

I know of no reports of bioluminescence in crinoids or asteroids except for deep-sea species (Herring 1974).

1.5.4 *Chemical deterrents*
Secondary metabolites are small organic molecules that affect other organisms and are not involved in basic metabolic pathways. One function in plants is as a chemical defense system that protects against predation and microbial infections (Iason 2005). They also have these functions in animals. As an antipredatory defense, they are characteristic of free-living organisms with a limited range of movement or limited control over their movements (Berenbaum 1995). Berenbaum noted this was the case for crinoids.

Parker (1881) gave an anecdotal account of *Crossaster papposus* being poisonous to cats but investigation of secondary metabolites in stellate echinoderms did not begin until the latter half of the twentieth century. Secondary metabolites with toxic, pharmaceutical and antimicrobial properties have been demonstrated (Singh et al. 1967, Stonik and Elyakov 1988, D'Auria et al. 1993, Hosttettmann and Marston 1995, Berenbaum 1995, Cimino and Ghiselin 2001, Amsler et al. 2001, Iorizzi et al. 2001). Secondary metabolites have been reported for ophiuroids (Amsler et al. 2001). There is variability in the presence and types of secondary metabolites (e.g. in crinoids, Rideout and Sutherland 1985) and in the physiological and behavioral responses of echinoderms to extracts (e.g. McClintock and Baker 1997).

Feeding deterrence assays are the usual method to demonstrate the presence of secondary metabolites (e.g., Rideout et al. 1979, Bryan et al. 1997, Amsler et al. 2001). Because predators can have resistance and tolerance at multiple levels (Segner and Braunbeck 1998), these assays to demonstrate toxicity are best done with ecologically irrelevant marine species (McClintock 1989). The demonstration that a number of reef fish feed on stalked crinoids from 400 m depth (McClintock et al. 1999) suggests defensive secondary metabolites are not present as the fish used are ecologically irrelevant and would not be expected to have evolved tolerance or resistance to toxins.

Lack of avoidance behavior to ecologically relevant predators by stellate echinoderms might be used as an indication of the presence of deleterious secondary metabolites and absence of tolerance and resistant by predators. Meyer (1985) found only 13 of 46 crinoid species had nocturnal activity and Schneider (1988) recorded diurnal behavior in crinoids that would expose them to fish predation.

1.5.5 *Aposematic warning*
The potential predator must be able to identify the prey with secondary metabolites. Having a defense when the predator cannot learn is not useful to the prey. Berenbaum (1995) noted in her review of chemical defenses that echinoderms have many conspicuously colored species. Bandaranayake (2006) concluded the co-occurrence of conspicuous coloration and noxious characteristics suggest aposematism is a common predator warning in many coastal marine invertebrate species. He listed the characteristics of aposematic coloration as bright colors, usually in banded or contrasting patterns, and classed echinoderms as one of the marine invertebrate groups with the most spectacular and most unusual color patterns.

A uniform color, unless glaring or conspicuous, may not be an aposematic warning but coincidental to another function (Bandaranayake 2006). Blumer (1960) identified a pigment, fringelite, in the upper Jurassic *Millericrinus* (= *Liliocrinus*). Wolkenstein et al. (2006) found it in the Middle Triassic crinoid *Carnallicrinus carnally* which belongs to another order. They suggested the occurrence of this pigment in fossil crinoids is a widespread feature. It is found throughout the body, even in the holdfast. Wolkenstein et al. (2008) suggested this pigment is closely related to recent crinoid pigments. Its occurrence throughout the body suggests it may have functioned as an antipathogen and not antipredatory and preceded anti-predator function.

It is strange aposematic coloration has not been studied in echinoderms, especially since Cuénot (1891) suggested many years ago the intense red color of *Echinaster sepositus* is aposematic coloration and indicates toxicity. *Echinaster* and other species of Echinasteridae have secondary metabolites (Stonik and Elyakov 1988, Iorizzi et al. 2001, Ivanchina et al. 2006). Extracts from *Henricia downeyae* have steroid glycosides that inhibit bacterial and fungal growth and deter fish feeding (Palagiano et al. 1996).

Clark (1915) observed fish generally avoid crinoids and suggested they recognized them as inedible either by sight or olfactory perception. Reports of fish preying on crinoids (Meyer and Macurda 1977, Meyer 1985) do not disprove Clark's observations but do indicate crinoid avoidance by fish is not universal. Reports of exposed diurnal crinoid species (Meyer 1973, Schneider 1988) suggest effective anti-predator mechanisms. Could the bright colors and patterns of crinoids be aposematic coloration? Australian coral reef crinoids, ophiuroids and asteroids often have bright and patterned colors while those of the southern coast typically do not (Clark 1921, 1938).

Aposematic coloration implies predators with color vision. Fish have color vision (Siebeck et al. 2008) and many coral reef fish also have UV perception (Siebeck and Marshall 2001). Brachyurans have only one or two spectral classes of photoreceptors (Cronin and Forward 1988, Weise 2002). They are either color blind or potential dichromates and would not respond to aposematic coloration. Most marine crustaceans have high affinity to blue-green wavelengths and some to violet/near ultraviolet (Johnson et al. 2002).

Does this mean stellate echinoderms have no warning deterrents for potential crustacean predators? Hancock (1955) noted the spider crab *Hyas graneus* attacked *Asterias rubens* but not *Crossaster papposus*. He suspected *C. papposus* might be toxic and placed pieces of the body of both species before the crab. All pieces of *A. rubens* were consumed but none of *C. papposus*. There is a major difference in coloration of the two species that suggests aposematic coloration in *C. papposus*. But the lack of color vision in the crab suggests it responded to a soluble chemical deterrent that was an olfactory warning. I know of no direct test of either aposematic coloration or olfactory warning systems in echinoderms.

Grobner (1988b) suggested bioluminescence by arms of unpalatable *Ophiopsila riisei* was an aposematic warning. Grobner (1989) acknowledged Guilford and Cahill (1989) were correct that he did not unequivocally show this was aposematic warning because the slopes of the learning curves of sighted and blind crabs were identical. An olfactory cue is possible. Grobner (1989) also noted his use of handling time was not the best measure of learnt aversion as the critical measure is consumption.

1.5.6 *Predator avoidance*
If defense by the previous mechanisms is not possible or adequate, predator avoidance is a final defense against predation. An activity period when the predator is not active and cryptic behavior when it is results in a tradeoff is between decreased activity and decreased predation. Diel rhythms associated with predator avoidance are found in many species of crinoids (e.g. Magnus 1967, Meyer and McCurda 1980, Vail 1987), ophiuroids (e.g. Hendler 1984, Sides 1985, Rosenberg and Lundberg 2004) and asteroids (e.g. Fenchel 1965, Ribi and Jost 1975, Thomassin 1976, McClintock and Lawrence 1982, Soliman et al. 1986, Keesing 1995). Cuénot (1891) suggested echinoderms may also be camouflaged to escape predation.

2 ARM REGENERATION

2.1 *Characteristics of regenerating arms*

Arm regeneration begins with production of a growing tip similar to that of intact arms and proceeds in the same manner (King 1898, Mann 1936). Salzwedel (1974) compared regenerating arms of *Amphiura filiformis* to intact arms of the same length and concluded regenerating arms are thin and become functional by thickening. Dupont and Thorndyke (2006) also described this increase in length followed by development of ossicles, tube feet and spines in *A. filiformis*. They also concluded it is adaptive for the arm to increase in length more rapidly than differentiation to functional recovery. Differentiation necessarily follows increase in length but full differentiation takes time. The question is whether a long arm that has not fully differentiated is functional. It appears the regenerating arm increases in length and, as it does so, differentiates and becomes functional. It seems likely growth and regeneration of arms is similar to that of meristematic growth of eudicots in which increase in length precedes differentiation.

Regenerating arms of *Luidia clathrata* also grow more quickly than intact arms but have little difference in dimensions from intact arms measured from the tip (Lawrence and Pomory, unpub.). Reports that the arm weight:length ratio of regenerating arms is less than that of intact arms in *Amphiura filiformis* (Salzwedel 1974), *Microphiopholis gracillima* (Stancyk et al. 1994) and *Astrobrachion constrictum* (Stewart 1996) are not surprising because of the greater development of the intact arm than the regenerating arm.

2.2 *Arm regeneration and amount of arm loss*

King (1898) began studies on the effect of amount of arm loss with her report containing a figure showing the rate of arm regeneration an individual *Asterias vulgaris* was indirectly related to the closeness of arm loss to the disc. She did not provide quantitative data to support this conclusion but did provide other figures showing the rate of regeneration of arms from the disc could be either equal or unequal.

A number of studies in the first quarter of the 20th century repeated her work. The results of these studies are suggestive but poor experimental design, poor or undescribed aquarium conditions, no or few replicates, lack of consideration of nutritional condition and no statistical analysis make their interpretations and conclusions unreliable. For example Schapiro (1914) stated regeneration is rapid at the base of the arm and slow at the tip in *Echinaster sepositus* and *Marthasterias glacialis* but provided no documentation. Zirpolo's (1921) study on the effect of position of arm loss in *Asterina gibbosa* had an inadequate number of replicates and too many variable conditions to draw conclusions. His report (Zirpolo 1926) with *Luidia ciliaris* described a single individual.

Mladenov (1983) reported *Florometra serratissima* regenerates arms from a long arm stump at about the same rate as one with a short stump. In a field study Vail (1989) found the rate of growth of arms of small intact *Oligometra serripinna* and regenerating arms of the same length is similar and both are greater than arm growth in large individuals. This interesting result indicates the similarity in growth of small intact arms and regenerating arms.

An indirect relation between the rate of arm regeneration and position of arm loss in terms of increase in arm length has been quantified for *Amphipholis filiformis* (Dupont and Thorndyke 2006) and *Ophionotus vitoriae* (Clark et al. 2007). This indirect relation between rate of arm regeneration and position of arm loss occurs in *Luidia clathrata* (Lawrence and Pomory, 2008) in terms of increase in arm length, dry weight and amount of organic matter. Mass as a measurement of regeneration provides information about production. Lawrence and Pomory (2008) found the difference in rate of arm regeneration according to position of arm loss corresponds to the rate of intact arm growth as the small *L. clathrata* reaches its asymptotic size (Dehn 1980) and of an arm regenerating from its base as it reaches its asymptotic size (Lawrence and Elwood 1991). This observation is like that of Vail (1989) and indicates differences in rate of arm regeneration and differentiation are under instantaneous positional control.

Amputation of arms at different positions in the same individual by Dupont and Thorndyke (2006) and Lawrence and Pomory (2008) eliminated possible effects of different amounts of nutrient reserves and ability to feed on regeneration. Dupont and Thorndyke (2006) stated the amount (length) of arm loss determines whether energy will be invested in growth and/or differentiation. An alternative explanation is the position of arm loss determines the rate of growth and differentiation because the mechanisms for regeneration depend on the amount of tissue present (Candia Carnevali 2006). Interesting aspects of this are that initiating the high rate of regeneration has no effect on intact arms and that the rate of regeneration slows as the regenerating arm reaches the asymptotic length regardless of position of arm loss.

The amount of arm loss can be interpreted in the original sense of King (1898), the amount of loss by an arm, or in an extended sense, the total amount of arm loss by several arms. The rate of arm regeneration in these different conditions has important implications for understanding regeneration in the individual and the mechanisms that control it. In contrast to analysis of the effect of position of arm loss on regeneration, it is necessary to use different individuals for analysis of the effect of number of arms lost. It is important to recognize arm loss may affect the ability to feed and make comparison of individuals with different numbers of arms lost more difficult. As with early studies on the effect of position of arm loss on regeneration, many studies are not convincing and must be critically analyzed.

Zeleny (1905) starved *Ophioglypha lacertosa* (= *Ophiura ophiura*) and reported the amount of regenerated arms increased with number of arms regenerating and suggested intact arms decreased regeneration. His analysis is not convincing. Ramsay et al. (2001) also found regeneration of arms in *Asterias rubens* fed in the laboratory initially decreased with an increase in number of arms lost but subsequently the amount of regeneration per arm became greatest in individuals with most lost arms. They stated this would be adaptive but did not suggest a mechanism.

Salzwedel (1974) reported the rate of regeneration per arm of *Amphiura filiformis* fed in the laboratory decreased with an increase in number of arms amputated but the total length and weight of regenerating arms per individual increased. His data are suggestive but the values and differences were small and the data were not analyzed statistically. Similar results have been reported for other ophiuroids starved in the laboratory (*Microphiopholis gracillima*, Fielman et al. 1991) and in the field (*Ophiophragmus filograneus*, Clements et al. 1994) and a crinoid in the field (*Florometra serratissima*, Mladenov 1983). Mladenov (1983) pointed out this cannot be explained by energy limitations because the total regeneration per individual increased with increasing number of regenerating arms but did not suggest a mechanism. An increase in total rate of arm regeneration with an increased number of arms lost would require increased use of nutrient reserves or greater allocation of absorbed nutrients to regeneration because a decrease in feeding would be expected. These observations that regeneration per arm changes with the number of arms lost indicate systemic rather than local control.

The third possibility, no effect of number of arms lost on regeneration per arm, has been reported. Morgulis (1909) stated the rate of arm regeneration in *Ophiocoma pumila* with one to three arms amputated was about the same as that of those with four to five arms amputated although the total amount regenerated per individual increased but his data are not adequate to support his conclusion. The lack of an effect has been reported for other ophiuroids in the field (*Microphiopholis gracillima*, Stancyk et al. 1994) or fed in the laboratory (*Ophiocoma scolopendrina*, Soong et al. 1997, *Amphiura filiformis*, Nilsson 1999) and an asteroid fed in the laboratory (*Leptasterias hexactis*, Bingham et al. 2000). These observations indicate local rather than systemic control of regeneration.

2.3   *Arm regeneration and nutrition*

Regeneration requires materials and energy. These can come from nutrient reserves as in the disc of *Microphiopholis gracilima* (Fielman et al. 1991) or from food. Nutritional studies require knowing the amount of food consumed. This is difficult with crinoids and ophiuroids but not with asteroids. *Luidia clathrata* allocates nutrients to the pyloric caeca and gonads in intact arms and not to regenerating arms if fed a maintenance level diet (Lawrence et al. 1986). Allocation is not only to production in the pyloric caeca and gonads but also to rapid regeneration of the arms if fed a high amount of food (Lawrence and Elwood 1991).

Both quality of food and quantity eaten affect regeneration. Comparison of rates of regeneration (e.g. Clarke et al. 2007 for ophiuroids) is difficult without knowledge of quality of food and quantity consumed. Prepared feeds provide great potential for investigating arm regeneration because quality can be controlled. Characteristics of regenerating *L. clathrata* fed a high protein:low energy or low protein:high energy diet differ (Lawrence et al., unpub.).

2.4   *Cost of arm loss and regeneration*

A cost to the organism from arm loss is expected and regeneration implies the benefit exceeds the cost. This has been little studied. Lawrence and Vasquez (1996) suggested a decrease in nutrient reserves and reproduction could be expected costs of arm loss. *Stichaster striatus* regenerating arms in the field had much less organic matter in the pyloric caeca

of intact arms (Lawrence and Larrain 1994) and *Ophiothrix fragilis* regenerating arms had less nutrient storage and gonad production (Morgan and Jangoux 2004). Seasonal reproduction is an additional factor. Pomory and Lawrence (2001) found no significant difference in body components of regenerating and intact *Ophiocoma echinata* during the winter. Stomach mass was less in regenerating individuals in summer. A final problem is availability of food. Pomory and Lares (2000) attributed the lack of differences in the amount of pyloric caeca and gonads in intact and regenerating *L. clathrata* in the field to low availability of food. High availability of food could decrease the cost.

A decrease in growth and production resulting from arm loss could be predicted from the allocation of nutrients to regeneration but inadequate attention has been given to the effect of a decrease in feeding capacity with loss of arms. This would certainly be expected for crinoids and ophiuroids but has not been considered or investigated. Arm loss decreases feeding and production in asteroids. Amputation of only the arm tips of *Pisaster ochraceus* reduced feeding for at least 16 weeks and resulted in decreased gonad production (Harrold and Pearse 1980). Arm loss in *Stichaster striatus* resulted in a major decrease in feeding that lasted at least six months and decreased growth and production by the pyloric caeca (Diaz-Guisado et al. 2006). *Heliaster helianthus* with autotomized arms similarly had decreased feeding that lasted at least six months and a decreased growth rate and pyloric caeca and gonad production (Barrios et al. 2008).

Maginnis (2006) pointed out research on the cost of autotomy and regeneration in general has compared intact individuals with regenerating individuals. This is certainly true for stellate echinoderms. She suggested an alternative comparison between individuals experiencing autotomy with or without regeneration. This would provide a better evaluation of the fitness resulting from regeneration. This phenotypic engineering would be feasible with stellate echinoderms. Morgulis (1909) produced individuals that did not regenerate the arm by cauterizing the radial nerve. Another possibility is altering the physiological state of the regenerating individual by hormone treatment. Hormones can be added to prepared feeds to manipulate regeneration as suggested by Maginnis (2006).

3   ARM LOSS AND LIFE-HISTORY STRATEGIES

Life history theory developed by plant biologists predicts loss of body parts followed by regeneration would be the strategy developed in species in which stress (low nutrient availability and/or ability to obtain nutrients) is low and disturbance (e.g. loss of body part) is high (Grime 2001). It is generally assumed plants are best able to recover from predation (disturbance) when growing in high resource conditions (low stress) (Hawkes and Sullivan 2001). This approach, considering stress as well as disturbance, has been applied to echinoderms (Lawrence 1991b). Species that have high frequency of arm loss (disturbance) must have high capacity for arm regeneration (productivity) either by high availability of food or ability to obtain food. The long life of *Amphiura filiformis* (Muus 1981) despite the high frequency of arm loss has been attributed to its ability to rapidly regenerate lost arms. This ability is probably related to the high availability of food where it occurs.

Blake (1989) noted the conflicting demands of flexibility and protection. Blake (1983) suggested predatory pressure could lead to structurally protected asteroids and consequently to feeders on small particles or sessile colonial animals. Lawrence (1991b) suggested adaptations reducing disturbance, including predation, would be predicted for species with low feeding capacity.

We can contrast the life-history strategies of *Pisaster* and *Asterias* species, closely related genera in the Asteriinae (Foltz et al. 2007, Perseke et al. 2008). *Asterias* species are active predators and scavengers (Menge 1979, Hulbert 1980, Sloan and Aldridge 1981, Dare 1982, Nickell and Moore 1991, Ramsay et al. 1997), which suggests a high capacity for production. *Pisaster ochraceus* is normally food limited, often to an extreme degree (Castilla and Paine 1987) and *Pisaster giganteus* has a slow rate of feeding (Harrold and Pearse 1987), which suggests a low capacity for production. These differences in ability to feed are correlated with a relative weak body wall in *Asterias* species and a robust one in *Pisaster* species (Fisher 1930). This in turn can be correlated with frequency of arm loss, high in *Asterias* species and low in *Pisaster* species (Lawrence 1992). Life-history strategy predicts stress tolerant species would not only be physically protected but chemically as well. As would be predicted, saponins from *Pisaster brevispinus* and *P. ochraceus* are more toxic than those from *Asterias forbesi* (Rio et al. 1965).

Meyer and McCurda (1977) suggested the appearance of durophagous teleosts during the Mesozoic led to the ecological restriction of stalked crinoids to deep water. This attributes the ecological isolation to disturbance alone. It does not consider stress, the low availability of food or the low ability to obtain food. Ameziane and Roux (1997) characterized the mesobathyal distribution of the continental margin where stalked crinoids are found as sheltered from hydrodynamic turbulences and low but not extreme food supply. This applies the concept of stress and disturbance

to their ecological distribution. If stalked crinoids have a low capacity to obtain food, they would have a low capacity to regenerate arms lost by predation. They would subsequently be restricted to areas where predation pressure is low.

The alternative question can be asked. Why are comatulids most common in shallow water (Gage and Tyler 1991, Ameziane and Roux 1997, Tokeshi 2002)? Why have they not exploited deep waters as they have shallow waters? Meyer (1985) reported frequency of arm loss of reef-dwelling crinoids ranging from 29 to 77%. The great activity of comatulids has been noted (Meyer and McCurda 1977, Meyer et al. 1984, Shaw and Fontaine 1990). The consequence of this is an ability to cope with disturbance but results in high requirement for food for their basic metabolic requirements. Comatulids have higher rates of metabolism than other crinoids groups (Baumiller and LaBarbara 1989). Gage and Tyler (1991) concluded the availability of particulate food in the deep sea is generally a significant limiting factor. This would mean comatulids there would be in a stressful situation.

Brisingid asteroids also are ecologically restricted to the deep sea (Hyman 1955). It is possible the same considerations applied to stalked crinoids also apply to them.

4 CONCLUSIONS

Arm loss and regeneration are integrally related and must be considered together. Many questions remain about the biology of arm loss and regeneration in stellate echinoderms. Although fundamentally alike, arm loss and regeneration of crinoids and ophiuroids differ in some ways from asteroids as a result of structural and functional differences in the arms. More attention must be given to the cost of arm loss and regeneration on functioning of the organism. Application of life-history theory can be used profitably to interpret arm loss and regeneration in terms of characteristics of the species and their environment.

Understanding arm loss and regeneration in stellate echinoderms at the organismal level is also important to understand their population biology and ecology and to have guidelines for investigating the molecular mechanisms involved.

NOTE ADDED IN PROOF: Since this paper was submitted, a review appeared that is relevant to Section 1.5.4 Chemical deterrents

Slattery, M. 2010. Bioactive components from echinoderms: Ecological and evolutionary perspectives. In: L.G. Harris, S.A. Böttger, C.W. Walker, M.P. Lesser (eds). *Echinoderms: Durham*. London: Taylor & Francis Group.

ACKNOWLEDGEMENTS

I thank William Ausich, Bill Baker, Larry Basch, Hans Hess, Tatsuo Oji, Rutger Rosenberg, Robert Scheibling, and Stephen Stancyk for information and Carlos Gaymer, Gordon Hendler, James McClintock, Christopher Pomory and Richard Turner for helpful comments on the manuscript.

REFERENCES

Allee, W.C., Park, O., Emerson, A.E. Park, T. & Schmidt, K.P. 1949. *Principles of Animal Ecology.* Philadelphia: W.B. Sanders, Company.

Alva, V. & Jangoux, M. 1990. Fréquence et causes presumée de la régénération brachiale chez *Amphipholis squamata* (Echinodermata, Ophiuroidea. In C. De Ridder, P. Dubois, M.-C. Lehaye, M. Jangoux (eds). *Echinoderm Research*: 147–153. Balkema: Rotterdam.

Ameziane, N. &. Roux. 1997. Biodiversity and historical biogeography of stalked crinoids (Echinodermata) in the deep sea. *Biodiversity Conservation.* 6: 1557–1570.

Amsler, C.D., Iken, K.B., McClintock, J.B. & Baker, B.J. 2001. Secondary metabolites from Antarctic marine organisms and their ecological implications. In J.B. McClintock & B.J. Baker (eds). *Marine Chemical Ecology*: 267–300. Boca Raton: CRC Press.

Anderson, J.M. 1962. Studies on visceral regeneration in sea-stars. I. Regeneration of pyloric caeca in *Henricia leviuscula* (Stimpson). *Biol. Bull.* 122: 321–342.

Anderson, J.M. 1965l Studies on visceral regeneration in sea-stars. II. Regeneration of pyloric caeca in Asteriidae, with notes on the source of cells in regenerating organs. *Biol. Bull.* 128: 1–23.

Aronson, R.B. 1987. Predation on fossil and Recent ophiuroids. *Paleobiology.* 13: 187–192.

Bandaranayake, W.M. 2006. The nature and role of pigments of marine invertebrates. *Nat. Prod. Rep.* 23: 223–255.

Barrios, J.V., Gaymer, C.F., Vásquez, J.A. & Brokordt, K.B. 2008. Effect of the degree of autotomy on feeding, growth, and reproductive capcity in the multi-armed sea star *Heliaster helianthus.* *J. Exp. Mar. Biol. Ecol.* 361: 21–27.

Basch, L.V. 1988. Bioluminescent anti-predator defense in a subtidal ophiuroid. In R.D. Burke, P.V. Mladenov, P. Lambert, R.L. Parsley. (eds). *Echinoderm Biology*: 503–515. Rotterdam: A.A. Balkema.

Baumiller, T.K. 2008. Crinoid ecological morphology. *Annu. Rev. Earth Planet. Sci.* 36: 221–249.

Baumiller, T.K. & LaBarbara, M. 1989. Metabolic raes of Caribbean crinoids (Echinodermata), with special reference to deep-0water stalked and stalkless taxa. *Comp. Biochem. Physiol.* 93 A: 391–394.

Berenbaulm, M.R. 1995. The chemistry of defense: theory and practice. *Proc. Natl. Acad. Sci. USA.* 92: 2–8.

Bingham, B.L., J. Burr, J. & Wounded Head, H. 2000. Causes and consequences of arm damage in the sea star *Leptasterias hexactis.* *Can. J. Zool.* 78: 596–605.

Birkeland, C., Dayton, P.K. & Engstrom, N.A. 1982. A stable system of predation on a holothurians by four asteroids and their top predator. *Aust. Mus. Mem.* No. 16: 175–189.

Blake, D.B. 1983. Some biological controls on the distribution of shallow water sea stars. *Bull. Mar. Sci.* 33: 703–712.

Blake, D.B. 1989. Asteroidea: functional morphology, classification and phylogeny. In M. Jangoux, J.M. Lawrence (eds) *Echinoderm Studies* 3: 179–223. Rotterdam: A.A. Balkema.

Blumer, M. 1960. Pigments of a fossil echinoderm. *Nature.* 188: 1100–1101.

Bourgoin, A. & Guillou. M. 1994. Arm regeneration in two populations of *Acrocnida brachiata* (Montagu) (Echinodermata: Ophiuroidea) in Douarnenez Bay, (Brittany: France): An ecological significance. *J. Exp. Mar. Biol. Ecol.* 184: 123–139.

Bowmer, T. & Keegan, B.F. 1983. Field survey of the occurrence and significance of regeneration *Amphiura filiformis* (Echinodermata: Ophiuroidea) from Galway Bay, west coast of Ireland. *Mar. Biol.* 74: 65–71.

Bryan, P.J., McClintock, J.B. & Hopkins, T.S. 1997. Structure and chemical defenses of echinoderms from the northern Gulf of Mexico. *J. Exp. Mar. Biol. Ecol.* 210: 173–186.

Buchanan, J.B. 1964. A comparative study of some features of the biology of *Amphiura filiformis* and *Amphiura chiajei* (Ophiuroidea) considered in relation to their distribution. *J. Mar. Biol. Ass. U.K.* 44: 565–576.

Campbell, A.C., Coppard, S., D'Abreo, C. & Tudor-Thomas, R. 2001. Escape and aggregation responses of three echinoderms to conspecific stimuli. *Biol. Bull.* 201: 175–185.

Candia Carnevali, M.D. 2006. Regeneration in echino-derems: repair, regrowth, cloning. *Inform. Syst. J.* 3: 64–76.

Castilla, J.C. & Crisp, D.J. 1970. Responses of *Asterias rubens* to olfactory stimuli. *J. Mar. Biol. Assoc. U.K.* 50: 829–847.

Castilla, J.C. & Paine, R.T. 1987. Predation and community organization on Easter Pacific, temperate zone, rocky intertidal shores. *Rev. Chileana Hist. Nat.* 60: 131–151.

Cimino, G. & Ghiselin, M.T. 2001. Marine natural products chemistry as an evolutionary narrative. In J.B. McClintock and B.J. Baker (eds) *Marine Chemical Ecology*: 115–154. Boca Raton: CRC Press.

Clark, H.L. 1915. The comatulids of Torres Strait; with special reference to their habits and reactions. *Pap. Dept. Mar. Biol. Carnegie Inst. Wash.* 8: 97–125.

Clark, H.L. 1921. The echinoderm fauna of Torres Strait: its composition and its origin. *Carnegie Institution of Washington, Pub.* 214.

Clark, H.L. 1938. Echinoderms from Australia. *Mem. Mus. Comp. Zool. Harvard College.* 55.

Clark, M.S., Dupont, S., Rossetti, H., Burns, G., Thorndyke, M.C. & Peck, L.S. 2007. Delayed arm regeneration in the Antarctic brittle star *Ophionotus victoriae.* *Aquat. Biol.* 1, 45–53.

Clements, L.A.J., S.S. Bell, & Kurdziel, J.P. 1994. Abundance and arm loss of the infaunal brittle star *Ophiophragmus filograneus* (Echinodermata: Ophiuroidea), with an experimental determination of regeneration rates in natural and planted seagrass beds. *Mar. Biol.* 121: 97–104.

Cronin, T.W. & Forward, Jr, R.B. 1988. The visual pigments of crabs. I. Spectral characteristics. *J. Comp. Physiol.* 162A: 463–478.

Cuénot, L. 1891. Études morphologiques sur les Échinodermes. *Arch. Biol.* 11: 313–680.

Cuénot, L. 1948. Anatomie, éthologie et systématique des échinoderms. In P. Grassé (ed) *Traité de zoologie*: 11: 3–272. Paris: Masson et Cie., Éditeurs.

Dare, P.J. 1982. Notes on the swarming behaviour and population density of *Asterias rubens* L. (Echinodermata:Asteroidea) feeding on the mussel, *Mytilus edulis* L. *J. Conseil* 40: 112–118.

D'Auria, M.V., Minale, L. & R. Riccio, R. 1993. Polyoxygenated steroids of marine origin. *Chem. Rev.* 93: 139–1895.

Deheyn, D., Mallefet, J. & Jangoux, M. 2000. Expression of bioluminescence in *Amphipholis squamata* (Ophiuroidea: Echinodermata) in presence of various organisms: a laboratory study. *J. Mar. Biol. Ass. U.K.* 80: 179–180.

Dehn, P.F. 1980 *Growth and reproduction in* Luidia clathrata *(Say) (Echinodermata: Asteroidea)*. Ph.D. dissertation, Tampa: University of South Florida.

Diaz-Guisado, D., Gaymer, C.F., Brokordt, K.B. & Lawrence, J.M. 2006. Autotomy reduces feeding, energy storage and growth of the sea star *Stichaster striatus*. *J. Exp. Mar. Biol. Ecol.* 338: 73–80.

Drolet, D., Himmelmann, J.H. & Rochette, R. 2004. Use of refuges by the ophiuroid *Ophiopholis aculeate*: contrasting effects of substratum complexity on predation risk from two predators. *Mar. Ecol. Prog. Ser.* 284: 173–183.

Dupont, S. & Thorndyke, M.C. 2006. Growth or differentiation? Adaptive regeneration in the brittlestar *Amphiura filiformis*. J. Exp. Biol. 209: 3873–3881.

Emson, R.H. & Wilkie, I.C. 1980. Fission and autotomy in echinoderms. *Oceanogr. Mar. Biol. Ann. Rev.* 28: 155–250.

Fenchel, T. 1965. Feeding biology of the sea star *Luidia sarsi* Duben & Koren. *Ophelia.* 2: 223–236.

Fielman, K.T., Stancyk, S.E., Dobson, W.E. & Clements, L.A.J. 1991. Effects of disc and arm loss on regeneration by *Microphiopholis gracillima* (Echinodermata:Ophiuroidea)in nutrient-free seawater. *Mar. Biol.* 111: 121–127.

Fisher, W.K. 1930. Asteroidea of the North Pacific and adjacent waters. Part 3. Forcipulata (concluded). *United States National Museum, Smithsonian Institution. Bulletin* 76.

Fleming, P.A., Muller, D., & Bateman, P.W. 2007. Leave it all behind: a taxonomic perspective of autotomy in invertebrates. *Biol. Rev.* 82: 481–510.

Foltz, D.W., Bolton, M.T., Kelley, S.P., Kelley, B.D. & Nguyen, A.T. 2007. Combined mitochondrial and nuclear sequences support the monophyly of forciculatacean sea stars. *Mol. Phylogen. Evol.* 43: 627–634.

Forbes, E. 1841. A history of British starfishes, and other animals of the class Echinodermata. London: John Van Voorst.

Fujita, T. & Ohta, S. 1989. Spatial structure within a dense bed of the brittle star *Ophiura sarsi* (Ophiuroidea: Echinodermata) in the bathyal zone off Ōtsuchi, northeastern Japan. 45, 389–300.

Gage, J.D. & Tyler, P.A. 1991. *Deep-sea biology: A natural history of organisms at the deep-sea floor.* Cambridge: Cambridge University Press.

Gahn, F.J. & Baumiller, T.K. 2005. Arm regeneration in Mississippian crinoids: evidence of intense predation pressure in the Paleozoic? *Paleobiology.* 31: 151–164.

Gaymer, C.F. & Himmelmann, J.H. 2008. A keystone predatory sea star in the intertidal zone is controlled by a higher-order predatory sea star in the subtidal zone. *Mar. Ecol. Prog. Ser.* 370: 143–153.

Gotto, R.V. 1963. Luminescent ophiuroids and associated copepods. *Ir. Nat. J.* 14: 137–139.

Grime, J.P. 2001. Plant strategies, vegetative processes, and ecosystem properties. Chichester: John Wiley & Sons, Ltd.

Grober, M.S. 1988a. Responses of tropical reef fauna to brittle-star luminescence (Echinodermata: Ophiuroidea). *J. Exp. Mar. Biol. Ecol.* 115: 157–168.

Grober, M.S. 1988b. Brittle-star bioluminescence functions as an apposematic signal to deter crustacean predators. *Anim. Behav.* 36: 493–501.

Grober, M.S. 1989. Bioluminescent aposematism: a reply to Guilford & Cahill. *Anim. Behav.* 37: 341–343.

Guilford, T. & Cuthill, I. 1989. Aposematism and bioluminescence. *Anim. Behav.* 37: 339–341.

Hancock, D.A. 1955. The feeding behavior of starfish on Essex oyster grounds. *J. Mar. Biol. Ass. U.K.* 34: 313–331.

Hancock, D.A. 1974. Some aspects of the biology of the sun-star *Crossaster papposus* (L.). *Ophelia.* 13: 1–30.

Harrold, C. &. Pearse, J.S. 1980. Allocation of pyloric caecum reserves in fed and starved sea stars, *Pisaster giganteus* (Stimpson): somatic maintenance comes before reproduction. *J. Exp. Mar. Biol. Ecol.* 48: 169–183.

Harrold, C. & Pearse, J.S. 1987. The ecological role of echinoderms in kelp forests. In: M. Jangoux, J.M. Lawrence (eds). *Echinoderm Studies.* 2: 137–233. Rotterdam: Balkema.

Hawkes, C.V. & Sullivan, J.J. 2001. The impact of herbivory on plants in different resource conditions: a met-analysis. *Ecology.* 82: 2045–2058.

Hendler, G. 1984. Brittlestsar color-change and phototaxis (Echinodermata: Ophiuroidea: Ophiocomidae). *P.S.Z.N.I.: Mar. Ecol.* 5: 379–410.

Herring, P.J. 1974. New observations on the bioluminescence of echinoderms. *J. Zool., London.* 172: 401–481.

Herring, P.J. 1995. Bioluminescent echinoderms: unity of function in diversity of expression? In R.H. Emson, A.B. Smith, A.C. Campbell (eds). *Echinoderm Research 1995*: 9–17. Rotterdam: Balkema.

Holland, N.D. 1991. Echinodermata: Crinoidea. In A.C. Giese, J.S. Pearse, V.B. Pearse (eds.) *Reproduction of Marine Invertebrates*: 6: 247–299. Pacific Grove: Boxwood Press.

Hopkins, T.S., Watts, S.A., McClintock, J.B. &. Marion, R.K. 1994. Contrasting size demographics, sub-lethal arm loss and arm regeneration in two populations of *Astropecten articulates* (Say) in the northern Gulf of Mexico. In B. David, A. Guille, J.P. Féral, M. Roux (eds). *Echinoderms through Time.* 311–316/. Rotterdam: Balkema.

Hosttettmann, K., &. Marston, A. 1995. Saponins. Cambridge: Cambridge University Press.

Hotchkiss, F.H.C. Churchill, S.E., Gelormini, R.G., Hepp, W.R., Rentler, R.J. & Tummarello, M.T. 1990. Events of autotomy in the starfish *Asterias forbesi* and *A. vulgaris*. In T. Yanigisawa, I. Yasumasu, C. Oguro, N. Suzki, T. Motokawa (eds). *Biology of Echinodermata*: 537–541. Rotterdam: Balkema.

Hulbert, A.W. 1980. The ecological role of *Asterias vulgaris* in three subtidal communities. In: M. Jangoux (ed.). *Echinoderms: Present and Past*: 191–196. Rotterdam: A.A. Balkema.

Hyman, L.H. 1955. *The Invertebrates: Echinodermata*. McGraw-Hill Book Company, Inc., New York.

Iason, G. 2005. The role of plant secondary metabolites in mammalian herbivory: ecological perspectives. *Proc. Nutritional Soc.* 64: 123–131.

Iorizzi, M., De Marino, S. & Zollo, F. 2001. Steroidal oligo-glycoside from the Asteroidea. *Curr. Org. Chem.* 5: 951–973.

Ivanchina, N.V.. Kicha, A.A., Kalinovsk, A.I., Dmitrenok, P.S., Dmitrenok, A.S., Chaikina, E.L., Stonik, V.A., Gavagnin, M. & Cimino, G. 2006. Polar steroidal compounds from the far eastern starfish *Henricia leviuscula*. *J. Nat. Prod.* 69: 224–228.

Johnson, M.L., Gaten, E. & Shelton, P.M.J. 2002. Spectral sensitivities of five marine decapods crustaceans and a review of spectral sensitivity variation in relation to habitat. *J. Mar. Biol. Ass. U.K.* 82: 835–842.

Keesing, J.K. 1995. Temporal patterns in the feeding and emergence haviour of the crown-of-thorns starfish *Acanthaster planci. Mar. Freshwat. Behav. Physiol.* 25: 209–232.

King, E. 1898. Regeneration in Aserias vulgaris. Arch. Entwicklungsmech. Organ. 7: 351–363.

Lawrence, J.M. 1991a. A chemical alarm response in *Pycnopodia helianthoides* (Echinodermata: Astseroidea). *Mar. Behav. Physiol.* 19: 39–44.

Lawrence, J.M. 1991b. Analysis of characteristics of echinoderms associated with stress and disturbance. In T. Yanagisawa, I. Yasumasu, C. Oguro, N. Suzuki, T. Motokawa (eds). *Biology of Echinodermata*: 11–26. Rotterdam: A.A. Balkema.

Lawrence, J.M. 1992. Arm loss and regeneration in Aster-oidea (Echinodermata). In: L. Scalera-Liaci and C. Canicatti (eds). *Echinoderm Research 1991*: 39–52. A.A. Balkema, Rotterdam.

Lawrence, J.M. & Ellwood, A. 1991. Simultaneous allocation of resources to arm regeneration and to somatic and gonadal production in *Luidia clathrata* (Say) (Echinodermata: Asteroidea). In T. Yanagisawa, I. Yasumasu, C. Oguro, N. Suzuki, T. Motokawa (eds) *Biology of Echinodermata*: 543–548. Rotterdam: A.A. Balkema.

Lawrence, J.M. & Larrain, A. 1994. The cost of arm autotomy in the starfish *Stichaster striatus. Mar. Ecol. Prog. Ser.* 109: 311–313.

Lawrence, J.M. & Pomory, C.M. (2008) Position of arm loss and rate of arm regeneration by *Luidia clathrata* (Echinodermata: Asteroidea). *Cah. Biol. Mar.* 49: 369–373.

Lawrence, J.M. & Vasquez, J. 1996. The effect of sublethal predation on the biology of echinoderms. *Oceanol. Acta.* 19: 431–440.

Lawrence, J.M., Klinger, T.S., McClintock, J.B., Watts, S.A., Chen, C.-P., Marsh, A. & Smith, L. 1986. Allocation of nutrient resources to body components by regenerating *Luidia clathrata* (Say) (Echinodermata: Asteroidea). *J. Exp. Mar. Biol. Ecol.* 102: 47–53.

Lawrence, J.M., Byrne, M., Harris, L., Keegan, B., Freeman, S. & Cowell, B.C. 1999. Sublethal arm loss in *Asterias amurensis, A. rubens, A. vulgaris* and *A forbesi* (Echinodermata: Asteroidea). *Vie Milieu*, 49: 69–73.

Lawrence, J.M., Cobb, J.C., Talbot-Oliver, T. & Plank, L.R. (in press) In: C.R. Johnson (ed). *Echinoderms in a changing world*. Rotterdam. A.A. Balkema.

Levin, A.V., Levina, E.V. & Levin, V.S. 1984. The relation of asteroids *Asterias amurensis* and *Distolasterias nipon* homogenates and chemical substances from far eastern starfishes. *Biol. Morya*. No. 5: 40–45. (in Russian).

Loo, L.O., Jonson, P.R., Sköld M. & Karlsson, O. 1996. Passive suspension feeding in *Amphiura filiformis* (Echinodermata: Ophiuroidea): feeding behavior in flume flow and potential feeding rate of field populations. *Mar. Ecol. Prog. Ser.* 139: 143–155.

Maginnis, T.L. 2006. The costs of autotomy and regeneration in animals: a review and framework for future research. *Behav. Ecol.* 17: 857–872.

Magnus, D.B.E. 1967. Ecological and ethological studies and experiments on the echinoderms of the Red Sea. *Stud. Trop. Oceanogr.* 5: 635–664.

Makra, A. & Keegan, B.F. 1999. Arm regeneration in *Acrocnida brachiata* (Ophiuroidea) at Little Killary, west coast of Ireland. *Biol. Environ.* 99B: 95–102.

Mallefet, J., Queby, F. & O'Hara, T. 2008. Distribution of luminescence in Ophiuroidea (Echinodermata). *Luminescence*. 23: 84.

Mallefet, J. & O'Hara, T.O. 2009. Ophiuroidea luminescence: diversity and development, a first analysis. Handbook. 13th Internat. Echionoderm Conference. p. 45.

Mann, H. 1936. Regeneratonsversuche an Seesternen und Schlangensternen. *Ber. d. Oberhess. f. Nat. u. Heilkunde N.F., Naturw. Abtg.* 17: 1–12.

Marrs, J., Wilkie, I.C., Sköld, M. Maclaren, W.J. & McKenzie, J.D. 2000. Size-related aspects of arm damage, tissue mechanics, and autotomy in the starfish *Asterias rubens. Mar. Biol.* 137: 59–70.

Mauzey, K.P., Birkeland C. & Dayton, P.K. 1968. Feeding behavior of asteroids and escape responses of their prey in the Puget Sound region. *Ecology*. 49: 603–619.

Mayo, P. & Mackie, A.M. 1976. Studies of avoidance reactions in several species of Britisih seastars (Echinodermata: Asteroidea). *Mar. Biol.* 38: 41–49.

Mazzone, F., Byrne, M. & Thorndyke, M.C. 2003. Arm autotomy and regeneration in the seastar *Coscinasterias muricata*. In J.-P. Féral, B. David (eds). *Echinoderm Research 2001*: 209–213. Lisse: Swets & Zeitlinger.

McCallum, H.I., Endean, R. & Cameron, A.M. 1989. Sublethal damage to *Acanthaster planci* as an index of predation pressure. *Mar. Ecol. Prog. Ser.* 56: 29–36.

McClintock, J.B. 1989. Toxicity of shallow-water Antarctic echinoderms. *Polar Biol.* 9:461–465.

McClintock, J.B. & Baker, B.J. 1997. A review of the chemical ecology of Antarctic marine invertebrates. *Amer. Zool.* 37: 329–342.

McClintock, J.B. & Lawrence, J.M. 1982. Photoresponse and associative learning in *Luidia clathrata* (Say) (Echinodermata: Asteroidea). *Mar. Behav. Physiol.* 9: 13–21.

McClintock, J.B., Baker, B.J., Baumiller, T.K. & Messing, C.G. 1999. Lack of chemical defense in two species of stalked crinoids: support for the predation hypothesis for Mesozoic bathymetric restriction. *J. Exp. Mar. Biol. Ecol.* 232: 1–7.

McClintock, J.B., Angus, P.A., Ho, C, Amsler, C.D. & Baker, B.J. 2008. A laboratory study of behavioral interactions of the Antarctic keystone sea star *Odontaster validus* with three sympatric predatory sea stars. *Mar. Biol.* 154: 10077–1084.

Menge, B.A. 1979. Coexistence between the seastars *Asterias vulgaris* and *A. forbesi* in a heterogeneous environment: a non-equlibrium explanation. *Oecologia.* 41: 245–272.

Meyer, D.L. 1973. Feeding behavior and ecology of shallow-water unstalked crinoids (Echinodermata) in the Caribbean Sea. *Mar. Biol.* 22: 105–129.

Meyer, D.L. 1985. Evolutionary implications of predation on Recent comatulid crinoids from the Great Barrier Reef. *Paleobiology.* 11: 154–164.

Meyer, D.L. &. Macurda, B.D., Jr. 1977. Adaptive radiation of the comatulid crinoids. *Paleobiology.* 3: 74–82.

Meyer, D.L. & Macurda, Jr., B.D. 1980. Ecology and distribution of the shallow-water crinoids of Palau and Guam. *Micronesica.* 16: 59–99.

Meyer, D.L., LaHaye, C.A., Holland, N.D., Arneson, A.C. & Strickler, J.R. 1984. Time-lapse cinematography of feather stars (Echiondermata: Crinoidea) on the Great Barrier Reef, Australia: demonstrations of posture changes, locomotion, spawning and possible predation by fish. *Mar. Biol.* 78: 179–184.

Mladenov, P.V. 1983. Rate of arm regeneration and potential causes of arm loss in the feather star *Florometra serratissima* (Echinodermata: Crinoidea). *Can. J. Zool.* 61: 2873–2879.

Morin, J.G. 1983. Coastal bioluminescence: patterns and functions. *Bull. Mar. Sci.* 33: 87–817.

Morgan, R. & Jangoux, M. 2004. Assessing arm regeneration and its effect during the reproductive cycle in the gregarious brittle-star *Ophiothrix fragilis. Cah. Biol. Mar.* 45: 277–280.

Morgulis, S. 1909. Regeneration in the brittle-star *Ophiocoma pumila*, with reference to the influence of the nervous system. *Proc. Amer. Acad. Arts Sci.* 44: 655–659.

Munday, B.W. 1993. Field survey of the occurrence and significance of regeneration in *Amphiura chiajei* (Echinodermata: Ophiuroidea) from Killary Harbrour (sic), west coast of Ireland. *Mar. Biol.* 115: 661–668.

Muus, K. 1981. Density and growth of juvenile *Amphiura filiformis* (Ophiuroidea) in the Øresulnd. *Ophelia.* 20: 153–168.

Nichols, D. 1994. Reproductive seasonality in the comatulid crinoids *Antedon bifida* (Pennant) from the English Channel. *Phil. Trans. R. Soc. Lond. B.* 343: 113–1344.

Nickell, T.D. & Moore, P.G. 1991. The behavioural ecology of epibenthic scavenging invertebrates in the Clyde Sea area: field sampling using baited traps. *Cah. Biol. Mar.* 32: 353–370.

Nilsson, H.C. 1999. Effects of hypoxia and organic enrichment on growth of the brittle stars *Amphiura filiformis* (O.F. Müller) and *Amphiiura chiajei* Forbes. *J. Exp. Mar. Biol. Ecol.* 237: 11–30.

Oji, T. 2001. Fossil record of echinoderm regeneration with special regard to crinoids. *Microscopy Res. Tech.* 55: 397–402.

Oji, T. & Okamoto, T. 1994. Arm autotomy and arm branching pattern as anti-predatory adaptations in stalked and stalkless crinoids. *Paleobiology.* 20: 27–39.

Palagiano, E., Zollo, F., Minale, L., Iorizzi, M., Bryan, P., McClintock, J.B. & Hopkins, T. 1996. Isolation of 20 glycosides from the starfish *Henricia downeyae*, collected in the Gulf of Mexico. *J. Nat. Prod.* 59: 348–354.

Parker, C. 1881. Poisonous qualities of the starfish (*Solaster papposus*). *The Zoologist.* 5(53): 214.

Perseke, M., Fritsch, G., Ramsch, K., Bernt, M., Merkle, D., Middendorf, M., Bernhard, D., Stadler, P.F. & Schlegel, M. 2008. Evolution of mitochondrial gene orders in echinoderms. *Mol. Phylogen. Evol.* 47: 855–864.

Pomory, C.M. & Lares, M.T. 2000. Rate of regeneration of two arms in the field and its effect on body components in *Luidia clathrata* (Echinodermata: Asteroidea). *J. Exp. Mar. Biol. Ecol.* 254: 211–220.

Pomory, C.M. & Lawrence, J.M. 2001. Arm regeneration in the field in *Ophiocoma echinata* (Echinodermata: Ophiuroidea): effects on body composition and its potential role in a reef food web. *Mar. Biol.* 139: 661–670.

Ramsay, K., Kaiser, M.J., Moore, P.G. & Hughes, R.N. 1997. Consumption of fisheries discards by benthic scavengers: utilization of energy subsides in different marine habitats. *J. An. Ecol.* 66: 884–896.

Ramsay, K., Kaiser, M.J. & Richardson, C.A. 2001. Invest in arms: behavioural and energetic implications of multiple autotomy in starfish (*Asterias rubens*). *J. Mar. Biol. Ass. U.K.* 50: 360–365.

Ribi, G. & Jost, P. 1975. Feeding rate and duration of daily activity of *Astropecten aranciacus* (Echinodermata: Asteroidea) in relation to prey density. *Mar. Biol.* 45: 249–254.

Rideout, J.A., Smith, N.B. & Sutherland, M.D. 1979. Chemical defense of crinoids by polyketide sulphates. *Experientia.* 35: 1273–1274.

Rideout, J.A. & Sutherland, M.D. 1985. Pigments of marine animals. XV Bianthrones and related polyketides from *Lamprometra palma gyges* and other species of crinoids. *Aust. J. Chem.* 38: 793–808.

Rio, G.J., Stempien, M.F., Jr., Nigrelli, R.F. & Ruggieri, G.D. 1965. Echinoderm toxins—I. Some biochemical and physiological properties of toxins from several species of Asteroidea. *Toxicon.* 3: 147–15.

Rosenberg, R. & Lundberg, L. 2004. Photoactivity patterns in the brittlestar *Amphipholis filiformis. Mar. Biol.* 145: 651–656.

Rosenberg, R. & Selander, E. 2000. Alarm signal response in the brittle star *Amphiura filiformis. Mar. Biol.* 136: 43–48.

Salzwedel, H. 1974. Arm-Regeneration bein *Amphiura filiformis* (Ophiuroidea). *Veröff. Inst. Meeresforsch. Bremerh.* 14: 161–167.

Schapiro, J. 1914. Über die Regenerationserscheinungen verschiedener Seesternarten. *Arch f. Entwicklungsmech. d. Org.* 38: 210–251.

Scheibling, R.E. & Metaxas, A .2008. Abundance, spatial distribution, and size structure of the sea star *Protoraster nodosus* in Palau, with notes on feeding and reproduction. *Bull. Mar. Sci.* 82: 221–235.

Schneider, J.A. 1988. Frequency of arm regeneration of comatulid crinoids in relation to life habit. In R. Burke, P.V. Mladenov, P. Lambert & R.L. Parsley (eds). *Echinoderm Biology*: 531–538. Rotterdam:. A.A. Balkema.

Segner, H. & Braunbeck, T. 1998. Cellular response profile to chemical stress. In G. Schüürmann & B. Markert (eds). *Ecotoxicology*: 521–569. New York: Wiley-Interscience.

Shaw, G.D. & Fontaine, A.R. 1990. The locomotion of the comatulid *Florometra serratissima* (Echinodermata: Cri-noidea) and its adaptive significance. *Can. J. Zool.* 68, 942–950.

Sides, E. 1985. Niche separation in three species of *Ophiocoma* (Echiondermata: Ophiuroidea) in Jamaica, West Indies. *Bull. Mar. Sci.* 36: 701–715.

Siebeck, U.E. & Marshall, N.J. 2001. Ocular media transmission of coral reef fish – can coral reef fish see ultraviolet light? *Vision Res.* 41: 133–149.

Siebeck, U.E., Wallis, G.M. & Literland, L. 2008. Colour vision in coral reef fish. *J. Exp. Biol.* 211: 354–360.

Singh, H., Moore, R.E. & Scheuer, P.J. 1967. The distribution of quinone pigments in echinoderms. *Experentia.* 23: 624–626.

Sköld, M. 1998. Escape responses in four epibenthic brittlestars. *Ophelia.* 49: 163–179.

Sköld, M. & Rosenberg, R. 1996. Arm regeneration frequency in eight species of Ophiuroidea (Echinodermata) from European sea areas. *J. Sea Res.* 35: 353–362.

Sloan, N.A & Aldridge, T.H. 1981. Observations on an aggregation of the starfish *Asterias rubens* L. in Morecambe Bay, Lancashire, England. *J. Nat. Hist.* 15: 407–418.

Sloan, N.A. & Northway, S.M. 1982. Chemoreception by the asteroid *Crossaster papposus* (L.). *J. Exp. Mar. Biol. Ecol.* 61: 85–98.

Solan, M. & Battle, E.J.V. 2003. Does the ophiuroid *Amphiura filiformis* alert conspecifics to the danger of predation through the generation of an alarm signal? *J. Mar. Biol. Ass. U.K.* 83: 1117–1118.

Solan, M. & Kennedy, R. 2002. Observation and quantification of in situ animal-sediment relations using time-lapse sediment profile imagery (t-SPI). *Mar. Ecol. Prog. Ser.* 228: 179–191.

Soliman, F.E.-S., Nojima, S. & Kikuchi, T. 1986. Daily activity patterns and their seasonal change in the sea star, *Asterina minor* (Hayashi) (Asteroidea: Asterinida). *Pub. Amakusa Mar. Biol. Lab.* 8: 143–171.

Soong, K., Shen, Y., Tseng, S.-H. & Chen, C.P. 1997. Regeneration and potential functional differentiation of arms in the brittlestar *Ophiocoma scolopendrina* (Lamarck) (Echinodermata: Ophiuroidea). *Zool. Stud.* 36: 90–97.

Stancyk, S.E., Golde, H.M., Pape-Lindstrom, P.A. & Dobson, W.E. 1994. Born to lose. I. Measures of tissue loss and regeneration by the brittlestar *Microphiopholis gracillima* (Echinodermata: Ophiuroidea). *Mar. Biol.* 118: 451–462.

Stewart, B. 1996. Sub-lethal predation and rate of regeneration in the euryalinid snake star *Astrobrachion constrictum* (Echinodermata: Ophiuroidea) in a New Zealand fiord. *J. Exp. Mar. Biol. Ecol.* 199: 269–283.

Stonik, V.A. & Elyakov, G.B. 1988. Secondary metabolites from echinoderms as chemotaxonomic markers. In P.J. Scheuer (ed). *Bioorganic Marine Chemistry*: Vol. 2. 43–86. Berlin: Springer-Verlag.

Swenson, D.P & McClintock, J.B. 1998. A quantitative assessment of chemically-mediated rheotaxis in the asteroid *Coscinasterias tenuispina*. *Mar. Fresh. Behav. Physiol.* 31: 63–80.

Thomassin, B.A. 1976. Feeding behavior of the felt-, sponge-, and coral feeding sea stars, mainly *Culcita schmideliana*. *Helgoländer wiss. Meeresunters.* 20: 51–65.

Tokeshi, M. 2002. Spatial distribution of a deep-sea crinoids *Pentametrocrinus tuberculatus* in the Izu-Ogasawara Arc, western Pacific. *J. Zool. Lond.* 258: 291–298.

Turner, R.L. 1974. Post-metamorphic growth of the arms in *Ophiophragmus filograneus* (Echinodermata: Ophiuroidea) from Tampa Bay Florida (USA). *Mar. Biol.* 24: 273–277.

Vail, L.1987. Diel patterns of emergence of crinoids (Echinodermata) from within a reef at Lizard Island, Great Barrier Reef, Australia. *Mar. Biol.* 93, 551–560.

Vail, L. 1989. Arm growth and regeneration in *Oligometra serripinna* (Carpenter) (Echinodermata: Crinoidea) at Lizard Island, Great Barrier Reef. *J. Exp. Mar. Biol. Ecol.* 130: 189–204.

Van Veldhuuizen, H.D. & Oakes, V.J. 1981. Behavioral responses of seven species of asteroids to the asteroid predator, *Solaster dawsoni*. *Oecologia.* 48; 214–220.

Viviani, C.A. 1978. Predación interespecífica, canibalismo y autotomía como mecanismo de escape en las especies de Asteroídea (Echinodermata) en el litoral del desierto del Norte Grande de Chile. Laboratorio de Ecología Marina, Universidad del Norte, Iquique.

Weise, K. 2002. Experimental systems in neurobiology. Springer: Berlin.

Wilkie, I.C. 1978. Arm autotomy in brittlestars (Echinodermata: Ophiuroidea). *J. Zool., Lond.* 186: 311–330.

Wilkie, I.C. 2001. Autotomy as a prelude to regeneration in echinoderms. *Microsc. Res. Tech.* 55: 369–396.

Witman, J.D., Genovese, S.J., Bruno, J.F., McLaughlin, J.W. & Pavin, B.I. 2003. Massive prey recruitment and the control of rocky subtidal communities on large spatial scales. *Ecol. Monogr.* 73: 441–462.

Wolkenstein, K., Gross, J.H., Falk, H. & Schöler, H.F. 2006. Preservation of hypericin and related polycyclic quinone pigments in fossil crinoids. *Proc. R. Soc. B.* 273: 451–456.

Wolkenstein, K., Gluchowski, E., Gross, J.H. & Marynowski, L. 2008. Hypericinoid pigments in millericrinids from the Lower Kimmeridgian of the Holy Cross Mountains (Poland). *Palaios.* 23: 773–777.

Woodley, J.D., Chornesky, Clifford, P.A., Jackson, J.B.C. et al. 1981. Hurricane Allen's impact on Jamaican coral reefs. *Science.* 214: 749–755.

Yee, A., Burkhardt, J. & Gilly, W.F. 1987. Mobilization of a coordinated escape response by giant axons in the ophiuroid *Ophiopterus papillosa*. *J. Exp. Biol.* 128: 2878–305.

Zeleny, C. 1905. Compensatory regulation. *J. Exp. Zool.* 2: 1–102.

Zirpolo, G. 1921. Ricerche sulla rigenerazione delle braccia di *Asterina fibbosa* Penn. *Pub. Staz. Zool. Napoli.* 3: 93–163.

Zirpolo, G. 1926. Sulla rigenerazione delle braccia di *Luidia ciliaris* Phil. *Boll. Soc. Nat. Napoli.* 37: 241–243.

Fossil echinoderms and palaeobiology

Echinoderms in a Changing World – Johnson (ed)
© 2013 Taylor & Francis Group, London, ISBN 978-1-138-00010-0

Taxonomy and palaeoecology of the genus *Linthia* (Echinoidea: Spatangoida) from Japan

K. Nemoto
Graduate School of Biological Science, Kanagawa University, Kanagawa Prefecture, Japan

K. Kanazawa
Department of Biological Sciences, Faculty of Science, Kanagawa University, Kanagawa Prefecture, Japan

ABSTRACT: All five species of the genus *Linthia* previously reported from Japan were examined. Biometrical and statistical analyses indicate that *L. nipponica* and *L. tokunagai* are the same species, a result supported by palaeoecological studies. From the Oligocene to early Pleistocene, Japanese *Lintha* species appear to have lived in similar habitats, i.e. in fine-grained sand in shallow neritic environments in the cold temperate zone, until the youngest species of the genus, *L. nipponica*, became extinct. They might have exclusively occupied an ecological niche because they are usually found predominantly in strata accompanied by no other echinoid species.

1 INTRODUCTION

The genus *Linthia* belonging to the family Schizasteridae of the order Spatangoida is an exceptional group with an extremely long history, having originated in late Cretaceous (Fischer 1966) and survived until the early Pleistocene. In the Palaeogene heart urchins of the genus *Linthia* were distributed worldwide (Kier & Lawson 1978) and appear to have given rise to some advanced schizasterids such as *Paraster* (McNamara & Philip 1980). In the Neogene, however, their distribution became restricted within the West Pacific, resulting in their extinction around Japan in the Pleistocene.

From the Tertiary and the Quaternary of Japan five *Linthia* species have been reported so far (Nisiyama 1968) and they were apparently a predominant species of the Japanese echinoid fauna during these times. In this study we examine the taxonomic relationships among the five *Linthia* species and the habitats where the primitive schizasterids thrived and disappeared.

2 *LINTHIA* FROM JAPAN

All five species previously reported from Japan are examined in this study (Figs. 1 and 2). Many specimens were collected during our field investigation and those deposited in Museums and Kanazawa University were also examined. Museum abbreviations used herein are: HUM, Hokkaido University Museum; TUM, Tohoku University Museum;
FMM, Fossa Magna Museum; TMNH, Togakushi Museum of Natural History; GM, Geological Museum; UMUT, University Museum, the University of Tokyo.

Linthia nipponica Yoshiwara, 1899 (Figs. 1.1–1.4)
Material. 39 specimens: TUM No. 5188, 8112, and 78350, UMUT CE18683-90, one specimen from TMNH, eight specimens from Kanazawa University, and 12 specimens collected in this study. All these specimens were used for biometric analysis.
Description (see also Yoshiwara 1899, Nisiyama 1968). Tests range from 33 mm to 74 mm in length and from 16 mm to 35 mm in height. Test widths are almost as large as test lengths. The test is rounded-cordiform and gently arched aborally. Posterior extremity of the test is truncated vertically. The widest portion is located approximately at the middle of test length and the tallest point is subcenral, immediately behind the apical disc. The frontal notch is distinct but shallow and about 17 mm wide and 1.5 mm deep for a specimen with 62 mm test length (Figs. 1.1–1.4.).

The petals are almost straight and somewhat sunken. In the specimen with 62 mm test length, the petal length is about 30 mm in the anterior and about 24 mm in the posterior petals, and the petal width at the middle of the petal length is about 6 mm in each petal. There are 36 pore pairs in the anterior and 27 in the posterior petal. The frontal ambulacrum is sunken and about 40 mm from adapical end to ambitus, and its width is about 12 mm at the middle point.

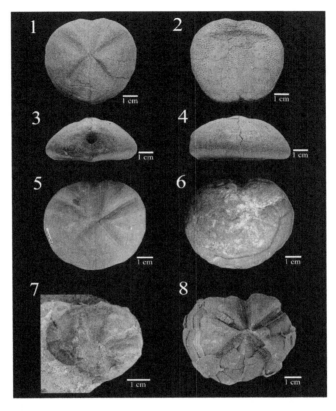

Figure 1. Five *Linthia* species from Japan. 1-4: *L. nipponica* from the Omma Fm. (1; aboral view, 2; adoral view, 3; posterior view, 4; right side view), 5: "*L. tokunagai*" from the Ogawa Fm. (TUM No. 23585 aboral view), 6: *L. praenipponica* from the Yamaga Fm. (TUM No. 35958 aboral view), 7: *L. yessoensis* probably from Hiragishi Fm. (HUM UHR16820 aboral view), 8: *L. boreasteria* from the Nissakutan Fm. (UHR09169 aboral view).

The peripetalous fasciole runs around the petals. The latero-anal fasciole branches from the peripetalous fasciole close behind the anterior petals and runs posteriorly slightly above the ambitus and below the periproct.

The peristome is depressed and the labrum is slightly projected downward. The anterior margin of the labrum is well arched anteriorly. The periproct is oval or almost circular in outline. It is located at upside of posterior extremity and visible in posterior view.

Occurrence. This species is the most common sea urchin fossil from the Pliocene to the early Pleistocene found around the central part of Japan. In the Nagano region the fossil record is also from the late Miocene.

Remarks. The holotype was not specified. This species seems to be the last species of the genus *Linthia*.

Linthia tokunagai Lambert, 1925 (Fig. 1.5)
Material. Two specimens: one specimen identified as this species probably by Nisiyama, deposited in TUM (No. 23585), and another one of TMNH.

Description. See Nisiyama (1968, p. 234–237).

Occurrence. The specimen deposited in the TUM was collected from the Ogawa Fm. (late Miocene) according to Nisiyama (1968), and another one deposited in TMNH from the Shigarami Fm. (middle Pliocene).

Remarks. The holotype was not specified and at present only two specimens definitely referred to this species are known. Lambert & Thiéry (1925, p. 520) originally described this species, based not on actual material but on the pictures shown in Tokunaga (1903, pl. 1, Figs. 6 and 7, and pl. 3, Fig. 1); they chose three specimens as *L. tokunagai* from the specimens described as *L. nipponica* by Tokunaga (1903). They mentioned that "*L. tokunagai*" differs from *L. nipponica* in terms of its shallow notch, shallow petals with curved and large pores, wide interporiferous zone, and short posterior petals. Thereafter, Nisiyama (1968) additionally suggested some criteria, which are discussed

Figure 2. Localities of *Linthia* examined in this study. 1, 2; *L. nipponica* and *L. tokunagai*; 3; *L. nipponica*, 4; L. *yessoensis*; 5; *L. praenipponica* (type specimen), 6; L. *boreasteria* (type specimen).

in section 3.1, below. With these taxonomic characters, however, it is impossible to discriminate between these species, as discussed below.

Linthia praenipponica Nagao, 1928 (Fig. 1.6)
Material. 14 specimens: seven specimens identified as this species by Nagao (TUM No. 35958, including the holotype), and seven specimens from GM (GSJF16627-1, GSJF16627-2, GSJF16628-1, GSJF16628-2, GSJF16629-1, GSJF16629-2, and GSJF16629-3).
Description. See Nagao (1928, p. 18–19).
Occurrence. This species was reported from the Kishima Fm. (late Oligocene) and Yamaga Fm. (middle Oligocene) of Kyushu (southwest Japan).
Remarks. Nagao (1928) mentioned that *L. praenipponica* closely resembles *L. nipponica*, but *L. praenipponica* can be distinguished from *L. nipponica* by its broader test, narrower posterior portion, sharper indentation of the anterior end, deeper notch, and more sunken petals. Nisiyama (1968) suggested that *L. praenipponica* can be distinguished from *L. tokunagai* by its deep notch, deep petals, and narrow posterior portion.

Linthia yessoensis Minato, 1950 (Fig. 1.7)
Material. Two specimens: HUM UHR16820 (referred to this species by Minato 1950) and UHR16821.
Description. See Minato (1950, p. 158).

Occurrence. Minato (1950) noted that the specimen he described was collected from the Shimizusawa region in the middle part of Hokkaido, where the Hiragishi Fm. (middle Eocene) is apparently distributed. Another specimen is known from the Shitakara Fm. (late Oligocene).
Remarks. The holotype was not specified. Except the two specimens examined here, no others referred to this species are known. Minato (1950) suggested that *L. yessoensis* can be differentiated from *L. praenipponica* because of its narrow peripetalous fasciole, which is slightly curved posteriorly in the interambulacrum (5), short and wide posterior petals, and low test. Nisiyama (1968) proposed other criteria to differentiate between these species; in *L. yessoensis*, large angles between anterior petals and between posterior petals, and very broad posterior petals.

Linthia boreasteria Nisiyama, 1968 (Fig. 1.8)
Material. Nine specimens probably belonging to this species (they are identified as *Linthia* sp., but were collected from the same formation where the holotype of this species was collected): HUM UHR09167 (three specimens), UHR09169, UHR09173, UHR09178 (two specimens), and UHR09183 (two specimens).
Description (see also Nisiyama, 1968). Test is somewhat elliptical rounded-cordiform. In UHR09169, test length is about 49 mm and test width is about 47 mm. The widest portion is at the middle of test length. The frontal notch is about 13 mm wide and 1.7 mm deep. The aboral and adoral surfaces of this specimen are not observable because these parts are partly broken or covered by matrix.
In UHR09178, the petals are almost straight and somewhat sunken; the anterior petals are about 29 mm long and the posterior ones are about 19 mm long, the petal width at the middle of the petal is about 7 mm in each petal. Because no intact anterior petal is preserved in any specimen, the number of the pore pairs was not determined, while that of the posterior petal is probably 20 in UHR09178. The frontal ambulacrum is sunken, but the precise form is uncertain because the anterior end is lost.
In the specimens examined, labrum, fascioles, and periproct are not observable, because they are interior molds and the oral sides and the posterior parts were not preserved. Nisiyama (1968) noted that the peristome is depressed and the labrum is projected slightly downward.
Occurrence. The specimen studied by Nisiyama (1968) and the specimens examined in this study were from the Nissakutan Fm. (Oligocene).
Remarks. The holotype specified by Nisiyama and deposited in Tohoku University was lost, according to Dr. Sato, a curator of the TUM (pers.

comm. 2008). Nisiyama (1968) suggested that *L. boreasteria* can be differentiated from *L. praenipponica* owing to its peripetalous fasciole curved anteriorly with a convex line in interambulacrum (5) and its large angle between anterior petals. He also noted that *L. boreasteria* is distinguishable from *L. yessoensis* in terms of its anteriorly curved peripetalous fasciole in interambulacrum (5) and its narrow posterior petals. However, these characters do not distinguish *L. boreasteria* from these species, as discussed below.

3 MORPHOLOGICAL COMPARISON IN *LINTHIA* SPECIES

Although Nisiyama (1968) suggested morphological features to distinguish these species from each other, the characters are based on one or a few specimens and the intraspecific variation was not taken into account. Here we biometrically analyze the taxonomic characters of *L. nipponica* and *L. tokunagai*, because the number of well-preserved specimens is adequate for the analysis. The morphologies of three other *Linthia* species are also examined.

3.1 *L. nipponica and L. tokunagai*

3.1.1 *Materials and methods*
Twenty-seven specimens of *L. nipponica*, two specimens of *L. tokunagai*, and 20 specimens probably belonging to *L. nipponica* are examined. The taxonomic characters analyzed are as follows (see also Fig. 3 and Table 1).

Petal length: distance between adapical and adradial ends of a petal along the perradial line of the petal, measured by using a string.

Petal width: width at the middle of a petal length, measured with a digital vernier caliper (CD-S15C). On some specimens that have lost their original petal shape by compression, the petal width was estimated for the reconstructed shape.

Angle between posterior petals: angle measured between perradial lines of posterior petals, measured with a digital microscope (VTX-600).

Width and depth of anterior notch: measured at the ambitus of a test with a digital microscope.

Each character was measured twice per specimen and the average of two measurements was used for analysis. For the length and width of a petal, the measurements of a pair, right and left, were averaged, if possible.

In addition to these characters, Nisiyama (1968) proposed another taxonomical character, depth of petals. But this character strongly depends on preservation and it is very difficult to estimate the original shape. For this reason, this character was excluded.

The measurements are assumed to be related to body size (test length), so that for petal width, notch width, and notch depth, the relationships to test length is described by allometric equations and shown on a log-log graph. For petal length, anterior petal length relative to posterior length is examined.

3.1.2 *Results*
3.1.2.1 Angle between posterior petals
Nisiyama (1968) suggested that the angle is about 65° for *L. nipponica* and about 80° for

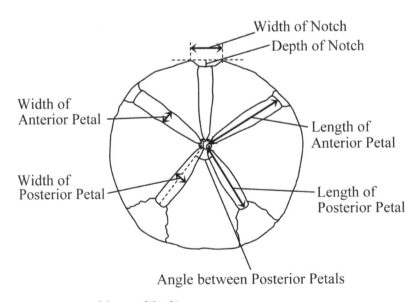

Figure 3. Measurement parts of the test of *Linthia*.

Table 1. Criteria to discriminate between *L. nipponica* and *L. tokunagai* suggested by Nisiyama (1968).

Taxonomic characters	*L. nipponica*	*L. tokunagai*
Po. petal/A. petal length	0.9 or more	0.7 or a little more
Width of petal	Narrower	Wider
Angle between Po. petals	About 65°	About 80°
Width of notch	Narrower	Wider
Depth of notch	Deeper	Shallower

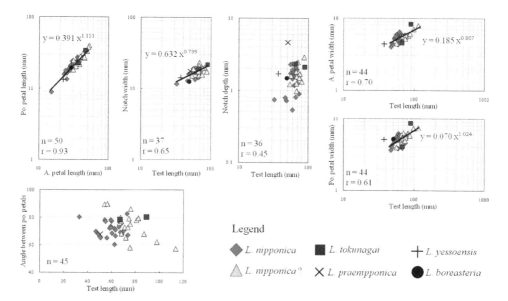

Figure 4. Allometric growth analysis of taxonomical characters that discriminate between *L. nipponica* and *L. tokunagai*, as suggested by Nisiyama (1968).

L. tokunagai. The measurements of *L. nipponica*, however, vary from 57° to 90° and the average is about 70°, while that of *L. tokunagai* is about 80° as described by Nisiyama (1968) (Fig. 4). It seems difficult to differentiate between these species using this character.

3.1.2.2 The ratio of posterior to anterior petal length
Lambert & Thiéry (1925) and Nisiyama (1968) noted that *L. tokunagai* has shorter posterior petals than *L. nipponica*. Figure 4, however, shows no such tendency. Nisiyama (1968) pointed out that the ratio of posterior to anterior petal length in *L. nipponica* is 0.9 or more, but no specimen showed such a ratio (Fig. 5).

3.1.2.3 Petal width
Nisiyama (1968) suggest that *L. tokunagai* has wider petals than *L. nipponica*. In two *L. tokunagai* examined, the measurement of the large specimen

is consistent with his suggestion, while that of the small one is not (Fig. 4).

3.1.2.4 Notch depth
Lambert & Thiéry (1925) and Nisiyama (1968) noted that *L. tokunagai* has a shallower notch than *L. nipponica*. In the specimens measured, however, there is no such tendency, and rather, *L. tokunagai* has a superficially deeper notch (Fig. 4).

3.1.2.5 Summary
These results demonstrate that the specimens of *L. nipponica* and *L. tokunagai* do not fall into two species by any character suggested by Nisiyama (1968).

3.2 *L. praenipponica, L. yessoensis, and L. boreasteria*

Although the specimens of *L. praenipponica*, *L. yessoensis*, and *L. boreasteria* were very poorly

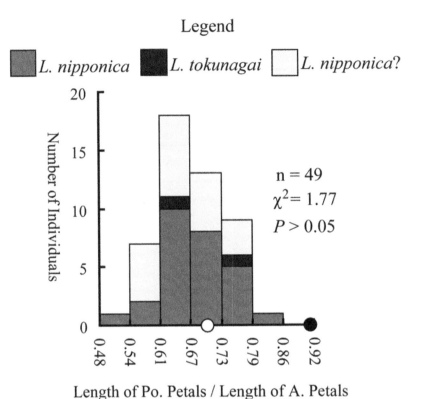

Figure 5. Frequency distribution histogram for the ratio of posterior to anterior petal length. This histogram shows a normal distribution and no peak corresponding to species differences. The values of *L. nipponica* and *L. tokunagai* suggested by Nisiyama (1968) are indicated as ● and O, respectively.

preserved and not suitable for biometric analysis, some characters are still measurable on the preserved parts of the tests, whereby taxonomic relationship can be discussed.

3.2.1 *L. praenipponica and L. nipponica*

Of the characters noted by Nagao (1928) to distinguish *L. praenipponica* from *L. nipponica*, the depth of notch appears to be the only one useful for this purpose. A specimen of *L. praenipponica* (Fig. 1.6) has a remarkably deep notch about 4.7 mm deep, and no specimens of *L. nipponica* have such a deep notch (Fig. 4). The other morphological features to characterize *L. praenipponica* are within the variation seen in *L. nipponica*.

3.2.2 *L. praenipponica and L. yessoensis*

Of the characters noted by Minato (1950) and Nisiyama (1968) to distinguish *L. yessoensis* from *L. praenipponica*, only three criteria, angle between posterior petals, ratio of posterior petal length to width, and width of posterior petals, are examinable because of poor preservation

of the material. The angle between posterior petals was about 62° for *L. yessoensis* and about 68° for *L. praenipponica*, but this result shows a tendency opposite to the criterion suggested by Nisiyama (1968). The length of the posterior petals of *L. yessoensis* was three times as long as its width in contrast to that of *L. praenipponica* (four times that of its width), as suggested by Minato (1950). The width of posterior petals was 5.1 mm in 37 mm long test for *L. yessoensis* and 5.2 mm in 50 mm long test for *L. praenipponica*, showing that *L. yessoensis* has significantly wider posterior petals than *L. praenipponica*, as noted by Nisiyama (1968).

3.2.3 *L. boreasteria*

L. boreasteria has a smaller angle between anterior petals (about 104°) than that of *L. praenipponica* (about 109°) and has wider posterior petals (about 7 mm in 70 mm wide test) than that of *L. yessoensis* (5.1 mm in 41 mm wide test). These results show a tendency opposite to the criteria suggested by Nisiyama (1968) and these criteria are useless for differentiating *L. boreasteria* from either *L. praenipponica* or *L. yessoensis*.

4 HABITATS OF *LINTHIA*

The habitats in which Japanese *Linthia* species lived are here deduced from our own field observations and previous palaeoenvironmental studies based on fossil mollusks and foraminifera and on sedimentary facies analysis.

4.1 *Field observations*

The following three formations were investigated: the Omma Fm. (Early Pleistocene, *L. nipponica* occurs), the Zukawa Fm. (Late Pliocene—Early Pleistocene, *L. nipponica* occurs), and the Kawazume Fm. (Late Miocene, *L. nipponica* occurs).

In these formations, fossil *Linthia* were usually found patchily concentrated in a fine-grained sand bed about 20–40 cm thick, which was presumably deposited by storm deposition. Many of the fossil specimens retain the whole test, while the tests do not show any particular orientation in a bed and their spines are lost in most cases. The fossil echinoids are often accompanied by fossils of extant molluscan species which live in fine-grained sand bottoms. This mode of occurrence indicates that the echinoid remains were buried in their habitats at depths shallower than storm wave base (Banno 2008).

4.2 *Paleoenvironment*

The palaeoenvironmental studies made on the formations where Japanese *Linthia* occur are summarized in Table 2.

4.2.1 *L. nipponica including L. tokunagai*

In the early Pleistocene Omma Fm., *L. nipponica* occurs from strata, for which depositional environment was inferred to be at depths about 10–20 m in a large bay in the cold temperate zone (ostracod analysis by Ozawa & Kamiya 2001).

In the late Pliocene to early Pleistocene Zukawa Fm., the habitat of *L. nipponica* was deduced to be in a shallow neritic zone less than 60 m deep (molluscan analysis by Matsuura 1985), and in the cold temperate zone (foraminifera analysis by Hasegawa 1979).

The early to middle Pliocene Shigarami Fm. is thought to have been deposited in a deep neritic or subneritic zone (molluscan analysis by Yano 1981 and Nagamori 1998, facies analysis by Yoshikawa 1996). However, the stratigraphic horizon where *L. nipponica* occurs is uncertain, so that the exact palaeoenviromental condition cannot be determined.

In the late Miocene Ogawa Fm., the depositional environment in which *L. nipponica* lived was inferred to be a shallow neritic zone at depths of 10 to 30 m in the cold temperate zone (molluscan analysis by Amano & Koike 1993).

4.2.2 *L. yessoensis*

The late Oligocene Shitakara Fm. is thought to have been deposited in a neritic zone (sedimentary facies and molluscan analysis by Minato 1954) and *L. yessoensis* would have lived in very fine- to fine-grained sand, as inferred from the matrix attached to the specimens examined.

4.2.3 *L. praenipponica*

The middle Oligocene Yamaga Fm. is thought to be formed in a shallow neritic zone (sedimentary facies analysis by Hayasaka 1991), and the late Oligocene Kishima Fm. is deduced to have been

Table 2. Habitats of Japanese *Linthia* species.

| Species | Age | Paleoenvironment | | | Formation examined |
		Water depth	Bottom sediment	Climate	
L. nipponica	e. Pleisto.	10 to 20 m (inner bay)	Fine-grained sand	Cold temp.	Omma Fm.
L. nipponica	l. Plio.—e. Pleisto.	Less than 60 m	Fine-grained sand	Cold temp.	Zukawa Fm.
L. nipponica (*L. tokunagai*)	e.—m. Plio.	Deep neritic (outer continental shelf) and maybe subneritic (continental sloop)	Sand and maybe mud	–	Shigarami Fm.
L. nipponica (*L. tokunagai*)	l. Mio.	10 to 30 m	Fine-grained sand	Cold temp.	Ogawa Fm.
L. yessoensis	l. Oligo.	Neritic	Very fine to fine-grained sand	–	Shitakara Fm.
L. praenipponica	l. Oligo.	Shallow neritic	–	–	Kishima Fm.
L. praenipponica	m. Oligo.	Deeper than storm wave base (about 30 m)	Very fine to fine grained sand	–	Yamaga Fm.

deposited in a neritic zone deeper than storm wave base (molluscan analysis by Inoue 1972). *L. praenipponica* seemingly inhabited very fine to fine-grained sand, as inferred from the matrix attached to the specimens examined.

5 DISCUSSION

5.1 *Taxonomy of Japanese Linthias*

The biometrical and statistical analyses show that there are no reasonable grounds to differentiate *L. tokunagai* from *L. nipponica* and rather indicate that they are the same species. In the Ogawa Fm., they were found from the same strata presumably deposited in the same environment, which also suggests that they are the same species. *L. tokunagai*, therefore, should be abandoned and included in *L. nipponica*, which was described earlier.

L. praenipponica and *L. yessoensis* are morphologically differentiated from each other and also from *L. nipponica*, and they should be treated as different species.

The criteria suggested by Nisiyama (1968) are useless to distinguish *L. boreasteria* from either *L. praenipponica* or *L. yessoensis*. In considering the stratigraphic and geographic distribution of *L. boreasteria* and *L. yessoensis*, they might be the same species.

5.2 *Ecological characteristics of Japanese Linthias*

From the Oligocene to early Pleistocene, Japanese *Linthia* species appear to have lived in similar habitats, i.e. in fine-grained sand in shallow neritic environments of cold temperate zones, until the youngest species of the genus, *L. nipponica*, became extinct. These all had similar test morphology to each other, indicating their similar modes of life.

In the strata where Japanese *Linthia* species occur, they are the predominant echinoid species and no other spatangoid urchin is usually found. They appear to have exclusively occupied these environments, and lack of competition might help explain how *Linthia* thrived for such a long time around Japan.

Early *Linthia* species from the Cretaceous to the Eocene in France are usually found in calcareous sand deposited in shallow waters. In the Palaeocene of Western Australia, *Linthia* occurs in a coarse calcarenite facies (McNamara & Philip 1980). In contrast, in the Eocene of Egypt (Carter & Hamza 1994) and in the upper Eocene of North America (Carter 1987), *Linthia* species are found in muddy sediment deposited in deep waters. These are the last records of *Linthia* in these regions.

Linthia's sub-globular form with gently arched aboral surface indicates that they may have been sand-dwellers rather than mud-dwellers (McNamara & Philip 1980, Kanazawa 1992). In the Tertiary of Japan *Linthia* were able to persist in their habitats in sandy bottoms in shallow waters, and this exceptional situation seems to provide a key to understanding the long history of Japanese *Linthia*. It is, however, still uncertain why such a situation could have existed around Japan. The last *Linthia* species, *L. nipponica*, became extinct when the late Pleistocene climate change began with the glacial eustatic sea-level fluctuation.

ACKNOWLEDGEMENTS

We are grateful to Dr Takuma Banno (Kanazawa Univ.) and Dr Ko Takenouchi (FMM) for their kind guidance during field investigations. They also provided us with many specimens examined in this study. Museum curators of HUM, TUM, TMNH, and UMUT kindly helped us access and observe the specimens deposited there. Two anonymous reviewers provided many helpful comments which improved the manuscript. We thank these persons heartily for their cooperation. This work was supported by Grant-in-Aid for Scientific Research (C) of JSPS (20540459).

REFERENCES

Amano, K. & Koike, K. 1993. Molluscan fauna from the Miocene Ogawa Formation in the western part of Nagano city, Nagano Prefecture. *Bulletin of Joetsu University of Education* 13: 287–297, pls. 1–2.

Banno, T. 2008. Ecological and taphonomic significance of spatangoid spines: relationship between mode of occurrence and water temperature. *Paleontological Research* 12: 145–157.

Carter, B. 1987. Paleogene echinoid distributions in the Atlantic and Gulf coastal plains. *Palaios* 2: 390–404.

Carter, B. & Hamza, F. 1994. Substrate preferences and biofacies distributions of Egyptian Eocene echinoids. *Palaios* 9: 237–253.

Fischer, A.G. 1966. Spatangoids. In Moore, R.C. (ed.). *Treatise on invertebrate paleontology. Part U. Echinodermata 3(2)*. Kansas: Geological Society of America and University of Kansas Press.

Hasegawa, S. 1979. Foraminifera of the Himi Group, Hokuriku province, central Japan. *The science reports of the Tohoku University. Second series, Geology* 49: 89–163, pls. 3–10.

Hayasaka, R. 1991. Sedimentary facies and environments of the Oligocene Ashiya Group in the Kitakyushu-Ashiya area, southwest Japan. *Journal of the Geological Society of Japan* 97: 607–619.

Inoue, E. 1972. Lithofacies fossil assemblages and sedimentary environment of Oligocene Kishima Formation in Karatsu coal field, northwest Kyushu, southwest Japan. *Reports, Geological Survey of Japan* no. 245: 1–68.

Kanazawa, K. 1992. Adaptation of test shape for burrowing and locomotion in spatangoid echinoids. *Palaeontology* 35: 733–750.

Kier, P.M. & Lawson, M.H. 1978. *Index of Living and Fossil Echinoids 1924–1970*. Washington: Smithsonian Institution Press.

Lambert, J. & Thiéry, P. 1925. *Essai de nomenclature raisonnée des échinides, fascicules 8 & 9*. Chaumont: Librairie L. Ferrière.

Matsuura, N. 1985. Successive change of the marine molluscan faunas from Pliocene to Holocene in Hokuriku region, central Japan. *Bulletin of the Mizunami Fossil Museum*: 71–158, pls. 32–42.

McNamara, K.J. & Philip, G.M. 1980. Australian Tertiary schizasterid echinoids. *Alcheringa* 4: 47–65.

Minato, M. 1950. On some Palaeogene fossils in Hokkaido. *Journal of the Geological Society of Japan* 56: 157–159.

Minato, M. 1954. The test remains of *Echinocardium cordatum* and the ecology of *Linthia yessoensis. Cenozoic Reseach*: 1–4.

Nagamori, H. 1998. Molluscan fossil assemblages and paleo-environment of the Pliocene strata in the Hokushin District, Nagano Prefecture, central Japan. *Earth Science* 52: 5–25.

Nagao, T. 1928. Palaeogene fossils of the island of Kyushu, Japan, part 2. *Science reports of the Tohoku Imperial University. 2nd series, Geology* 12: 11–140, pls. 1–17.

Nisiyama, S. 1968. The echinoid fauna from Japan and adjacent regions, part 2. *Palaeontological Society of Japan, Special Papers*: 1–491, pls. 19–30.

Ozawa, H. & Kamiya, T. 2001. Palaeoceanographic records related to glacio-eustatic fluctuations in the Pleistocene Japan Sea coast based on ostracods from the Omma Formation. *Palaeogeography, Palaeoclimatology, Palaeoecology* 170: 27–48.

Tokunaga, S. 1903. On the fossil echinoids of Japan. *Journal of the College of Science Imperial University of Tokyo* 17: 1–27, pls. 1–4.

Yano, T. 1981. Neogene fossil fauna from the Dojiri basin, northern part of Nagano Prefecture. *Abstracts: Annual meeting of the geological society of Japan* 88: 121.

Yoshikawa, H. 1996. Sedimentary facies and environments of early Pliocene lower Shigarami Formation in Takafu syncline area, northern fossa magna. *The Journal of the Sedimentological Society of Japan*: 47–58.

Yoshiwara, S. 1899. On some new fossil echinoids of Japan. *Journal of the Geological Society of Japan* 6: 1–4.

Echinoderms in a Changing World – Johnson (ed)
© *2013 Taylor & Francis Group, London, ISBN 978-1-138-00010-0*

Development and functional morphology of sutural pores in Early and Middle Cambrian gogiid eocrinoids from Guizhou Province, China

R.L. Parsley
Department of Earth and Environmental Sciences, Tulane University, New Orleans, Louisiana, USA

ABSTRACT: Sutural pores (epispires) are arguably the most primitive respiratory structures in echinoderms. While wide spread throughout the Echinodermata, they are especially common in the polyphyletic "Homalozoa" and Blastozoa. Complete ontogenies are known for three well preserved gogiid eocrinoid genera from Guizhou Province, China and they provide excellent information on pore development. In the Lower Cambrian gogiid *Guizhoueocrinus* small circular pores first develop late in the juvenile stage and considerably earlier in younger genera (Middle Cambrian *Sinoeocrinus* and *Globoeocrinus*) across sutures in a narrow band directly under the transverse food groove. Here they quickly mature into oval pores with a distinct raised rim. Subsequently, juvenile pores develop across the theca or may show an intermediary stage of pore emplacement just above the stalk in the basal and near basal thecal plates. In some specimens, large unpored patches on the upper theca may persist until full maturity at which point the theca evenly fills out with pores. Mature Guizhou gogiids have large pores commonly shared between three thecal plates. Pore geometry indicates that some resorption and redeposition has occurred. Pores first appear close to the ambulacra and dorsal nervous tissue associated with the brachioles. Second generation pores open near the stem and suggests presence of an aboral nerve center. Generally distributed third generation pores appear to be for overall respiration. Exceptional preservation in several specimens of *Guizhoueocrinus* indicates that podia or tissue blisters did not protrude into the slipstream but pockets of tissue extend into the body cavity from the edges of the pores. This is observable in some specimens where pockets have been everted probably due to gas filling the body cavity from postmortem putrefaction. The pockets of tissue taper and their edges anchor on the internal margins of the pores. Intake and exhaust of seawater was probably propelled by cilia. This internal sutural pore structure appears basal to more advanced respiratory structures such as decked pores (*Rhopalocystis*), cothurnopores (*Cothurnocystis*), diplopores, (diploporids), fistulipores (fistuliporids) and pore rhombs (rhombiferans).

1 INTRODUCTION

Ontogeny in gogiid eocrinoids is complex and varies considerably between genera. Features such as number and morphology of brachioles, morphology and symmetry of the associated ambulacra, emplacement and growth of thecal plates, and emplacement and growth of stalk plates are all distinct at certain growth stages. Many of these features can be seen in Figures 1–4. Perhaps no feature is more distinct in the various morphological stages than the emplacement and growth of the sutural pores.

Sutural pores or epispires are the most primitive respiratory structures in the echinoderms and make their appearance in the Lower Cambrian. They are circular to elliptical openings between two thecal plates and in large mature specimens in some genera they are a triangular opening between three plates. Sutural pores are common in a wide range of early (Cambrian) echinoderms such as

Trochocystites Barrande, 1887 (Homostelea) and *Ceratocystis* Jaekel, 1901 (Stylophora); but, they are especially common and diverse in gogiids (Eocrinoidea). All sutural pores are probably not the same, they may have evolved several times, but those in the Blastozoa and "Homalozoa" which apparently serve only a respiratory function are probably homologous.

Recently numerous essentially complete Early and Middle Cambrian gogiids, preserved as high fidelity molds in fine grained siliciclastics have been collected in Guizhou Province, in southwest China. The Balang Formation near Kaili City has produced hundreds of specimens of the new Lower Cambrian genus *Guizhoueocrinus* Zhou et al. 2007, and the basal Middle Cambrian Kaili Formation near Balang in Taijiang County has produced over 1500 specimens of *Sinoeocrinus* and about 4000 specimens of the new genus *Globoeocrinus* Zhao et al. 2008 = *Sinoeocrinus globos* Zhao et al. 1999 *nomen nudum*. Each of these three genera is

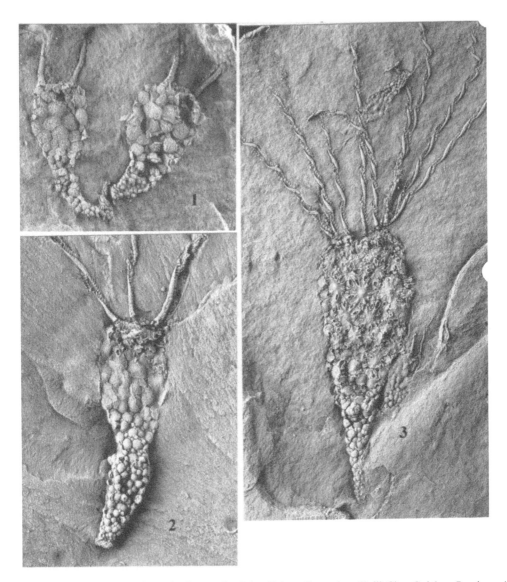

Figure 1. *Guizhouoecrinus yui* from the Lower Cambrian Balang Formation, Kaili City, Guizhou Province. All specimens are latex casts. (A) Pair of poreless juveniles; TH = 4 mm for both specimens; J—15—5—b. (B) Sutural pores developed under ambulacra; some pores show signs of maturity with their oval outlines and well defined rims; TH = 9 mm; K—WB—229. (C) Mature specimen with well developed mature oval sutural pores over the upper part of the theca; some pores in the distal part of the theca just above the stalk are well developed in (commonly) four plate rosettes; TH = 19 mm; KW—0—8.

monospecific and each is collected in an essentially biocoenosic setting. Such paleoecologic conditions are not known for Cambrian eocrinoids in western North America or Bohemia, the other regions where Cambrian gogiids are well represented (Robison 1965, Sprinkle 1973).

The metric used in discussing ontogeny is Thecal Height (TH) and is the distance from the summit of the theca and the top of the stalk. Measurements are significant only within each of the monospecific genera but between genera, stage (early juvenile, late juvenile, early mature, mature, late mature) heights are roughly comparable.

All specimens illustrated in this paper are housed in the Paleontological Museum, Guizhou University, Guiyang, Guizhou Province, China.

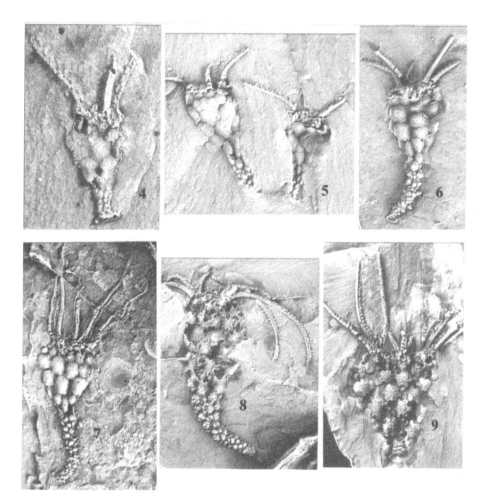

Figure 2. *Sinoeocrinus lui* from the basal Middle Cambrian, Kaili Biota, Kaili Formation, Balang, Guizhou Province, China. All specimens are latex casts. (A) Poreless juvenile; TH = 1.4 mm; GTBM 15335. (B) Pair of juveniles with sutural pores developed under the ambulacra; TH = 3.5 mm; GTBM—95319. (C) Advanced juvenile with well developed sutural pores under the ambulacra and scattered pores on thecal plates just above the stalk; TH = 3.5 mm; GTBM 953689. (D) Late juvenile/early adult with nascent pores developing on all thecal plates under the sub-ambulacral pores; TH = 9 mm; GTBM 932206. (E) Mature specimen with some sutural pores becoming triangular and formed by three plates, the rest are mature oval pores with prominent rims developed between two plates; TH = 18 mm; GTBM—101332. (F) Very mature (gerontic?) specimen with most sutural pores triangular and developed between three thecal plates; TH = 21 mm. GM 9-41582.

2 ONTOGENY AND EXTERNAL MORPHOLOGY OF GOGIID SUTURAL PORES

2.1 *Early Cambrian gogiid from Guizhou Province*

Guizhoueocrinus (Early Cambrian) is highly complex and has an elaborate ontogeny. Sutural pores do not appear in early stages of juvenile ontogeny (TH = 2–6 mm) (Fig. 1A) and appear in late stages (TH ≥ 6 mm) in a irregular narrow band between plates in the two or three irregularly constituted plate circlets under the ambulacra (Fig. 1B). By TH = 8 mm it is common for pores to become oval through peripheral plate growth and develop a prominent narrow and elevated perimeter wall. By TH ≥ 11 mm oval sutural pores are common in most specimens but in a few, pore emplacement is late and pore emplacement at this thecal height may be just beginning.

The second stage of pore emplacement begins at the top of the stalk with the formation of a

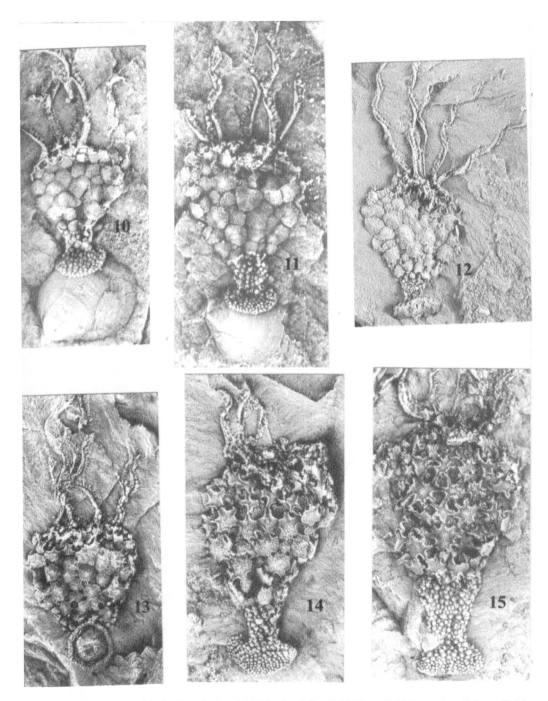

Figure 3. *Globoeocrinus globulus* from the basal Middle Cambrian Baili Biota, Kaili Formation, Balang, Guizhou Province, China. All specimens are latex casts. (A) Poreless juvenile specimen, TH = 2.5 mm; GTBM 911035. (B) Large juvenile specimen with sutural pores developed under the ambulacra; TH = 5 mm; GTBM 93007. (C) Small mature specimen with large non-pored area under well developed subambulacral pores. TH = 6.5 mm; GTBM 951760. (D) Mature specimen with sutural pores developed on edges of most thecal plates; TH = 7 mm; GTBM 142585. (E) Specimen with sutural pores ranging from incipient to mature; TH = 8 mm. GTBM 943436. (F) Advanced mature specimen with fully developed triangular sutural pores formed by three plates; TH = 11 mm; GTBM 943147.

2.2 *Pores in Middle Cambrian gogiids from Guizhou Province*

The two basal Middle Cambrian gogiids from Guizhou Province are somewhat different morphologically; *Sinoeocrinus* has a similar theca and stalk/attachment disk to that of *Guizhoueocrinus* but differs in having straight brachioles. *Globoeocrinus* has a globular theca and a shorter stalk made up of small platelets that resemble the early juvenile stages of *Guizhoueocrinus* and *Sinoeocrinus*. Mature specimens of *Globoeocrinus* resemble *Gogia ojenai* Durham, 1979 from the Latham Shale, Marble Mountains, in Southern California.

Both of the Guizhou Middle Cambrian gogiids are morphologically simpler than *Guizhoueocrinus* and have a simpler ontogeny. While features such as stalk and thecal plating are paedomorphic relative to *Guizhoueocrinus*, the emplacement of sutural pores is paramorphic, i.e. more mature aspects are propelled into earlier stages of ontogeny.

In *Sinoeocrinus* sutural pores are absent in the smallest juveniles, TH = 1.4 to 2 mm (Fig. 2A). In small juveniles, TH = 2 to *ca.* 3.5 to 4 mm and, as in *Guizhoueocrinus,* are restricted to a narrow band under the ambulacra (Fig. 2B). They also mature rapidly (develop peripheral rims and become elliptical) during early stages of thecal development. At *ca.* TH = 6 mm sutural pores appear sporadically in thecal plates just above the stalk (Fig. 2C). The "basal" series of thecal plates can be identified as the plates with sutural pores on the upper and lateral faces but none on the bottom. This may be an arbitrary differentiation between theca and stalk but it is consistent and is probably correct. At *ca.* TH ≥ 7 mm sutural pores begin to develop sporadically over the rest of the theca and at TH = 10 to 12 mm sutural pores are spread over the theca in various stages of development (Fig. 2D). At TH ≥ 12 mm pores develop into complex structures, including widening through resorption, and the concomitant development of high and thick peripheral rims. At TH = ≥17 mm (to 23+ mm) sutural pores continue to enlarge and commonly acquire a triangular shape and are shared by three thecal plates (Figs. 2E and 2F). These are the largest most complex sutural pores observed in gogiids.

The second basal Middle Cambrian gogiid *Globoeocrinus* co-occurs with and probably is about three times more abundant than *Sinoeocrinus*. It is the most abundant of all known species. Like *Sinoeocrinus* sutural pores appear under the ambulacra from TH = 2.5 to 5 mm (Figs. 3A and 3B). As in the other species they acquire a mature geometry as early juveniles. From TH = 5–7 mm pores randomly appear over the entire theca and

Figure 4. *Guizhoueocrinus yui* from the Lower Cambrian Balang Formation, Kaili City, Guizhou Province. Latex Cast. Specimen with preserved soft-part everted respiratory pockets that extend into the theca from the internal margins of the sutural pores; TH = 17 mm; KMS4 15.

circular pore surrounded by (commonly) four thickened plates to form a raised rosette around the pore. These pores are always circular and do not develop into elliptical or oval outlines. This stage can begin as early as TH = 6 to 8 mm and by TH = 11 mm normal pores (between two plates) are forming in the thecal plates above the stalk. By TH = 12 mm formation of pores over the rest of the theca is common (Fig. 1C). In some specimens emplacement of thecal pores is retarded and, for example, a specimen at TH = 16 mm may closely resemble a specimen at TH = 9 mm. This retardation seems to be due to large areas that are devoid of pores. This condition has been observed up to TH = 16 mm, after this point even distribution of pores (especially on the upper two thirds) on the theca, TH = 16 to 28+ mm, appears to be a uniform condition.

Pores enlarge, apparently by resorption and redeposition of stereom and the peripheral wall becomes relatively taller and broader. This condition is especially present in mature specimens TH = <15 mm and in contrast to Middle Cambrian gogiid species enlargement to form triangular pores formed in the corner areas of three plates is generally not present.

in most cases bypass the stage of emplacement just above the stalk. In early mature specimens (TH = 7 to 11 mm) a varying sized non-pored area (involving varying numbers of plates) on the upper half of the theca is commonly present (Fig. 3C). James Sprinkle, University of Texas (personal comm., 2006) has observed similar conditions in North American specimens of *Gogia* and is of the opinion that these poreless areas are restricted to the posterior of the theca. This cannot be confirmed in *Globoeocrinus* and there is a possibility that it may simply be retarded development as is seen in *Guizhoueocrinus*. In specimens TH = >11 to 15+ mm. well developed pores are generally distributed over the theca and large ovoid pores formed by three or four thecal plates are common on the upper half of the theca (Figs. 3E and 3F).

In keeping with the heterochronous tendencies in eocrinoids: *Rhopalocystis* Ubaghs, 1963 from the Upper Tremadocian of Morocco has large offset, generally hexagonal, thecal plates and round sutural pores with elevated rims; these characters are quite similar to those of a late juvenile gogiid. Significant differences in the ambulacra, subsurface decking in the pores and the holomeric stalk certainly bespeaks significant morphologic advancement and placement in its own family, Rhopalocystidae Ubaghs, 1968. However, the overall aspects of thecal outline, plating and pores suggests its derivation from an overall paedomorphic gogiid.

2.3 *Order of pore emplacement*

Very small juvenile specimens of the genera discussed in this paper do not have sutural pores and it is assumed that respiratory exchange is carried out across the integument and through the stereom. As these organisms grow the distance that oxygen will diffuse does not meet the needs of the animal and sutural pores develop to alleviate respiratory needs. Because sutural pores are emplaced in phases and in rather specific areas it is assumed that they are positioned to alleviate areas of high metabolic activity.

As indicated above, sutural pores have essentially three series of emplacement: first, under the ambulacra; second, just above the stalk in the lowermost thecal plates; and third, general emplacement over the rest of the theca. The first two series occur in narrow bands and suggest that they are servicing local needs on the theca's interior and the third series opens in a more random fashion over the rest of the theca suggests that they are enhancing respiratory needs as the volume of the theca increases as the result of plate growth and addition. The pattern of pore emplacement suggests that internal fluid circulation is not very

efficient and/or respiratory exchange across the plate stereom was inadequate for metabolic needs. The first series always occurs in juveniles and in the Guizhou genera mature into oval-elliptical pores with well developed rims by the earliest stages of maturity. It is likely that these first stage pores provided increased respiratory exchange for the dorsal nerve ring, feeding apparatus and gonads.

The second series of pore emplacement just above the junction of the stalk with the theca in *Guizhoueocrinus* and *Sinoeocrinus* suggests enhanced respiratory exchange for the aboral nerve plexus. (This series of pore emplacement may in fact argue for the presence of a crinoid-like nerve plexus).

The final phase of pore emplacement appears to be for general respiratory enhancement and in mature specimens mature oval/elliptical pores are commonly rather evenly spread over the entire theca. In large specimens for each species the "open or pore space" outlined by the pore openings is relatively significantly greater than in juvenile specimens. And, while it has not been measured it is probable that the area of pore opening is in proportion to the internal volume of the theca and the stalk in late juvenile and mature specimens.

3 INTERNAL SOFTPART ANATOMY OF GOGIID SUTURAL PORES

A few specimens of *Guizhoueocrinus* (Lower Cambrian) have been found with an unusual form of soft-part preservation where the sutural pores are covered with peaked conical covers (Fig. 4). The only other recorded case of covers over the pores in gogiids was reported by Ubaghs and Vizcaïno (1990) on several small specimens of *Gogia (Alaniscystis) andalusae* from the Lower Cambrian of the Sierra Morena Oriental, Andalusia, Spain. These coverings over the pore openings resemble a pair of conjoined bubbles and the pair has the shape and profile of an exploded airbag on a steering wheel. In their view, respiratory exchange took place through the surface of these lightly calcified covers above the surface of the theca. The pore covers are preserved in the same manner as those in *Guizhoueocrinus*; they are external molds. My best specimen, like the Spanish specimens, was casted with latex to obtain an in-life aspect. The Guizhou specimens are larger and seemingly better preserved than the Spanish specimens and results in a rather different interpretation of the covers.

I view the covers as being anchored along the internal margin of the pore margin and not along the upper or outer margin of the pore nor up on the marginal rim. The covers or pockets are in fact everted and, in living specimens, extended into and not out of the body. The everted pockets are most common on the upper and ambital portions of the theca. They are parabolic in profile and elliptical in cross section. Their length or depth (distance of penetration into the theca) appears to be about the same as the length of the pore opening. Because the pocket is essentially parabolic in profile the area (A) of the parabola can be calculated by the equation

$$A = 2ld/3$$

where l = the length, d = the depth, and by observation $l \approx d$. Doubling the area of a parabola would serve as a proxy for the approximate area of respiratory surface (rs) for each pocket.

$$Ars = 2ld/3 \times 2$$

This model is approximate because the additional area from the ellipsoid curvature of the pockets (and not known in the non-everted state) has not been included in the calculations.

As the pores grew, the size of the pockets respiratory surface area increased greatly. For example, a sutural pore 0.5 mm in length has 0.33 mm² of respiratory surface; a sutural pore 1 mm in length has about 1.33 mm² of respiratory surface. Respiratory area increases by a factor of 4 in this model for each doubling of the pore length. This marked increase in respiratory area may be directly related to the fact that pores are added rather abruptly, along with pronounced thecal thickening (and increase in volume), in early mature specimens of all species and show a tendency to mature into ovoid openings with well developed rims over a small thecal height increase. The abrupt increase in thecal volume is compensated for by rapid increase in respiratory surface area.

Ubaghs and Vizcaïno (1990) asserted that the epispires were covered by a thin calcareous cover. In *Guizhoueocrinus* there is no evidence for rigid or semi-rigid calcification of the pockets (they are after all, everted) but it is reasonable to assume that spicular support may have been present to prevent collapse of the pocket but with little loss to respiratory efficiency (Paul 1976).

The mechanism for this rare preservational mode would appear to be caused by *post mortem* putrifacting gas collecting inside the upper part of the theca and everting the pockets. This requires special conditions to prevent the escape of internally produced gas (rapid burial in copious fine grained material of sufficient mass to seal in produced gas?) that would normally readily dissipate into the environment. The accumulated gas would force the pockets out while the covering sediment and decay began the process of flattening the theca.

3.1 Probable functioning of the respiratory pockets

How oxygen rich water was drawn into the pocket and oxygen depleted water expelled is speculative, but a ciliary pumping mechanism seems reasonable. There is also no evidence for internal linkage with the water vascular system. Efficiency of respiratory exchange would have been enhanced if there was at least some circulation of body fluids. (For general discussion on respiratory efficiency in "cystoid" blastozoans see Paul 1972, 1976). There is a distinct probable pairing of characters in blastozoans relative to body fluid circulation, where sutural pore structures (epispires, diplopores, pore rhombs) are associated with a tightly closed theca due to the presence of tight-valved anal pyramids, tightly fitting roof-like cover plates over the mouth and sometimes a small pyramid over the gonopore. A circulation/respiratory mechanism, under these conditions, may be generated by peristalsis of the gut where internal pressure is above ambient, by virtue of effective plugging of the principal thecal openings and oxygen carrying currents generated by peristalsis of the gut. Body fluids are pumped past the internal surfaces of the pockets where oxygen/carbon dioxide exchange takes place (see Parsley 1990). Because of the specific areas of pore emplacement in juveniles it seems likely that circulation was limited, at least in the early ontogenetic stages, and the general positioning of pores in late juveniles and mature specimens suggests an overall more efficient circulation of body fluids and a more respiratory exchange system.

4 CONCLUSIONS

Guizhoueocrinus from the Lower Cambrian Balang Formation is the earliest eocrinoid known from complete specimens and in sufficient quantity to document its ontogeny. Sutural pores appear under the ambulacra in late juvenile stages and develop into mature elliptical openings with well developed rims by early maturity. The emplacement of pores occurs much earlier in the Middle Cambrian genera *Sinoeocrinus* and *Globoeocrinus* where the earliest pores are paramorphically emplaced in the early juvenile stage, commonly at about half of the thecal height as emplacement in *Guizhoueocrinus*.

This stage is followed by emplacement at the base of the theca (poorly defined in *Globoeocrinus*) and followed by random emplacement of pores over the rest of the theca. In large mature specimens pores and marginal rims are large, increasingly ovoid and apparently were modified by resorption and redeposition.

Several specimens of *Globoeocrinus* show exceptional preservation and demonstrate that the sutural pores are probably not basal perimeters for calcareous bubble-like sacs extending out from the theca as suggested by Ubaghs and Vizcaïno (1990) but actually extend into the theca as internal pocket-like structures. Circulation in the pockets was probably ciliary and their apparent rapid growth was driven by the rapid increase in respiratory area as the structure grew.

It is quite likely that sutural pore pocket structure is ancestral to more complex respiratory structures. Diplopores and pore-rhombs, in the blastozoans; cothurnopores, in the stylophorans, are parsimonious probable modifications of a sutural pore pocket.

ACKNOWLEDGMENTS

Yuanlong Zhao, Guizhou University, Guiyang made the material available for study. This study was made possible through grants from the National Science Foundation S.G.E.R. 0207292 and the McWilliams Fund, Earth and Environmental sciences, Tulane University.

REFERENCES

Barrande, J. 1887. Classe des échinoderms. Ordre des cystidées. In Système Silurien du centre de la Bohême. Pt.1. 7:233p, Privately published, Prague.

Durham, J. 1979. A Lower Cambrian *eocrinoid. Journal of Paleontology* 52:195–199.

Jaekel, O. 1901. Uber Carpoideen; eine neue Klasse von Pelmatozoen. Zeitschrift der Deutschen Geologischen Gesellschaft 52: 661–677.

Parsley, R.L. 1990. *Aristocystites,* a recumbent diploporid (Echinodermata) from the Middle and Late Ordovician of Bohemia, (ČSSR). *Journal of Pale ontology* 64 (2): 278–293.

Paul, C.R.C. 1972. Morphology and Function of exothecal pore-structures in cystoids. *Palaeontology,* 72 (1):1–28.

Paul, C.R.C. 1976. Respiration rates in primitive (fossil) echinoderms. *Thalassia Jugoslavica* 12 (1): 277–286.

Robison, R.A. 1965. Middle Cambrian eocrinoids from western North America. *Journal of Paleontology* 39: 355–364.

Sprinkle, J. 1973. Morphology and evolution of blastozoan echinoderms. *Mus. Comparative* Zool., Harvard University, Special publication. 284p.

Ubaghs, G. 1963. Rhopalocystis destombsei n.g., n.sp. Eocrinoïde de l'Ordovician inférieur (Tremadician supérior) du Sud marocain. *Serv. Geol.* Maroc., Notes & Mém. 23 (172): 25–40.

Ubaghs, G. 1968. Eocrinoids. In R.C. Moore (ed) Treatise on Invertebrate Paleontology Pt. S, Echinodermata1 (2): S455–S495.

Ubaghs, G. & Vizcaïno, D. 1990. A new eocrinoid from the lower Cambrian of Spain. *Palaeontology* 33 (1): 249–256.

Zhao, Y., Yuan, J., Zhu, M., Yang, R., Guo, Q., Huang, Y. & Pan, Y. 1999. A progress report on research on the early Middle Cambrian Kaili biota, Guizhou, PRC. Acta Palaeontologica Sinica 38 (Sup.): 10–14.

Zhao, Y., Parsley, R. & Peng, J. 2007. Early Cambrian eocrinoids from Guizhou Province, South China. Palaeogeography, Palaeoclimatology, Palaeoecology 254: 317–327.

Zhao, Y., Parsley, R. & Peng, J. 2008. Basal Middle Cambrian short stalked eocrinoids from the Kaili Biota: Guizhou Province, China. Journal of Paleontology 82: 415–422.

Echinoderms in a Changing World – Johnson (ed)
© *2013 Taylor & Francis Group, London, ISBN 978-1-138-00010-0*

Comparison of asteroid and ophiuroid trace fossils in the Jurassic of Germany with resting behavior of extant asteroids and ophiuroids

Y. Ishida
Kamiogi Suginami-ku, Tokyo, Japan

M. Röper
Bürgermeister-Müller-Museum, Solnhofen, Germany

T. Fujita
National Museum of Nature and Science, Tokyo, Japan

ABSTRACT: Two different size star-shaped trace fossils are described from the Upper Jurassic Hienheim Formation in Germany. The large and small traces were assigned to *Asteriacites quinquefolius* and *Asteriacites lumbricalis,* respectively. To identify the producer of *A. quinquefolius* and to examine the producing process, resting traces of living asteroids, *Astropecten scoparius* and some deep-sea starfishes were observed in an aquarium and *in situ,* respectively. In an aquarium, *A. scoparius* buried itself shallowly, keeping its arms in a pentamerous symmetrical position. When it moved from the resting position, the three front arms of the moving asteroid bulldozed the substrate in front of these arms. Consequently, a star-shaped like depression was left behind with four radiating wider and sub-triangular arm furrows tapering toward the tip and one straight or indistinct sub-triangular depression left by the bulldozing starfish. Similar shaped resting traces were also observed on the deep-sea floor. The sizes of traces were more or less the same as the producing animals. The producers were observed to move from the resting position in the direction of the indistinct depression. *Asteriacites quinquefolius* was very similar to these resting traces of living asteroids, and its producer is suggested to be an asteroid. On the other hand, *Asteriacites lumbricalis* is interpreted to originate from ophiuroids based on the previous experimental study of living ophiuroids. The resting traces of living ophiuroids (four well preserved depressions and an ill-defined one) were much wider than the producing animals. The size and shape of trace fossils suggest that the producers may belong to an ophiuroid species found from the Late Jurassic around Ried. The producers were observed to move from the resting position in the opposite direction of an ill-defined depression remained. These studies show that there are important differences in the morphology and in the production of asteroid and ophiuroid traces.

1 INTRODUCTION

The ichnogenus *Asteriacites* is a star-shaped trace fossil and includes five ichnospecies. Producers and their producing process have been so far discussed for *A. lumbricalis* and *A. quinquefolius* (Seilacher 1953, Mángano et al. 1999, Ishida et al. 2004). *Asteriacites lumbricalis* (Schlotheim 1820) has been found from the Ordovician to the Cenozoic of Europe, USA, South America, Greenland and Japan (Oppel 1864, Seilacher 1953, Lewarne 1964, Chamberlain 1971, Goldring & Stephenson 1972, Hakes 1977, Müller 1980, Hess 1983, Dam 1990, West & Ward 1990, Mikulas 1990, Twitchett & Wignall 1996, Twitchett 1999, Mángano et al. 1999, Wilson & Rigby 2000, Bell 2004, Ishida et al. 2004). *Asteriacites lumbricalis* was interpreted as a resting trace of asteroids by some workers (Oppel 1864, Lewarne 1964, Bell 2004), however, many examples have been interpreted as ophiuroid traces (Seilacher 1953, Chamberlain 1971, Mikulas 1990, West & Ward 1990, Mángano et al. 1999, Wilson & Rigby 2000). Ishida et al. (2004) observed the producing behavior of extant ophiuroids and interpreted *A. lumbricalis* from the Lower Triassic of Japan as an ophiuroid trace. *Asteriacites quinquefolius* (Quenstedt 1876) has been found from the Carboniferous to Triassic of Europe and USA (Seilacher 1953, Hakes 1977, Twitchett & Wignall 1996). *Asteriacites quinquefolius* was interpreted as an asteroid trace based on the comparison of the morphologies with traces made by living asteroids in an aquarium (Seilacher 1953), but details of the actual producing behavior by extant asteroids have been scarcely studied.

Two sizes of star-shaped trace fossils have been recovered from the Upper Jurassic Upper Hienheim Formation, near Hienheim in Germany (Röper & Rothgaenger 1998). In the present study, we examined the morphology of these trace fossils and ascribed the smaller one to *Asteriacites lumbricalis* and the larger one to *Asteriacites quiquefolius*. We also observed the behavior of extant asteroids both in an aquarium as well as *in situ*, and compared their traces with the trace fossils. This was done in order to determine the producers of the fossils and to clarify the processes involved in trace production.

2 MATERIALS AND METHODS

Two sizes of star-shaped trace fossils were found in an outcrop at Ried near Hienheim (east Bavaria) in Germany (Fig. 1C). These outcrops consist of rocks belonging to the Upper Heinheim Formation. This formation is composed mainly of fine grained limestone intercalated with muddy limestones. The rocks contain ophiuroids, asteroids, mollusks, brachiopods and other fossils (Röper & Rothgaenger 1998). The geologic age of the Upper Heinheim Formation corresponds to the Lower Tithonian (Upper Jurassic) based on

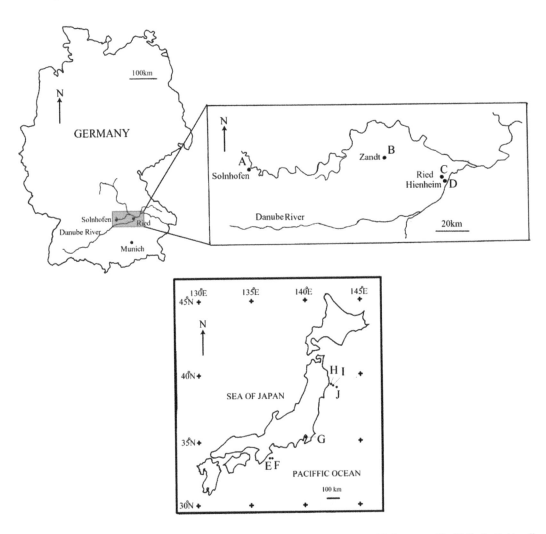

Figure 1. Sampling localities of the trace fossils (C), fossil ophiuroids (A–D) and living asteroids (E–J). A, *Ophiurella speciosa*; B, *Geocoma carinata* (Fig. 7A); C, *Asteriacites lumbricalis* (Fig. 2), *Asteriacites quinquefolius* (Fig. 3), *Sinosura kelheimense* (Fig. 7B); D, *Ophiopetra lithographica*; E, Asteroidea sp. (Fig. 5B); F, Astropectinidae sp. (Fig. 5G), star-shaped traces (Fig. 5G–I); G, *Astropecten scoparius* (Fig. 4); H, Astropectinidae sp. (Fig. 5E); I, *Ceramaster japonicus* (Fig. 5C), Asteroidea sp. (Fig. 5F); J, *Thrissacanthias* sp. (Fig. 5D), Benthopectinidae sp. (Fig. 5A).

ammonite biostratigraphy (Röper & Rothgaenger 1995, 1998). The morphology of the star-shaped trace fossils was analyzed by tracing them onto transparent paper placed upon the fossil-bearing sediment. The small sized trace fossils were morphologically compared with the Upper Jurassic ophiuroids recovered from Solnhofen to Hienheim (Fig. 1A–D). The large sized trace was compared to the, as of yet, unidentified asteroid specimens (Röper and Rothgaenger 1998). The traces and fossils are stored at the Bürgermeister-Müller-Museum, Solnhofen, Germany.

One specimen of *Astropecten scoparius* Valenciennes (arm length 7 cm) was collected from Tokyo Bay (Fig. 1G). Its behavior was observed on a coarse sandy substrate in a 1 kL aquarium in the Tokyo Sea Life Park. *In situ* observations on the traces and behavior of extant asteroids were made on silty bottoms off central to northern Japan in the Pacific by a deep-sea camera during cruises of the R.V. Tansei-maru of the

Ocean Research Institute, University of Tokyo, at the following stations (see Fig. 1 for the positions); St. I, KT-84-9, 418–463 m deep, July 1984; St. J, KT-84-9, 1214–1406 m deep, July 1984; St. E, KT-86-6, 427–442 m deep, May 1986; St. F, KT-86-6, 923–861 m deep, May 1986; St. H, KT-87-5, 152–153 m deep, May 1987.

3 RESULTS

3.1 *Trace fossils*

Two size classes of star-shaped trace fossils were found in fine grained limestone. A total of 26 smaller-sized traces were discernible in 3 separate blocks: 16 in concave epirelief (Fig. 2A, C) and 10 in convex hyporelief (Fig. 2B, D). The density of the trace fossils calculated from the three blocks (Fig. 2E–G) was 187 individuals/m². Many traces are isolated, while others are in contact with

Figure 2. *Asteriacites lumbricalis* from the Hienheim Formation, Germany. A and C, concave epirelief; B and D, convex hyporelief. Sketches of E and F correspond to photographs C and D respectively, and sketch G shows the other block in concave epirelief. Arrows in sketches show the presumed moving directions of the producers. Scale bars = 1 cm.

one another. The traces have five radiating slender, sub-triangular impressions tapering toward the tips and a central sub-circular impression. One of the five arm impression is conspicuously shorter and shallower with more ill-defined edges than the others. This arm impression is straight or gently curved, and shows faint transverse striations. The diameter of the central area is 5.0–8.1 mm (mean 6.2 mm, $n = 16$), the longest arm impression measures 7.6–17.5 mm (mean 12.6 mm, $n = 16$), and the width at the arm base is 1.9–3.0 mm (mean 2.3 mm, $n = 16$). The morphology agrees well with the descriptions of *Asteriacites lumbricalis* (Seilacher 1953, Chamberlain 1971, Hakes 1977, Twitchett & Wignall 1996, Mángano et al. 1999, Wilson & Rigby 2000, Bell 2004, Ishida et al. 2004).

One large-size star-shaped trace fossil was preserved in concave epirelief on the bedding plane (Fig. 3). It has five radiating wide, sub-triangular depressions tapering toward the tips and faint transverse striations. One of the five radiating depressions is shorter and shallower than the others with an ill-defined edge. The radius of the central area is 16 mm. The longest arm furrow measures 53 mm, and the width at the arm base is 12 mm. The deepest part of the depression is near the central area with a depth of 5 mm. The morphology agrees well with the description of *Asteriacites quinquefolius* (Seilacher 1953, Hakes 1977).

3.2 Aquarium observations of extant asteroid species

In the aquarium, *Asteropecten scoparius* often rested without moving. When resting, the asteroid slowly buried its body vertically and shallowly into the substrate (Fig. 4A, B). The arms were radially symmetrical during resting periods. When the asteroid started to move from the resting position, the pentamerally symmetric posture was kept. The three front arms of the moving asteroid bulldozed the substrate in front of these arms (Fig. 4C, D). Consequently a star-shaped depression was left behind with four radiating wider and sub-triangular arm furrows tapering toward the tip and one straight or indistinct sub-triangular depression left by the bulldozing starfish (Fig. 4E).

3.3 In situ observations of extant asteroid species

Star-shaped traces and the behavior of asteroids very similar to those recorded in the aquarium

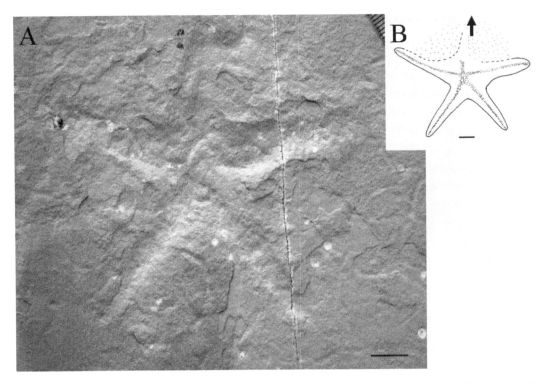

Figure 3.   Photograph (A) and sketch (B) of *Asteriacites quinquefolius* from the Hienheim Formation. Arrow in sketch B shows the presumed moving direction of a producer. Scale bars = 1 cm.

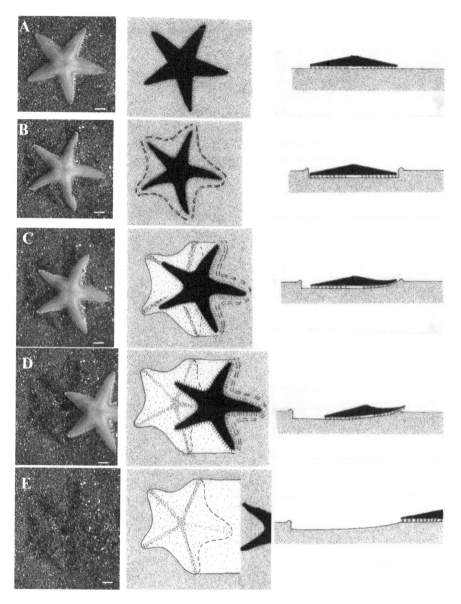

Figure 4.   Production of resting trace by *Astropecten scoparius* in an aquarium. A and B, burrowing into the substrate to rest; C and D, escaping from the substrate; E, resting trace. Left column, upper view of the asteroid; middle column, schematic diagram in horizontal view; right column, schematic diagram in vertical view. Scale bars = 1 cm.

were also observed on the deep-sea floor (Fig. 5). Several specimens of asteroid shallowly buried themselves in the silty substrate (Fig. 5A–E). When moving from the resting position, they bulldozed the substrate in front of themselves in the direction of movement (Fig. 5F). Star-shaped traces were observed with four radiating wider and sub-triangular arm furrows and one ill-defined arm depression (Fig. 5G–I).

4   DISCUSSION

The studied ichnospecimens of *Asteriacites lumbricalis* (Fig. 2A–D) are interpreted to be made by ophiuroids because of the morphological similarities to living ophiuroid resting traces with disc depression as shown by Ishida et al. (2004). The arm and disc furrows of the resting traces of the extant ophiuroids were much wider than the arms

Figure 5. *In situ* observations of asteroids by deep-sea camera. Resting postures (A–E, G), moving behavior from the resting position (F) and resting traces (G–I). A, Benthopectinidae sp. (St. J); B, Asteroidea sp. (St. E); C, *Ceramaster japonicus* (St. I); D, *Thrissacanthias* sp. (St. J); E, Astropectinidae sp. (St. H); F, Asteroidea sp. (St. I); G, Astropectinidae sp., resting trace of asteroid (St. F); H–I, resting traces of asteroids (St. F). Large arrows in F–I show the presumed moving directions of the producers. Small arrows in G–I show an ill-defined depression of the five arm furrows. Scale bars = 5 cm.

and discs of the producing ophiuroids (Ishida et al. 2004: Fig. 4). The size ranges of four species of fossil ophiuroids ($n = 17$) found from the Upper Jurassic around Ried were measured and compared with those of the present ichnospecimens of *Asteriacites lumbricalis* ($n = 16$) (Fig. 6). The size ranges of *Geocoma carinata* and *Sinosura kelheimense* (Fig. 7) were within the size range of *Asteriacites lumbricalis* suggesting that the producers of *A. lumbriclis* may possibly be one or both of these ophiuroids. Ishida et al. (2004) clarified that the producer of the ophiuroid resting traces moved from the resting position in the opposite direction of an ill-defined depression. Following these observations, the directions of movement of the present producers were nearly uniform judging from the asymmetric morphologies of the trace fossils (Fig. 2E–G).

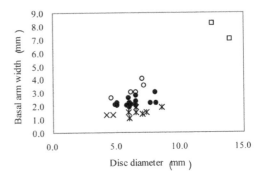

Figure 6. Comparison of the disc diameter and the basal arm width of *Asteriasites lumbricalis* from Ried and of the fossil ophiuroids around Ried.
●, *Asteriasites lumbricalis*; □, *Ophiurella speciosa*; ○, *Ophiopetra lithographica*; ×, *Geocoma carinata*; ∗, *Sinosura kelheimense*.

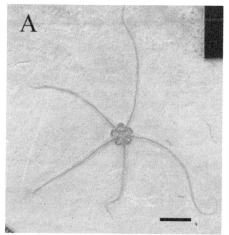
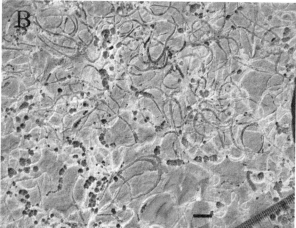

Figure 7. Fossil ophiuroids from the Upper Jurassic around Ried. A, *Geocoma carinata* from Zandt (Fig. 1B); B, *Sinosura kelheimense* from Ried (Fig. 1C). Scale bars = 1 cm.

The ichnospecimen *Asteriacites quinquefolius* (Fig. 3) is morphologically very similar to the resting traces of living asteroids (Figs. 4 and 5), and its producer is thus interpreted to be an asteroid. The body size and the ratio of the radius to the interradius of *Asteriacites quinquefolius* (R/r 3.3, arm length 53 mm) were similar to those of the unidentified asteroid fossils from the Upper Jurassic of Ried and Hienheim (R/r 2.9–3.2, arm length 12.5–48.5 mm, *n* = 3; Röper and Rothgaenger 1998: Figs. 139, 142 and 143), suggesting that the producers of *Asteriacites quinquefolius* may be these unidentified asteroids.

Based on the analysis of the production of the resting trace of living asteroids, the producers of asteroid resting traces were observed to move in the direction of an ill-defined depression of five arm depressions (Figs. 4 and 5). The direction of movement of the producer of the trace fossil can be estimated upon comparison to the asymmetric morphology of the trace (Fig. 3B).

ACKNOWLEDGEMENTS

We would like to thank the staff of the R.V. Tansei-maru for their kind help in sampling. Many thanks go to H. Arai and the staff of Tokyo Sea Life Park, H. Kohtsuka of the University of Tokyo and T. Kato for their kind help with aquarium experiments. Thanks are due to the members of the trace fossil study Group for their comments on the trace fossils. Thanks are extended to an anonymous referee for improving this manuscript.

REFERENCES

Bell, C.M. 2004. Asteroid and ophiuroid trace fossils from the Lower Cretaceous of Chile. *Palaeontology* 47: 51–66.

Chamberlain, C.K. 1971. Morphology and ethology of trace fossils from the Ouachita mountains, southeast Oklahoma. *J. Paleont.* 45: 212–246, pls. 29–32.

Dam, G. 1990. Palaeoenvironmental significance of trace fossils from the shallow marine Lower Jurassic Neill Klinter Formation, East Greenland. *Palaeogeogr. Palaeoclimatol. Palaeoecol.* 79: 221–248.

Goldring, R. & Stephenson, D.G. 1972. The depositional environment of three starfish beds. *Neues Jahrb. Geol. Paläont. Mh.* 1972: 611–624.

Hakes, W.G. 1977. Trace fossils in Late Pennsylvanian cyclothems, Kansas. In T.P. Crimes & J.C. Harper (eds), *Trace fossils 2*; 209–226. Seel Hous Press: Liverpool.

Hess, R. 1983. Das Spurenfossil *Asteriacites* im klastischen Permoskyth (Prebichl-Schichten) der südlichen Admonter Schuppenzone (Ostalpen) und seine paläogeographische Bedeutung. *Neues Jb. Geol. Paläont. Mh.* 1893: 513–519.

Ishida, Y., Fujita, T. & Kamada, K. 2004. Ophiuroid trace fossils in the Triassic of Japan compared to the resting behavior of extant brittle stars. In T. Heinzeller & J.H. Nebelsick (eds), *Echinoderms München*: 433–438. Taylor & Francis Group: London.

Lewarne, G.C. 1964. Starfish traces from the Namurian of County Clare, Ireland. *Palaeontology* 7: 508–513.

Mángano, M.G., Buatois, L.A., West, R.R. & Maples, C.G. 1999. The origin and palaeoecologic significance of the trace fossil *Asteriacites* in the Pennsylvanian of Kansas and Missouri. *Lethaia* 32: 17–30.

Mikulas, R. 1990. The ophiuroid *Taeniaster* as a tracemaker of *Asteriacites*, Ordvician of Czechoslovakia. *Ichnos* 1: 133–137.

Müller, A.H. 1980. Zur Ichnologie und Lebensweise triadischer Asterozoa. *Freiberger Forschungsheft* C 357, S: 69–76, 3 Bilder, 2 Bildtafeln.

Oppel, A. 1864. Über das Lager von Seesternen im Lias und Keuper. *Jh. Ver. vaterl. Naturk. Wttbg.* 20: 206–212.

Quenstedt, F.A. 1876. *Petrefaktenkunde Deutschlands. 4. Die Asteriden und Encriniden nebst Cysti und Blastoideen.* Fuess, Leipzig: 742pp.

Röper, M. & Rothgaenger, M. 1995. Eine neue Fossillagerstätte in den Ostbayerischen Oberjura-Plattenkalken bei Brunn/Oberpfalz. *Jb. Mitt. Freunde der Bayer. Staatssamml. Pal. Hist. Geol.* München 23: 32–46.

Röper, M. & Rothgaenger, M. 1998. *Die Plattenkalke von Hienheim.* Eichendorf Verlag: 1–110.

Schlotheim, E.F. von. 1820. *Die Petrefactenkunde auf ihrem jetzigen Standpunkte durch die Beschreibung seiner Sammlung versteinerter und fossiler Überreste des Thier- und Pflanzenreiches der Vorwelt erläutert;* Becher (Gotha): 437pp, 15pls.

Seilacher, A. 1953. Studien zur Palichnologie II. Die fossilen Ruhespuren (Cubichnia). *Neues Jb. Geol. Paläont., Abhandlungen* 98: 87–124.

Twitchett, R.J. 1999. Palaeoenvironments and faunal recovery after the end-permian mass extinction. *Palaeogeogr. Palaeoclimatol. Palaeoecol.* 154: 27–37.

Twitchett, R.J. & Wignall, P.B. 1996. Trace fossils and the aftermath of the Permo-Triassic mass extinction: evidence from northern Italy. *Palaeogeogr. Palaeoclimatol. Palaeoecol.* 124: 137–151.

West, R.R. & Ward, E.L. 1990. *Asteriacites lumbricalis* and a protasterid ophiuroid. 321–327. In A.J. Boucot (ed), *Evolutionary paleobiology of behavior and coevolution*: 725pp. Elsevier: Amsterdam.

Wilson, M.A. & Rigby, J. 2000. *Asteriacites lumbricalis* von Schlotheim 1820: Ophiuroid trace fossils from the Lower Triassic Thaynes Formation, Central Utah. *Ichnos* 7: 43–49.

Morphology

Echinoderms in a Changing World – Johnson (ed)
© 2013 Taylor & Francis Group, London, ISBN 978-1-138-00010-0

Stereom differentiation in spines of *Plococidaris verticillata*, *Heterocentrotus mammillatus* and other regular sea urchins

N. Grossmann & J.H. Nebelsick
Institute of Geosciences, Tübingen, Germany

ABSTRACT: Regular sea urchin spines (mainly *Plococidaris verticillata* and *Heterocentrotus mammillatus*) are studied with respect to their general structure, surface sculpturing and internal stereom morphology; variations of stereom densities are also measured and recorded. Differences occur not only in basic architecture, but also in the distribution of specific stereom types within the spines. The primary verticulate spines of the cidaroid *P. verticillata* contain a central medulla surrounded by a radiating layer both consisting of galleried stereom, which is then enclosed by a final microperforate cortex. The large massive primary spines of *H. mammillatus* also show a galleried medulla followed, however, by a labyrinthic radiating layer which is regularly interrupted by dense growth lines. The spine surface shows a complex longitudinally orientated microstructure. This investigation is part of a more general study concerning the biomimetic potential of sea urchin skeleton by analyzing their morphological, mineralogical and mechanical properties.

1 INTRODUCTION

The focus of the study presented here is to map the distribution, variations and gradients of stereom types within different representative primary spines of regular echinoid and correlate these to spine function. This study is part of a wider investigation on the biomimetic potential of sea urchin skeletons (Nickel et al. 2008, Presser 2009a, b, Grossmann 2010). The field of biomimetics (for overview see Buschan 2009) generally deals with the analysis of biological phenomena and their potential transfer to technical applications (for example Barthlott & Neinhuis 1997 and Neinhuis & Barthlott 1997 for the application of plant surface water repulsive phenomena—the "Lotus-Effect"—in industrial products). There have, however, been few studies concerning the biomimetic potential of the echinoderm skeleton, despite their potential in various fields of biomechanics, material sciences and nanotechnology (see Weber et al. 1971, Hiratzka et al. 1979, Currey 1999, Oaki & Imai 2006, Vecchio et al. 2007).

Spines belong to the most striking features of sea urchins. It is no surprise that the immense morphological diversity of these spines is related to the various functions such as defense, locomotion and feeding which they fulfill (e.g. Strathmann 1981, Denny & Gaylord 1996). Common to all echinoderm skeletal elements, spines consist of stereom, a three dimensional meshwork of trabeculae which thereby encompasses a network of interconnecting pore spaces. The stereom itself is constructed by high Magnesium Calcite with Mg concentration ranging from 2 to 12 mol% (Magdans & Gies 2004, Magdans 2005) and contains ca. 0.1 weight% organic macromolecules (Weiner et al. 2000). This pore space is filled by living tissue, the stroma, during life, consisting of among others, sclerocytes, phagocytes, and collagen fibers (for detailed analysis of the stroma in spines see Märkel & Röser 1983a, b, Märkel et al. 1989 and Ameye et al. 2001).

The microstructure of echinoderms has long been of interest (e.g. Müller 1854, Stewart 1871, Stock et al. 2004, Xiao et al. 2007). The different "Bauplans" of several sea urchin spines were compared by Hesse (1900), who classified them accordingly, and in doing so suggested that more recently evolved sea urchin spines are more simply organized than those of ancient taxa. Based on echinoid stereom, Smith (1980) differentiated and defined stereom microstructures such as rectilinear, galleried, labyrinthic, fascicular, perforated or even imperforate. These stereom types differ in general structure, orientation as well as the size and density relationships of struts to pore spaces. Labyrinthic sterom, for example, indicates a disorganized arrangement of trabeculae while galleried stereom shows long, connected, parallel galleries resulting in regularly arranged pores of uniform size. Stereom morphology can be related to different functions including substrates for muscle tissue insertion, load bearing surfaces or filling material.

A typical cidaroid spine (for example that of *Plococidaris verticillata*) can be separated into the base including milled ring, base and acetabulum, and the shaft including the collar or collorette and tip or summit (terminology following Mortensen 1928). The spine base is compact and has a smooth or perforated surface. The shaft itself (except for that of the collar area) is covered by a polycrystalline cortex that can show sculpturing and can be encrusted by organisms (David et al. 2009). The medulla is the central internal part of a spine surrounded by the radiating layer, followed by the cortex. Cidaroids occur in a very wide range of environments including both hard and soft substrates and use their spines for locomotion and defense.

Regular echinoids belonging to the family Echinoida (including the investigated *Heterocentrotus mammillatus*) show a different type of spine construction. The core of these spines consists of the medulla, following by a radiating layer which extends to the surface. These spines are covered by an epithelium which inhibits encrustation. Spines of the Echinoida can show discrete "growth lines" which usually occur at regular intervals and completely enclose the previous growth stage. A well known example for Echinoida spines are those of *Heterocentrotus mammillatus*, a common tropical Indo-Pacific echinoid which can inhabit high wave energy reef crests. One obvious function of these spines is to wedge the echinoid into reef crevices during the day (for detailed studies on the biology and morphology see Dotan 1990a, b).

The primary spines of *Heterocentrotus mammillatus* have long attracted interest due to their structure and the presence of distinct growth lines. Their potential for biomimetic research has also been outlined (e.g. Weber 1969, Hasenpusch 2000). The origin of growth lines is still under discussion with various explanations. Authors including Deutler (1926), Weber (1969), Heatfield (1971), Pearse & Pearse (1975) and Dotan & Fishelson (1984) suggested that the lines are due to periodic differences in growth rates. Ebert (1967, 1984, 1986, 1988) in contrast postulated that growth lines are the result of regeneration following traumatic autotomy, which can occur in distinct manner within the spines (for further information see Wilkie 2001, Maginnis 2006 and Fleming et al. 2007). A further possibility postulated by Ebert (1984) is that the growth lines result from reconciling the different growth trajectories of the test, ambulacral plates and spines.

2 MATERIALS & METHODS

The following sea urchin species were chosen due to (1) the wide variation of stereom microstructure shown by the spines, (2) the large spine size

Table 1. Stereom interpretations (after Smith 1980).

Stereom coarseness ($\bar{A}$)		Trabecular density ($\bar{A}/\bar{t}$)	
<25 μm	Coarse	<1	Compact
10–25 μm	Medium	1–2	Dense
<10 μm	Fine	2–4	Open

which was conducive to biomimetic experiments and (3) the relative ease of keeping specimens alive in aquarium: *Plococidaris verticillata* (Lamarck 1816) belonging to the cidaroids and *Heterocentrotus mammillatus* (Linnaeus 1758) belonging to the family Echinometridae. Additional spine material was derived from the cidaroids *Eucidaris metularia* (Lamarck 1816), *Prionocidaris baculosa* (Lamarck 1816), *Phyllacanthus imperialis* (Lamarck 1816) and the echinometrid *Echinometra mathaei* (Blainville 1825). All of the above mentioned species were kept in several artificial seawater aquaria (300 L). The aquaria were fully equipped to reflect a reef ecosystem with other marine invertebrates, algae and fishes. Investigated spines were later either completely removed from the test or cut near the milled ring from living specimens. Living specimens and isolated spines were obtained from aquarium supply houses, additional spines originate from collections housed at the Institute of Geosciences, University of Tübingen.

Spines were cleaned using ultrasonic distilled water bath and hydrogen peroxide (H_2O_2). Sputtering for Scanning Electron Microscopy (SEM) was done with gold, platinum or carbon in a Sputter Coater (BAL-TEC Model SCD 005/CEA 035). SEM was completed at the Institute of Geosciences in Tübingen (LEO Model 1450 VP). Critical-point-drying (cpd) and subsequent SEM (Hitachi S 800) studies were completed at the Max-Planck-Institute for development biology in Tübingen. Before cpd the spines were cut off the living sea urchin and then thrown in a solution of 2.5% Glutaraldehyde/PBS at 6°C for several hours. Stereom pore measurements and designations (see Table 1) follow Smith (1980): average maximum pore diameter ($\bar{A}$) and average minimum trabecular thickness (t) between adjacent pores were obtained from an area 200 μm × 200 μm. $\bar{A}$ is an indicator for the stereom coarseness, $\bar{A}/(t)$ ratio is a measure of trabecular density (Smith 1980). Stereom measurements were attained using Adobe Photoshop CS3.

3 RESULTS

3.1 *Cidaroids*

Primary spines of *Plococidaris verticillata* reach sizes of up to 20 mm with a test diameter of about 30 mm. The external morphology of aboral primary spines is characterized by up to four successive whorls of sharp ridges, and culminates in a crown shape structure at the tip of a spine. Oral primary spines lack whorls.

Randomly arranged smaller thorns are distributed between the whorls. Additionally, fine longitudinal striations cover the spine reaching over the whorls and stretch from the milled ring to the tip. Other cidaroids show a wide variety of surface sculpturing as seen in Figure 1A & B for *Eucidaris metularia* and *Prionocidaris baculosa*. *E. metularia* has rounded nodules on the spine (see Cutress, 1965 for *Eucidaris tribuloides*). The surface on the outer side of the spines is perforated and the pores are arranged around the nodules. Internally the spine is separated

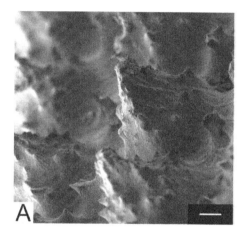

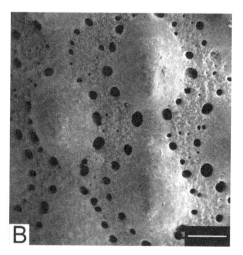

Figure 1. Surface structures of different sea urchin spines A) *Prionocidaris baculosa* (scale bar = 40 μm), B) *Eucidaris metularia* (scale bar = 200 μm).

into the medulla, the radiating layer and cortex. *Plococidaris* spines do not show growth lines. Stereom measurements are summarized in Table 2.

The medulla consists of galleried stereom with the main axis running parallel to the length of the spine (Fig. 2). The width of the medulla ranges in size from the base to the tip of a spine with a width of about 450 μm. The amount of pore space increases from the base onwards, reaches its maximum value half way to the tip, and then decreases in size. Stereom coarseness is compact at the base, the shaft is dense/medium and the tip shows a fine coarseness. The pores of the medulla cover about 20 to 25% volume. The average maximum diameter is 20 μm.

The radiating layer is also galleried, but with the main axis inclined to the medulla. The coarseness changes from the base (coarse) to the tip (fine) and the density from compact to dense/compact. The pore space shows values from 16 to 20%. The average maximum diameter of pores is 18 μm (shaft) to 9 μm (tip). The cortex, the outermost layer, is irregular perforated, and consists of a thick calcite layer. It includes several canals (diameter: 10 μm) and can be encrusted on the outer surface by various organisms. The thickest part of the cortex is at the tip and base (~80 μm = <1% diameter of spine), with flanks showing values of about 60 to 70 μm.

Spine tissue investigations through critical-point-drying of another cidaroid *Prionocidaris baculosa* showed different cells (e.g. sclerocytes and phagocytes) in the cavities of the stereom, as well as collagen fibers (Fig. 3A & B). In general the inner pore walls are covered by a thin ciliated epithelium (~2 μm).

3.2 *Echninoida*

Primary spines of *Heterocentrotus mammillatus* are large and robust reaching lengths of 100 mm (test width of 90 mm) and become flattened towards the peristome. The spine surface is characterized by parallel arrangement of distinct struts (ca. 100 μm wide) separated by spaces ranging in width from

Table 2. Stereom features of *Plococidaris*.

Structure	Stereom	Position	Ā (μm)	7 (μm)
Medulla	Galleried	Centre	17.1	15.6
		Tip	9.7	12.4
r. layer	Galleried	Centre	18.1	17.5
		Tip	12.4	12.2
Cortex	Micro-perforate	Centre	69.2 Ø	
		Tip	83.8 Ø	

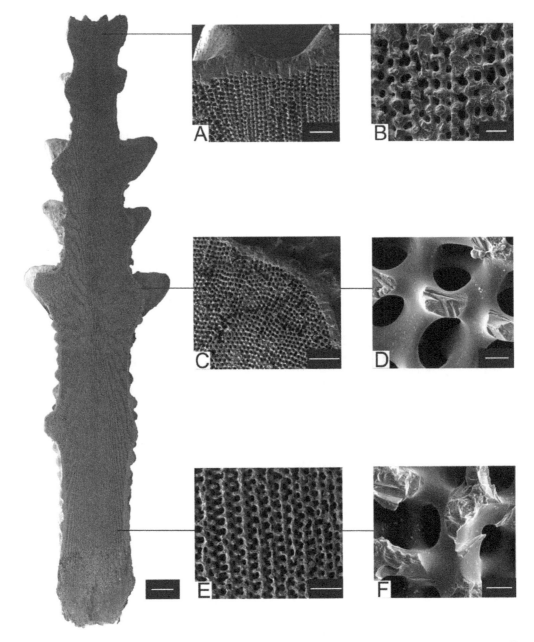

Figure 2.   Compiled SEM picture of a longitudinal section of a section of a primary spine of *Plococidaris verticillata* (length 25 mm). Stereom details: A & B) the tip with cortex and part of the medulla, C & D) the radiating layer and E & F) the medulla. Scale bars: spine: 600 µm; A = 100 µm, B = 20 µm, C & E = 200 µm; D & F = 10 µm.

10 to 30 µm (Fig. 4). The struts are wedge shaped in cross section. Their outer surfaces are generally smooth but support fine transverse striations. Spines of *Echinometra mathaei* also show these features (Fig. 4).

Internally, the spine is separated into the medulla, the radiating layer and the growth lines. The most striking feature of the internal morphology is the alternation of labyrinthic stereom and the much denser stereom, representing the growth

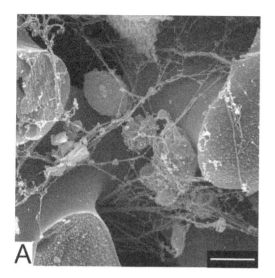

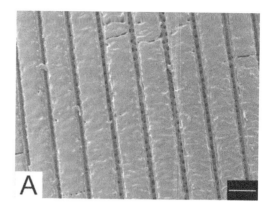

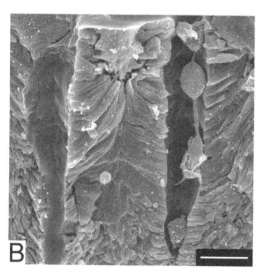

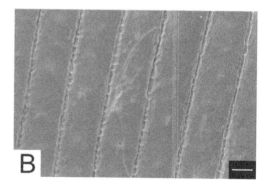

Figure 4. Surface structure of spines of A) *Echinometra mathaei* (scale bar = 100 μm), B) *Heterocentrotus mammillatus* spine (scale bar = 40 μm).

Figure 3. Internal view of a broken *Prionocidaris baculosa* spine following critical-point-drying. A) spine stereom (scale bar = 4 μm), B) cortex canal (scale bar = 10 μm).

lines. In longitudinal view, the growth lines consist of imperforate stereom. The average diameter of these growth lines ranges from 20 to 40 μm depending on the position in the spine (except at the base). The growth lines are thickest at the spine base (80 to 100 μm) and tips (~41 μm) and thinnest along the flanks (~32 μm) (see Fig. 5). The stereom coarseness of the medulla is fine/medium and with average maximum pore diameters of around 15 μm (see Table 3). The size of

each pore ranges from 70 to 100 μm². The pore space of the medulla varies slightly along the length of the spine (base 18%; shaft 19%; tip 18%). The trabecular density ranges from compact/dense to compact. The stereom coarseness of the radiating layer is fine to medium (from the base to the tip). The average pore diameter is 18 μm. The trabeculardensity is fine compact to dense. At each tip of a spine the stereom of the medulla spread out in a fountain-like manner. The stereom coarseness of the pores at the base ranges from fine to medium in comparison to the rest of the spine (Figs. 5E & F). The density is compact to dense with a pore space of about 12%. The spine is covered by a thin epithelium which precludes encrustation during life.

4 DISCUSSION & FUTURE WORK

Sea urchin spines such as those from *Heterocentrotus mammillatus* are of a special interest for

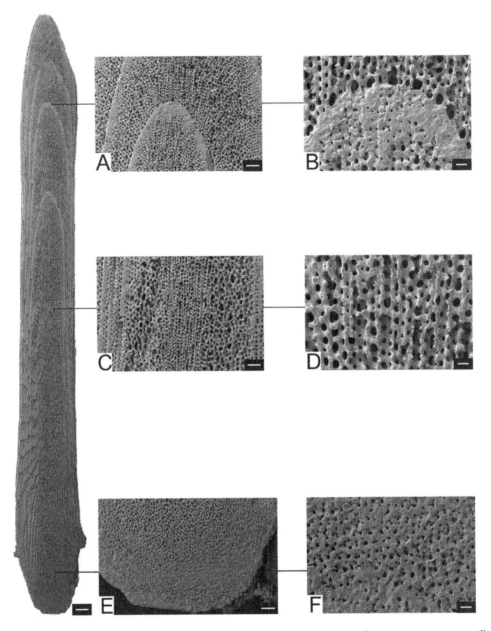

Figure 5. Compiled SEM picture of a longitudinal section of a primary spine of *Heterocentrotus mammillatus* (length 20 mm). Stereom details: A & B) growth lines, C & D) medulla & radiating layer and E & F) the base. Scale bars: spine = 300 μm; A, C, E = 100 μm; B, D, F = 20 μm.

biomimetic research as they are very stable and lightweight. The stereom definitions from Smith (1980) can be applied to the stereom of the investigated spines. A major difference, however, is the scale of stereom structures present. The stereom structures described by Smith (1980) are fairly restricted in extent within individual plates, where the studied stereom types extend along the whole length of the spines which can reach sizes over 10 cm.

The major differences between spine types of cidaroids and echinoida were confirmed by the present study. Both have complex layered structures consisting of different stereom types and

Table 3. Stereom features of *Heterocentrotus*.

Structure	Stereom	Position	Ā (μm)	t̄ (μm)
Medulla	Galleried	Base	12.5	14.7
		Centre	13.1	12.6
		Tip	12.2	14.2
r. layer	Labyrinthic	Base	7.3	10.0
		Centre	19.8	23.7
		Tip	13.7	13.8

orientation. The following stereom types were found: (1) galleried, (2) labyrinthic, (3) microperforate and (4) imperforate. Differences exist not only in the stereom types, but also in the measured values of density and trabecular thickness within the separate stereom types.

Cidaroid spines contain differently oriented galleried stereom (thus differentiating the medulla from the radiating layer) and a microperforate cortex. Investigated echinoida spines show more highly differentiated stereom types containing galleried in the medulla, labyrinthic stereom in the stereom and imperforate growth lines which regularly alternate with the previously mentioned layers.

To what extent these stereom differences represent adaptation to ecological factors will depend on future planed studies within the present project dealing with mechanical stiffness and impact behavior. Differences in density and trabecular thickness within the stereom types show for example that the stereom at the base of the spines of both species is denser than the rest of the stereom. This can be correlated with higher loading pressures than the rest of the spine. This obviously has to be seen also within the context of soft connecting tissues which can spread the loading pressure from the spine to the test as a whole. Further details such as the splaying of labyrinthic stereom at the end of the spines of *Heterocentrotus* (also seen within the spines just below the growth lines) may also be an adaptation to impact pressure at the tip (or former tip) of the spines. Further functions that have to be taken into account are the supply and maintenance of organic soft tissue on the surface of the Echinoida (such as *Heterocentrotus*). The surface characteristics of cidaroid spines are also important as an adequate substrate for encrustation which may be important in predatory interactions.

Future work will include: (1) a correlation between function of spines and their microstructure, (2) investigating more species from other families, (3) exploring evolutionary aspects of spine morphology, (4) exploring the biomimetic potential of spines which ideally reduce volume and weight without losing structural integrity,

and (5) the implementation of high resolution microtomographic CT methods.

ACKNOWLEDGMENTS

We would like to thank our partners within the biomimetic working group (University of Tübingen: Klaus Nickel, Christoph Berthold, Volker Presser, Stefanie Schultheiss and Melanie Keuper; Institute of Textile Technology and Process Engineering Denkendorf: Thomas Stegmaier, Achim Vohrer and Hermann Finckh), as well as Hartmut Schulz and Peter Fittkau for technical assistance at the SEM Lab (University of Tübingen) and Jürgen Berger (MPI Tübingen) for the critical-point-drying. This project is funded by the Landesstiftung Baden-Württemberg.

REFERENCES

Ameye, L., De Becker, G., Killian, C., Wilt, F., Kemps, R., Kuypers, S. & Dubois, P. 2001. Proteins and saccharides of the sea urchin organic matrix of mineralization: characterization and localization in the spine skeleton. *Journal of Structural Biology* 134: 56–66.

Barthlott, W. & Neinhuis, C. 1997. Purity of the sacred lotus or escape from contamination in biological surfaces. *Planta* 202: 1–8.

Buschan, B. 2009. Biomimetics. Philosophical Transactions of the Royal Society A 367: 1443–1444.

Currey, J.D. 1999. The design of mineralised hard tissues for their mechanical functions. *Journal of Experimental Biology* 202: 3285–3294.

Cutress, B.M. 1965. Observations on growth in *Eucidaris tribuloides* (Lamarck), with special reference to the origin of the oral primary spines. *Bulletin of Marine Science* 15: 797–834.

David, B., Stock, S.R., De Carlo, F, Hétérier, V & De Ridder, C. 2009. Microstructures of Antarctic cidaroid spines: diversity of shapes and ectosymbiont attachments. *Marine Biology* 156: 1559–1572.

Denny, M. & Gaylord, B. 1996. Why the sea urchin lost its spines: hydrodynamic forces and survivorship in three echinoids. *Journal of Experimental Biology* 199: 717–729.

Deutler, F. 1926. Über das Wachstum des Seeigelskeletts. *Zoologische Jahrbücher* 48: 119–200.

Dotan, A. 1990a. Population structure of the echinoid *Heterocentrotus mammillatus* (L.) along the littoral zone of south-eastern Sinai. *Coral Reefs* 9: 75–80.

Dotan, A. 1990b. Distribution of regular sea urchins on coral reefs near the south-eastern tip of the Sinai Peninsula, Red Sea. *Israel Journal of Zoology* 37: 15–29.

Dotan, A. & Fishelson, L. 1984. Morphology of spines of Heterocentrotus mammillatus. In B.D. O'Connor (ed.), Echinodermata: Proceedings of the fifth international echinoderm conference, Galway 24–29.9.1984: 253–260. Rotterdam: Balkema.

Ebert, T.A. 1967. Growth and repair of spines in the sea urchin *Strongylocentrotus purpuratus* (Stimpson). *Biological Bulletin* 133: 141–149.

Ebert, T.A. 1984. The non-periodic nature of growth rings in echinoid spines. In B.D. O´Connor (ed.), *Echinodermata: Proceedings of the International Echinoderm Conference, Galway September 24–29.9.1984*: 261–267. Rotterdam: Balkema.

Ebert, T.A. 1986. A new theory to explain the origin of growth lines in sea urchin spines. *Marine Ecology-Progress Series* 34: 197–199.

Ebert, T.A. 1988. Growth, regeneration and damage repair of spines of the slate pencil sea urchin *Heterocentrotus mammillatus* (L.) (Echinodermata: Echinoidea). *Pacific Science* 42(3–4): 160–172.

Fleming, P.A., Muller, D. & Bateman, P.W. 2007. Leave it all behind: a taxonomic perspective of autotomy in invertebrates. *Biological Review* 82: 481–510.

Grossmann, N. 2010. Stereom differentiation in sea urchin spines under special consideration as a model for a new impact protective system. Unpubl. PhD Thesis, University of Tübingen, Germany.

Hasenpusch, W. 2000. Die Stachel der Griffelseeigel. *Mikrokosmos* 89(1): 23–27.

Heatfield, B.M. 1971. Growth of the calcareous skeleton during regeneration of spines of the sea urchin, *Strongylocentrotus purpuratus* (Stimpson): a light and scanning electron microscopic study. *Journal of Morphology* 134(1): 57–90.

Hesse, E. 1900. Die Mikrostruktur der fossilen Echinoideenstachel und deren systematische Bedeutung. *Neues Jahrbuch für Mineralogie, Geologie und Palaeontologie* 13: 185–264.

Hiratzka, L.F., Goeken, J.A., White, R.A. & Wright, C.B. 1979. In vivo comparison of replamineform, silastic, and bioelectric polyurethane arterial grafts. *Archives of Surgery* 114: 698–702.

Magdans, U. & Gies, H. 2004. Single crystal structure analysis of sea urchin spine calcites: systematic investigations of the Ca/Mg distribution as a function of habitat of the sea urchin and the sample location in the spine. *European Journal of Mineralogy* 16(2): 261–268.

Magdans, U. 2005. Mechanismen der Biomineralisation von Calcit am Beispiel von Seeigelstacheln: Untersuchung der Wechselwirkung zwischen Sorbat-Molekülen und Calcit-Wachstumsgrenzflächen mit Oberflächen-Röntgenbeugung und numerischen Simulationen. Unpubl. Dissertation, Ruhr Universität Bochum.

Maginnis, T.L. 2006. The costs of autotomy and regeneration in animals: a review and framework for future research. *Behavioral Ecology* 17(5): 857–872.

Märkel, K. & Röser, U. 1983a. The spine tissues in the echinoid *Eucidaris tribuloides*. *Zoomorphology* 103: 25–41.

Märkel, K. & Röser, U. 1983b. Calcite-resorption in the spine of the echinoid *Eucidaris tribuloides*. *Zoomorphology* 103: 43–58.

Märkel, K., Röser, U. & Stauber, M. 1989. On the ultrastructure and the supposed function of the mineralizing matrix coat of sea urchins (Echinodermata, Echinoida). *Zoomorphology* 109: 79–87.

Mortensen. T. 1928. A monograph of the Echinoidea. I. Cidaroidea. Copenhagen: Reitzel.

Müller, J. 1854. Über den Bau der Echinodermen. *Abhandlungen der königlichen Akademie der Wissenschaften* 44: 123–219.

Nickel, K.G., Presser, V., Schultheiß, S., Berthold, C., Kohler, C., Nebelsick, J.H., Grossmann, N., Stegmaier, T., Finckh, H. & Vohrer, A. 2008. Seeigelstachel als Modell für stoffdurchlässige Einschlagschutzsysteme. In Kesel, A.B. & Zehren, D. (eds.), *Bionik-Patente aus der Natur, Tagungsbeiträge*: 26–37. B-I-C: Bremen.

Neinhuis, C. & Barthlott, W. 1997. Characterization and distribution of water-repellent, self-cleaning plant surfaces. *Annals of Botany* 79: 667–677.

Oaki, Y. & Imai, H. 2006. Nanoengineering in echinoderms: the emergence of morphology from nanobricks. *Small* 2(1): 66–70.

Pearse, J.S. & Pearse, V.B. 1975. Growth zones in the echinoid skeleton. *American Zoologist* 15: 731–753.

Presser, V., Schultheiß, S., Berthold, C. & Nickel, K.G. 2009a. Sea urchin spines as a model system for permeable, light-weight ceramics with graceful failure behavior. Part I. Mechanical behavior of sea urchin spines under compression. *Journal of Bionic Engineering* 6: 203–213.

Presser, V., Kohler, C., Živcová, Z., Berthold, C., Nickel, K.G., Schultheiß, S., Gregorovà, E. & Pabst, W. 2009b. Sea urchin spines as a model-system for permeable, light-weight ceramics with graceful failure behavior. Part II. Mechanical behavior of sea urchin spine inspired porous aluminum oxide ceramics under compression. *Journal of Bionic Engineering* 6: 357–364.

Smith, A.B. 1980. Stereom microstructure of the echinoid test. *Special Papers in Palaeontology* 25.

Stewart, C. 1871. On the minute structure of certain hard parts of the genus cidaris. *Quarterly Journal of Microscopical Science* s2–11(41): 51–55.

Stock, S.R., Ignatiev, K., Veis, A, Almer, J.D., De Carlo, F. 2004. Microstructure of sea urchin teeth studied by multiple x-ray modes. In: Nebelsick J.H.& Heinzeller P. (eds.), *Echinoderms: München*: 359–364. München: Balkema.

Strathmann, R.R. 1981. The role of spines in preventing structural damage to echinoid tests. *Palaeobiology* 7(3): 400–406.

Vecchio, K.S., Zhang, X., Massie, J.B., Wang, M. & Kim, C.W. 2007. Conversion of sea urchin spines to Mg-substituted tricalcium phosphate for bone implants. *Acta Biomaterialia* 3: 785–793.

Weber, J.N. 1969. Origin of concentric banding in the spines of the tropical echinoid *Heterocentrotus*. *Pacific Science* 23: 452–466.

Weber, J.N., White, E.W. & Lebiedzik, J. 1971. New porous biomaterials by replication of echinoderm skeletal microstructures. *Nature* 233: 337–339.

Weiner, S., Addadi, L. & Wagner, H.D. 2000. Materials design in biology. *Materials Science and Engineering C* 11: 1–8.

Wilkie, I.C. 2001. Autotomy as a prelude to regeneration in echinoderms. *Microscopy research and technique* 55: 369–396.

Xiao X., de Carlo F. & Stock S.R. 2007. Practical error estimation in zoom-in and truncated tomography reconstructions. *Review of Scientific Instruments* 78: 063705-063707.

Echinoderms in a Changing World – Johnson (ed)
© *2013 Taylor & Francis Group, London, ISBN 978-1-138-00010-0*

Estimating echinoid test volume from height and diameter measurements

L.F. Elliott, M.P. Russell & J.C. Hernández

Biology Department, Villanova University, Pennsylvania, USA

ABSTRACT: The purple sea urchin, *Strongylocentrotus purpuratus*, occurs along the western coast of North America, and shows considerable phenotypic plasticity. Test diameter is the parameter used to quantify and compare urchin populations among habitats. However, diameter fails to account for variations in shape and height:diameter ratio. Volume is a superior metric of urchin size. Two formulae based on height and diameter have been used to estimate volume, but neither were compared with independent estimates of volume to assess their accuracy and precision. We empirically determined the volume of 572 urchins to estimate volume independent of height and diameter. The samples were collected from 16 sites and encompassed the size and shape spectrum. From our analysis we developed a volume formula using height and diameter that accounts for shape differences. We show that functional interpretations of allometric differences can change if volume instead of diameter is used to gauge size.

1 INTRODUCTION

Test diameter is the standard measure to quantify sea urchin size in both field and laboratory studies. For example, diameter is used to construct and compare size frequency distributions (Swan 1961, Ebert 1968, Russell 1987, Ebert 1992). It is also the measure used in growth studies (e.g., Fuji 1967, Ebert 1977, Ebert 1980a, 1980b, Russell 1987, Ebert & Russell 1992, Meidel & Scheibling 1998, Vadas et al. 2002). Studies quantifying allometric relationships have used diameter as the measure of overall size (Ebert 1980b, Russell 1998, Hagen 2007). The functional significance of relatively larger demipyramids (jaws) in Aristotle's Lantern is based on the allometric relationship of jaw to test diameter. Larger jaws indicate low food availability and greater allocation to the food processing apparatus—Aristotle's Lantern (Ebert 1980b, Black et al. 1984, Levitan 1991). Analyzing large scale demographic patterns and viability studies also use test diameter to compare populations (Lester 2007, Ebert 2010).

The purple sea urchin, *Strongylocentrotus purpuratus*, occurs along the west coast of North America from Alaska to Baja California, Mexico (Ricketts et al. 1988). It plays a key role in structuring algal assemblages and higher trophic levels (Pearse 2006). Intertidal *S. purpuratus* show high site fidelity as tagged sea urchins will remain in the same tidepool for at least one year (Russell 1987). On rocky substrates that are soft enough to be eroded, purple sea urchins can be found living in cavities or "pits". The foraging activity and spine abrasion produce these cavities and purple urchins are capable of altering the micro-topography of the substrate (Fewkes 1890). The substrate in turn drastically alters test growth. Purple sea urchins grown under laboratory conditions in pits and on flat surfaces have different shapes that mimic the morphological diversity found in the field (Fig. 1). In addition, Grupe (2006) found differences in the

Figure 1. Two purple sea urchin tests from a laboratory study by Hernandez & Russell (2010). The test on the top was from a sea urchin grown in a pit and the one of the bottom from a sea urchin grown on a flat surface. If these urchins are quantified by diameter they would be considered the same size.

shapes of purple sea urchins in the same tidepools. Urchins in pits had a greater height:diameter (H:D) ratio than urchins not in pits.

The linear measures of diameter and height are insufficient to describe urchin size. Weight is a better measure because it accounts for shape variation and H:D. However, weight can be impractical to use in field studies. In laboratory studies "drain time" differences can affect the values of weight recorded. When weight has been used in field studies (e.g., Lauzon-Guay & Scheibling 2007) a regression of weight-to-diameter is established for a subsample. Then all other data recorded for size are diameter and these values then converted to weight.

Here we evaluate two formulae that use diameter and height to estimate echinoid test volume— the oblate spheroid and the one used by Vasseur (1952). To assess these models we developed and tested an independent method to measure internal test volume. We used this method to quantitatively compare the two models and produced a modified formula for estimating volume for purple sea urchins using diameter and height.

2 MATERIALS AND METHODS

2.1 *Field sites and sample preparation*

We collected sea urchins from 16 field sites (Table 1) during the low tide between 18 May and 16 July 2007. Our focus was to obtain a wide size range and we did not distinguish microhabitat type, e.g., pits versus non-pits. The sites encompassed a variety of substratum types (sandstone, mudstone, and granite) as well as sea urchin shapes.

Samples were frozen and shipped to the lab at Villanova where they were processed by draining the body fluid, removing the gonads, and soaking the skeletal elements, including Aristotle's lantern, in 6% sodium hypochlorite (bleach) overnight. Bleached samples were rinsed in tap water and allowed to soak for ~12h before drained and air-dried. Some tests disarticulated in the bleaching process (Table 1) and for those that were intact we recorded diameter and height with knife-edge, digital vernier calipers (Fig. 1). For each sample we measured the length of one demipyramid from Aristotle's lantern (jaw) using a dissecting microscope fitted with an ocular micrometer.

2.2 *Internal volume estimates*

We developed a method to estimate the internal volume of cleaned echinoid tests that is independent of measures of diameter and height. The concept is simple—fill the intact test with a substance of known density. Volume can then be calculated by multiplying the inverse of filler density by the difference in weight of the filled and empty test. Fluids would prove impractical because of the porous nature of the test so we settled on finely graded sand.

Aquarium sand was sifted repeatedly through 600 μm mesh to produce a matrix that resembled hour-glass sand. In 50 independent trials we determined the density of the matrix by measuring the weight of a known volume, 10 ml in a 10 ml

Table 1. Field sites and sample sizes. Site name, coordinates, and dates collected in 2007. Sample size collected (N) and number processed for volume estimates (n) differ because some individuals disarticulated in the preparation process.

Site	Latitude N	Longitude W	Date	N	n
Palmerston	50°36′15.3″	128°16′39.8″	17 Jun	51	46
Bamfield	48°44′43.1″	125°07′36.4″	14 Jun	47	36
Boiler	44°49′53.1″	124°03′34.5″	19 May	50	41
Gregory Pt	43°20′02.3″	124°22′38.5″	16 May	31	25
Cape Blanco	42°50′26.5″	124°33′49.4″	17 May	50	44
Whiskey	42°13′12.5″	124°22′48.3″	18 May	50	22
Devil's gate	40°24′15.5″	124°23′28.6″	30 Jun	50	18
Bruhl	39°36′16.2″	123°47′22.1″	29 Jun	53	29
Arena cove	38°54′58.1″	123°42′47.1″	01 Jul	50	43
Bodega	38°19′09.7″	123°04′28.9″	02 Jul	51	23
Palomarin	37°55′49.9″	122°44′54.1″	17 Jul	48	41
Bean hollow	37°13′36.1″	122°24′39.7″	17 Jul	58	54
Garapata	36°20′07.4″	121°56′03.7″	15 Jul	50	49
Partington	36°10′26.6″	121°41′47.3″	16 Jul	51	48
Cambria	35°34′41.1″	121°07′04.0″	03 Jul	21	9
White's Pt	33°42′54.0″	118°19′10.0″	04 Jul	52	44

graduated cylinder. We used the average of these trials as the density of the sand.

The peristomal openings of cleaned tests were sealed with parafilm and the test/parafilm were then weighed (±.001 g). The sand-matrix was funneled through surgical tubing into the test through the aboral opening (the periproct and madreporite were removed). Residual sand was gently brushed from the exterior surfaces and the test/parafilm/sand-matrix weighed. Volume was then determined using the equation:

$$\text{Volume} = \frac{1}{d} \times \left(\text{Test}_{\text{Filled}} - \text{Test}_{\text{Empty}}\right) \qquad (1)$$

where d is density of the sand and $\text{Test}_{\text{Filled}}$ and $\text{Test}_{\text{Empty}}$ are the two measures of weight.

To evaluate precision and potential size bias of this method, 6 individuals were selected that represented the range of sizes in our sample (test diameter measurements in mm): the smallest (11.08), the largest (77.55), and four intermediate samples (69.97, 56.06, 38.44, 21.54). The volume estimate method was applied five times to each of these tests and the Coefficient of Variation (CV) calculated for each sample.

All intact tests (n = 572) were then evaluated for volume estimates using this method. These data provided a measure of volume independent of test height and diameter.

2.3 Evaluation of volume formulae

We used our independent measures of volume based on the amount of sand-matrix to assess the two formulae that estimate volume from height and diameter. The first was a formula introduced by Vasseur (1952):

$$\text{Volume} = \text{Diameter}^2 * \frac{\text{Height}}{2} \qquad (2)$$

The second is the formula for an oblate spheroid first used by Ebert (1988):

$$\text{Volume} = \frac{4}{3}\pi \times \left(\frac{1}{2} \times \text{Diameter}\right)^2 \times \frac{1}{2}\text{Height} \qquad (3)$$

This formula is mathematically equivalent to the one used by Middleton et al. (1998).

$$\text{Volume} = \eta\frac{\pi}{6}\text{Diameter}^3 \qquad (4)$$

where η is the average height-to-diameter ratio of a large sample.

3 RESULTS

3.1 Volume estimates from sand-matrix

All error estimates are reported as ± s. The mean density of the sifted sand matrix was 1.5175 ± 0.0207 g/ml (n = 50). One s is equivalent to a volume estimate of 0.01364 ml. The smallest sea urchin in our sample (test diameter = 11.08 mm) produced a mean volume estimate of 0.299 ml ± 0.00346 (n = 5) which is less than 5% of 1 s of the volume. Figure 2 plots the CV for each of the 6 tests where repeated estimates (n = 5) of volume were recorded. The slope of the line is not significantly different from zero (t = 0.17, p = .8735).

Figure 3 plots the calculated values of volume based on height and diameter (equations 2 and 3) against the measured volume estimate. Both the Vasseur and Oblate Spheroid formulae underestimate test volume and the difference increases as volume increases. Visual inspection of this plot shows the underestimates of the Vasseur formula are greater. A paired t-test evaluating the difference between these estimates shows a significant difference (t = −28, p < .0001). Based on these results the Vasseur formula was eliminated from further analyses.

A formula that is both unbiased and precise would produce a plot of residuals evenly distributed around zero for all volume sizes. A regression of the residuals of the Oblate Spheroid formula against Volume (Fig. 4) produces a negative slope that is significantly different from zero (r^2 = .258, slope = −0.052, t = −14.05, p < .0001) and an intercept that is not significantly different from zero

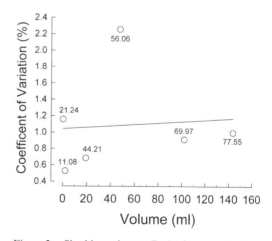

Figure 2. Size bias estimates. Each of 6 sea urchin tests spanning the range of sizes in our sample was measured five times for volume and the resulting CV plotted against volume. The diameters (mm) for each test are listed for each point. r^2 = .007 for the regression line.

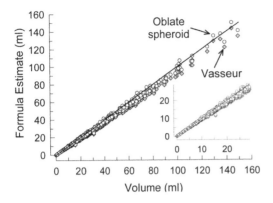

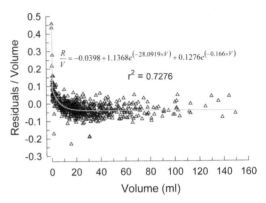

Figure 3. Volume estimates. For each sea urchin, volume was estimated independent of test height and diameter measurements by filling the empty test with sand of known mass (x-axis). The two formulae, Oblate Spheroid (open circles) and Vasseur (filled diamonds), were applied to the values for height and diameter for each individual (n = 572) and plotted on the y-axis. The plotted line of slope = 1 indicates a perfect fit between the calculated and measured values of volume.

Figure 5. Residuals standardized by volume (n = 572). The exponential decay function that defines the best fit (gray) line is displayed on the plot along with the r^2 value (R = Residual and V = Volume) Multiplying both sides of this equation by V yields a size-specific correction for the Oblate Spheroid equation to estimate volume.

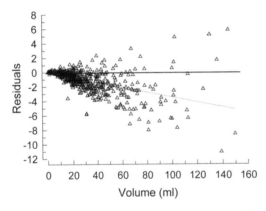

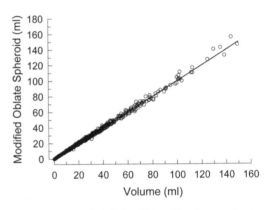

Figure 4. Oblate Spheroid residuals plot. As expected the range of the residuals increases with increasing volume. There is a significant negative trend (gray line) and the slope indicates that as urchin size increases the underestimate of the Oblate Spheroid formula also increases (n = 572). The black line has a slope and intercept of 0.

Figure 6. Modified Oblate Spheroid volume estimates (n = 572). The correlation between the calculated volume (y-axis) and measured value (x-axis) is clear. The black line defines a slope = 1 and is not significantly different from the regression line for these data.

(t = −0.034, p = .58). This trend shows significant bias and as size increases the underestimate of volume of the Oblate Spheroid formula also increases.

When the residuals are standardized by urchin size (residuals/volume) the distribution of points appears to be a negative exponential decay (Fig. 5). The equation that defines this function can be used to modify the Oblate Spheroid formula. Multiplying the right side of the equation in Figure 5 by Volume (V) yields a size-specific

correction. A Modified Oblate Spheroid equation then is:

$$Volume = Eq.3 - Correction \qquad (5)$$

where Eq. 3 is the Oblate Spheroid (equation 3 above) and *Correction* is the right side of the function in Figure 5 multiplied by V. In the *Correction* term one simply substitutes the Oblate Spheroid equation (equation 3) wherever V appears to generate volume estimates from diameter and height measurements.

Figure 6 plots the volume estimates of the Modified Oblate Spheroid equation (5) against the measured values of volume from filling the tests with sand. The slope (= 1.0045, t = 1.79, p = .08) and intercept (= –0.054, t = –0.55, p = .58) of the regression are not significantly different from 1.0 and zero respectively.

Repeating the same analysis of the residuals for the Oblate Spheroid from Figure 4 yields a different result for the Modified Oblate Spheroid equation (Fig. 7). These residuals are equally distributed around zero and the linear regression for these data yields a slope (t = 1.79, p = .07) and intercept (t = –.55, p = .58) not significantly different from zero.

3.2 *Allometry analysis*

To assess the biological significance of using volume rather than diameter as the metric of overall size we used the samples from this study in an allometric analysis. Sea urchins with lower food availability have relatively larger jaw sizes and it has been suggested that this relationship is functional and could be used to assess resource availability (Ebert 1980b, Black et al. 1984, Levitan 1991). In these previous studies the standard measure of overall size has been diameter. Here we compare the results of two allometric analyses on the same samples—one using diameter, and the other volume, to compare jaw sizes among field sites. We restricted the analyses to the 12 sites from Table 1 that had estimates of volume n ≥ 25.

We did not use an ANCOVA to compare jaw sizes because the untransformed and ln-transformed data violated the assumption of homogeneity of slopes with both diameter (ln-transformed, $F_{11,472} = 2.61$, p = .003) and volume (ln-transformed,

$F_{11,472} = 2.92$, p = .0009) as the covariates. Instead we compared the ratio of jaw size to overall size using ANOVA. We used the cube root of volume estimates from the Modified Oblate Spheroid equation so the units between the analyses were standardized. The untransformed data violated the assumption of normality (jaw:diameter – W = .973, p = .0002; jaw:volume$^{1/3}$ – W = .976, p = .003) whereas the ln transformed data did not (ln[jaw:diameter] – W = .985, p = .520; ln[jaw:volume$^{1/3}$] – W = .982, p = .220).

Both ANOVAs revealed significant differences in relative jaw sizes among field sites (ln[jaw:diameter] – $F_{11,484} = 18.17$, p < .0001; ln[jaw:volume$^{1/3}$] – $F_{11,484} = 25.21$, p < .0001). Significant differences for comparisons among all pairs using Tukey-Kramer HSD for both analyses are plotted in Figure 8.

These two analyses yield very different results. First, there is a significant difference in the rankings of the field sites (Kendall's τ = .757, p = .0006). Although some of the pair-wise comparisons of sites are consistent between the analyses, e.g., Bruhl Bay and Gregory Point (no difference in both analyses), and Arena Cove and Partington (different in both analyses) others are not. In the Cape Blanco and Bean Hollow comparison there is no significant difference in the ln(jaw:diameter) comparison but in the ln(jaw:volume$^{1/3}$) comparison there is (Fig. 8). Of the 66 possible pair-wise comparisons between field sites, 27 show a difference between the two analyses—significantly different in one but not the other.

4 DISCUSSION

Sea urchins that have similar test diameters can have dissimilar overall sizes due to differences in test height and shape (Fig. 1). The purple sea urchin displays a high degree of morphological variation and phenotypic plasticity. Some of this variation may be induced by the micro-topography of the substrata as sea urchins found in pits compared to those on flat surfaces show taller H:D ratios (Grupe 2006, Hernandez & Russell 2010). Although the standard metric for comparing purple sea urchins and other echinoid populations is test diameter, measures that account for test shape variation are superior because they more accurately reflect overall size. For example, there is a 1% difference in diameter between the tests in Figure 1 (the bottom test is slightly larger); however there is a 37% difference in height. Using the Modified Oblate Spheroid formula to calculate volume reveals a 25% difference in overall size between these two individuals. Classifying these urchins by diameter would result in scoring them equivalent

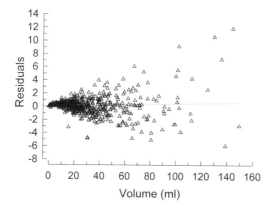

Figure 7. Modified Oblate Spheroid residuals (n = 572). The residuals are evenly distributed around zero for all sizes of the sea urchins indicating this formula yields an unbiased estimate of volume. The slope and intercept of the regression line (gray) that fit these data are not significantly different from 0.

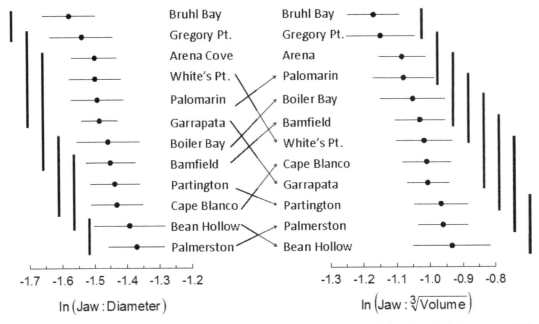

Figure 8. Allometry analysis. Results of two ANOVA analyses comparing the relative sizes of the jaw among field sites. The means of ln(Jaw:Diameter) and ln(Jaw:Volume$^{1/3}$) ± s are plotted for each field site for each analysis. The vertical lines on the outer edge of each plot connect sites that are not significantly different for that analysis. The arrows in the middle connect the same field sites between the analyses and help visualize the change in the rankings.

but volume more accurately reflects the real difference in overall size. Although we chose an extreme example to visually illustrate our point (Fig. 1), significant differences in biological interpretations can result by using diameter versus volume as illustrated in the allometric analysis (Fig. 8).

Vasseur (1952) used equation 2 in an allometric comparison of growth of Aristotle's lantern in *S. droebachiensis* and *S. pallidus* from different populations in Norway. Ebert (1988) used the Oblate Spheroid formula (equation 3) in an interspecific study of several sea urchins to evaluate the effects of allometric ontogenetic changes on volume. In a study of the deep-sea echinoid *Echinus affinus* Middleton et al. (1998) used equation 4 to model growth. Equation 4 yields a volume estimate from diameter only but depends on a constant, η, the average H:D from a large sample. These studies did not measure volume directly but based their estimate of volume on measures of diameter or height and diameter. Here we show that both equations 2 and 3 consistently underestimate internal volume (Fig. 3) and that for *S. purpuratus* the Modified Oblate Spheroid equation provides an unbiased estimate of volume using the linear dimensions of height and diameter (Figs. 5–7). To produce similar equations for other species requires applying the methodology outlined in this study.

Our sites spanned a large portion of the geographic range and encompassed a variety of different substrates like soft mudstone and sandstone at Palomarin, Gregory Pt., and Arena Cove, to hard granite at Garrapata and Bodega (Table 1). This wide spatial and habitat range made it more likely that we captured the morphological diversity that produces differences in overall size from differences in shape (Fig. 1). Although the parameters we generated in this study are specific for *S. purpuratus* the methods we developed could be applied to other species.

Although tedious, our method to assess internal volume produced precise estimates that were not biased with increasing size (Fig. 2). This method was successful in part because we painstakingly refined a sand matrix by repeated sifting until it was uniformly dense. This matrix had the consistency of hour-glass sand and once it was refined it was relatively easy to manipulate and pour into an urchin test that had the peristomal opening sealed with parafilm. After the tests were cleaned and dried and the sand matrix prepared, the processing of individual samples proceeded quickly (~5 minutes each) allowing us to generate data for 572 sea urchins.

Any measure of the internal volume of the test will vary with the relative curvature and position of the peristomal membrane. This membrane changes position with the movement of Aristotle's lantern during feeding activities (Ellers & Telford 1992). Our method standardized the measures of volume by fixing the position where the peristomal membrane would occur by sealing the oral opening with parafilm perpendicular to the oral edge of the test height measurement. We have observed the curvature of the peristomal membrane extend both distally and proximally to this plane so standardizing it yielded consistent observations across our samples.

In summary, the approach outlined here could be applied to other echinoids to develop species-specific parameters for the Modified Oblate Spheroid method to estimate volume. The parameters we provide here could also be used to re-evaluate previous allometric studies that provide data on height and diameter.

ACKNOWLEDGEMENTS

We received support from an NSF grant to MPR and a Villanova Undergraduate Research Award to LFE. We are grateful for the laboratory assistance of Nathan Haag, Sam Morelli, Ray Costa, Michelle Harris, Victoria Garcia and Aimee Lee Russell. This study benefited from discussions with Drs. Bill Fleischman, Peter Petraitis, and Thomas Ebert. The results of this study fulfilled the requirements for a Senior Thesis in Biology at Villanova University for LFE.

REFERENCES

Black, R., C. Codd, D. Hebbert, S. Vink, & J. Burt. 1984. The functional significance of the relative size of Aristotle's Lantern in the sea urchin, *Echinometra mathaei* (de Blainville). *Journal of Experimental Marine Biology and Ecology* 77:81–97.

Ebert, T.A. 1968. Growth rates of the sea urchin *Strongylocentrotus Purpuratus* related to food availability and spine abrasion. *Ecology* 49:1075–1091.

Ebert, T.A. 1977. An experimental analysis of sea urchin dynamics and community interactions on a rock jetty. *Journal of Experimental Marine Biology and Ecology.* 27:1–22.

Ebert, T.A. 1980a. Estimating parameters in a flexible growth equation, the Richards function. *Canadian Journal of Fisheries and Aquatic Sciences.* 37: 687–692.

Ebert, T.A. 1980b. Relative growth of sea urchin jaws: An example of plastic resource allocation. *Bulletin of Marine Science.* 30: 467–474.

Ebert, T.A. 1988. Allometry, design and constraint of body components and of shape in sea urchins. *Journal of Natural History* 22:1407–1425.

Ebert, T.A. 2010. Demographic patterns of the purple sea urchin *Strongylocentrotus purpuratus* along a latitudinal gradient, 1985–1987. *Marine Ecology Progress Series* 406:1057–120.

Ebert, T.A., & Russell, M.P. 1992. Growth and mortality estimates for red sea urchin *Strongylocentrotus franciscanus* from San Nicolas Island, California. *Marine Ecology Progress Series* 81:31–41.

Ellers, O. 1993. A mechanical model of growth in regular sea urchins: predictions of shape and a developmental morphospace. *Proceedings of the Royal Society of London* 254:123–129.

Ellers, O., & M. Telford. 1992. Causes and consequences of fluctuating coleomic pressure in sea urchins. *Biological Bulletin* 182:424–434.

Fewkes, J.W. 1890. On Excavations made in rocks by sea-urchins. *The American Naturalist* 24:1–21.

Fuji, A. 1967. Ecological studies on the growth and food consumption of Japanese common littoral sea urchin, *Strongylocentrotus interdedius* (A. Agassiz). *Memoirs of the Faculty of Fisheries Hokkaido University* 15:83–160.

Grupe, B.M. 2006. Purple sea urchins (*Strongylocentrotus purpuratus*) in and out of pits: the effects of micro-habitat on population structure, morphology, growth and mortality. University of Oregon.

Hagen, N.T. 2007. Enlarged lantern size in similar-sized, sympatric, sibling species of Strongylocentrotid sea urchins: from phenotypic accommodation to functional adaptation for durophagy. *Marine Biology* 153:907–924.

Hernández, J.C., & Russell, M.P. 2010. Substratum cavities affect growth-plasticity, allometry, movement and feeding rates in the sea urchin *Strongylocentrotus purpuratus*. *Journal of Experimental Biology* 213:520–525.

Lauzon-Guay, J.-S., & Scheibling, R.E. 2007. Behaviour of sea urchin *Strongylocentrotus droebachiensis* grazing fronts: food-mediated aggregation and density dependent facilitation. *Marine Ecology Progress Series* 329:191–204.

Lester, S.E., Gaines, S.D., & Kinlan, B.P. 2007. Reproduction on the edge: large-scale patterns of individual performance in a marine invertebrate. *Ecology* 88:2229–2239.

Levitan, D.R. 1991. Skeletal changes in the test and jaws of the sea urchin Diadema antillarum in response to food limitation. *Marine Biology* 111:431–435.

Meidel, S.K. & R.E. Scheibling. 1998. Size and age structure of subpopulations of sea urchins *Strongylocentrotus droebachiensis* in different habitats. In M. Telford & R. Mooi (eds), *Proceedings of the Ninth International Echinoderm Conference*: 737–742.A.A. Balkema: Rotterdam.

Middleton, D.A.J., Gurney, W.S.C., & Gage, J.D. 1998. Growth and energy allocation in the deep-sea urchin *Echinus affinis*. *Biological Journal of the Linnean Society* 64:315–336.

Pearse, J.S. 2006. Ecological role of purple sea urchins. *Science* 314:940–941.

Ricketts, E.F., Calvin, J., & Hedgpeth, J.W. 1988. Between Pacific Tides. Stanford University Press, Stanford, CA.

Russell, M.P. 1987. Life history traits and resource allocation in the purple sea urchin *Strongylocentrotus purpuratus* (Stimpson). *Journal of Experimental Marine Biology and Ecology* 108:199–216.

Russell, M.P. 1998. Resource allocation plasticity in sea urchins: rapid, diet induced, phenotypic changes in the green sea urchin, *Strongylocentrotus droebachiensis*. *Journal of Experimental Marine Biology and Ecology* 220:1–14.

Swan, E.F. 1961. Some Observations on the Growth Rate of Sea Urchins in the Genus Strongylocentrotus. *Biological Bulletin* 120:420–427.

Vadas, R.L., Sr., Smith, B.D., Beal, B., & Dowling, T. 2002. Sympatric growth morphs and size bimodality in the green sea urchin (Strongylocentrotus droebachiensis). *Ecological Monographs* 72:113(120).

Vasseur, E. 1952. Geographic variation in the Norwegian sea urchins, *Strongylocentrotus droebachiensis* and *S. pallidus*. *Evolution* 6:87–100.

Echinoderms in a Changing World – Johnson (ed)
© 2013 Taylor & Francis Group, London, ISBN 978-1-138-00010-0

Stone canal morphology in the brachiolaria larva of the asterinid sea star *Parvulastra exigua*

V.B. Morris
School of Biological Sciences A12, University of Sydney, NSW, Australia

P. Cisternas
Department of Anatomy and Histology, University of Sydney, NSW, Australia

R. Whan
Electron Microscope Unit, University of Sydney, NSW, Australia

M. Byrne
Department of Anatomy and Histology, University of Sydney, NSW, Australia

ABSTRACT: The morphology of the stone canal and the pore canal in late brachiolaria larvae of *Parvulastra exigua* is described from the series of sections through the larvae obtained by laser-scanning confocal microscopy. The stone canal connects primary podium D of the hydrocoele with the pore canal, close to where the pore canal opens into the anterior coelom. The pore canal opens externally at the hydropore. The stone canal and the pore canal are the aboral components of the water vascular system. The connection of the stone canal to a primary podium of the hydrocoele is a means of naming the primary podia of the larva, according to a defined system.

1 INTRODUCTION

The stone canal in the adult asteroid is the structure that connects the ring canal of the water vascular system to the pore canals of the madreporite, which connect with the exterior (Hyman 1955). The stone canal is attached to the wall of the axial sinus and opens aborally into the ampulla which lies below the madreporic pore canals and which is part of the axial sinus (Gemmill 1914).

In the larva, the stone canal has been described as forming on the posterior wall of the anterior coelom (MacBride 1896) or as a groove in the left hydrocoele (Hyman 1955). The pore canal, which later branches forming the pores of the madreporite, develops before the stone canal as an outgrowth from the anterior coelom that fuses with the ectoderm forming an opening, which is the hydropore (MacBride 1896). In *Parvulastra exigua*, the pore canal is present in early brachiolaria larvae whereas the stone canal is not seen until a late stage of brachiolarian development (Morris et al. 2009). In *Asterias rubens*, which has a feeding larval stage, the stone canal forms five to six days before metamorphosis (Gemmill 1914). The first note of a clear distinction between the pore canal and the stone canal is attributed to Ludwig (MacBride 1896 p. 388).

Here, we describe the morphology of the pore canal and the stone canal in late brachiolaria larvae of the asterinid *P. exigua*. We add to a previous report of the stone canal given in the account of the development of the primary podia in *P. exigua* (Morris et al. 2009). Larval development in *P. exigua* is abbreviated: the brachiolarian stage is reached without trace of a preceding bipinnarian feeding-stage and the adult rudiment starts to form within a few days of fertilization (Byrne 1995). The observations were made in a confocal laser-scanning microscope from whole-mounts of larvae made autofluorescent by glutaraldehyde fixation (Morris et al. 2009).

The primary podia of *P. exigua* have been named (Morris et al. 2009) according to the Carpenter system as defined by Hyman (1955). In this system, the echinoderm rays are named from A to E in clockwise order in oral view where ray A is defined as lying opposite the interradius housing the madreporite, which is thus the interradius between rays C and D. The connection of the stone canal to primary podium D that we are reporting for *P. exigua* has application in identifying the ray homologies between the echinoderm classes, a problem addressed by Hotchkiss (1998).

The stone canal and the pore canal are early stages in the development of the coelomic connection between the water vascular system and the external environment. They are the aboral components of the water vascular system in contrast to the hydrocoele and its derivatives, which are the oral components of the system. Conceptually, they are a phylotypic part of the echinoderm body plan with some claim to homology with the proboscis pore of hemichordates, a homology noted in the early literature (Gemmill 1914).

2 METHODS

Adults of *P. exigua* were collected from the intertidal zone of rocky shores near Sydney, NSW, Australia and were spawned and fertilized as described by Byrne (1995). The embryos and larvae were reared in small dishes in filtered sea water at temperatures between 19°C and 21°C. They were fixed in 2.5% (v/v) glutaraldehyde (Pro-SciTech) in filtered sea water for 1–2 hours, dehydrated in an ethanol series to 100% ethanol, then cleared in 2:1 (v/v) benzyl benzoate/benzyl alcohol (Sigma). They were fixed at 7 days and 10 days after fertilization. For viewing, the larvae were mounted, in the clearant, in a chamber between glass coverslips that sealed a hole cut through a conventional microscope slide.

The mounted larvae were viewed in the Olympus FluoView 1000 laser-scanning system (version 1.7.1.0) attached to an Olympus IX81 inverted microscope. Each larva was excited with a 633 nm helium-neon laser with the emission collected from 645–745 nm. A stack of images was collected along the Z axis (the Z stack) with a pixel dimension of 1.24×1.24 μm at a slice thickness of 1.10 μm averaged over two frames with a 20x UplanApo objective lens NA 0.7 in a 512×512 pixel array, 12 bits/pixel. In the Z stack, the larva was sectioned in the XY plane; sections of the larva in the XZ and YZ planes were constructed from the Z stack. The resolution in the XZ and YZ planes was lower, 2.592 μm, compared with a resolution of 0.386 μm in the sections from the Z stack itself, a property of confocal imaging (Cox 2007).

3 RESULTS

The morphology of the pore canal, the stone canal and their connections in the brachiolaria larva of *P. exigua* is shown in the series of selected sections of one 7 day larva and one 10 day larva in Figures 1 and 2, respectively. The pore canal connects the hydropore to the anterior coelom while the stone canal branches from the pore canal and connects to primary podium D. A variant of this morphology is shown in Figure 3.

3.1 *The morphology in Figure 1*

The hydropore which lies on the larval dorsal surface just left of the midline opens into a thick-walled channel, the pore canal (Fig. 1A). The pore canal connects with the anterior coelom which lies over the anterior end of the archenteron (Fig. 1B-G and I-M). Close to where the lumen of the pore canal opens into the anterior coelom, a bud is seen on the pore canal (Fig. 1C,I). The bud, in deeper sections, transforms into a thick-walled channel, the stone canal (Fig. 1D,E,F and J,K,L). The stone canal connects to primary podium D (Fig. 1G,M).

3.2 *The morphology in Figure 2*

The profile of the pore canal, seen in a section below the dorsal ectoderm of the larva (Fig. 2A), tracks anteriorly (Fig. 2B) before opening into the anterior coelom (Fig. 2C). The separate profile of the stone canal has turned to the larval left (Fig. 2D) and its wall joins the posterior wall of podium D (Fig. 2E) in the region where the lumen of podium D joins the lumen of podium C (Fig. 2E,F). Podium C is dorsal with respect to podium D. The order of the series of sections is from larval dorsal to larval ventral.

3.2.1 *The point of branching of the pore canal and the stone canal*

In Figure 2G-O, the view is in a plane orthogonal to that in Figure 2A-F and is of a series of transverse sections of the larva constructed from the same Z stack with larval dorsal uppermost and larval left on the left of each panel. The hydropore (Fig. 2G) leads into the pore canal (Fig. 2H). The profile of the pore canal extends ventrally (Fig. 2I) before the separate profile of the stone canal is seen (Fig. 2J,K). The pore canal joins the anterior coelom which is on its right in the panel, and in the larva (Fig. 2L). Thus, the junction of the stone canal and the pore canal is close to where the pore canal opens into the anterior coelom. In the enlarged views (Fig. 2M-O), the stone canal proceeds leftward, towards the hydrocoele, to join podium D.

3.3 *The morphology in Figure 3*

A variant of the morphology of the pore canal and the stone canal is shown in Figure 3. The hydropore on the larval dorsal surface is irregularly shaped (Fig. 3A,B). Two thick-walled channels lead internally from the hydropore (Fig. 3C,D). Both channels open into the anterior coelom (Fig. 3E,F and H,I). The larger of the channels becomes a separate profile in deeper sections (Fig. 3G,J) and it connects with primary podium D (Fig. 3K-N and O-Q).

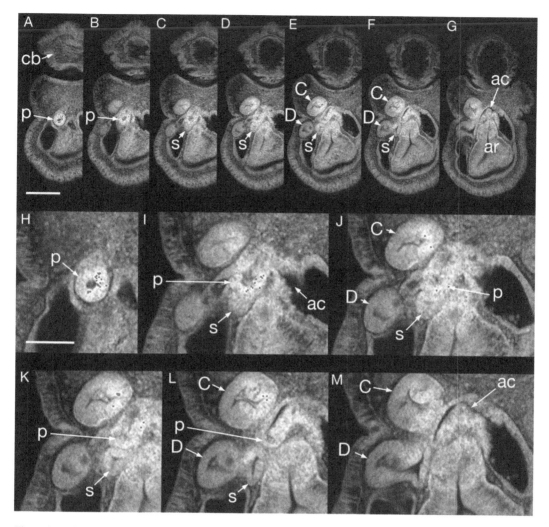

Figure 1. The paths of the pore canal and the stone canal in a 7 day brachiolaria larva (described in text), larval dorsal views, anterior at top of panel. A-G, selected sections from the Z stack. H-M, enlarged views from the Z stack. ac, anterior coelom; ar, archenteron; cb, central brachium; C, primary podium C; D, primary podium D; p, pore canal; s, stone canal. Scale bars 50 μm and 100 μm.

3.4 *Number of observations*

Thirteen larvae were examined. In ten of these, the stone canal connected with primary podium D. In two of these ten, the connection to podium D was close to the junction between the bases of podium D and podium C. In the remaining three larvae, the connection was unclear.

4 DISCUSSION

We report evidence of the connection of the stone canal in the larval stages of *P. exigua* to primary podium D, with instances of the connection being close to the junction between primary podia C and D. This evidence is in agreement with the description by Gemmill (1914 p. 257) of the inner end of the stone canal in *Asterias rubens* to be between pouches I and II of the hydrocoele and in *Solaster endeca* to be opposite pouch II. In *Asterina gibbosa*, MacBride (1896 p. 362) describes the stone canal as running from the second lobe of the hydrocoele, which is the same structure as pouch II in *A. rubens*. The primary podia C and D in *P. exigua* (Morris et al. 2009) lie in similar positions to pouches I and II in *A. rubens*. The pouches in *A. rubens* are named from I to V in oral view in clockwise order, with pouch I in an anterior-dorsal

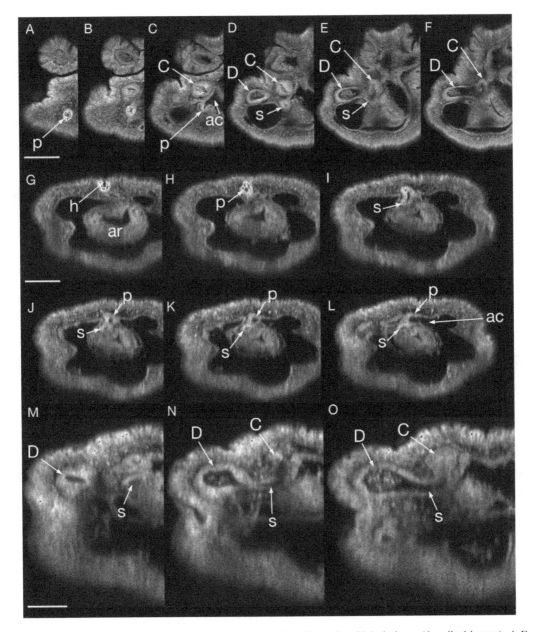

Figure 2. The paths of the pore canal and the stone canal in a 10 day brachiolaria larva (described in text). A-F, selected sections from the Z stack, larval dorsal views, anterior at top of panel. G-O, selected transverse sections of the larva (constructions from the Z stack of the XZ plane), larval dorsal uppermost, viewed from the posterior with larval left on left of panel. ac, anterior coelom; ar, archenteron; C, primary podium C; D, primary podium D; h, hydropore; p, pore canal; s, stone canal. Scale bars 50 μm and 100 μm.

position (Gemmill 1914). Primary podium C of *P. exigua* is thus in a similar position to pouch I of *A. rubens*.

The primary podia in *P. exigua* are named according to the Carpenter system, as defined by Hyman (1955) such that the location of the madreporite defines the CD interradius. In the absence of a madreporite in *P. exigua* at the bra-chiolarian stage that we observed, the following two sources of evidence were used to name the

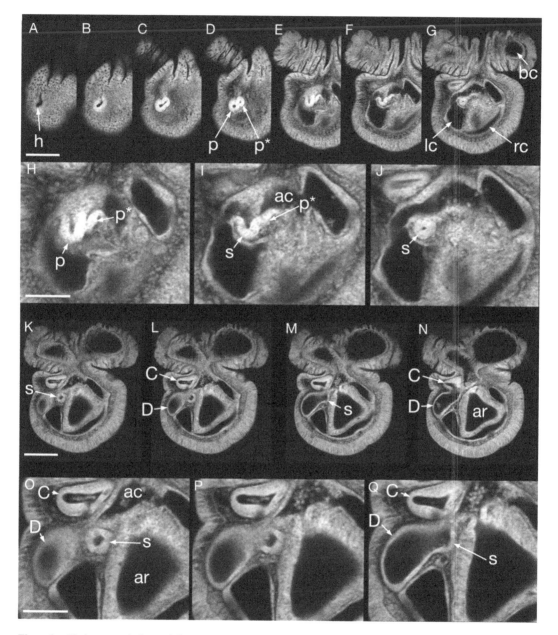

Figure 3. Variant morphology of the pore canal and the stone canal in a 7 day brachiolaria larva (described in text), larval dorsal views, anterior at top of panel. A-G and K-N are selected sections from the Z stack. H-J and O-Q are enlarged views of portions of the sections in A-G and K-N respectively. ac, anterior coelom; ar, archenteron; bc, brachial coelom; C, primary podium C; D, primary podium D; h, hydropore; lc, left coelom; p, pore canal; p*, small pore canal; rc, right coelom; s, stone canal. Scale bars 50 μm and 100 μm.

primary podia. The first, given previously (Morris et al. 2009), is from *A. rubens* where the nascent madreporite lies between the larval arm rudiments I and II which become aligned later with the hydrocoele pouches I and II (Gemmill 1914).

The second is from *Asterias forbesi* where the axial complex, which contains the stone canal and ends at the madreporite, lies in the interbrachial septum of the CD interradius (Hyman 1955 p. 283). Thus, when the madreporite which is a defined standard for naming the primary podia and the rays that develop from them cannot be observed, the inner connection of the stone canal to the hydrocoele, or the ring canal, might be used to name the primary podia, particularly in echinoderm larvae.

The stone canal in the brachiolaria larva of *P. exigua* is described as a branch from the pore canal, close to where the pore canal joins the anterior coelom. The stone canal is thus open to the anterior coelom. The descriptions in the early literature vary with MacBride (1896 p. 362), for *A. gibbosa*, writing that the epithelium of the stone canal becomes continuous with that of the pore canal, but also that the conjoined tubes are open to the anterior coelom. For *A. rubens* Gemmill (1914) describes the stone canal and the pore canal opening independently into the anterior coelom. The importance of the connection of both canals to the anterior coelom is that, in later development, the anterior coelom forms the axial sinus to whose inner wall the stone canal is attached. Thus, the ampulla that lies between the funnel-like opening of the stone canal and the pore canals of the madreporite above it in the adult, a region that Gemmill (1914) called the ampullary region of the axial sinus, probably has its developmental origin from the region of the anterior coelom where the stone canal and the pore open in the larva.

The variant form of stone canal morphology in Figure 3 might be a consequence of the earlier time at which the pore canal forms in brachiolaria larvae, the smaller canal in Figure 3 perhaps being the pore canal that formed early. We assume, in this larva, that the later-formed stone canal has failed to make the connection with this early pore canal and instead developed another path to the exterior. This could be the outcome of a mistake in the signalling mechanism that guides the stone canal epithelium to the junction with the epithelium of the pore canal. The variant morphology supports the separate nature of the pore canal and the stone canal: in forming early, the pore canal is a larval structure, most likely with a role in early larval functions, whereas the stone canal is a structure of the adult.

ACKNOWLEDGEMENTS

We acknowledge the facilities and the scientific and technical assistance from staff of the Australian Microscopy & Microanalysis Facility at the Electron Microscope Unit, The University of Sydney.

REFERENCES

Byrne, M. 1995. Changes in larval morphology in the evolution of benthic development by *Patiriella exigua* (Asteroidea: Asterinidae), a comparison with the larvae of *Patiriella* species with planktonic development. *Biol. Bull.* 188: 293–305.

Cox, G. 2007. *Optical imaging techniques in cell biology*. Boca Raton, FL: CRC Press, Taylor & Francis.

Gemmill, J.F. 1914. The development and certain points in the adult structure of the starfish *Asterias rubens*, L. *Phil. Trans. Roy. Soc. B* 205: 213–294.

Hotchkiss, F.H.C. 1998. A "rays-as-appendages" model for the origin of pentamerism in echinoderms. *Paleobiology* 24: 200–214.

Hyman, L.H. 1955. *The invertebrates: Echinodermata* IV. New York: McGraw-Hill.

MacBride, E.W. 1896. The development of *Asterina gibbosa*. *Q.J. Microsc. Sci.* 38: 339–411.

Morris, V.B., Selvakumaraswamy, P., Whan, R. & Byrne, M. 2009. Development of the five primary podia from the coeloms of a sea star larva: homology with the echinoid echinoderms and other deuterostomes. *Proc. Roy. Soc. B* 276: 1277–1284.

Echinoderms in a Changing World – Johnson (ed)
© *2013 Taylor & Francis Group, London, ISBN 978-1-138-00010-0*

Arm damage and regeneration of *Tropiometra afra macrodiscus* (Echinodermata: Crinoidea) in Sagami Bay, central Japan

R. Mizui & T. Kikuchi

Graduate School of Environment and Information Sciences, Yokohama National University, Japan

ABSTRACT: A natural population of the comatulid crinoid *Tropiometra afra macrodiscus* was monitored in Sagami Bay, central Japan. Serious damage to the arms of these crinoids was observed after a strong typhoon struck Sagami Bay. A rapid regeneration rate of 17.6 cm year^{-1} was observed among these crinoids. This rate was more than twice the value of the normal growth rate (8.1 cm year^{-1}). Microscopic observations revealed that only 20% of breakages had occurred at syzygial articulations. The pattern of arm loss was obviously different from that caused by predator attacks in other autotomic species. A negative relationship was observed between the length of stumps and that of regeneration. Our results suggest that *T. a. macrodiscus* controls the rate of regeneration of each arm in order to maintain efficient feeding and for rapid recovery of reproductive ability.

1 INTRODUCTION

The feather star *Tropiometra afra macrodiscus* (Hara)—a comatulid crinoid—is found from Hong Kong to central Japan and in the Bonin Islands (Clark 1947, Kogo 1998) and is one of the most common shallow-water species in southern Japan (Kogo 2006). In Sagami Bay, this species is the largest in size and one of the most common feather star species in the rocky shallow-water macro benthic community. They have only ten arms with a length of up to 40 cm long; these arms are whippy and considerably stout but cannot be curled. Seldom are the arms cut off even when strongly touched during collection (Utinomi and Kogo 1965). They are diurnal and observed to extend their arms perpendicular to the substrate where exposed to currents. *T. a. macrodiscus* is not capable of swimming; they use thick and large cirri for their locomotion. Therefore, their locomotory ability is poor. Although the species is relatively large in size, their ecology and/or behavior are still not well understood.

In order to clarify their behavioral pattern, long-term *in situ* observations of four young individuals of *T. a. macrodiscus* were carried out on an artificial reef off Manazuru Peninsula, Sagami Bay from October 2005. During the period of observation, a strong typhoon struck the artificial reefs (September, 2007) and four young individuals and one adult individual were observed to lose parts of every arm. Such serious damage was not observed to have occurred to any other individuals or species during the observation period. In general, the

damage or loss of arms of feather stars occurs due to predation by fish or by other invertebrates (Mladenov 1983, Schneider 1988, Vail 1989, Messing 1997). There is currently little information on the damage caused to comatulids by natural hazards. After the typhoon had passed, *in situ* observations were continued for a further six months to obtain information on the patterns of regeneration of damaged arms of *T. a. macrodiscus*.

Here, we report information on the growth rate of *T. a. macrodiscus*, the extent of arm damage caused by the heavy typhoon event, and the mechanism of recovery.

2 MATERIALS AND METHODS

This study was conducted on an artificial reef deployed on a sandy bottom of 20 m depth off the Manazuru Peninsula in the northwestern part of Sagami Bay (Fig. 1). The network of artificial reefs consisted of 10 steel-frame blocks. The block on which individuals of *T. a. macrodiscus* were studied was 5 m wide and 10 m long and had six pinnacles. All of the observed individuals were distributed on one of the pinnacles, which extended to a height of 5.1 m above the sea bottom (Fig. 1-d). *In situ* monitoring every one or two months between December 2005 and March 2008 was carried out by SCUBA diving. The distribution, growth, and regeneration of four young and one adult individual were monitored.

During SCUBA diving monitoring, distributional points and the longest arm length of each individ-

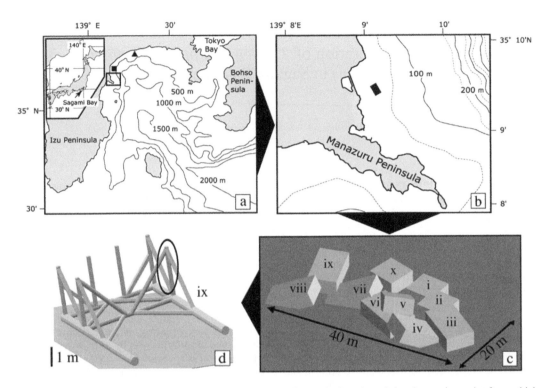

Figure 1. Location of the study site. a. The black triangle indicates the location of the observation point from which wave height data were recorded. The black box indicates the location of the observation point from which the current velocity data were recorded. b. Location of the artificial reef off Manazuru Peninsula, Sagami Bay. c. Schematic illustration of ten artificial reef blocks on the sandy bottom at a depth of 20 m. d. A detailed illustration of artificial reef block ix. The circle shows the region where the observed individuals of *T. a. macrodiscus* were distributed.

ual (measured with a ruler to the nearest 1.0 cm) were recorded. Clear digital images were taken to map distributional positions on the artificial reef block, growth and regeneration of damaged arms. To avoid disturbance, we decided not to use tags to identify individuals. However, since the locomotory ability of *T. a. macrodiscus* is very limited and there were no other individuals of this species on the same block, we could easily and accurately identify individuals by their arm lengths, morphology, body color, and distributional positions on the block.

Three young and one adult of *T. a. macrodiscus* were collected and fixed in 80% ethanol in the laboratory after monitoring. The position of arm breakage, the length of the stump (distance from radial to the point of breakage), and the length of the regenerated portion of each arm (measured with a pair of calipers to the nearest 1.0 mm) were recorded under microscopic observation to determine the pattern of arm loss and the relationship between the length of the stump and that of the regenerated arm.

Data on the wave height and the current velocity recorded during the typhoon were obtained from the observatory stations of the National Research Institute for Earth Science and Disaster Prevention (NIED) and the Kanagawa Prefectural Fisheries Research Institute, respectively. The NIED station is located approximately 22 km northeast and the latter station approximately 4 km north of the artificial reef.

3 RESULTS

3.1 *Distribution and growth*

Two young individuals of *T. a. macrodiscus*—one brown (Y1) in color, the other dark brown (Y2)—were found on the top of the artificial reef located at a depth of 15.9 m (4.1 m from the bottom) in October 2005. The length of the longest arm of Y1 was 9 cm and that of Y2 was 8 cm. One Adult Individual (AD) was located next to Y1 and Y2.

In March 2006, two further individuals—one light brown (Y3) in color, the other dark brown (Y4)—were found 1.8 m below Y1 and Y2 (17.7 m depth)., The length of the longest arm in both Y3 and Y4 was 11 cm. These four young individuals showed continuous growth until August 2007 (Fig. 2). During this period, Y3 and Y4 were observed to change their positions on the reef. In August 2006, Y3 was found 0.9 m above and Y4 found 0.6 m below the respective positions at which they had initially been observed. Y3 was observed to ascend only once in several months, and was finally found at the top of the reef next to Y1 and Y2 in January 2007. In contrast, Y4 ascended and descended repeatedly, and from June 2007 was found at a depth of 17 m (1.1 m below the other three individuals). Active migration of Y1, Y2, and AD was not observed during the observation period. By August 2007, the length of the longest arm of Y1, Y2, Y3, and Y4 had increased to 24 cm, 26 cm, 25 cm, and 20 cm, respectively (Fig. 2). The longest arm length of AD was 34 cm. The estimated average growth rate determined on the basis of the longest arm length in Y1, Y2, Y3, and Y4 between October 2005 and August 2007 was 8.1 cm year^{-1}.

In October 2006, Y1, Y2, and Y4 were observed with eggs and Y3 and AD had sperm in the genital pinnules. By January 2007, eggs and sperm had been released.

3.2 Typhoon event and damage to arms

Between September 6th and 7th 2007, Sagami Bay was struck by a strong typhoon. This typhoon caused the worst damage to the coastal area of Sagami Bay in the last ten years. Wave heights had reached 6 m before noon on September 6th; the maximum wave height recorded at 2100 hours was over 7 m. The current velocity at 10 m depth was greater than 0.7 m sec^{-1} (1.36 kt) between 0200 and 0400 hours on September 7th; the maximum current speed was 0.823 m sec^{-1} (1.60 kt) at a time after 0300 hours.

Eight days later, on September 15th, the arms of all individuals were observed to have been damaged. In particular, Y1 and Y4 had lost all their arms. Y2, Y3 and AD had retained some arms; however, these arms were severely damaged. By October 13th, these latter individuals had each lost a part of all their arms. Such extensive damage had not previously been observed in the population of *T. a. macrodiscus* living on this artificial reef.

After the typhoon had passed, some of the kelp and soft corals that had previously been distributed on the artificial reef had completely disappeared. However, some the dominant species of feather stars on the artificial reef, such as *Oxycomanthus* spp. (multiarmed species capable of curling their arms), and all individuals of *T. a. macrodiscus* (except Y1, Y2, Y3, Y4 and AD), which were distributed near the bottom or in the area surrounding the artificial reef, escaped serious injury.

3.3 Recovery from damage and average regeneration rate

One month after the typhoon, regeneration was observed in some arms of the monitored individuals (Fig. 3). The time at which regeneration commenced differed among individuals. Furthermore, when observed in December 2007, the regeneration was observed to proceed differently among the arms of each individual (Fig. 4).

All the damaged arms of the monitored individuals showed continuous regeneration. The arms of Y2 and Y4 had almost completely regenerated by March 2008, six months after the typhoon had passed. The estimated average regeneration rate determined on the basis of the longest arm length in Y1, Y2, Y3, and Y4 between October 2007 and March 2008 was 17.6 cm year^{-1} (Fig. 2). This rate was more than twice the value of the normal growth rate.

3.4 Breakage and regeneration of arms

Under microscopic observation, all breakage positions could be recognized by differences in

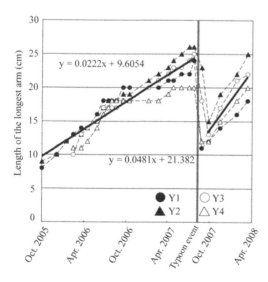

Figure 2. Changes in the length of the longest arm of Y1, Y2, Y3, and Y4 during the study period. The solid line represents the regression line of the average length of the longest arms. The average growth rate observed between Oct 2005 and Aug 2007 was 8.1 cm year^{-1}. The average growth (regeneration) rate observed between Oct 2007 and Mar 2008 was 17.6 cm year^{-1}.

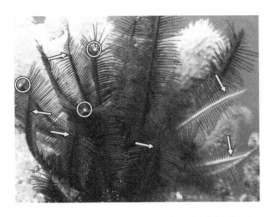

Figure 3. The appearance of Y2, Y3, and AD in October 2007. Both regenerated (arrows) and non-regenerated (circles) arms can be observed.

Figure 4. The appearance of Y4 in December 2007. The arrows show the positions at which breakage occurred. Regeneration was observed to proceed differently among the arms.

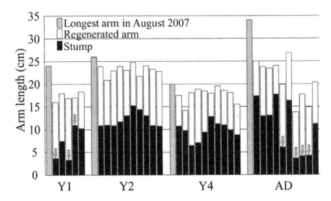

Figure 5. The length of the longest arm in August 2007 and that of the regenerated arms and stumps in Y1, Y2, Y4, and AD. The arrows indicate where breakage occurred at syzygial articulations.

Table 1. The length of the regenerated arm of Y1, Y2, Y4, and AD.

Individual	Number of arms	Length of regenerated arm (mm)			
		Average	Maximum	Minimum	Total
Y1	5	100.6	135	60	503
Y2	10	109.4	130	73	1094
Y4	10	79	117	44	790
AD	10	103	138	63	1030

thickness and shade of color between the stumps and regenerated arms. The positions of breakage, lengths of stumps and regenerated arms were different for each arm (Fig. 5). In the case of Y1, five damaged arms were monitored, excluding one arm observed for the development of genital pinnules and four arms that had been broken after the typhoon. Seven breakages (20%) in two individuals were observed at syzygial articulations (Fig. 5). Six of these breakages occurred relatively close to the base.

The lengths of the regenerated arm were also observed to differ between each of the arms (Fig. 5). Y1, Y2, and AD were similar in terms of the average, maximum, minimum, and total lengths of the regenerated arms (except total length

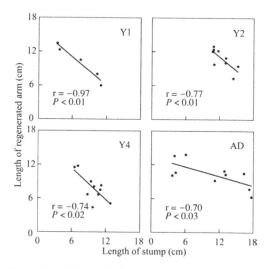

Figure 6. Relationship between length of regenerated arm and length of stump in Y1, Y2, Y4, and AD.

of regenerated arms in Y1) (Table 1). Compared with these individuals, the extent of regeneration in Y4 was observed to be poor. In Y1, Y2, Y4, and AD, a strong negative relationship was observed (r = −0.97, −0.77, −0.74, and −0.70, respectively) between the length of the stump and that of the regenerated arm (Fig. 6).

4 DISCUSSION

The details of early and post-larval developmental stages of *T. a. macrodiscus* are still unknown; however, the results of the present study and those of an ongoing survey indicate that the spawning season of *T. a. macrodiscus* is between November and December. By taking the observed growth rate into consideration, it appears that the four young individuals— Y1, Y2, Y3, and Y4—monitored in the present study were fertilized between November and December 2004, and that their age was 2 years and 9 or 10 months as of August 2007. The arm lengths of the largest specimen of *T. a. macrodiscus* on the artificial reef were greater than 35 cm. Therefore, *T. a. macrodiscus* would appear to take a minimum of 4 years to reach its maximum size after fertilization. Further, Y1, Y2, Y3, and Y4 were observed with eggs or sperm in the genital pinnules in October 2006. It is presumed that *T. a. macrodiscus* reaches sexual maturity at the age of two years.

In August 2007, the longest arm length of Y4 was obviously shorter than that of the other

three young individuals. Y1 and Y2, which were distributed at the top of the artificial reef block, were exposed to multidirectional current during the observation period (Fig. 3). Y3 migrated to the top of the reef next to Y1 and Y2. On the other hand, the Y4 specimen was positioned on the vertical surface of the artificial reef block; this specimen occurred below the top surface of the reef and was exposed to weaker current as compared to the specimens distributed at the top of the artificial reef (Fig. 4). Such a distribution was unfavorable for efficient feeding. This suggests that the feeding environment of the habitat makes a difference to growth. Vail (1989) has also mentioned that it is possible that conditions at a particular locality influence rate of growth.

In general, feather stars autotomize their arms in order to reduce damage when attacked by predators (Messing 1997). Mladenov (1983) has reported that the most frequent points of breakage in regenerating arms are observed at the syzygial arm articulations, which are known to be specialized for purposes of autotomy. Additionally, nearly all the regenerated arms were observed to be proximally truncated near the visceral mass (Schneider 1988). However, autotomic phenomena caused by predator attacks have not been reported from *T. a. macrodiscus* yet. The pattern of arm loss observed in this study is obviously different from that observed to be caused by predator attacks in other autotomic species. The points of breakage had not occurred at the syzygial arm articulations and had not been as near the visceral mass as previously reported.

Initial arm lengths and body sizes were observed to differ between individuals. However, during regeneration (204 days), the total regenerated arm lengths of Y1 (the values were doubled as only five arms were observed), Y2, and AD were similar (1006, 1094, and 1030 mm, respectively). This indicates that the rate of regeneration of *T. a. macrodiscus* does not change with increasing body size or age (at least older than three years of age). On the other hand, the length of the regenerated arms of Y4 was observed to be approximately 20% shorter than that of the others. This indicates that, in addition to growth, the feeding environment of the habitat makes a difference in regeneration.

The negative relationship between the length of the stump and that of the regenerated arm observed in Y1, Y2, Y4, and AD indicates that this was a means by which *T. a. macrodiscus* attempted to reduce the gap among the lengths of its arms and to preferentially regenerate the proximal arm by controlling the rate of regeneration in each arm. If the lengths of their arms differ, a change is induced in the direction of the current, thus reducing its feeding efficiency. When the number of arms in

Oxycomanthus japonicus increases by autotomy of a single arm and the subsequent regeneration of a pair of arms, the next such increase tends to occur at a certain distance away from the previous pair (tends to not occur in neighbouring radii) (Shibata and Oji 2003). Shibata and Oji explained that the reason underlying such behavior of *O. japonicus* is to ensure that the density of arms remains relatively constant for maintaining an efficient filtration fan for feeding. Genital pinnules exist on the proximal half of arms in *T. a. macrodiscus*. Increasingly greater number of eggs and sperms are contained near the base of the pinnules. Thus, preferentially regenerating proximal parts of a greater number of arms favors the recovery of reproductive ability than regenerating fewer arms fully. Mladenov (1983) discussed that arms are important for not only feeding but also for reproductive structures; therefore, the existence of strong selective pressure for the prompt regeneration of lost arms could be expected in crinoids.

In conclusion, a rapid rate of regeneration in *T. a. macrodiscus* was observed during recovery from natural disasters such as typhoons. This regeneration rate does not change with body size or age. The feeding environment will exert some effects on the difference in regeneration rate similar to those exerted on the growth rate. Our results also indicate that *T. a. macrodiscus* controls the rate of regeneration in each of its arms to ensure an efficient feeding ability and rapid recovery of its reproductive activity. Most feather stars can survive with curling and/or cutting off its arms when attacked by predators. On the other hand, *T. a. macrodiscus* appears to be able to survive by rapidly regenerating its long robust arms. The discovery of individuals of this species lacking arms may thus signify the occurrence of a severe natural disturbance.

ACKNOWLEDGMENTS

We would like to thank National Research Institute for Earth Science and Disaster Prevention (NIED) and Kanagawa Prefectural Fisheries Research Institute Sagami-Bay Experimental for permission to use their data of wave height and current velocity. We are grateful to Dr. T. Fujita and Dr. V.S. Kuwabara for reading the manuscript and making valuable suggestions.

REFERENCES

Clark, A.H. 1947. A monograph of existing crinoids 1(4b). *Bulletin of the United States National Museum* 82: 1–473.

Kogo, I. 1998. Crinoids from Japan and its adjacent waters. *Osaka Museum of Natural History, Special Publications* 30: 1–148.

Kogo, I. 2006. Comatulid Fauna (Echinodermata: Crinoidea: Comatulida) of Sagami Sea and a Part of Izu Islands, Central Japan. *Memoirs of the National Science Museum, Tokyo* 41: 223–246.

Messing, C.G. 1997. Living comatulids. *Paleontological Society Papers* 3: 3–30.

Mladenov, P.V. 1983. Rate of arm regeneration and potential causes of arm loss in the feather star *Florometra serratissima* (Echinodermata: Crinoidea). *Canadian Journal of Zoology* 61: 2873–2879.

Schneider, J.A. 1988. Frequency of arm regeneration of comatulid crinoids in relation to life habit. In R.D. Burke, P.V. Mladenov, P. Lambert, and R.L. Parsley (eds), *Echinoderm Biology*: 531–538. Rotterdam: Balkema.

Shibata, T. F. and Oji, T. 2003. Autotomy and arm number increase in *Oxycomanthus japonicus* (Echinodermata, Crinoidea). *Invertebrate Biology* 122(4): 375–379.

Utinomi, H. and Kogo, I. 1965. On some crinoids from the coastal sea of Kii Peninsula. *Publications of the Seto Marine Biological Laboratory* 13(4): 263–286.

Vail, L. 1989. Arm growth and regeneration in *Oligometra serripinna* (carpenter) (Echinodermata: Crinoidea) at Lizard Island, Great Barrier Reef. *Journal of Experimental Marine Biology and Ecology* 130: 189–204.

Taxonomy

Echinoderms in a Changing World – Johnson (ed)
© *2013 Taylor & Francis Group, London, ISBN 978-1-138-00010-0*

Comparisons of ophiactid brittle stars possessing hemoglobin using intronic variation

A.B. Christensen & E.F. Christensen
Biology Department, Lamar University, Beaumont, TX, USA

ABSTRACT: The sequences of two hemoglobin genes were used to examine the relationship between three species of ophiactid brittle star. cDNA for the genes was sequenced from a Texas ophiactid. This information was then used to amplify genomic DNA from *Ophiactis simplex, O. rubropoda*, and the Texas ophiactid. While the protein coding sequences were the same for all three groups, differences were found in the sequences of two of the introns. The sequences for these genes further support the close evolutionary relationship of these species, but are inadequate to further resolve it.

1   INTRODUCTION

Analysis of a fragment of the COI gene suggests that populations of the brittle star *Ophiactis simplex, O. rubropoda*, and a Texas ophiactid share a recent evolutionary history and may be in the process of speciation (Christensen et al. 2008). All of the populations included in the study produce the respiratory pigment hemoglobin that is contained in coelomocytes present in the water vascular system (*O. simplex*: Christensen 1998, *O. rubropoda*: Ruppert and Fox 1988, Texas ophiactid: Christensen 2004). While hemoglobins of various organisms are thought to share a common evolutionary origin (Hardison 2001), the amino acid sequence of closely related species typically share less than 50% homology (Kapp et al. 1995). Hemoglobin genes of vertebrates and many invertebrates typically have a two intron—three exon structure. The position of these two introns is conserved, occurring at positions B12 (intron 1) and G7 (intron 2) of the protein (Hardison 1998). The sea cucumber *Caudina arenicola* has a third intron occurring in the codon for amino acid nine (Kitto et al. 1998). An investigation into the nucleotide sequence of the hemoglobin gene and its structure was undertaken to further examine the relationship between these three closely related ophiactid species.

2   MATERIALS AND METHODS

cDNA from Texas individuals was made using reverse transcription. The hemoglobin gene was PCR amplified using primers for *Hemipholis elongata* hemoglobin (Christensen et al. in prep)

and 5′ RACE (Ozawa et al. 2006) and sequenced using Big Dye Terminator v3.1 sequencing kit on an ABI 3100 Gene Analyzer. Genomic DNA was isolated using a modified Chelex protocol (Mederios-Bergen et al. 1995) from: 5 *O. simplex* Long Beach, CA; 5 *O. simplex* San Diego, CA; 5 TX ophiactid Port Aransas, TX; 3 TX ophiactid South Padre Island, TX (SPI); and 6 *O. rubropoda* from Georgetown, SC. Using the cDNA sequence, primers were generated for regions suspected of possessing introns. These regions were PCR amplified and sequenced as above. Comparisons of sequences were made using the align two sequences function of BLAST (NCBI).

3   RESULTS

Two different hemoglobin genes were identified from cDNA: hemoglobin A (HbA) codes for 158 amino acids and hemoglobin B (HbB) codes for 144 amino acids. The two proteins share a 59% homology with one another. The protein coding sequence was identical in all individuals sampled, regardless of location of origin.

Both genes possess the two introns typical of many hemoglobins. In HbA intron 1 splits the codon for amino acid 42 and intron 2 occurs between the codons for amino acids 116 and 117 (Fig. 1). In HbB these introns occur in the codon for amino acid 33 and between codons for amino acids 107 and 108, respectively (Fig. 2). Intron 2 of HbA is 314 bp in length and was invariant among all individuals surveyed; however intron 2 of HbB had three variants, ranging in size from 361 to 380 bp (Fig. 2). The most common form, A, appears in all species. The next common variant, B,

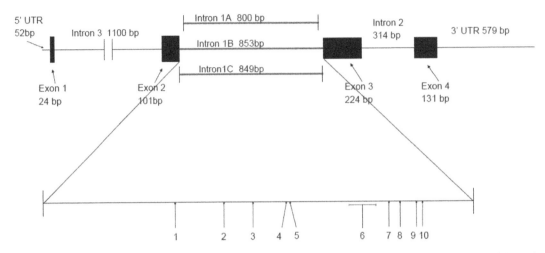

Figure 1. Gene structure of hemoglobin A *Ophiactis simplex/rubropoda*. The blowout shows the relative positions of the indels of intron 1. Size and variant in which indels occur are: 1 = 1 bp, C; 2 = 1 bp, A,C; 3 = 3 bp, A,B; 4 = 1, B; 5 = 6 bp, C; 6 = 53 bp, A; 7 = 1 bp, C; 8 = 1 bp, B, C; 9 = 1 bp, C; and 10 = 1 bp, A,B.

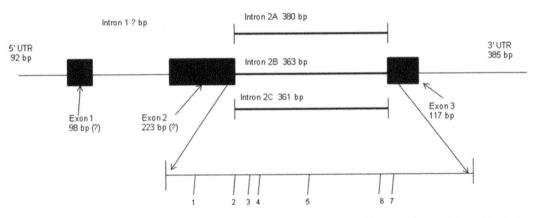

Figure 2. Gene structure of hemoglobin B of *Ophiactis simplex/rubropoda*. The blowout of intron 2 shows the relative positions of the indels. Size and variant in which indels occur are: 1 = 6 bp, B,C; 2 = 9, B,C; 3 = 1 bp, C; 4 = 1 bp, C; 5 = 14 bp, A; 6 = 3 bp, B,C; and 7 = 15 bp, B,C. The sequence and size of intron 1 has not been determined yet.

Table 1. Distribution of genotypes among the different population sampled.

Locality	HbA intron 1A	HbA intron 1B	HbA intron 1C	HbB intron 2A	HbB intron 2B	HbB intron 2C
San Diego, CA	3	1	0	4	1	0
Long Beach, CA	1	3	0	4	1	0
Port Aransas, TX	4	0	1	4	0	0
SPI, TX	0	0	3	2	0	1
George-town, SC	1	0	1	3	3	0

has 5 indels of 3–15 nucleotides each and appears in the two most distant species, *O. simplex* (CA) and *O. rubropoda* (SC) (Table 1). The third variant C appeared in one individual of the TX ophiactid (SPI); it is the same as the second variant with two additional single base deletions. Intron 1 of HbA has three variants, ranging from 800 to 851 bp (Fig. 1), two of which possess a 50 nucleotide insert not seen in the most common form. Variant B has only been detected in *O. simplex* (CA), while variant C occurs in both Texas and *O. rubropoda* (SC) (Table 1). There are no unique genotype combinations that appear in a single population. HbA has a third intron of approximately 1100 bp in between the codons for amino acids eight and nine (Fig. 1); as the sequencing of HbB is incomplete, the presence and position of a third intron is unknown at this time.

4 CONCLUSIONS

The gene structure of HbA is similar to that of the sea cucumber, *C. arenicola* (Kitto et al. 1998). This suggests a common evolutionary origin for echinoderm hemoglobins.

The lack of differences in the protein coding regions of HbA and HbB suggests a very recent evolutionary past for these ophiactid brittle stars. When combined with the COI data, these populations represent shallowly separated species with a common ancestry despite the geographical distances. The variations in the intronic sequence appear to be due to ancestral mutational events that occurred before their separation and that there has been no drift in the populations. However, due to the small sample sizes, we are unable to further resolve their relationships.

REFERENCES

Christensen A.B. 1998. The properties of the hemoglobins of *Ophiactis simplex* (Ophiuroidea, Echinodermata). *Am Zool* 38:120A.

Christensen A.B. 2004. A new distribution record and notes on the biology of the fissiparous brittle star *Ophiactis simplex* (Ophiuroidea, Echinodermata) in Texas. *Texas J. Sci.* 56:175–179.

Christensen, A.B., E.F. Christensen and D.W. Weisrock. 2008 Population genetic structure of North American *Ophiactis* spp. possessing hemoglobin. *Marine Biology* 154:755–763.

Hardison, R. 1998. Hemoglobins from bacteria to man: evolution of different patterns of gene expression. *J. Exp. Biol.* 201:1099–1117.

Hardison, R.C. 2001. Organization, evolution and regulation of the globin genes. In: *Disorders of Hemoglobin: Genetics, Pathophysiology and Clinical management.* M.H. Steinberg, B.G. Forget, D.R. higgs, and R.L. Nagel eds. Pp95–116. Cambridge University Press.

Kapp, O.H., L. Moens, J. Vanfleteren, C.A.N. Trotman, T. Suzuki, and S.N. Vinogradov. 1995. Alignment of 700 globin sequences: Extent of amino acid substitution and its correlation with variation in volume. *Protein Science* 4:2179–2190.

Kitto, G.B., P.W. Thomas, and M.L. Hackert. 1998. Evolution of cooperativity in hemoglobins: what can invertebrate hemoglobins tell us? *J. Exp. Zool.* 282:120–126.

Medeiros-Bergen, D., N.T. Perna, J.A. Conroy, and T.D. Kocher. 1998. Identification of ophiuroid post-larvae using mitochondrial DNA. In: Mooi R, Telford M (eds) *Echinoderms: San Francisco, Proceedings of the 9th International Echinoderm Conference*, Balkema, Rotterdam, The Netherlands pp 399–404.

Ozawa, T., H. Kishi, and A. Muraguchi. 2006. Amplification and analysis of cDNA generated from a single cell by 5'-RACE: application to isolation of antibody heavy and light chain variable gene sequences from single B cells. *Bio. Tech.* 40:469–478.

Ruppert E.E., and R.S. Fox (eds) 1988. *Seashore animals of the southeast.* University of South Carolina Press.

Population biology

Echinoderms in a Changing World – Johnson (ed)
© *2013 Taylor & Francis Group, London, ISBN 978-1-138-00010-0*

Field and laboratory growth estimates of the sea urchin *Lytechinus variegatus* in Bermuda

M.P. Russell
Biology Department, Villanova University, Villanova, PA, USA

T.A. Ebert
Oregon State University, Corvallis, OR, USA

V. Garcia
Biology Department, Villanova University, Villanova, PA, USA

A. Bodnar
Bermuda Institute of Ocean Science, St. George's, Bermuda

ABSTRACT: We conducted a tagging study of *Lytechinus variegatus* for 1 year (2005–2006) in Bermuda. At two sites, Flatts Inlet (n = 248) and Emily's Bay (n = 116), we collected all individuals, recorded test diameters, injected them with calcein, and released them. We also held a laboratory sample (n = 119) of tagged urchins in a tank stocked periodically with seagrass. In 2006, samples were collected, skeletal elements cleaned, and demipyramids (jaws) from Aristotle's lanterns measured and examined for the tag. We recovered 505 sea urchins with 11 tagged and 21 with 4 tagged from Flatts Inlet and Emily's Bay respectively. In the lab only one individual was not tagged. The jaw-test allometry indicated food-limitation in the laboratory sample. Early growth is rapid and individuals between 25–35 mm can grow as large as 60 mm in one year. Subsequent growth is slower and individuals > 70 mm may be 5–7 years old.

1 INTRODUCTION

Measuring growth rates and estimating age and longevity are essential to quantifying life history and understanding population dynamics (Ebert 1999). Fluorochrome tagging techniques have been very useful in many echinoid studies (e.g. Kobayashi and Taki 1969, Taki 1971, 1972a, b, Märkel 1975, Pearse and Pearse 1975, Taki 1978, Ebert 1980, Russell et al. 1998, Lamare and Mladenov 2000, Russell and Meredith 2000, Johnson et al. 2001). However, these methods have not been used to study *Lytechinus variegatus*, a common, subtropical and tropical sea urchin. This species is an abundant and dominant member of many seagrass communities found in the Gulf of Mexico, Caribbean, as well as tropical and subtropical Atlantic (Moore 1963, Camp 1973, Vadas 1982, Beddingfield and McClintock 1994, Greenway 1995, Hendler et al. 1995, Junqueira et al. 1996, Watts 2007).

Previous laboratory studies suggest that early growth rates are high. Pawson and Miller (1982) reported that newly settled individuals (0.42 mm) reach 8.4 mm in 36 weeks, and 23 mm in 84 weeks.

Based on size distributions, Beddingfield and McClintock (2000) suggested that *L. variegatus* is fast growing and short lived, reaching a maximum size in 3–4 years.

This type of rapid development and early sexual maturation make *L. variegatus* a good candidate for commercial and laboratory use (Watts et al. 2007). Although there is no current commercial fishery for *L. variegatus*, further work quantifying its life history is important for potential management purposes.

Other growth studies of *L. variegatus* have used a combination of size distribution data and growth bands in skeletal plates (Moore 1963, Camp 1973, Beddingfield and McClintock 2000, Hill et al. 2004). Interpreting skeletal growth lines as chronometers is fraught with logistic and conceptual issues (see Russell and Meredith 2000 for review) and more recent work on *L. variegatus* skeletal growth lines indicates that they are not related to age (Hill et al. 2004).

The most reliable method for quantifying growth and assessing longevity in sea urchins is mark and recapture studies using fluorochrome tagging (see Ebert 2007 for review). Here we report the results of a field and laboratory fluorochrome-tagging study of *L. variegatus* in Bermuda.

2 MATERIALS AND METHODS

We collected sea urchins from two field sites in Bermuda for the mark recapture study. In 2005, 248 individuals were collected and tagged from Flatts Inlet (32°10'22.48"N 64°44'12.97"W—Nov. 14) and 116 from Emily's Bay (32°22'11.16" N 64°40'14.60"W—Nov. 17). Additionally, 117 sea urchins were collected and tagged from an adjacent site at Flatts Inlet (Nov. 18) and kept in the laboratory for the duration of the experiment. The two smallest individuals from the Flatts Inlet collection of Nov. 14 were added to the lab sample (test diameters 9.53 and 10.78 mm).

The laboratory samples were kept at the Bermuda Institute of Ocean Sciences in a concrete tank 7 m long, 0.5 m deep, and 0.8 m wide. The bottom of the tank was lined with coral rocks. Periodically the tank was stocked with blades of the seagrass, *Thalassia testudinum* and flushed to remove fecal and detritus material. Ambient seawater was continuously filtered and pumped into one end of the tank and drained through a mesh-covered standpipe at the opposite end.

In 2005 we injected all sea urchins through the peristomal membrane with a calcein solution (0.25 g in 500 ml of sea water). In both 2005 and 2006 we recorded test diameters with knife-edge, digital vernier calipers.

In 2006 we collected all sea urchins from the mark-recapture sites: Flatts Inlet on Nov. 4 and Emily's Bay on Nov. 7. These samples were processed within 24 h by draining the body fluid, removing the gonads, and soaking the skeletal elements, including Aristotle's lantern, in 6% sodium hypochlorite (bleach) overnight. Bleached samples were rinsed in hot tap water and allowed to soak for ~12 h before drained, air-dried, and shipped to Villanova. Lab samples were processed on Nov. 8.

From each sea urchin recovered in 2006 we measured the length of one demipyramid of Aristotle's lantern (hereafter referred to as "jaw") using a dissecting microscope fitted with an ocular micrometer. The jaw was illuminated with ultraviolet light (wavelength 470 nm) to detect the calcein mark on both the oral and aboral ends. If tagged, the growth increments were recorded for each jaw. See Lamare and Mladenov (2000) for illustration of jaw growth quantification using fluorochrome markers.

Previous work (e.g. Ebert and Russell 1993) has shown that the four parameter Tanaka function (Ebert 1999, Tanaka 1982, Tanaka 1988) is best suited to analyzing echinoid mark and recapture data where S_t is size at time t:

$$S_t = \frac{1}{\sqrt{f}} \ln\left[2f(t-c) + 2\sqrt{f^2(t-c)^2 + fa} \right] + d \quad (1)$$

The Tanaka function can be expressed as a difference equation (Ebert 1999, Ebert and Russell 1993) where size of the jaw at one year (S_{t+1}) is a function of original jaw size (S_t):

$$S_{t+1} = \frac{1}{\sqrt{f}} \ln\left| 2G + 2\sqrt{G^2 + fa} \right| + d \quad (2)$$

where $\quad G = \frac{E}{4} - \frac{fa}{E} + f \quad (3)$

and $\quad E = e^{\sqrt{f}(S_t - d)} \quad (4)$

A nonlinear regression (SAS) analysis of the difference equation was used to estimate the growth parameters, a, d, and f. Parameter c of the Tanaka function was calculated from equation 5.

$$c = \frac{a}{E} - \frac{E}{4f} \quad (5)$$

Measures of jaw growth used in the Tanaka function analyses were converted to centimeters to allow the nonlinear regression to converge on a solution. To convert the jaw growth parameters into test diameter growth, the relationship between test diameter, D, and jaw length, J, was established using the allometric relationship:

$$D = \alpha J^\beta \quad (6)$$

The parameters α and β were calculated from Model II linear regressions of ln transformed measurements of both tagged and untagged sea urchins (Ebert and Russell 1994).

3 RESULTS

More than twice as many sea urchins were recovered from the field at Flatts Inlet in 2006 (n = 505) than in 2005 (n = 248) and only 11 were tagged (Fig. 1a). In contrast, there were only 21 sea urchins recovered from Emily's Bay in 2006 but four were tagged (Fig. 1b).

After one year we recovered 108 sea urchins from the tank in the laboratory (Fig. 1c); however, one of these individuals was deformed with a height approximately equal to the diameter and excluded from growth analyses. During the course of the study the remains of 11 individuals were recovered

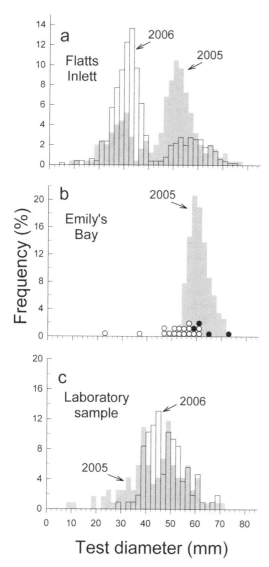

Figure 1. Size distributions of field and laboratory samples in 2005 (shaded bars) and 2006 (open bars). a) Flatts Inlet. Both 2005 (n = 248) and 2006 (n = 505) show a strong bimodal distribution. b) Emily's Bay. In 2005, n = 116. Only 21 sea urchins were recovered in 2006 so these individuals are plotted along the x-axis as not-tagged (open circles, n = 17) and tagged (filled circles, n = 4). c) Laboratory sample. In 2005, n = 119. All individuals recovered from the tank at the end of one year were tagged (n = 107).

at various times. It was not possible to determine the exact date or the source of mortality for these individuals and they were also excluded from growth analyses. Of the original 119 put into the tank 118 were tagged with calcein. The individual that was

not tagged was recovered less than two weeks after tagging. We restricted the laboratory growth analysis to the 107 non-deformed individuals that survived the entire year.

Figure 2 compares the test-jaw allometric relationships among Flatts Inlet, Emily's Bay, and the laboratory sample. An ANCOVA of ln(Jaw) using ln(diameter) as the covariate of all the samples recovered from Flatts Inlet and Emily's Bay violates the assumption homogeneity of slopes (F = 6.30, P = 0.012). Because Emily's Bay lacks the smaller urchins found at Flatts Inlet (Figs. 1a,b) the analysis was repeated restricting the samples to the size range of urchins that were found tagged at both sites (>58 mm). The results of this ANCOVA met the homogeneity of slopes assumption (F = 0.37, P = 0.55) and showed no difference between the sites (F = 3.14, P = 0.081). Therefore the data for Emily's Bay and Flatts Inlet were combined in subsequent analyses and referred to as field samples (Fig. 2).

The jaw-test allometric relationship for the laboratory sample is

$$D = 2.00J^{1.26} \qquad (7)$$

and for the field:

$$D = 4.24J^{1.02} \qquad (8)$$

The Walford plot in Figure 3 shows the original jaw size on the x-axis and size in 2006 on the y-axis. The only overlap between the field and laboratory samples occurs in the larger sizes and the Tanaka function regressions (equation 2, Table 1) show substantially greater growth rates in the field.

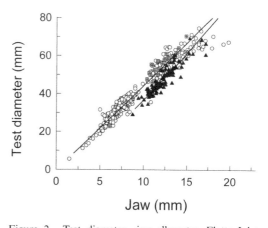

Figure 2. Test diameter—jaw allometry. Flatts Inlet (open circles, n = 502) and Emily's Bay (shaded crossed squares, n = 21) were combined to produce the upper allometric regression line. The filled triangles (n = 107) represent the laboratory sample (lower regression line).

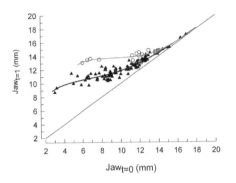

Figure 3. Walford plot of jaw growth. $\text{Jaw}_{t=0}$ is length of jaw in 2005 and $\text{Jaw}_{t=1}$ is final size when recovered in 2006. The diagonal line of slope = 1 represents no growth. Flatts Inlet (open circles, n = 11) and Emily's Bay (shaded crossed squares, n = 4) were combined to produce upper Tanaka growth curve. The laboratory sample (closed triangles, n = 107) is the lower Tanaka growth curve.

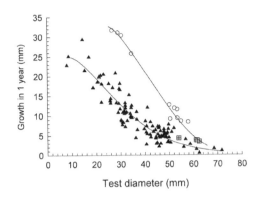

Figure 4. Test diameter growth rate estimates from tagged individuals. Flatts Inlet (open circles, n = 11) and Emily's Bay (shaded crossed squares, n = 4) were combined to produce the upper growth curve and the laboratory sample (closed triangles, n = 107) is the lower one.

Table 1. Results of nonlinear regression analyses fitting the Tanaka function parameters to the jaw growth data (equation 2). The parameter estimates are for jaw size in centimeters.

Field sample—Emily's Bay and Flatts Inlet

Source	SS	df	MS
Model	29.5127	3	9.8376
Error	0.0206	12	0.0017
Total	29.5333	15	

Parameter	Estimate	se	95% confidence interval (±)
a	0.2361	0.1819	0.3963
d	0.6544	0.1574	0.3429
f	56.6483	33.3734	72.7517
Model	160.7000	3	53.57810

Laboratory sample

Source	SS	df	MS
Error	0.2484	103	0.00241
Total	161.0000	106	

Parameter	Estimate	se	95% confidence interval (±)
a	1.5575	0.3137	0.6223
d	0.0276	0.0645	0.1279
f	17.3973	2.8134	5.5797

Figure 4 uses the allometric regressions (equations 7 and 8) to convert jaw growth estimates into test growth. These plots indicate rapid growth for small sea urchins (<30 mm test diameter) but slow rates for larger individuals (>50 mm) and slower growth rates in the >60 mm size range. If the first mode in Figure 1a is the 0–1 year class then using the upper growth curve in Figure 4 would result in individuals >70 mm in the 4–5 year class or older.

Finally, during this study we observed a spawning event in the laboratory sample on June 8, 2006. Most of the individuals climbed near the surface along the sides of the tank and released copious amounts of gametes turning the water murky for several hours. We recovered embryos from the water and observed development for several days but were not equipped to bring the larvae through to settlement.

4 DISCUSSION

There was a low recapture rate of tagged sea urchins in the field, 15 of the 354 originally tagged. This low rate was not due to failure of the tagging method as all the sea urchins in the laboratory sample were successfully tagged. Mortality cannot be ruled out as a contributing factor for the low return rate in the field. Although we did not observe high mortality in the laboratory sample, other sources may be at play in the field, e.g. predators or storms. Many large individuals were recovered from both field sites. In the >60 mm size range (Fig. 1) we retrieved 46 sea urchins from Flatts Inlet and 5 (out of 21—3 that were tagged) from Emily's Bay. Given the observed growth rates (Fig. 4) these larger individuals are probably a collection of age classes (see below). One conclusion that is consistent with the tagging pattern, size distributions, and growth rates is that this species is mobile and the low return rates could be due to immigration and emigration into and out of the study sites.

Previous field studies on *L. variegatus* growth have used either size distributions (Moore et al. 1963, Camp et al. 1973, Engstrom 1982, Beddingfield and McClintock 2000) or skeletal growth rings (Camp et al. 1973, Hill et al. 2004) and none have inferred individuals older than 4 years. The validity of using growth rings, or growth bands, as chronometers for sea urchins, has been widely debated. Validation studies using fluorchrome markers have shown that growth rings are inaccurate indicators of age for some species (Pearse and Pearse 1975, Ebert 1988, Russell and Meredith 2000). Moore et al. (1963) found that *L. variegatus* lacked growth rings whereas Hill et al. (2004) found ill-defined rings that were possibly related to food availability. This study is the first to successfully recapture fluorochrome-tagged *L. variegatus* after one year in the field.

The laboratory sample was collected in the inlet immediately seaward and adjacent to the Flatts Inlet field site. There is no reason to assume that these sea urchins had a different jaw-test allometric relationship at the start of the experiment in 2005. However, after one year these individuals had a significantly different relationship than either of the field samples (Fig. 2). A common allometric relationship (equation 8) described the Emily's Bay and Flatts Inlet sea urchins and these sites are on different parts of the island separated by tens of kilometers. Previous work on several echinoid species has shown that food limitation produces a shift in resource allocation that results in relatively larger jaws (Ebert 1980). Our laboratory sea urchins were limited to a diet of seagrass and Beddingfield et al. (1998) showed that *L. variegatus* grew faster and had higher survival when fed a more diverse diet than seagrass. To illustrate the dramatic difference in jaw sizes between the field and laboratory samples one can compare a mid-size sea urchin of 40 mm test diameter. The field sample jaw size would be 9.05 mm (equation 8) whereas the laboratory jaw size would be 10.69 mm (equation 7). This 18% difference was produced in just one year. The lower growth rates we observed and the difference in jaw allometry strongly indicates that the laboratory sample was food limited.

Despite the higher level of resource allocation to jaws in the food-limited laboratory sample, overall jaw growth was higher in the field (Fig. 3). Annual test growth inferred from the jaw-test allometry also showed faster growth rates in the field (Fig. 4). Using this annual growth rate curve and the size distributions from Flatts Inlet, it is possible to estimate the age ranges of the largest individuals to get an approximation of longevity. Assume that the first mode in the size distributions (~30 mm test diameter) of both the 2005 and 2006 populations

from Flatts Inlet represent the 0–1 year age class. Applying the curve from Figure 4 would yield a test size of ~60 mm for the 1–2 year age class. If this calculation was repeated three more times then individuals ~70 mm would be in the 4–5 year age class.

The laboratory samples indicate that there is a relatively high degree of variation in growth in smaller (<40 mm) and mid-size (45–55 mm) sea urchins. The field samples also show variation in this latter size category. However, as is true for other species (Ebert 2007) variation in growth is much smaller in the larger size classes and for sea urchins > 68 mm the estimated growth increment is less than 2.0 mm per year. These patterns indicate that the larger size classes and particularly the larger modes in the size distributions (Fig. 1) represent a collection of ages.

The size distribution of Flatts Inlet in both 2005 and 2006 is bimodal (Fig. 1a). Given the growth patterns we observed and the early post-settlement growth and age estimates of Pawson and Miller (1982), we infer that the smaller mode in each year represents recruitment for that year and the larger mode is a collection of year classes. Flatts Inlet is a narrow channel and the only above-water connection to the sea for Harrington Sound. Although the daily tidal fluctuation range is not great (~1 m) the tidal currents through the channel are intense and often produce standing waves (*pers. obs.*). The patch of seagrass bed that we sampled in the inlet is adjacent to the narrowest connection to Harrington Sound. Any competent larvae produced in the Sound and transported seaward on an ebb tide, or entrained in water currents in the inlet on a flowing tide, would be concentrated and have the opportunity to settle at our field site. A similar recruitment "hot-spot" was observed by Sewell and Watson (1993) in the channel into Boca del Infierno, a semi-enclosed bay on Vancouver Island in British Columbia. We suggest that this unique physiographic situation may explain the distinctive first-year classes for this site and predict that future sampling efforts would produce similar bimodal distributions (Fig. 1a).

The strikingly different size structure in Emily's Bay suggests a very different scenario. No sea urchins in the first-year-class range were found in 2005 and only two individuals < 40 mm were there in 2006 (Fig. 1b). Therefore larval recruitment seems to be low or non-existent. However, the size range of the four marked individuals recovered in 2006 is consistent with the idea that the larger size mode is a collection of age classes.

Two future avenues of investigation that would provide further data to address the ideas presented here are: use of the fluorochrome tagging to assess movement patterns (Dumont et al. 2004) and deployment of settlement collectors to test the idea that Flatts Inlet is a hot-spot for recruitment.

ACKNOWLEDGEMENTS

We appreciate the Grants-in-aid program at the Bermuda Institute of Ocean Science, the Villanova Biology Department, and NSF for financial and logistic support. In Bermuda, J. Ward (Director of Conservation Services) and T. Sleeter (Director of Environmental Protection) approved this study. G. Gamba, J. Hopper, M. Talag, J. Quinn, and C. Crowder assisted with the field and laboratory work and N. Dollahon assisted with microscopy.

REFERENCES

Beddingfield, S.D. & McClintock, J.B. 1998. Differential survivorship, reproduction, growth and nutrient allocation in the regular echinoid *Lytechinus variegatus* (Lamarck) fed natural diets. *Journal of experimental marine biology and ecology* 226: 195–215.

Beddingfield, S.D. & McClintock, J.B. 1994. Environmentally-induced catastrophic mortality of the sea urchin *Lytechinus variegatus* in shallow seagrass habitats of St. Joseph's Bay. *Bulletin of marine science* 55: 235–240.

Beddingfield, S.D. & McClintock, J.B. 2000. Demographic characteristics of *Lytechinus variegatus* (Echinoidea: Echinodermata) from three habitats in a north Florida Bay, Gulf of Mexico. *Marine ecology* 21: 17–40.

Camp, D.K., Cobb, S.P. & Breedveld, J.F. 1973 Overgrazing of seagrasses by a regular urchin, *Lytechinus variegatus*. *Bioscience* 23: 37–38.

Dumont, C.P., Himmelman J.H. & Russell, M.P. 2004. Size-specific movement of green sea urchins *Strongylocentrotus droebachiensis* on urchin barrens in eastern Canada. *Marine Ecology Progress Series* 276:93–101.

Ebert, T.A. 1980. Relative growth of sea urchin jaws: an example of plastic resource allocation. *Bulletin of marine science* 30: 467–474.

Ebert, T.A. 1988. Calibration of natural growth lines in ossicles of two sea urchins, *Strongylocentrotus purpuratus* and *Echinometra mathaei*, using tetracycline. In R.D. Burke, P. Mladenov, P. Lambert & R.L. Parsley (eds), *Echinoderms: Proceedings of the Sixth International Echinoderm Conference*: 435–444. A.A. Balkema: Rotterdam.

Ebert, T.A. 1999. Plant and animal populations: methods in demography. San Diego: Academic Press.

Ebert, T.A. 2007. Growth and survival of post-settlement sea urchins. In: Lawrence JM (ed.) *Edible Sea Urchins: Biology and Ecology*. Elsevier.

Ebert T.A. & Russell, M.P. 1993. Growth and mortality of subtidal red sea urchins (*Stongylocentotus franciscanus*) from San Nicolas Island, California, USA: problems wi th models. *Marine Biology* 117:79–89.

Ebert, T.A., & Russell, M.P. 1994. Allometry and Model II Non-linear regression. *Journal of Theoretical Biology* 168:367–372.

Engstrom, N.A. 1982. Immigration as a factor in maintaining populations of sea urchin *Lytechinus variegatus* (Echinodermata: Echinoidea) in seagrass beds on the southwest coast of Puerto Rico. *Studies on geotropically fauna and environment*. 17: 51–50.

Greenway, M. 1995 Trophic relationships of macrofauna within a Jamaican seagrass meadow and the role of the Echinoid *Lytechinus variegatus* (Lamarck). *Bulletin of marine science* 56: 719–736.

Hendler, G., Miller J.E., Pawson, D.L & Kier, P.M. 1995. *Sea stars, sea urchins, and allies. Echinoderms of Florida and the Caribbean*. Washington, DC, Smithsonian Institution Press.

Hill, S.K., Aragona, J.B. & Lawrence, J.M. 2004. Growth bands in test plates of the sea urchins *Arbacia punctulata* and *Lytechinus variegatus* (Echinodermata) on the Central Florida Gulf Coast Shelf. *Gulf of Mexico science* 1: 96–100.

Johnson, A.S., Ellers O., Lemire J., Minor, M. & Leddy H. 2001. Sutural loosening and skeletal flexibility during growth: determination of drop-like shapes in sea urchins. *Proceedings of the Royal Society of London*. 269: 215–220.

Junqueira, A.O.R., Ventura, C.R.R., Carvalho, A.L. & Schmidt A.J. 1996. Population recovery of the sea urchin *Lytechinus variegatus* in a seagrass flat (Araruama Lagoon, Brazil): the role of recruitment in a disturbed environment. *Invertebrate reproduction and development* 31: 143–150.

Kobayashi, S. & Taki J. 1969. Calcification in sea urchins. I. A tetracycline investigation of growth of the mature test in *Strongylocentrotus intermedius*. *Calcified Tissue Research*. 4: 210–223.

Lamare, M.D. & Mladenov, P.V. 2000. Modeling somatic growth in the sea urchin *Evechinus chloroticus* (Echinoidea:Echinometridae). *Journal of experimental marine biology and ecology* 243: 17–43.

Märkel, K. 1975. Wachstum des coronarskeletes von *Paracentrotus lividus* Lmk. (Echinodermata, Echinoidea). *Zoomorphologie*. 82: 259–280.

Moore, H.B., Jutare, T., Bauer, J.C. & Jones, J.A. 1963. Biology of *Lytechinus variegatus*. *Bulletin of Marine science*. 13: 23–53.

Pawson, D.L. & Miller, J.E. 1982. Studies of genetically controlled phenotypic characters in laboratory-reared *Lytechinus variegatus* (Lamarck) (Echinodermata: Echinodea) from Bermuda and Florida. J.M. Lawrence (ed.), *Echinoderms: Proceedings of the International Conference*, Tampa Bay. A.A. Balkema, Rotterdam: 1982.

Pearse, J.S. & Pearse, V.B. 1975. Growth zones in the echinoid skeleton. *American Zoologist*. 15: 731–753.

Russell, M.P., Ebert, T.A. & Petraitis, P.S. 1998. Field estimates of growth and mortality of the green sea urchin, *Stronglyocentrotus droebachiensis*. *Opihelia* 48: 137–153.

Russell, M.P. & Meredith, R.W. 2000. Natural growth lines in echinoid ossicles are not reliable indicators of age: a test using *Stongylocentrotus droebachiensis*. *Invertebrate Biology* 119: 410–420.

Sewell, M.A. & Watson, J.C. 1993. A "source" for asteroid larvae?: recruitment of *Pisaster ochraceus, Pycnopodia helianthoides* and *Dermasterias imbricata* in Nootka Sound, British Columbia. *Marine biology* 117: 387–398.

Taki, J. 1971. Tetracycline labelling of test plates in Strongylocentrotus intermedius. Scientific Reports of Hokkaido Fisheries Experimental Station. 13: 19–29.

Taki, J. 1972a. A tetracycline labelling observation of growth zones in the jaw apparatus of *Strongylocentrotus intermedius*. *Bulletin of the Japanese Society of Scientific Fisheries*. 38: 181–188.

Taki, J. 1972b. A tetracycline labelling observation of growth zones in the test plate of *Strongylocentrotus intermedius*. *Bulletin of the Japanese Society of Scientific Fisheries*. 38: 117–121.

Taki, J. 1978. Formation of growth lines in test plates of the sea urchin, *Strongylocentrotus intermedius*, reared with different algae. *Bulletin of the Japanese Society of Scientific Fisheries*. 44: 955–960.

Tanaka, M. 1982. A new growth curve which expresses infinitive increase. Amakusa marine biology lab. Kyushu University 6: 167–177.

Tanaka, M. 1988. Eco-physiological meaning of parameters of ALOG growth curve. Amakusa marine biology lab. Kyushu University 9: 13–106.

Vadas, R.L., Denchel, T. & Ogden, J.C. 1982. Ecological studies on the sea urchin, *Lytechinus variegatus*, and the local algal-seagrass communities of the Miskito Cays, Nicaragua. *Aquatic biology* 14: 109–125.

Watts, S.A., McClintock, J.B. & Lawrence, J.M. 2007. Ecology of Lytechinus In: Lawrence JM (ed.) *Edible Sea Urchins: Biology and Ecology*. Elsevier.

Echinoderms in a Changing World – Johnson (ed)
© *2013 Taylor & Francis Group, London, ISBN 978-1-138-00010-0*

Potential use of production and biomass for life-history comparisons of sea urchins

T.A. Ebert

Oregon State University, Corvallis, OR, USA

ABSTRACT: Comparing life histories has used measures of growth or a combination of growth and survival. A useful relationship is between the growth-rate constant, K, of the Brody-Bertalanffy growth model, and the instantaneous mortality rate, M. Asymptotic size, S_∞, and K have been combined as KS_∞ in different forms but this is not useful because small K and large S_∞ can have the same product as large K and small S_∞. Life history comparisons are more difficult with other growth models. Biomass and production provide a currency that is independent of any particular growth model but require measurements in units of mass or energy so models with linear dimensions need conversion. Analysis is illustrated using the red sea urchin *Strongylocentrotus franciscanus*. In addition to P/B, production of spawn mass relative to total production is a measure of relative allocation and adds an extra dimension to life-history comparisons of sea urchins.

1 INTRODUCTION

Comparison of life histories is important because patterns provide insights into the evolution of species as well as possible constraints to management of those of commercial importance. The salient features of a life history are the age at first reproduction, the scheduling of reproductive episodes and numbers of offspring at each time, and the probabilities of survival from one reproductive episode to the next. These traits are consequences of resource allocation to morphology, physiology, and behavior. For many species including echinoderms, number of offspring at each reproductive event is a function of size so as individuals grow they also produce more gametes. Growth becomes important in the trade-off between current reproduction and investment in growth to increase reproductive output at the next reproductive event. Given suitable variation, such a trade-off will be selected if it increases fitness more than just increasing current gamete production.

One approach to comparison of life histories has been to focus on attributes of growth and survival or some combination of these (e.g. Beverton & Holt 1956, Ebert 1975, Robertson 1979, Charnov 1993). In such comparisons, growth has been modeled using the Brody-Bertalanffy equation, which is a special case of the Richards function (Eq. 1),

$$S_t = S_\infty(1 - be^{-Kt})^{-n},\qquad(1)$$

where S_t is size at time t, S_∞ is asymptotic size, K is a growth rate constant, n is a shape parameter, and

$$b = \frac{S_\infty^{-1/n} - S_0^{-1/n}}{S_\infty^{-1/n}}\qquad(2)$$

with S_0 size at $t = 0$.

When $n = -1$, Eq. 1 is the Brody-Bertalanffy equation and there are several ways of comparing species. One is to use K and an estimate of the instantaneous mortality rate, M,

$$N_t = N_0 e^{-Mt},\qquad(3)$$

where N_0 is initial number of individuals and N_t is number remaining at time t.

A graph of M vs. K shows a positive relationship for a variety of species (Beverton & Holt 1956, Shine & Charnov 1992, Charnov 1993) including sea urchins (Ebert 1975, 2007). The meaning is clear: species that approach maximum size slowly (small K) also tend to have low mortality (small M).

Another approach has been to focus just on growth characteristics and use the product of K and S_∞, called ω by Gallucci & Quinn (1979) and used in studies of marine invertebrates including sea urchins (Appeldoorn 1983, Duineveld & Jenness 1984). The instantaneous growth rate is

$$\frac{dS}{dt} = K(S_\infty - S_t),\qquad(4)$$

so when $S_t = 0$, the growth rate is KS_∞, the maximum. Pauly (1991) added an exponent to S_∞ and used a log-transformation,

$$\exp(\phi') = KS_\infty^2, \tag{5}$$

or

$$\phi' = \ln K + 2\ln S_\infty. \tag{6}$$

The exponent 2 was added so size would be related to area rather than just a linear measurement. It is clear that ω and ϕ' are similar in how K and S_∞ are combined. Both of these indices require the Brody-Bertalanffy model and both include all combinations of K and S_∞ that give the same value of ω or ϕ' such as a large and slow-growing species and a small and fast growing one. Species that grow rapidly and have a small maximum size really have a very different life history from species that grow slowly and attain a large size. For this reason alone ω or ϕ' are not appropriate measures for life-history comparisons. But there are other reasons as well.

If the Richards function has $n \neq -1$ such as in the logistic ($n = +1$) or Gompertz ($n = \pm\infty$) functions, the growth rate is

$$\frac{dS}{dt} = nKS_t \left(1 - S_\infty^{-\frac{1}{n}} S_t^{\frac{1}{n}} \right) \tag{7}$$

and there is no constant to pair with S_∞ to yield a single value like ω or ϕ'. The maximum growth rate may not be at $S_0 = 0$ and so combining K and S_∞ to produce a single number for comparison is not reasonable. Comparisons also are difficult if the Tanaka function (Tanaka 1982, 1988) is the appropriate model for growth. The question is, how can comparisons be made when different growth functions are required? The goal of the present work is to explore the use of production, production/biomass, and relative allocation to somatic vs. gamete production for possible use in comparing life-histories.

Estimates of production and biomass have been used for a variety of purposes, mostly for ecosystem studies, but including life histories of species (Mann 1969, Zaika 1970, Robertson 1979). The relationship with lifespan is not surprising. If the survival rate is independent of age and just somatic production and biomass are used, P/B is the same as M, the instantaneous mortality coefficient (Allen 1971). The average lifespan is $1/M$. The calculation of production and biomass, however, must include spawn mass and so the ratio no longer retains the simple relationship with M.

Gonad development for echinoderms has tended to focus on application of a gonad index (e.g. Bennett & Giese 1955, Bishop & Watts 1994, Lester et al. 2007) and only infrequently are dissections done covering a wide size range and

presented as gonad weight as a function of diameter or total wet weight (e.g. Baker 1973, Miller & Mann 1973, Tegner & Levin 1983, Ebert & Russell 1994, Ebert et al. 2011). Occasionally, researchers have drained Perivisceral Fluid (PVF) before weighing (e.g. Kramer & Nordin 1975), which, given the inverse relationship between PVF and gonad mass, makes any calculations using total wet weight for comparisons impossible. Both production and biomass calculations require the conversions from linear measures of size to ash-free dry weight, joules, or grams of carbon. Ash-free dry weights are used here.

2 METHODS

2.1 Determination of ash-free dry weights

The approach to ash-free determinations was developed using dissections made during the mid to late 1970s (Ebert 1982, 1988). Diameter, height, and total wet weight were measured and then sea urchins were dissected into body components: body wall, lantern, gut, gonad, perivisceral fluid, and gut contents. The organic content of body parts was determined indirectly by subtraction of water and ash from total wet weight. Body parts dried at 80°C provide one estimate of the amount of water lost. Additional water is present that is tightly bound to the salt present in the body tissues and this moisture is not lost until the tissue is heated to higher temperatures such as during ashing. The general formula for estimating ash-free content, AF, is

$$AF = W - (H_{80} + H_{500} + A) - C \tag{8}$$

where AF is ash-free dry weight, W is wet weight of body part, H_{80} is water lost at 80°C, H_{500} is additional non-organic loss at 500°C, A is ash weight, and C is calcite weight for lantern and body wall.

The additional loss at 500°C was based on weight loss of sea water dried at 80°C and then ashed. It may be additional water held by salt and not removed at 80°C or it may be sodium and potassium volatilized if the furnace temperature exceeds 550°C (Grove et al. 1961).

2.2 Calculation of production and biomass

Red sea urchins were collected 26 February 2008 at Orford Reef in southern Oregon (42° 47′N, 124° 35′W) and dissected 4 March 2008 at the Oregon Institute of Marine Biology. Total wet weight, diameter, and height were measured and then individuals were dissected into component parts.

Other than the perivisceral fluid, all parts were dried at 80°C and weighed to the nearest 0.01 g. Subsequently, the body wall and lantern were treated with sodium hypochlorite bleach to remove tissue, soaked in fresh water for at least 24 hours, rinsed, dried, and weighed to determine the $CaCO_3$ fractions. Gonad production and spawn mass could not be determined from these dissections but provide conversions that can be used with published values of seasonal changes in gonad size as a function of diameter (Baker 1973, Kramer & Nordin 1975).

The calculation for production and biomass were done for a cohort but the result is the same as for a stable and stationary population of many cohorts (Allen 1971). Production, P, is

$$P = G\overline{B} \tag{9}$$

where $\overline{B}$ is mean biomass during each period and G is the mean instantaneous relative growth rate of an individual of weight w (Crisp 1971). The Tanaka growth function is,

$$D_t = \frac{1}{\sqrt{f}} \ln\left[2f(t-c) + 2\sqrt{f^2(t-c)^2 + fa} \right] + d, \tag{10}$$

where

a = a parameter related to maximum growth rate, which is approximately $1/\sqrt{a}$,
c = age at which growth rate is maximum,
d = a parameter that shifts the body size at which growth is maximum,
f = a measure of the rate of change of the growth rate,

$$c = \frac{a}{E} - \frac{E}{4f}, \tag{11}$$

and

$$E = exp\left(\sqrt{f}\left(D_0 - d\right)\right), \tag{12}$$

where D_0 is the size at time 0.

The instantaneous growth rate of diameter is

$$\frac{dD}{dt} = \frac{1}{\sqrt{f(t-c)^2 + a}}. \tag{13}$$

Further details of the Tanaka equation are given in Tanaka (1988) and Ebert et al. (1999).

Previously published Tanaka growth parameters for *S. franciscanus* were used (Ebert 2008): $f = 0.1943$, $d = 6.9458$, $a = 0.1996$, and $c = 2.1084$ with an initial diameter, $D_0 = 1.48$ cm.

Growth in diameter must be changed to growth in ash-free dry weight, AF, which is an allometric relationship

$$AF = \alpha D^\beta \tag{14}$$

or

$$\ln AF = \ln\alpha + \beta \ln D, \tag{15}$$

$$\frac{d\ln AF}{dD} = \frac{\beta}{D} \tag{16}$$

and

$$G = \frac{d\ln AF}{dt} = \frac{\beta}{D} \cdot \frac{1}{\sqrt{f(t-c)^2 + a}}, \tag{17}$$

the mass-specific growth rate (Crisp 1971). Annual survival rates used previously published values (Ebert et al. 1992, 1999, Ebert 2008). For the smallest size, 1.48 cm, survival was 0.363 yr^{-1}. After one year estimated diameter is 2.69 cm and for this and all larger individuals a survival rate of 0.97 yr^{-1} was used. Calculations of production and biomass for *S. franciscanus* followed the approach used by Brey et al. (1995) for *Sterechinus neumayeri*.

Somatic cohort production, P_s, is

$$P_s = \sum_{i=0}^{\omega} N_i M_i G_i, \tag{18}$$

where N_i is the mean number of individuals in the age class i to $i+1$, M_i is the mean ash-free mass of individuals in the age class i to $i+1$, G_i is the mass-specific growth rate, and the summation is from age-i equals 0 to a maximum age, ω. Because gamete (spawn) mass, M_g, starts at zero each year, production is

$$P_g = \sum_{i=0}^{\omega} N_i M_{gi} \tag{19}$$

Biomass is concentrated at the midpoint of the time interval and for the somatic portion

$$B_s = \sum_{i=0}^{\omega} N_i M_i \tag{20}$$

and gamete (spawn) biomass is the same as gamete production $B_g = P_g$. Total production is $P_s + P_g$ and total biomass is $B_s + B_g$. The sums of somatic and gamete production and biomass over all sizes were used to calculate production/biomass. Gamete production divided by total production is a measure of relative allocation to reproduction.

3 RESULTS

3.1 Ash-free dry weight

A 0-intercept regression was used to describe the relationship between dry and ash weight of sea salt and periviseral fluid, PVF, dried at 80°C (Fig. 1). Slopes of the two lines are different ($F_{1,13} = 16.722$, $P = 0.001$) and the difference between the slopes is the organic content of the PVF. For sea salt, S, dry weight at 80°C is ash, A, plus non-organic loss at 500°C, H_{500}.

$$S = A + H_{500}. \qquad (21)$$

so

$$A + H_{500} = 1.1756A \qquad (22)$$

For the periviseral fluid, ash-free weight, AF, as a function of dry weight is

$$PFV\ AF = 1.2092 - 1.1756 = 0.0336 \text{ dry wt (g)}. \quad (23)$$

Dry vs. wet weight for PVF (Fig. 2) used natural logarithms and slope was fixed at 1.0. Multiplying

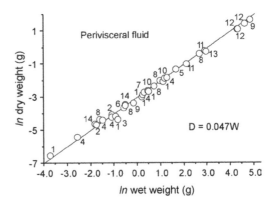

Figure 2. ln Dry weight vs. ln wet weight for periviseral fluid, N = 34; slope was fixed at 1.0; $r^2 = 0.994$, the intercept = -3.0576 with 95% limits of -3.1122—3.0030 so the fixed proportion of wet weight that is dry weight is 0.047; Areas are A, Australia; E, Enewetak Atoll; H, Hawaii; I, Israel; Z, Zanzibar; 1 (H) *Colobocentrotus atratus*; 2 (H) *Cyrtechinus verruculatus*; 3 (Z) *Diadema setosum*; 4 (E) *Echinostrephus aciculatus*; 5 (I) *Echinothrix calamaris*; 6 (E) *Echinothrix diadema*; 7 (E) *Echinometra mathaei*; 8 (A) *Heliocidaris erythrogramma*; 9 (E) *Heterocentrotus trigonarius*; 10 (I) *Lovenia elongata*; 11 (A) *Phyllacanthus parvimensis*; 12 (A) *Salmacis belli*; 13 (E) *Tripneustes gratilla*; 14 (I) *Tripneustes gratilla elatensis*.

the conversion of wet, W, to dry by dry to organic (Eq. 23) is

$$PVF\ AF = 0.0336 \times 0.047 = 0.00158W. \qquad (24)$$

An error was introduced because of the procedure used to estimate ash weight. Tissues were dried in aluminum foil pans to determine water loss. For ashing, tissue was removed from the pan, weighed, and placed in a crucible. The entire dry mass was never recovered from the aluminum pan because some dried body fluid always remained. The error was corrected by calculating % water loss for the entire tissue sample and applying this percentage to the tissue to be ashed to estimate original wet weight on the assumption that this sample lost water at the same rate as the entire original sample. Corrected values for water loss, H_{80}, in gut and gonad samples were compared (Fig. 3). Slopes were homogeneous ($F_{1,39} = 0.0157$, $P = 0.901$) and the relationship is

$$A = 0.0346H_{80}. \qquad (25)$$

with Eq. 8, ash-free weight, AF, can be estimated from water lost at 80°C (H_{80}), wet weight (W), and calcite weight (C). H_{500} in gut and gonad is

$$H_{500} + A = 1.1756 \times 0.0346H_{80}, \qquad (26)$$

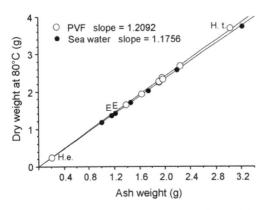

Figure 1. Relationship between estimates of ash weight of sea water and periviseral fluid based on samples dried at 80°C for 48 hours and ashed at 500–580°C; sea salt samples from Enewetak Atoll (September 1977) labeled E; other sea salt samples from Brisbane, Australia, (February 1976); one PVF sample, *Heterocentrotus trigonarius* (H.t) from Enewetak Atoll; *Heliocidaris erythrogramma* (H.e.) from Hastings Pt, NSW, all other PVF samples are *Salmacis belli* from Moreton Bay, Qld, Australia; 95% confidence limits for the slopes are 1.1681–1.1831 (sea salt) and 1.198–1.220 (PVF).

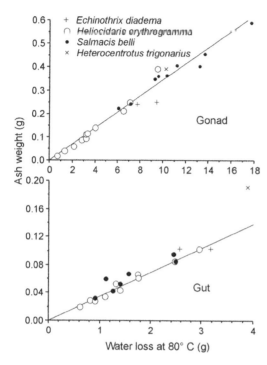

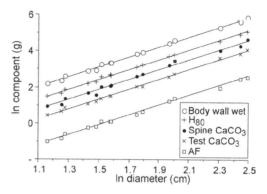

Figure 4. Components of the body wall of *Strongylocentrotus franciscanus* with a common slope of 2.730; H_{80} is water loss during drying at 80°C, spine and test $CaCO_3$ are calcite weights following bleaching; AF is ash-free dry weight.

Figure 3. Ash vs. water lost at 80°C (H_{80}); slopes for gut and gonad are not different; dissections of *Heliocidaris erythrogramma* and *Salmacis belli* were in Australia; *Echinothrix diadema* and *Heterocentrotus trigonarius* were dissected at Enewetak Atoll; note scale differences for gonad and gut.

$$H_{500} + A = 0.04064H_{80}, \qquad (27)$$

$$AF = W - (H_{80} + 0.04064H_{80}) - C, \qquad (28)$$

so

$$AF = W - 1.04064H_{80} - C. \qquad (29)$$

The *AF* content can be estimated using Eqs. 24 and 29 with wet weight for the perivisceral fluid and wet and dry weights for other body parts together with $CaCO_3$ in the body wall and lantern.

3.2 Production and biomass of Strongylocentrotus franciscanus

Calculation of production and biomass of *Strongylocentrotus franciscanus* was done in two parts. The first was conversion of the dissection data for urchins from Orford Reef to ash-free dry weight. The second part of the analysis was to determine size-specific spawn mass.

Components of the body wall as functions of diameter (Fig. 4) have similar slopes

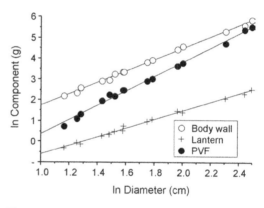

Figure 5. Wet weights of body components of *Strongylocentrotus franciscanus* showing different allometric exponents (slopes); Body wall (2.7356); Lantern (2.0705); PVF (3.4315).

($F_{4,65} = 0.666$, $P = 0.62$) indicating that water content and ash-free components do not change with size. Variation with size was greatest for the spine calcite component, which is reasonable because spine damage and repair vary more than damage and repair to the test.

There are allometric changes during growth and other body components have different allometric exponents (Fig. 5). Aristotle's lantern becomes smaller relative to the body wall and the perivisceral fluid fraction increases. As a consequence of these relative changes, the sum of ash-free components for body wall, lantern, gut, and PVF as a function of test diameter (Fig. 6) shows an allometric exponent that is different from the slope for ash-free dry weight of the body wall (Fig. 4).

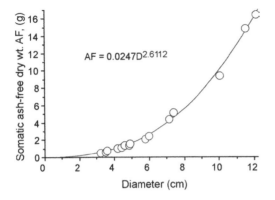

Figure 6. Somatic ash-free dry weight (g) vs. diameter (cm) for *Strongylocentrotus franciscanus* from Orford Reef, Oregon.

Table 1. Estimate of the correction value C for diameter when gonads begin to develop in *Strongylocentrotus franciscanus* using maximum gonad wet weight in grams (G) and test diameter in centimeters (D); data from Baker (1973, N = 44) Kramer & Nordin (1975, N = 70) and this study (N = 15); the model is $\ln G = A + \beta \ln(D - C)$; $r^2 = 0.94$.

Parameter	Estimate	Lower <95%>	Upper
A	−0.7848	−1.3413	−0.2283
β	2.4650	2.2353	2.6947
C	2.7022	2.4724	2.9319

Gonad production and spawn mass could not be determined from the Orford Reef dissections but provide conversions for published values of seasonal changes in gonad size. The initial step was to estimate the size when gonads begin to develop. The Orford reef sample from late February was combined with March data from Albert Head, Vancouver Is. (Kramer & Nordin 1975) and data from southern California (Baker 1973) to estimate the size when gonads begin to develop (Table 1). The estimate for size is 2.70 cm and was used to correct diameter for analysis of maximum and minimum gonad size first in estimating a common allometric exponent, β, and then for individual estimates of the allometry coefficient, α. (Fig. 7).

Data from Kramer & Nordin (1975) are wet weights but the required relationship is ash-free dry weight. A 0-intercept regression was used to determine the conversion constant using Eq. 29 and H_{80} for the sea urchins collected at Orford Reef (Fig. 8).

The results of dissections of *S. franciscanus* coupled with analysis of the Kramer & Nordin (1975) gonad data were used to calculate production and biomass of a cohort. The Tanaka equation (Eq. 10) was used with published parameters (Ebert 2008) to determine size at age with ω = 200 years. Numbers in a cohort (Eq. 3) used published values (Ebert et al. 1992, 1999). Using the conversion values of wet weight to ash-free dry weight (Eqs. 24 and 29) and diameter to ash-free weight (Figs. 6 and 8), mean biomass was calculated for each mean age and production was calculated (Eqs. 9 and 17). The sum of production and biomass for each age class is the cohort production and biomass (Eqs. 18–20). These sums (Table 2) show that most production and biomass are concentrated in spawn production.

Individual somatic and spawn production (Fig. 9) show changing allocation with size.

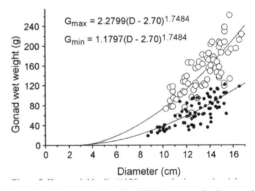

Figure 7. Kramer & Nordin (1975) max and min gonad weights vs. diameter with correction of 2.70 cm for when gonads begin to develop.

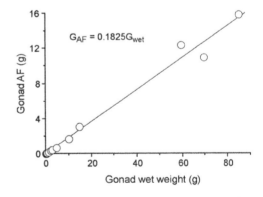

Figure 8. Gonad ash-free dry weight vs. gonad wet weight (g) for *Strongylocentrotus franciscanus* from Orford Reef, Oregon, USA; $r^2 = 0.989$, 95% limits for the constant are 0.1716–0.1734.

Allocation to somatic and spawn is about equal between 7 and 8 cm, which is an age of about four to five years. Past this, an ever increasing fraction of production is directed to spawn. Peak somatic production is at about 6 cm (≈3–4 yr) followed by

Table 2. Ash-free dry weight production and biomass for a cohort of the red sea urchin *Strongylocentrotus franciscanus*.

Attribute	Ash-free dry weight (g)
Ps (somatic production)	12.88
Pg (gamete production)	143.43
Bs (somatic biomass)	406.34
Bg (gamete biomass)	143.43
Pg/(Pg+ Ps)	0.92
ΣP (total production)	156.31
ΣB (total biomass)	549.77
P/B	0.28

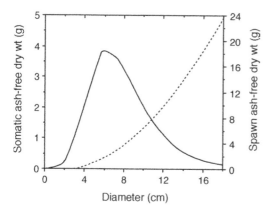

Figure 9. Relation between test diameter and individual production of *Strongylocentrotus franciscanus*; solid line is somatic production and dashed line is spawn production.

a decline, whereas spawn production continues to increase with size out to the largest individuals (Fig. 7).

4 DISCUSSION

Comparison of production and production/biomass across species avoids problems associated with use of different growth models. Furthermore, separation of production of spawn and somatic tissue permits comparisons of resource allocation of surplus energy (Ware 1980) that is an important component of theoretical considerations in life-history evolution (e.g. Perrin & Sibly 1993, Engen & Sæther 1994, Kozlowski 1996, Heino & Kaitala 1999). When tagging data are available, sea urchins display indeterminate growth (e.g. McShane et al. 1997, Ebert et al. 1999, 2008) and this may be the usual condition for echinoids. Resolving the problem of determinate vs. indeterminate growth

in echinoderms in general is an important part of testing life-history theories as presented by Heino & Kaitala (1999).

The large fraction of production devoted to gametes in *Strongylocentrotus franciscanus* and *Sterechinus neumayeri* (Table 3) is understandable in the context of very long life where growth becomes very slow. One theory for the evolution of indeterminate growth assumes that mortality increases with increasing reproductive investment (Sibly et al. 1985). In *S. franciscanus*, however, reproduction continues to increase with size (Figs. 7 and 9) but survival appears to remain unchanged (Ebert 2008).

If successful production of a new individual that lives sufficiently long to reproduce is unlikely, ever increasing allocation to reproduction and long life are major components of a bet-hedging strategy. *Sterechinus antarcticus*, however, poses a problem. The fraction of production devoted to gametes is only 0.48 but the instantaneous mortality rate, M, is 0.07 yr^{-1}, similar to *S. franciscanus* and *S neumayeri* and P/B is the lowest of any species in Table 3 that includes spawn mass in calculations. Estimation of spawn production for *S. antarcticus* may have been a problem or this species has evolved with very different selective forces. With current data it is not possible to know.

There are some methodological problems in analysis of *S. franciscanus* data. Growth and survival estimates are based on 903 tagged individuals and a size structure with N = 1345 (Ebert 2008), which probably are sufficient but there are no ash-free determinations of body parts. Body fluid of sea urchins is isotonic with sea water (Binyon 1966) and so the estimates of ash-free weights are expected to follow the determinations for other echinoid species presented here although error is introduced due to differences in salinity across sites. Low salinities occur off the Oregon coast. Mean annual coastal salinity at Cape Arago, Oregon, was 32.5‰ (Wyatt et al. 1965) and so salt content of tissues would be low. Highest salinities for dissections in Figure 2 occur at Eilat, Israel, with reported values of 41‰ (Paldor & Anati 1979). Salinities at other areas where dissections were done range from 34‰ to 34.5‰ (Milford & Church 1977, Atkinson et al. 1981). Salinity was not measured prior to dissections but Figure 2 shows similar dry weight measurements for 14 species collected from different geographic areas and so with current data, salinity does not appear to be a major source of error in analysis but should be measured in further studies because it can be used to adjust Eq. 29.

Species in Table 3 generally follow a pattern of decreasing fraction of gamete production with increasing M yr^{-1}. The pattern can be understood as

Table 3. Production/biomass and instantaneous mortality, M yr^{-1}, estimates; $Pg/(Pg + Ps)$ is spawn production, Pg, divided by total production.

Species	Area	P/B	$Pg/(Pg + Ps)$	M yr^{-1}	Author
Moira atrops	Florida, USA	1.24*	0.30	0.84[†]	Moore & Lopez 1966
Strongylocentrotus droebachiensis	Nova Scotia, Canada	0.80	0.19	0.41[†]	Miller & Mann 1973
Parechinus angulosus	South Africa	0.59	0.18**	0.12	Greenwood 1980
Sterechinus neumayeri	Antarctica	0.44–0.45	0.96–0.98	0.06	Brey et al. 1995
Strongylocentrotus franciscanus	Washington, USA	0.28	0.92	0.03	This study
Sterechinus antarcticus	Antarctica	0.12	0.48	0.07	Brey 1991
Strongylocentrotus pallidus	Arctic	0.07	§	0.08	Bluhm et al. 1998

*T. Brey (pers, comm.), **Brey et al. (1995), [†]Ebert (1975), [§]Bluhm et al. (1998) did not include spawn.

indicating that with higher mortality, fewer reproductive episodes occur as individuals approach the size of the largest individuals in a population. With low mortality, there are more reproductive episodes while somatic production is low. The sample in Table 3 is small and more species are required, which raises additional problems.

There are problems with lack of standard methods. Growth estimates in Table 3 come from size-frequency distributions (Moore & Lopez 1966, Greenwood 1980) natural growth lines (Miller & Mann 1973, Brey 1991, Bluhm et al. 1998), and tagging (Brey et al. 1995, Ebert et al. 1999, Ebert 2008). Resolution of natural lines after growth becomes very slow has not been achieved (e.g. Brey et al. 1995, Shelton et al. 2006) and bias in estimating P/B when growth of the largest individuals is not known. There are problems with determining spawn production because in most studies only a gonad index is reported and so size at which gonads begin to develop can not be determined nor can the allometric exponent relating gonad size to weight or diameter. An approximation of spawn mass as a percentage of maximum gonad size has been used. For example, Brey et al. (1995) used 60% for *S. neumayeri* based on Pearse & Giese (1966). Brockington et al. (2001), however, found very little change in gonad size over a period of two years and considered the estimate of gonad production used by Brey et al. (1995) to be too high. Based on low mortality and the published growth curve, the estimate of spawn production by Brey et al. (1995) seems more reasonable. Brockington et al. (2001), however, bring up the problem of measuring spawn production when there is no single spawning event during a year. Obviously more studies of gonad production are needed with dissections over a wide size range.

The validity of using production, biomass and relative spawn mass for life history comparisons can not be established by the small number of species in Table 3 although from a theoretical standpoint applicability is clear. The problem is collecting sufficient data for a large number of species to examine patterns. Critical to analysis are estimates of growth and survival. These are best obtained by chemical tagging so slow growth of large individuals can be determined. Estimating survival requires large numbers of measured individuals to determine size structure and also good tag returns. Calculation of production requires additional effort with respect to dissections that include monthly gonad samples as well as other body components, both wet and dry and done so confidence limits can be estimated (Brey 1990). The single major problem in expanding the species list in Table 3 is monthly gonad dissections that include all sizes and not converted into a gonad index.

Assembling all necessary data is a challenge. Production estimation, however, adds an additional level to analysis of life histories with respect to allocation and has the potential to become a new invariant (Charnov 1993) in the exploration of life histories and so deserves additional attention.

ACKNOWLEDGEMENTS

Work on growth and survival of red sea urchins was with S. Schroeter, J. Dixon, and A. Bradbury with support from the Pacific States Fishery Commission and Washington state employees. Work at Enewetak Atoll was supported the US Dept. of Energy administrated through the University of Hawaii. Sea urchins were collected at Orford Reef by T. Foley of the *F/V Mach I* and transported and maintained at the Oregon Institute of Marine Biology by S. Groth and B. Miller, Oregon Dept of Fish & Wildlife. Portions of the work were funded by the Ocean Sciences Division Biological Oceanography of the US National Science Foundation grants OCE75-10442 and OCE-0623934.

REFERENCES

Allen, K.R. 1971. Relation between production and biomass. *Journal of the Fisheries Research Board of Canada* 28: 1573–1581.

Appeldoorn, R.S. 1983. Variation in the growth rate of *Mya arenaria* and its relationship to the environment as analyzed through principal components analysis and the ω parameter of the von Bertalanffy equation. *Fishery Bulletin U.S.* 81: 75–84.

Atkinson, M., Smith, S.V. & Stroup, E.D. 1981. Circulation in Enewetak Atoll lagoon. *Limnology and Oceanography* 26: 1074–1083.

Baker, S.L. 1973. *Growth of the red sea urchin* Strongylocentrotus franciscanus *(Agassiz) in two natural habitats.* MS Thesis, San Diego: San Diego State University.

Bennett, J. & Giese, A.C. 1955. The annual reproductive and nutritional cycles in two western sea urchins. *Biological Bulletin* 109: 226–237.

Beverton, R.J. & Holt, S.J. 1956. A review of methods for estimating mortality rates in exploited fish populations, with special reference to sources of bias in catch sampling. *Conseil Permanent International pour l'Exploration de la Mer. Rapports et Procè-Verbaux des Réunions* 140: 67–83.

Binyon, J. 1966. Salinity tolerance and ionic regulation. In: R.A. Boolootian (ed.) *Physiology of echinodermata*: 359–377. New York: Wiley & Sons.

Bishop, C.D. & Watts, S.A. 1994. Two-stage recovery of gametogenic activity following starvation in *Lytechinus variegatus* Lamarck (Echinodermata: Echinoidea). *Journal of Experimental Marine Biology and Ecology* 177: 27–36.

Bluhm, B.A., Piepenburg D. & von Juterzenka, K. 1998. Distribution, standing stock, growth, mortality and production of *Strongylocentrotus pallidus* (Echinodermata: Echinoidea) in the northern Barents Sea. *Polar Biology* 20: 325–334.

Brey, T. 1990. Confidence limits for secondary production estimates: application of the bootstrap to the increment summation. *Marine Biology*. 106: 503–508.

Brey, T. 1991. Population dynamics of *Sterechinus antarcticus* (Echinodermata: Echinoidea) on the Weddell Sea shelf and slope, Antarctica. *Antarctic Science* 3: 251–256.

Brey, T., Pearse, J., Basch, L. McClintock, J. & Slattery, M. 1995. Growth and production of *Sterechinus neumayeri* (Echinoidea: Echinodermata) in McMurdo Sound, Antarctica. *Marine Biology* 124: 279–292.

Brockington, S., Clarke, A. & Chapman A.L.G. 2001. Seasonality of feeding and nutritional status during the austral winter in the Antarctic sea urchin *Sterechinus neumayeri*. *Marine Biology* 139: 127–138.

Charnov, E.L. 1993. Life History Invariants: Some Explorations of Symmetry in Evolutionary Ecology. New York: Oxford University.

Crisp, D.J. 1971. Energy flow measurements. In N.A. Holme & A.D. McIntyre (eds.) *Methods for the study of marine benthos*: 197–279. Oxford: Blackwell.

Duineveld, G.C.A, & Jenness, M.I. 1984. Differences in growth rates of the sea urchin *Echinocardium cordatum* as estimated by the parameter ω of the von Bertalanffy equation applied to skeletal rings. *Marine Ecology Progress Series* 19: 65–72.

Ebert, T.A. 1975. Growth and mortality in postlarval echinoids. *American Zoologist* 15: 755–775.

Ebert, T.A. 1982. Longevity, life history, and relative body wall size in sea urchins. *Ecological Monographs* 52: 353–394.

Ebert, T.A. 1988. Allometry, design and constraint of body components and shape in sea urchins. *Journal of Natural History* 22: 1407–1425.

Ebert, T.A. 2007. Growth and survival of postsettlement sea urchins. In: J.M. Lawrence (ed.) *Edible sea urchins: Biology and ecology*: 95–134. Amsterdam: Elsevier.

Ebert, T.A. 2008. Longevity and lack of senescence in the red sea urchin *Strongylocentrotus franciscanus*. *Experimental Gerontology* 43: 734–738.

Ebert, T.A., Dixon, J.D., & Schroeter, S.C., 1992. Experimental outplant of cultured juvenile red sea urchins, *Strongylocentrotus franciscanus* in California. Final Technical Report FG-0182. Sacramento: California Department of Fish and Game.

Ebert, T.A., Dixon, J.D., Schroeter, S.C., Kalvass P.E., Richmond, N.T., Bradbury, W.A., & Woodby, D.A., 1999. Growth and mortality of red sea urchins *Strongylocentrotus franciscanus* across a latitudinal gradient. *Marine Ecology Progress Series* 190: 189–209.

Ebert, T.A., Hernandez, J.C. & Russell, M.P. 2011. Problems of the gonad index and what can be done: Analysis of the purple sea urchin *Strongylocentrotus purpuratus*. *Marine Biology* 153: 47–58.

Ebert, T.A. & Russell, M.P. 1994. Allometry and Model II nonlinear regression. *Journal of Theoretical Biology* 168: 367–372.

Ebert, T.A., Russell, M.P., Gamba, G. & Bodnar, A. 2008 Growth, survival, and longevity estimates for the rock-boring sea urchin *Echinometra lucunter lucunter* (Echinodermata, Echinoidea) in Bermuda. *Bulletin of Marine Science* 82: 381–403.

Engen, S. & Sæther, B.-E. 1994. Optimal allocation of resources to growth and reproduction. *Theoretical Population Biology* 46: 232–248.

Gallucci, V.F. & Quinn, T.J. 1979. Reparameterizing, fitting, and testing a simple growth model. *Transactions of the American Fisheries Society* 108: 14–25.

Greenwood, P.J. 1980. Growth, respiration and tentative energy budgets for two populations of the sea-urchin *Parechinus angulosus* (Leske). *Estuarine and Coastal Marine Science* 10: 347–368.

Grove, E.L., Jones, R.A. & Mathews, W. 1961. The loss of sodium and potassium during the dry ashing of animal tissue. *Analytical Biochemistry* 2: 221–230.

Heino, M. & Kaitala, V. 1999. Evolution of resource allocation between growth and reproduction in animals with indeterminate growth. *Journal of Evolutionary Biology* 12: 423–429.

Kozlowski, J. 1996. Optimal allocation of resources explains interspecific life-history patterns in animals with indeterminate growth. *Proceedings of the Royal Society of London Series B* 263: 559–566.

Kramer, D.E. & Nordin, D.M.A. 1975. Physical data from a study of size, weight and gonad quality for the red sea urchin [*Strongylocentrotus franciscanus* (Agassiz)] over a one-year period. *Fisheries Research Board of Canada Manuscript Report Series* 1372: 1–91.

Lester, S.E., Gaines, S.D. & Kinlan, B.P. 2007. Reproduction on the edge: large-scale patterns of individual performance in a marine invertebrate. *Ecology* 88: 2229–2239.

Mann, K.H. 1969. The dynamics of aquatic ecosystems. *Advances in Ecological Research* 6: 1–81.

McShane, P.E. & Anderson, O.F. 1997. Resource allocation and growth rates in the sea urchin *Evechinus chloroticus* (Echinoidea; Echinometridae). *Marine Biology* 128: 657–663.

Milford, S.N. & Church, J.A. 1977. Simplified circulation and mixing models of Moreton Bay, Queensland *Australian Journal of Marine and Freshwater Research* 28: 23–34.

Miller, R.J. & Mann, K.M. 1973. Ecological energetics of the seaweed zone in a marine bay on the Atlantic coast of Canada. III. energy transformations by sea urchins. *Marine Biology* 18: 99–114.

Moore, H.B. & Lopez, N.N. 1966. The ecology and productivity of *Moira atropos* (Lamarck). *Bulletin of Marine Science* 16: 648–667.

Paldor N. & Anati D.A. 1979. Seasonal variations of temperature and salinity in the Gulf of Eilat (Aqaba). *Deep-Sea Research* 26: 661–672.

Pauly, D. 1991. Growth performance in fishes: Rigorous description of patterns as a basis for understanding causal mechanisms. *Aquabyte* 4: 3–6.

Pearse, J.S. & Giese, A.C. 1966. Food, reproduction and organic constitution of the common Antarctic echinoid *Sterechinus neumayeri* (Meissner). *Biological Bulletin* 130: 387–401.

Perrin, N. & Sibly, R.M. 1993. Dynamic models of energy allocation and investment. *Annual Review of Ecology and Systematics* 24: 379–410.

Robertson, A.I. 1979. The relationship between annual production:biomass ratios and lifespans for marine macrobenthos. *Oecologia* 38: 193–202.

Shelton, A.O., Woodby, D.A., Hebert, K., & Witman, J.D. 2006. Evaluating age determination and spatial patterns of growth of the red sea urchins in southeast Alaska. *Transactions of the American Fisheries Society* 135: 1670–1680.

Shine, R. & Charnov, E.L. 1992. Patterns of survivorship, growth and maturation in snakes and lizards. *American Naturalist* 139: 1257–1269.

Sibly, R, Calow, P. & Nichols, N. 1985. Are patterns of growth adaptive? *Journal of Theoretical Biology* 112: 553–574.

Tanaka, M. 1982. A new growth curve which expresses infinite increase. *Publications from the Amakusa Marine Biological Laboratory* 6: 167–177.

Tanaka, M. 1988. Eco-physiological meaning of parameters of ALOG growth curve. *Publications from the Amakusa Marine Biological Laboratory* 9: 103–106.

Tegner, M.J. & Levin, L.A. 1983. Spiny lobsters and sea urchins: analysis of a predator-prey interaction. *Journal of Experimental Marine Biology and Ecology* 73: 125–150.

Ware, D.M. 1980. Bioenergetics of stock and recruitment. *Journal of the Fisheries Research Board of Canad.* 37: 1012–1024.

Wyatt, B. Still, R. & Haag, C. 1965. Surface temperature and salinity observations at Pacific Northwest shore stations: for 1963 and 1964. *Data Report No. 21. Ref. 65–20. Oregon State University*: 1–18 http://hdl.handle.net/1957/6570.

Zaika, V.Y. 1970. Relationship between the productivity of marine mollusks and their life-span. *Oceanology* 10: 547–552.

Echinoderms in a Changing World – Johnson (ed)
© *2013 Taylor & Francis Group, London, ISBN 978-1-138-00010-0*

Natural *vs* artificial: Quantifying settler and juvenile echinoderms on cobble substrate

L.B. Jennings & H.L. Hunt

Department of Biology, University of New Brunswick, Saint John, New Brunswick, Canada

ABSTRACT: Juvenile and settler echinoderms can be difficult to quantify as they are often cryptic, but must be sampled to examine recruitment and early post-settlement events. This study examined the use of artificial collectors (Astroturf) compared to sampling the natural substrate to quantify settler and juvenile sea urchins and sea stars on cobble substrate in the NW Atlantic. More sea urchin settlers accumulated on Astroturf on the benthos than suspended in the water and more again were found among cobbles in pans than any other surface. There was no difference in the number of sea star settlers on various Astroturf collectors although more were on the artificial turf than the natural substrate. Juvenile sea urchin abundance did not significantly differ between three different methods: cobbles in pans, collecting cobbles from the benthos, and suction sampling. The best collection method depends on the question of study, because the substrate displaying the greatest abundance and best precision varied.

1 INTRODUCTION

Recruitment of most benthic invertebrates, including echinoderms, is dependent on larval supply, settlement and early post-settlement events (Hunt & Scheibling 1997). However, it is hard to observe these processes in the field because juvenile and settler echinoderms are small and often cryptic. To date, various sampling methods have been used to better understand settlement and early post-settlement events.

Settlement occurs when organisms leave the water column to start life on the benthos and undergo metamorphosis (Rodriguez et al. 1993). Individuals are often only a few hundred microns in size. Since settlers are small and often attach to substrate that cannot be removed, such as corals or bedrock, collector devices are often used to quantify them. Collectors have been used to quantify recruitment of echinoderms for over 70 years (Loosanoff 1964) and are useful in collecting different classes and species of echinoderms (Gabaev 1998, Balch & Scheibling 2000). They provide a standardized substrate that is logistically easy to manipulate. Many different styles of collectors have been used, including oyster shells (Loosanoff 1964), light diffuser plastic (Harrold et al. 1991), biospheres (Keesing et al. 1993), scrub brushes (Ebert et al. 1994), and plastic turf (Balch & Scheibling 2000). Artificial turf has become popular for quantifying settlement because it is easily obtainable, non-destructive, and easy to manipulate in the

field and analyze in the lab (Harrold et al. 1991). Plastic grass (Astroturf) has been used around the world as an artificial settling surface for echinoderms (Balch & Scheibling 2000, Lambert & Harris 2000, Lamare & Barker 2001). However, the number of individuals on a collector such as Astroturf indicates that larvae are competent to settle but does not necessarily reflect settlement on natural substrate. Larvae of many species, including sea urchins and barnacles, may preferentially settle on the artificial substrate when settlement on the natural substrate is low or vice versa (Pineda & Caswell 1997, Lamare & Barker 2001). Therefore, collectors should be considered as an index of settlement supply until they are compared to settlement on the natural substrates of the area.

To examine early post-settlement processes, good estimates of both settlement and abundance a few months later are needed. While using artificial collectors can help quantify the supply of settlers, recruitment patterns on collectors are likely to be biased by effects of the collectors on water currents, predators and resources. For example, Harris et al. (1994) found that survival of juvenile sea urchins was greater on plastic turf than on nearby rocks and attributed differences to predator abundance. Therefore, other methods should be sought to determine the number of young recruits. Often this is done by examining settlers on natural substrates by assessing the area within quadrats *in situ* (e.g. Tomas et al. 2004). However, small individuals (<10 mm) may be overlooked using this method (but see Pearse & Hines 1987, Lamare & Barker

2001, Tomas et al. 2004). In the Bay of Fundy where juveniles grow slowly (Robinson & Macintyre 1997), field observations will likely miss juveniles up to a few years old. Other methods used include suction sampling on bedrock (with >85% efficiency) (McNaught 1999) and collecting substrate to examine in the lab (Lopez et al. 1998).

This study compared several artificial and natural substrate collector types as methods to quantify echinoderm settlers and evaluated techniques for quantifying abundance of juvenile echinoderms. Three common echinoderm species found in the North West Atlantic Ocean were examined: the sea urchin *Strongylocentrotus droebachiensis* and the sea stars *Asterias rubens* and *A. forbesi*. *A. rubens* and *A. forbesi* are known to hybridize (Harper & Hart 2007) and are morphologically indistinguishable as young juveniles; thus the two species were therefore grouped together for this study. *S. droebachiensis* is a commercially exported species and is known to be ecologically important through its ability to change kelp forests into rocky barrens (Scheibling 1996, Andrew et al. 2002). The objectives of this study were (1) to compare various artificial and natural substrate settlement collectors for sea urchin and sea star settlement at a cobble substrate site; (2) to evaluate various collection methods of juvenile (~3 month old) sea urchins at the same site.

2 METHODS

2.1 *Study site*

This study took place in the Bocabec Cove region of Passamaquoddy Bay, Bay of Fundy, in New Brunswick, Canada (Fig. 1). The rocky subtidal

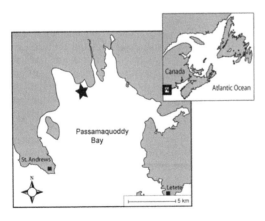

Figure 1. Map of Passamaquoddy Bay on the east coast of Canada. Black star indicates field site (45°08.507'N 67°00.105'W).

site was chosen based on the presence of adult sea cucumbers, sea urchins, sea stars and brittle stars, indicating that it represented a suitable habitat for echinoderms. Previous work on the settlement and recruitment of these species has been done at this site (Jennings & Hunt 2010). The mean water depth was approximately 10 m at high tide and 4 m at low tide, and there was no strong current (average current in this area is <0.09 m sec^{-2} and max current is <0.20 m sec^{-2}; Chang, pers. comm.).

2.2 *Settlement*

Two experiments were performed to compare collection of settling echinoderms. Experiment 1 was undertaken in 2005 and compared three types of Astroturf collectors. Experiment 2 was done in 2007 and 2008 and compared two of the Astroturf collectors with two types of natural substrate collection.

2.2.1 *Experiment 1: Astroturf collectors*

In 2005, three different types of Astroturf collectors were compared to determine the effect of collector placement (suspended in the water column or flat on the ground) and exclusion of macrofaunal predators (1 cm mesh over the turf) on the number of settlers. The suspended Astroturf was in the form of a pipe collector, designed after those used by Balch & Scheibling (2000). A 47 × 20 cm piece of Astroturf, the artificial settling substrate, was inserted into a 15 cm diameter PVC pipe 20 cm in length. Each pipe was suspended about 1 m above the bottom between a float and cinder blocks and had a vane on the top of the pipe to keep it facing into the current. 1 cm mesh lids covered the ends of the pipes to prevent large predators from entering.

The second and third types of Astroturf settlement collector had the Astroturf attached to tiles. Each collector was comprised of a 15 × 15 cm ceramic tile bolted to a brick with a Vexar mesh screen between the two pieces to prevent predators from entering via the bottom. Astroturf (15 × 15 cm) was attached to the tiles. Tiles either had no cage around them, or were covered with a 1 cm mesh cage 10 cm in height to test the effect of excluding macro-predators.

Four replicates of each of the three types of Astroturf collectors (pipe, tile caged, tile uncaged) were deployed at the site from June to October 2005 along a transect 20 m in length that was parallel to the shore at a constant depth. The collectors were placed in four groups, each of which had one of each type of collector, separated by 3–5 m. All Astroturf was pre-soaked for at least 10 days in filtered sea water to grow a biofilm, indicated by a slimy layer. The Astroturf was changed approximately every two weeks by SCUBA divers. Each piece of Astroturf was placed in a plastic bag in situ to minimize the loss of organisms upon ascent. Upon return to the lab, the Astroturf was soaked in 50% EtOH for at least 10 min to initiate the release of echinoderm tube feet (Balch et al. 1999, Balch & Scheibling 2000). The sample was then rinsed through a 180-μm sieve with fresh water and preserved in 80% EtOH. All echinoderms were identified, counted and measured using a dissecting microscope.

2.2.2 *Experiment 2: Astroturf collectors and natural substrate collection*

To compare settlement on the artificial turf to that on natural substrate, seven replicates of each type of method were deployed in July and retrieved in September 2007, placed every 1 m along a 21 m transect in a random order. Note that sea urchin settlement in 2007 occurred earlier than expected based on previous years and occurred before the collectors were deployed. Therefore the experiment was re-run in June/July in 2008 to compare methods for sea urchin settlement.

Two types of artificial collectors were used: the pipe collectors (as described above) and the tile collectors without cages (as described above except the size of the Astroturf was 47 × 20 cm and there was no mesh between the brick and the tile). Two different collection methods using natural substrate were also examined. The first was cobbles (the natural substrate) set out in plastic pans (32 × 22 cm). Approximately 25 rocks covered with dead coralline algae and soaked to ensure a biofilm were placed in each pan. The second settler collection method used was suction sampling with a manual bilge pump as in McNaught (1999). 25 × 25 cm quadrats were randomly placed off the transect (to ensure the location had not been disturbed while changing the other collectors). All cobbles in each quadrat were placed in sampling bags. The area within each quadrat was then suction sampled using a manual bilge pump into a 500 μm mesh bag. The Astroturf was processed as described for experiment 1 and the echinoderms were preserved in 80% EtOH. The pans of cobbles and the suction samples were frozen until they could be examined and the number of echinoderms counted.

2.3 *Recruitment*

Abundances of young juvenile (≤3 mm) sea urchins were quantified in October 2007 using three different methods: suction sampling, collection of cobbles, and colonization of pans of cobbles. The methods quantified abundances along a 20 m transect (different from the collector transect, but at the same depth). Suction sampling was done

with a manual bilge pump as described for settlement for five 25 × 25 cm quadrats randomly placed along the transect. All cobbles in the quadrat were placed in a sampling bag and the area within the quadrat was then suction sampled using a manual bilge pump into a 500 μm mesh bag. Collection of cobbles alone was done in five different 25 × 25 cm quadrats along the same transect. All the cobbles were removed and placed in a sampling bag, but the area within the quadrats was not suction sampled. Five pans of rocks of the same type used to quantify settlement were put out in late June before settlement occurred and were left for 4 months to measure recruitment. The samples were kept frozen until they could be examined and the sea urchins counted.

2.4 *Statistical analyses*

While the collectors were changed biweekly, only the weeks occurring within the peak settlement periods were included in the analysis as periods of low settlement tended to mask differences between collector types. If peak settlement was spread out over more than one biweekly period then the data was pooled for those two periods. The peak settlement for sea urchins was in July and very early August in 2005 (two collection dates) and the end of June and early July in 2008 (one collection date). The peak settlement for sea stars was early to mid August in 2005 (two collection dates) and mid July to mid August in 2007 (two collection dates). The numbers of settlers m^{-2} were compared among the various collector types separately for urchins and sea stars using 1 factor ANOVAs. Variances were homogeneous (experiment 1 O'Brien [0.5] test $F_{2,9} \leq 1.43$, $p \geq 0.29$; experiment 2 O'Brien [0.5] test $F_{3,24} \leq 1.62$, $p \geq 0.21$). The numbers of sea urchin young of the year m^{-2} (settled during the previous summer and approximately 3 months old) were compared among different methods of collection using a 1 factor ANOVA. Sea star recruitment was not analyzed as no individuals were present in the samples. Variances were homogeneous (O'Brien [0.5] test $F_{2,12} = 1.16$, $p = 0.3470$). Student-Newman-Keuls's (SNK) tests were used as post-hoc tests when needed for all ANOVAs. Coefficients of variation (standard deviation/mean × 100%) were also calculated to assess precision among collectors. ANOVAs were performed using the program JMP 5.0.1.2. ©2003 SAS Institute Inc.

3 RESULTS

3.1 *Settlement*

The method that collected the most settlers differed between sea urchins and sea stars. When the three types of Astroturf collectors were compared in 2005, significantly more sea urchins were collected on the tile caged and tile uncaged collectors than in the pipe collectors (Fig. 2, ANOVA, $F_{2,9} = 10.31$, p = 0.0047, SNK post-hoc test).

When Astroturf was compared to the natural substrate in 2008, more sea urchins were found in the pans of rocks than in any other treatment (Fig. 3, ANOVA, $F_{3,24} = 22.48$, p < 0.0001, SNK post-hoc test). However, suction sampling did not result in significantly different density of settler sea urchins than using Astroturf, either flat on the ground or suspended in a pipe (Fig. 3). The coefficients of variation were lower for the natural sub-

strate, both in the pan and suction samples (29% and 37%, respectively), than the Astroturf (43% suspended and 65% on the benthos) (Fig. 4).

When the number of settler sea stars was compared on artificial turf in 2005, there was no significant difference between the numbers of individuals found in the pipe collectors, tile caged collectors and the tile uncaged collectors (Fig. 5, ANOVA, $F_{2,9} = 3.20$, p = 0.0890). The tile uncaged collectors had a higher coefficient of variation (78%) than the other two collector types (Fig. 4). When artificial turf was compared to natural substrate, significantly more sea stars were found on the Astroturf regardless of whether it was suspended in a pipe or flat on the ground on tiles, than on the natural substrate, both in a pan and suction sampled (Fig. 6, ANOVA, $F_{3,24} = 23.46$, p < 0.0001, SNK post-hoc test).

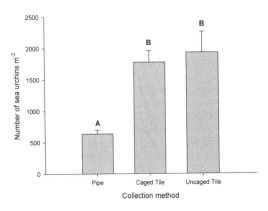

Figure 2. Mean number of sea urchins m^{-2} (+1 SE) settling on three different Astroturf collectors at a rocky subtidal site during the peak settlement period in 2005 (July and early August). Letters that are different indicate significantly different results.

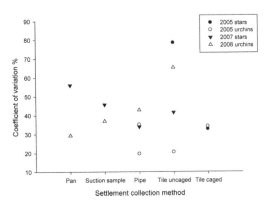

Figure 4. Coefficient of variation of number of settlers versus collection method for sea urchins (*Strongylocentrotus droebachiensis*) (open shapes) and sea stars (*Asterias* spp.) (closed shapes).

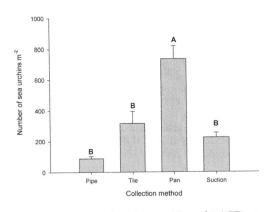

Figure 3. Mean number of sea urchins m^{-2} (+1 SE) settling on four different collector types at a rocky subtidal site during the peak settlement period in 2008 (end of June and early July). Pipe and tile are artificial turf collectors (Astroturf) and pan and suction are natural substrate (cobbles). Letters that are different indicate significantly different results.

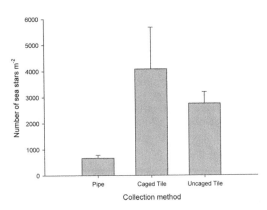

Figure 5. Mean number of sea stars m^{-2} (+1 SE) settling on three different Astroturf collectors at a rocky subtidal site during the peak settlement period in 2005 (early to mid August).

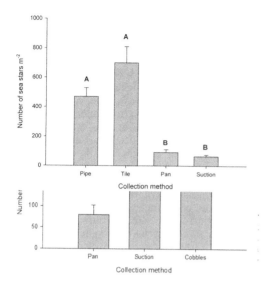

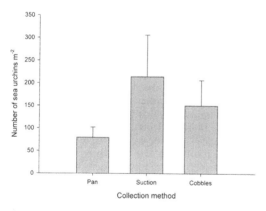

Figure 6. Mean number of sea stars m^{-2} (+1 SE) settling on four different collector types at a rocky subtidal site during the peak settlement period in 2007 (mid July to mid August). Pipe and tile are artificial turf collectors (Astroturf) and pan and suction are natural substrate (cobbles). Letters that are different indicate significantly different results.

Figure 7. Mean number of juvenile sea urchins (1.3–3.0 mm) m^{-2} (+1 SE) quantified using three different collection methods at a rocky subtidal site in October 2007.

3.2 Recruitment

No juvenile sea stars were found in any of the samples collected to compare methods. There were no significant differences in the number of young of the year sea urchins collected among the three different methods (Fig. 7, ANOVA, $F_{2,12} = 1.12$, p = 0.3575). However, the pan collection method had a lower coefficient of variation (64%) than the collection of cobbles alone (96%) or suction sam-

pling (84%). Most of the sea urchins in the suction samples were found on the cobbles collected prior to using the bilge pump suction sampler with only one or two sea urchins (5–15%) per sample being found in the sediment collected by the bilge pump.

4 DISCUSSION

Not all methods of collecting settler sea urchins and sea stars performed equally well at this cobble-bottom site, and the results differed between taxonomic groups. Sea urchin settlers were found in greater numbers on the natural substrate in pan collectors than on the Astroturf. Laboratory experiments have found greater sea urchin settlement on natural substrates than on light diffuser plastic (Harrold et al. 1991, Lamare & Barker 2001) or glass (Rowley 1989), although Harrold et al. (1991) noted that the opposite trend occurred in the field, where more settled on the plastic. There are likely other factors, including the hydrodynamics around the settling surface, that interact with substrate type and influence the collector type the competent larvae will settle on (Mullineaux & Butman 1991). The coefficient of variation was less for the collection methods using natural substrate, indicating that they have greater precision. Therefore, if the question of interest is to quantify potential settlement of sea urchins, then putting out trays of cobbles should be considered as it resulted in higher numbers of sea urchins and greater precision than Astroturf at this site. However, if the question is to quantify how many very recently settled sea urchins are actually on the substrate at a site, then suction sampling (or at least collecting the cobbles as very few sea urchins were in the suction sample bags) should be used at this site. However, this method is not appropriate for substrates that cannot be destructively sampled. The time between sampling periods can be adjusted to compromise between logistics and early post-settlement processes.

A greater abundance of recently settled sea urchins was found on Astroturf that was lying on the substrate than on turf suspended in the water column (although the difference was not significant in 2008). It is possible that sea urchin larvae that are near the benthos are more likely to settle than those higher in the water column perhaps due to hydrodynamic and boundary layer differences. The sea urchin S. droebachiensis has been previously noted to settle more on collectors that were near the substrate than those suspended (Harris et al. 1994, Balch et al. 1998). The current at our study site was quite low (average in area < 0.09 m sec^{-2}), but higher currents may result in more dissimilar quantities on suspended and benthic turf as there would be greater differences between the near bottom and free stream velocities. Therefore, site characteristics, such as hydrodynamics need to be considered when choosing a method. Suspended settlement collectors may be useful if they can be deployed and retrieved from the surface, eliminating some difficulties with the logistics of field work, but benthic settlement collectors tend to give a higher estimate of larval competent to settle. While temporal patterns were not explicitly examined, the same trends occurred over time. Sea urchin settlement tended to be higher on tiles than in pipe collectors both in 2005 and 2008.

Asterias spp. settlers were found in greater numbers on plastic turf than on rocks. We have only found a few juveniles at this site on the natural substrate (cobbles with coralline algae) over the past five years and no sea stars were found in the recruitment samples in this study. It is possible that *Asterias* spp. are undergoing high early post-settlement mortality in the first few days at this site (although no bodies have been found) or that they are settling elsewhere in the area where more food is available (e.g. in the mussel beds shallower in the subtidal or on the nearby scallop farm). Higher abundances of sea stars have been noted in kelp beds rather than on rocky barrens (Balch & Scheibling 2000), suggesting that sea stars may not be settling on the natural substrate at this site because of the lack of kelp or other suitable substrates. Our results for sea star settlement suggest that, if the question of interest concerns the supply of competent larvae, then Astroturf would be a good substrate for use in a collector, but may overestimate settlement on natural cobble substrate at this site. The trends of settlement on Astroturf were similar between years for sea stars as well, with no significant differences between tiles and pipe collectors both in 2005 and 2007. While sea urchins had a difference in supply of settlers between suspended and benthic substrate and sea stars did not in experiment 1 (2005), caution should be used when comparing the results of experiment 2 between groups of organisms as they were done in different years (2007 and 2008).

There was no significant difference between the tile collectors that had 1 cm mesh cages over them and those that did not for either taxonomic group. This suggests that any major predators of newly settled sea urchins and sea stars were less than 1 cm and not deterred by the cage. Scheibling & Robinson (2008) found that crabs 0.9–2.4 cm (carapace width) and shrimp 0.2–4.2 cm (carapace length) had high predation rates on newly settled sea urchins. Animals of these sizes would be able to fit through 1 cm mesh. Since smaller mesh sizes may affect water flow, it is very difficult to separate actual settlement and early post-settlement processes.

Many studies examine recruitment as early post-settlement processes remove individuals from the populations. Since observing these animals in the field may not be possible due to their size, other methods of collection were examined to quantify recruitment. There were no significant differences between methods of collecting juvenile sea urchins three months after settlement between the three methods: colonization of pans of rocks, suction sampling, and collection of cobbles alone. While the pans of cobbles had the lowest coefficient of variation, they are not necessarily the best method for sampling recruitment. The pans require two field days, one to put out and one to collect as compared to one field day for the other methods. Also, since timing of the settlement peak does vary between years (Jennings & Hunt 2010), there is also the risk of putting the pans out after settlement occurs, resulting in not finding any juveniles in the samples. Collecting just the cobbles requires only one field day and less time underwater and in the lab than does suction sampling. In California, collection of cobbles resulted in sampling sea urchins as small as 0.4 mm (Rowley 1989) and in New Brunswick, this method has sampled 0.35 mm sea urchins (pers. obs.). This method will only be useful on cobble substrates where destructive sampling can occur. Other sites should consider suction sampling or pans of rocks.

The abundance of settler sea urchins was greater in the pans than in the suction sampled area, while that of recruits was greater in the suction sampled area than the pans (although not significantly). This is more likely due to differences in post-settlement processes rather than differences in settlement. *S. droebachiensis* has higher settlement on live algae than on dead algae (such as in the pans in this study) (Lambert & Harris 2000). However, the natural substrate also hosts predators and competitors which may reduce the density of settler sea urchins in a few hours. Recent settlers of some benthic species experience up to 90% mortality in the first few days (Gosselin & Qian 1997). The settler sea urchins may have had less early post-settlement mortality in the pans than on the natural substrate because other organisms had not yet immigrated, perhaps being hindered by the plastic walls of the pan. The pans were out longer for the recruitment study, giving other organisms enough time to immigrate on to these cobbles and predate on or compete with the small juvenile sea urchins. Future studies should examine the settlement methods on shorter time scales to determine if they differ in settlement or very early post-settlement processes. More research also is needed on the predatory and competitive interactions between small juvenile sea urchins and other benthic animals.

4.1 Conclusions

The results of this study suggest different sampling methods should be used based on taxa and question of interest. At the study site in the Bay of Fundy, pans of cobbles are recommended to quantify the supply of settler sea urchins, Astroturf should be used to estimate the abundance of potential sea star settlers, and collecting natural cobble substrate should be used to quantify recently settled and juvenile sea urchins. While these methods work best for this particular site, other sites may require different methods due to substrate or hydrodynamic differences.

ACKNOWLEDGEMENTS

We would like to thank Marie-Josée Maltais, Monica Shaver, Bryan Morse, Julie Ellsworth, Mary-Claire Sanderson, and Mack Sprague and the Navy Island Dive Company for all their help with field work. We would also like to thank Jeff Houlahan and Myriam Barbeau and an anonymous reviewer for their comments on this manuscript. This project was funded by a NSERC Discovery Grant to HLH and a NSERC Postgraduate Scholarship to LBJ.

REFERENCES

Andrew, N.L., Agatsuma, Y., Ballesteros, E., Bazhin, A.G., Creaser, E.P., Barnes, D.K.A., Botsford, L.W., Bradbury, A., Campbell, A., Dixon, J.D., Einarsson, S., Gerring, P.K., Herbert, K., Hunter, M., Hur, S.B., Johnson, C.R., Juinio-Menez, M.A., Kalvass, P., Miller, R.J., Moreno, C.A., Palleiro, J.S., Rivas, D., Robinson, S.M.L., Schroeter, S.C., Steneck, R.S., Vadas, R.L., Woodby, D.A. & Xiaoqi, Z. 2002. Status and management of world sea urchin fisheries. *Oceanography and Marine Biology: an Annual Review* 40: 343–425.

Balch, T., Hatcher, B.G. & Scheibling, R.E. 1999. A major settlement event associated with minor meteorologic and oceanographic fluctuations. *Canadian Journal of Zoology* 77: 1657–1662.

Balch, T. & Scheibling, R.E. 2000. Temporal and spatial variability in settlement and recruitment of echinoderms in kelp beds and barrens in Nova Scotia. *Marine Ecology Progress Series* 205: 139–154.

Balch, T., Scheibling, R.E., Harris, L.G., Chester, C.M. & Robinson, S.M.C. 1998. Variation in settlement of *Strongylocentrotus droebachiensis* in the northwest Atlantic: Effects of spatial scale and sampling method. In R. Mooi & M. Telford (eds), *Echinoderms: San Francisco*: 555–560. Rotterdam: Balkema.

Ebert, T.A., Schroeter, S.C., Dixon, J.D. & Kalvass, P. 1994. Settlement patterns of red and purple sea urchins (*Strongylocentrotus franciscanus* and *S. purpuratus*) in California, USA. *Marine Ecology Progress Series* 111: 41–52.

Gabaev, D.D. 1998. Some aspects of the ecology of young echinoderms settling on artificial substrata. In R. Mooi & M. Telford (eds), *Echinoderms: San Francisco*: 31–33. Rotterdam: Balkema.

Gosselin, L.A. & Qian, P.-Y. 1997. Juvenile mortality in benthic marine invertebrates. *Marine Ecology Progress Series* 146: 265–282.

Harper, F.M. & Hart, M.W. 2007. Morphological and phylogenetic evidence for hybridization and introgression in a sea star secondary contact zone. *Invertebrate Biology* 126: 373–384.

Harris, L.G., Rice, B. & Nestler, E.C. 1994. Settlement, early survival and growth in a southern gulf of Maine population of *Strongylocentrotus droebachiensis* (Muller). In B. David, A. Guille, J.-P. Feral & M. Roux (eds), *Echinoderms through Time*: 701–706. Rotterdam: Balkema.

Harrold, C., Lisin, S., Light, K.H. & Tudor, S. 1991. Isolating settlement from recruitment of sea urchins. *Journal of Experimental Marine Biology and Ecology* 147: 81–94.

Hunt, H.L. & Scheibling, R.E. 1997. Role of early post-settlment mortality in recruitment of benthic marine invertebrates. *Marine Ecology Progress Series* 155: 269–301.

Jennings, L.B. & Hunt, H.L. 2010. Settlement, recruitment and potential predators and competitors of juvenile echinoderms in the rocky subtidal zone. *Marine Biology* 157: 307–316.

Keesing, J.K., Cartwright, C.M. & Hall, K.C. 1993. Measuring settlement intensity of echinoderms on coral reefs. *Marine Biology* 117: 399–407.

Lamare, M.D. & Barker, M.F. 2001. Settlement and recruitment of the New Zealand sea urchin *Evechinus chloroicus*. *Marine Ecology Progress Series* 218: 153–166.

Lambert, D.M. & Harris, L.G. 2000. Larval settlement of the green sea urchin, *Strongylocentrotus droebachiensis*, in the southern Gulf of Maine. *Invertebrate Biology* 119: 403–409.

Loosanoff, V.L. 1964. Variations in time and intensity of setting of the starfish, *Asterias forbesi*, in Long Island Sound during a twenty-five year period. *Biological Bulletin* 126: 423–439.

Lopez, S., Turon, X., Montero, E., Palacin, C., Duarte, C.M. & Tarjuelo, I. 1998. Larval abundance, recruitment and early mortality in *Paracentrotus lividus* (Echinoidea). Interannual variability and plankton-benthos coupling. *Marine Ecology Progress Series* 172: 239–251.

McNaught, D.C. 1999. The indirect effects of macroalgae and micropredation on the post-settlement success of the green sea urchin in Maine. University of Maine.

Mullineaux, L.S. & Butman, C.A. 1991. Initial contact, exploration and attachment of barnacle (*Balanus amphitrite*) cyprids settling in flow. *Marine Biology* 110: 93–103.

Pearse, J.S. & Hines, A.H. 1987. Long-term population dynamics of sea urchins in a central California kelp forest: rare recruitment and rapid decline. *Marine Ecology Progress Series* 39: 275–283.

Pineda, J. & Caswell, H. 1997. Dependence of settlement rate on suitable substrate area. *Marine Biology* 129: 541–548.

Robinson, S.M.C. & Macintyre, A.D. 1997. Aging and growth of the green sea urchin. *Bulletin of the Aquaculture Association of Canada* 97: 56–60.

Rodriguez, S.R., Ojeda, F.P. & Inestrosa, N.C. 1993. Settlement of benthic marine invertebrates. *Marine Ecology Progress Series* 97: 193–207.

Rowley, R.J. 1989. Settlement and recruitment of sea urchins (*Strongylocentrotus* spp.) in a sea urchin barren ground and a kelp bed: are populations regulated by settlement or post-settlement processes? *Marine Biology* 100: 485–494.

Scheibling, R.E. 1996. The role of predation in regulating sea urchin populations in eastern Canada. *Oceanologica Acta* 19: 421–430.

Scheibling, R.E. & Robinson, M.C. 2008. Settlement behaviour and early post-settlement predation of the sea urchin *Strongylocentrotus droebachiensis*. *Journal of Experimental Marine Biology and Ecology* 365: 59–66.

Tomas, F., Romero, J. & Turon, X. 2004. Settlement and recruitment of the sea urchin *Paracentrotus lividus* in two contrasting habitats in the Mediterranean. *Marine Ecology Progress Series* 282: 173–184.

Echinoderms in a Changing World – Johnson (ed)
© 2013 Taylor & Francis Group, London, ISBN 978-1-138-00010-0

Situational cannibalism in *Luidia clathrata* (Echinodermata: Asteroidea)

J.M. Lawrence
Department of Biology, University of South Florida, Tampa, FL, USA

J.C. Cobb
Fish and Wildlife Research Institute, Saint Petersburg, FL, USA

T. Talbot-Oliver & L.R. Plank
University of South Florida, Tampa, Florida, USA

ABSTRACT: *Luidia clathrata* held one week in an aquarium did not interact with each other. Amputation of an arm from an individual resulted in immediate attack by intact individuals. Severed arms were ingested by intact individuals. Attack was stopped after amputating an arm from all individuals. Attack did not occur when the individuals were exposed to bright light. *Luidia clathrata* in aquaria also scavenged decaying individuals. Individuals staked to the bottom after arm amputation in the field produced two responses from nearby intact individuals: some attacked the wounded individuals and ingested severed arms while others did not. Cannibalism by *L. clathrata* depends on the state of the individual, whether the individual is intact or wounded and probably on hunger level. Intact individuals do not release incitants for feeding but wounded individuals do. The incitants are probably amino acids in the coelomic fluids and tissue. Cannibalism probably does not occur in the field because of the high capacity for movement and because individuals are dispersed.

1 INTRODUCTION

Elgar and Crispi (1992) defined cannibalism as killing and consumption of all or part of a conspecific individual. Sibling cannibalism has been reported for brooding starfish (Byrne 1996, Bosch & Slattery 1999, Roediger & Bolton 2008) and juveniles may prey upon newly settled larvae (*Mediaster aequalis*, Birkeland et al. 1971). But starfish are unusual because cannibalism by adults can be sublethal, consumption of part of an individual (the arm) without killing. Sublethal cannibalism occurs in numerous species: *Asterias forbesi* (Galtsoff & Loosanoff 1939, Menge 1979), *Asterias rubens* (Hancock 1955, Vinberg 1967, Anger et al. 1977, Menge 1979, Hurlburt 1980, Harris et al. 1998, Witman et al. 2003), *Asterias amurensis* (Fukuyama & Oliver 1985), *Pycnopodia helianthoides* (Greer 1961, Mauzey et al. 1968, Paul & Feder 1975), *Meyenaster gelatinosus* (Dayton et al. 1977, Viviani 1978), *Heliaster helianthus* (Gaymer & Himmelman 2008), *Oreaster reticulatus* (Scheibling 1982), *Crossaster papposus* (Hancock 1958, Himmelman 1991), *Leptasterias hexactis* (Bingham et al. 2000), *Acanthaster planci* (Moran 1986), *Astropecten articulatus* (Wells et al. 1961) and *Luidia magellanica*, (Viviani 1978).

Cannibalism has never been reported for *Luidia clathrata* although field populations have been observed for over thirty years. A chance observation in the laboratory showed *L. clathrata* is cannibalistic on individuals with an amputated arm. Investigation of this situational cannibalism in the laboratory and field provides insight into the control mechanisms.

2 MATERIAL AND METHODS

2.1 Species and site

Luidia clathrata occur in protected waters on sand and shell hash from Virginia to southern Brazil. It is an active predator that feeds on infauna (McClintock and Lawrence 1985). The population studied occurs in Tampa Bay, Florida (27°58′15.23″N, 82°36′33.41″W) at 2–3 m depth on fine sand mixed with shell fragments and mud (Klinger 1979). Individuals in the population studied are dispersed, usually 1 or 2 m apart (Lawrence, Cobb, Talbot-Oliver, Plank, unpub.). Individuals in the laboratory do not overlap although they may touch. They have been maintained together without food for up to two months without showing cannibalism (Lawrence, unpub.).

2.2 Laboratory observations

Luidia clathrata were collected on three occasions (9 December 1982, 3 October 1988, 15 November 2008) and transported to the laboratory. Ten individuals were placed into aquaria (50 cm × 25 cm × 30 cm) with recirculating, filtered sea water and maintained for one week without food. On these three occasions, an individual was removed from an aquarium and one or two arms amputated. The wounded individual and the arms were returned to the aquarium and the response of the intact individuals observed. Subsequently, in two trials arms were amputated from all individuals and their behavior observed.

In a second experiment, individuals from the 2008 collection that had been held for one week without food were taken from other aquaria and two placed in each of 5 containers (32 cm × 18 cm × 15 cm). After 4 h, they were placed under bright light. One arm was amputated from an individual from each container and the individual and arm returned to the container. Observations of all individuals were recorded on digital or 8 mm video for five minutes.

In a third experiment, a decaying individual was placed in an aquarium with ten intact individuals that had been held without food for one week.

2.3 Field observations

Field observations were made on 15 November 2008 at the collection site during slack tide between 9:50 and 10:45 a.m. The sky was overcast. Four stations with two or more intact *Luidia clathrata* within a one m radius were selected for trials. For each trial an intact *L. clathrata* was collected from outside the station and staked in the center of the station. Three arms were amputated from this individual and placed next to it. The response of intact individuals was recorded digitally for five to ten minutes.

3 RESULTS

3.1 Laboratory observations

In each of the three trials when an individual with amputated arms was returned to the aquarium, the intact individuals immediately became extremely active and attacked the severed arms and wounded individual. On one occasion an intact individual managed to place its mouth on the amputated stump but the wounded individual escaped. The severed arms were ingested. In two trials no attacks occurred after arms had been amputated from all individuals.

In three trials intact individuals exposed to bright light did not attack the wounded individual

or severed arm. Attack in the other two trials was not immediate. Intact individuals moved quickly to the decaying individual and began ingestion.

3.2 Field observations

The responses of intact *Luidia clathrata* near the staked individual with amputated arms are summarized in Figure 1.

Intact individuals responded to the staked individuals in two basic ways: 1) movement toward the staked individual and arms, sometimes followed by contact with the staked individual and ingestion of the severed arm, 2) movement that was not toward the staked individual and severed arms.

4 DISCUSSION

Luidia clathrata shows no cannibalistic behavior on conspecific individuals in the laboratory or field unless an individual is wounded or decaying. It is likely both wounded and decaying individuals are releasing the same organic incitants and the only difference is whether the individual is alive or dead. Amino acids common to animal tissue induce feeding in *L. clathrata* (McClintock et al. 1984) and are probably the stimuli in both situations. *Pycnopodia heliantoides* apparently detects different feeding incitants from intact and damaged prey as they bypass intact prey encountered moving to a damaged one (Brewer & Konar 2005). *Odontaster validus* can distinguish between fed and starved conspecifics (Kidawa 2001).

Cannibalistic behavior by *L. clathrata* is suppressed by bright light or trauma of arm amputation. *L. clathrata* is active during the night (McClintock & Lawrence 1981) and during the day only when light is reduced by cloud cover (pers. obs.).

Flight is the usual response of asteroids to conspecific tissue and/or body fluid (*Crossaster papposus*, Sloan and Northway 1982, Himmelmann 1991, *Pycnopodia helianthoides*, Lawrence 1991, *Coscinasterias tenuispina*, Swenson and McClintock 1998, *Asterias amurensis*, Levin et al. 1984, Fukuyama & Oliver 1985, *Distolasterias nipon*, Levin et al. 1984), presumably a defense against predation. Because of the number of *L. clathrata* in the aquarium, it was not possible to know if individuals varied in their response to the introduction of a wounded individual. However, both fleeing and attack responses to a wounded individual were observed in the field. Flight is a defense against predation. Trade-offs existing between predation risk and energy intake affect foraging behavior of animals (Lima & Dill 1990). Starvation increases foraging in *L. clathrata* (McClintock & Lawrence 1985). The flight/attack responses of *L. clathrata* to a wounded conspecific

Trial	Behavioral Schematic	Time	Observations
1	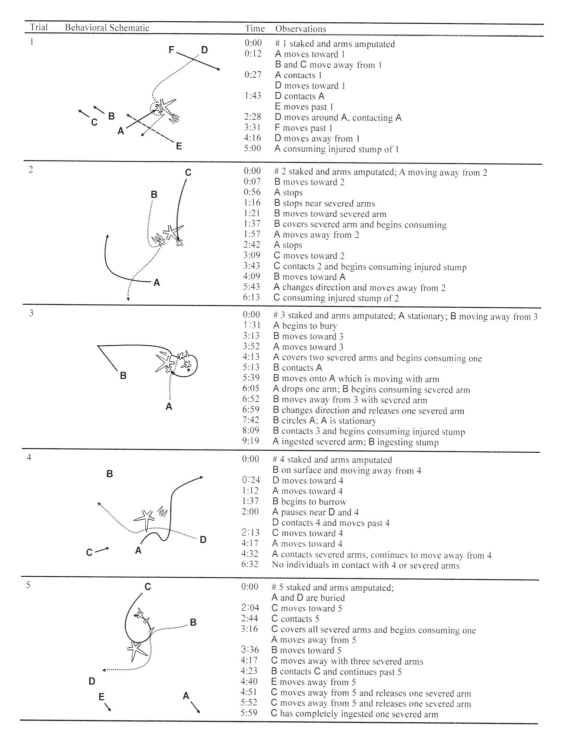	0:00	# 1 staked and arms amputated
		0:12	A moves toward 1
			B and C move away from 1
		0:27	A contacts 1
			D moves toward 1
		1:43	D contacts A
			E moves past 1
		2:28	D moves around A, contacting A
		3:31	F moves past 1
		4:16	D moves away from 1
		5:00	A consuming injured stump of 1
2		0:00	# 2 staked and arms amputated; A moving away from 2
		0:07	B moves toward 2
		0:56	A stops
		1:16	B stops near severed arms
		1:21	B moves toward severed arm
		1:37	B covers severed arm and begins consuming
		1:57	A moves away from 2
		2:42	A stops
		3:09	C moves toward 2
		3:43	C contacts 2 and begins consuming injured stump
		4:09	B moves toward A
		5:43	A changes direction and moves away from 2
		6:13	C consuming injured stump of 2
3		0:00	# 3 staked and arms amputated; A stationary; B moving away from 3
		1:31	A begins to bury
		3:13	B moves toward 3
		3:52	A moves toward 3
		4:13	A covers two severed arms and begins consuming one
		5:13	B contacts A
		5:39	B moves onto A which is moving with arm
		6:05	A drops one arm; B begins consuming severed arm
		6:52	B moves away from 3 with severed arm
		6:59	B changes direction and releases one severed arm
		7:42	B circles A; A is stationary
		8:09	B contacts 3 and begins consuming injured stump
		9:19	A ingested severed arm; B ingesting stump
4		0:00	# 4 staked and arms amputated
			B on surface and moving away from 4
		0:24	D moves toward 4
		1:12	A moves toward 4
		1:37	B begins to burrow
		2:00	A pauses near D and 4
			D contacts 4 and moves past 4
		2:13	C moves toward 4
		4:17	A moves toward 4
		4:32	A contacts severed arms, continues to move away from 4
		6:32	No individuals in contact with 4 or severed arms
5		0:00	# 5 staked and arms amputated;
			A and D are buried
		2:04	C moves toward 5
		2:44	C contacts 5
		3:16	C covers all severed arms and begins consuming one
			A moves away from 5
		3:36	B moves toward 5
		4:17	C moves away with three severed arms
		4:23	B contacts C and continues past 5
		4:40	E moves away from 5
		4:51	C moves away from 5 and releases one severed arm
		5:52	C moves away from 5 and releases one severed arm
		5:59	C has completely ingested one severed arm

Figure 1. Schematic of behavioral responses of *Luidia clathrata* to a staked individual and three amputated arms. Time reported in minutes:seconds following the amputation of arms. Intact starfish for each trial are represented by capital letters. Individuals with amputated arms are referred to by trial number. Attempted cannibalism is represented by ※. Cannibalistic feeding was not observed in trial 4.

individual in the field may be mediated by the hunger state of the individual. Scavenging gastropods also show flight or feeding on newly dead conspecific individuals, which is suggested to depend on hunger state (McKillup & McKillup 1995).

Behaviors leading to feeding involve movement toward food, initiation of feeding and then feeding (Lindstedt 1971). For non-visual animals such as *L. clathrata* these would require chemicals released by the food and receptors specific for them. Intact *L. clathrata* is preyed upon by the closely related *Luidia senegalensis* (Avery, pers. comm.). This indicates intact *L. clathrata* releases a compound that results in predation by *L. senegalensis* that does not result in cannibalism by *L. clathrata*. The difference in response must be that *L. senegalensis* has a receptor for this compound while *L. clathrata* does not.

Prey vulnerability, age, developmental stage, size, morphology and population density are factors leading to cannibalism (Dong & Polls 1992). None of these seem applicable to *L. clathrata* in the field. Rather, as Sloan and Northway (1982) proposed for *Crossaster papposus,* dispersal in the field suggests *L. clathrata* has intraspecific avoidance related to prey distribution that can be overcome when conspecifics are wounded.

REFERENCES

Anger, K., Rogal, U., Schriever, G. & Valentin, C. 1977. In situ investigation on the echinoderm *Asterias rubens* as a predator of soft bottom communities in the western Baltic Sea. *Helgoländer wiss. Meeresunters.* 29: 439–459.

Bingham, B., Burr, J. & Wounded Head, H. 2000. Causes and consequences of arm damage in the sea star *Leptasterias hexactis. Can. J. Zool.* 78: 596–605.

Birkeland, C., Chia, F.-S. & Strathmann, R.R. 1971. Development, substratum selection, delay of metamorphosis and growth in the seastar, *Mediaster aequalis* Stimpson. *Biol. Bull.* 141: 99–108.

Bosch, I. & Slattery, M. 1999. Costs of extended brood protection in the Antarctic sea star, *Neosmilaster georgianus* (Echinodermata: Asteroidea). *Mar. Biol.* 134: 449–459.

Brewer, R. & Konar, B. 2005. Chemosensory responses and foraging behaviour of the seastar *Pycnopodia helianthoides. Mar. Biol.* 147: 789–795.

Byrne, M. 1996. Viviparity and intragonad cannibalism in the dimunitive sea star *Patiriellal vivipara* and *P. parvivipara* (family Asterinidae). *Mar. Biol.* 125: 551–567.

Dayton, P.K., Rosenthal, R.J., Mahen, L.C. & Antezana, T. 1977. Population structure and foraging biology of the predaceous Chilean asteroid *Meyenaster gelatinosus* and the escape biology of its prey. *Mar. Biol.* 39: 361–370.

Dong, G. & Polls, G.A. 1992. Major factors affecting cannibalism. In: M.A. Elgar & B.J. Crispi (eds.): 13–37. *Cannibalism: ecology and evolution.* Oxford University Press: New York.

Elgar, M.A. & Crispi, B.J. (eds.). 1992. *Cannibalism: ecology and evolution.* Oxford University Press: New York.

Fukuyama, A.K. & Oliver, J.S. 1985. Sea star and walrus predation on bivalves in Norton Sound, Bering Sea, Alaska. *Ophelia.* 24: 17–36.

Galtsoff, P.S. & Loosanoff, V.L. 1939. Natural history and method of controlling the starfish *Asterias forbesi* (Desor). *Bull. Bur. Fish.* 49: 75–132.

Gaymer, C.F. & Himmelman, J.H. 2008. A keystone predatory sea star in the intertidal is controlled by a higher-order predatory sea star in the subtidal zone. *Mar. Ecol. Prog. Ser.* 370: 143–153.

Greer, D.L. 1961. Feeding behavior and morphology of the digestive system of the sea star *Pycnopodia heliaanthoids* (Brandt) Stimpson. M.S. thesis. University of Washington: Seattle.

Hancock, D.A. 1955. The feeding behavior of starfish on Essex oyster beds. *J. Mar. Biol. Ass. U.K.* 34: 313–331.

Hancock, D.A. 1958. Notes on starfish on an Essex oyster bed. *J. Mar. Biol. Ass. U.K.* 37: 565–589.

Harris, L.G., Tyrrell, M. & Chester, C.M. 1998. Changing patterns for two sea stars in the Gulf of Maine. In: R. Mooi & M. Telford (eds): 243–248. *Echinoderms: San Francisco.* Balkema: Rotterdam.

Himmelman, J.H. 1991. Diving observations of subtidal communities in the northern Gulf of St. Lawrence. *Can. Spec. Pub. Fish. Aquat. Sci.* 113: 319–332.

Hurlburt, A.W. 1980. The functional role of *Asterias vugaris* Verrill (1866) in three subtidal communities. Ph.D. thesis, Univ. New Hampshire: Durham.

Kidawa, A. 2001. Antarctic starfish, *Odontaster validus,* distinguish between fed and starved conspecifics. *Polar Biol.* 24: 408–410.

Klinger, T.S. 1979. A study of sediment preference and its effect on distribution in *Luidia clathrata* Say (Echinodermata: Asteroidea). MS thesis. University of South Florida: Tampa.

Lawrence, J.M. 1991. A chemical alarm response in *Pycnopodia helianthoides* (Echinodermata: Asteroidea). *Mar. Behav. Physiol.* 19: 39–44.

Levin, A.V., Levina, E.V. & Levin, V.S. 1984. The reaction of asteroids *Asterias amurensis* and *Distolasterias nipon* to homogenates and chemical substances from far eastern starfishes. *Biol. Morya.* 5: 40–45. (in Russian)

Lima, S.L. & Dill, L.M. 1990. Behavioral decisions made under the risk of predation—a review and prospectus. *Can J. Zool.* 68: 619–640.

Lindstedt, K.J. 1971. Chemical control of feeding behavior. Comp. *Biochem. Physiol.* 39: 553–581.

Mauzey, K.P., Birkeland, C. & Dayton, P.K. 1968. Feeding behavior of asteroids and escape responses of their prey in the Puget Sound region. *Ecology.* 494: 603–619.

McClintock, J.B. & Lawrence, J.M. 1981. An optimization study on the feeding behavior of *Luidia clathrata* Say (Echinodermata: Asteroidea). *Mar. Behav. Physiol.* 7: 263–275.

McClintock, J.B. & Lawrence, J.M. 1985. Characteristics of foraging in the soft-bottom benthic starfish *Luidia clathrata* (Echinodermata: Asteroidea): prey selectivity, switching behavior, functional responses and movement patterns. *Oecologia.* 66: 291–298.

McClintock, J.B., Klinger, T.S. & Lawrence, J.M. 1984. Chemoreception in *Luidia clathrata* (Echindermata: Asteroidea): qualitative and quantitative responses to low molecular weight compounds. *Mar. Biol.* 84: 47–52.

McKillup, S.C. & McKillup, R.V. 1995. The responses of intertidal scavengers to damaged conspecifics in the field. *Mar. Fresh. Behav. Physiol.* 27: 49–57.

Menge, B.A. 1979. Coexistence between the seastars *Asterias vulgaris* and *A. forbesi* in a heterogeneous environment: a non-equilibrium explanation. *Oecologia.* 41: 245–272.

Moran, P.J. 1986. The *Acanthaster* phenomenon. *Oceanogr. Mar. Biol. Ann. Rev.* 24: 379–480.

Paul, A.J. & Feder, H.M. 1975. The food of the sea star *Pycnopodia helianthoides* (Brandt) in Prince William Sound, Alaska. *Ophelia.* 14: 15–22.

Roediger, L.M. & Bolton, T.F. 2008. Abundance and distribution of South Australia's endemic sea star, *Parvulastra parvivipara* (Asteroidea: Asterinidae). *Mar. Freshwat. Res.* 59: 205–213.

Scheibling, R.E. 1982. Feeding habits of *Oreaster reticulatus* (Echinodermata: Asteroidea). *Bull. Mar. Sci.* 32: 504–510.

Sloan, N.A. & Northway, S.M. 1982. Chemoreception by the asteroid *Crossaster papposus* (L.). *J. Exp. Mar. Biol. Ecol.* 61: 85–98.

Swenson, D.P. & McClintock, J.B. 1998. A quantitative assessment of chemically-mediated rheotaxis in the asteroid *Coscinasterias tenuispina*. *Mar. Freshwat. Behav. Physiol.* 31: 63–80.

Vinberg, T.L. 1967. Biology of the nutrition of *Asterias rubens* L. in the littoral of the White Sea. *Zool. Zh.* 69: 929–931. (in Russian).

Viviani, C.A. 1978. Predación interespecífica, canibalismo y autotomía como mecanismo de escape en las especies de Asteroídea (Echinoodermata) en el litoral del desierto del Norte Grande de Chile. Laboratorio de Ecología Marina, Universidad del Norte, Iquique.

Wells, H.W., Wells, M.J. & Gray, I.E. 1961. Food of the sea-star *Astropecten articulates*. *Biol. Bull.* 120: 265–271.

Witman, J.D., Genovese, S.J., Bruno, J.F., McLaughlin, J.W. & Pavin, B.I. 2003. Massive prey recruitment and the control of rocky subtidal communities on large spatial scales. *Ecol. Monogr.* 73: 441–462.

Echinoderms in a Changing World – Johnson (ed)
© *2013 Taylor & Francis Group, London, ISBN 978-1-138-00010-0*

Aspects of the biology of an abundant spatangoid urchin, *Breynia desorii* in the Kimberley region of north-western Australia

J.K. Keesing & T.R. Irvine
CSIRO Wealth from Oceans National Research Flagship, CSIRO Marine and Atmospheric Research, Wembley, Australia

ABSTRACT: This is the first report on the biology of the spatangoid urchin *Breynia desorii*, which was found to be very abundant in the remote Kimberley region of north-western Australia. During our survey in June 2008 *B. desorii* were found at densities of up to 220,000 per hectare but more commonly at 100s to 1000s per hectare. A high proportion of urchins collected were found to be in spawning and post spawning condition. Sexes were in equal proportion overall, although there was a bias towards males in urchins 85 mm and larger. Ripe females had much larger gonads than ripe males. Gonad colour was found to be a poor indicator of sex but a good indicator of the fully spawned condition. The size range of urchins sampled was between 20 and 97 mm in length and maturity is thought to occur between 45 and 54 mm. Size frequency distributions were found to vary between locations. At Gourdon Bay and Perpendicular Head populations consisted of large adults with modal sizes of 70–74 mm and 60–64 mm respectively while the population at James Price Point consisted almost solely of a single cohort of immature juveniles with a modal size of 25–29 mm. Although based on limited information we suggest these small animals are a cohort of 1 year olds and that the adult populations at the other sites are accumulations of urchins aged 3+ years. Detailed information on size, wet weight, dry weight and ash free dry weight of both *B. desorii* and *Lovenia elongata* is also provided and using literature values for rates of bioturbation we discuss the potential for these species to play a key role in coastal ecosystems in north-eastern Australia. Extensive industrial developments planned for this region mean that it is important to gather empirical data on this species; in particular its role in sediment biogeochemistry, bentho-pelagic coupling and maintaining diversity in soft sediment habitats.

1 INTRODUCTION

The burrowing and bioturbating behaviour of macrobenthic animals in soft sediment habitats plays an important role in the biogeochemistry of marine ecosystems (Haese 2004). Spatangoid or heart urchins can occur in high densities in sediment dominated habitats (e.g. Ursin 1960, Ferber & Lawrence 1976, Lohrer et al. 2005). As a result, their burrowing and grazing activities can constitute an important ecological consequence for the environment (e.g. Lohrer et al. 2004, 2005, Vopel et al. 2007). The spatangoid urchin *Breynia desorii* was recently found to form extensive populations, sometimes at high densities of many thousands per km of towed video in north-western Australia (Fry et al. 2008). Thus *B. desorii* may play an important ecological role in the shallow marine ecosystem of north-western Australia and yet there have been no studies of the biology and ecology of this species. McNamara (1982) reports *B. desorii* as occurring along the full length of Australia's tropical and temperate Indian Ocean coastline from Lucky Bay near Esperance in the south to Darwin in the north, from intertidal to 140 m. This is the first reported study of population characteristics and biology of *B. desorii* and includes notes on *Lovenia elongata*, another spatangoid urchin which is also common in this region.

2 MATERIALS AND METHODS

The study was carried out on and near the Dampier Peninsula in the north-western region of Australia known as the Kimberley. Nine sites within 5 locations were surveyed between 9th and 29th June 2008 (see Table 1). Heart urchins were collected using an epibenthic sled with a mouth opening of 1.5 m wide by 0.5 m high towed at 1–2 knots behind a 20 m fishing vessel. A video camera was mounted on the sled to record the habitat being dredged. Sites chosen were based on knowledge from other video surveys being undertaken in the region (Fry et al. 2008) and the tow distance of each sled varied from 45 m to 281 m as calculated

Table 1. GPS coordinates of the nine epibenthic sleds.

Location	Operation	Latitude	Longitude
Gourdon bay	2873	18.381°S	121.871°E
	2904	18.383°S	121.845°E
James price point	1811	17.365°S	122.065°E
	2037	17.420°S	122.075°E
	2200	17.476°S	122.090°E
South of perpendicular head	5515	16.915°S	122.334°E
Perpendicular head	714	16.670°S	122.626°E
Packer island	278	16.525°S	122.782°E
	487	16.565°S	122.730°E

from the ships Global Positioning System. Heart urchins collected in the sled were weighed and where densities were high a large sample of randomly selected individuals was measured along the greatest length axis.

At Gourdon Bay 30 urchins were maintained in individual 2 litre buckets and injected through the test with 1–2 ml of 0.5 M Potassium Chloride to assess if any spawning response could be induced. Also at Gourdon Bay a sample of each size class of *Breynia desorii* was dissected by removing the top of the test and assessed visually and/or microscopically for sex, gonad size, colour and condition. Gonad size for each animal was ascribed to a size class of 5 = >25 mm, 4 = 20–25 mm, 3 = 15–19 mm, 2 = 10–14 mm, 1 = 5–9 mm or 0 = <5 mm by measuring the size of the largest gonad. Gonads were judged as being ripe if large round oocytes or milky sperm flowed from them when dissected during macroscopic examination.

For animals whose sex and gonad condition could not be determined from dissection, a sample of gonad was taken and preserved in 95% ethanol for histological examination and later sectioned at 5 microns and stained with Haemotoxylin and Eosin. Very small gonads were whole mounted and stained with Diff Quik Giemsa. Gonads examined microscopically were staged as Ripe (having a large proportion of mature oocytes or spermatozoa), Partly spawned (having vacant or partly collapsed lumen with immature oocytes or spermatids/spermatogonia) or Fully spawned (having collapsed lumen few or no mature or mid stage oocytes or sperm cells). Because a full size range of urchins could not be collected from Gourdon Bay, samples from James Price Point and Perpendicular Head were also included in the analysis of sex ratio and maturity using the same methods.

A full size range of urchins (across multiple sites) was also collected to determine length, wet weight, dry weight and ash free dry weight. Wet weight was a blotted wet weight, dry weight was

determined after drying to constant weight at 80 degrees Celsius and ash free dry weight was determined after ashing at 500 degrees Celsius for 4 hours. To investigate the resulting length—weight relationship difference in the two species, the morphological components of 20 *Breynia desorii* and 10 *Lovenia elongata* were measured separately. Length and total wet weight were measured of each individual before being dissected into test, gonad and ingested sand components. For each component dry weight was determined after drying to constant weight at 80 degrees Celsius and ash free dry weight was determined after ashing at 500 degrees Celsius for 3 hours. The percent of organic content was determined as the AFDW/dry weight, and the proportion of each component was calculated as a percentage of total weight (test + gonad + sand). Differences in component proportions between the 2 species were determined by t-tests for samples with equal variance.

Algal content of ingested sediment was examined in a further 11 *Breynia desorii*. 10 ml of sediment closest to the mouth was removed from the gut and pigments extracted in 10 ml 90% acetone overnight in a freezer. The supernatant was poured off, filtered through a syringe filter and after suitable dilution read in a Turner fluorometer before and after acidification with 2 drops 1.2 M hydrochloric acid. The sediment was then dried to constant weight at 80 degrees Celsius and weighed. Concentrations of chlorophyll *a* and phaeopigments per gram dry sediment were calculated as in Knap et al. (1996).

3   RESULTS

3.1   *Epibenthic sled sampling*

Nine sleds targeting soft bottom habitats occupied by heart urchins were carried out. The summary details for these sled operations are shown in Table 2. Four species of heart urchins were collected, of which *Breynia desorii* (Fig. 1A) and *Lovenia elongata* (Fig. 1B) were the most abundant. All four species co-occurred in only one sled, at the site south of Perpendicular Head, attributing to a total density of 961 heart urchins per hectare. Gourdon Bay and James Price Point had the greatest densities of heart urchins, with averages of 9659 (±419 S.E.) and 74,022 (±73,907 S.E.) per hectare respectively.

3.2   *Size frequency*

Breynia desorii from Gourdon Bay ranged in size from 46 to 97 mm with a mean size of 70.1 mm (±0.2 SE) and with 91.2% of animals between 60 and 79 mm in size (Fig. 2). The population sampled from Perpendicular Head also comprised

Table 2. Number of individuals, total weight and density of heart urchins collected in epibenthic sled operations.

Location	Operation	Species	Count	Total wet weight g	Density urchins per hectare
Gourdon bay	2873	*Breynia desorii*	251	22590	9766.5
Gourdon bay	2873	*Lovenia elongata*	8	318	311.3
Gourdon bay	2904	*Lovenia elongata*	134	7506	3182.9
Gourdon bay	2904	*Breynia desorii*	255	22780	6057.0
James price point	1811	*Breynia desorii*	1	50	66.2
James price point	2200	*Breynia desorii*	5435*	80000	221836.7
James price point	2037	*Breynia desorii*	2	1	163.9
South of perpendicular head	5515	*Breynia desorii*	9	1460#	508.5
South of perpendicular head	5515	*Lovenia elongata*	1		56.5
South of perpendicular head	5515	*Rhynobrissus hemiasteroides*	1		56.5
South of perpendicular head	5515	*Echinolampas ovata*	6	196	339.0
Perpendicular head	714	*Breynia desorii*	124	7800	3850.9
Packer island	278	*Breynia desorii*	7	1126	482.8
Packer island	487	*Breynia desorii*	2	300	88.1

*This count was calculated from the 80 kg sample with an average weight of individuals measured from a subsample; #Weight was measured for the pooled sample of *B. desorii, L. elongata* and *R. hemiasteroides.*

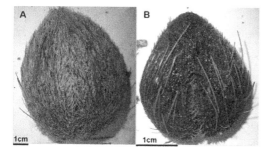

Figure 1. Dominant heart urchins (A) *Breynia desorii* (B) *Lovenia elongata.*

predominantly large animals. Sizes ranged from 40 to 90 mm with a mean size of 64.2 mm and with 71.8% of animals between 60 and 69 mm in size (Fig. 2). At Perpendicular Head the most abundant size class was 60–64 mm which was 10 mm smaller than at Gourdon Bay where the most abundant size class was 70–74 mm. The size distribution of the population sampled at James Price Point was much smaller. Sizes ranged from 20 to 52 mm with a mean size of 27.2 mm and with 61.6% of animals between the most abundant size class of 25–29 mm (Fig. 2). Specimens of *Lovenia elongata* from Gourdon Bay ranged in size from 59 to 74 mm total length with a mean size of 66.5 mm and with 48.8% of animals falling in the 65–69 mm size class (Fig. 3).

3.3 Sex ratio and maturity

Table 3 shows the summary data for analysis of gonad condition and sample sizes for each size class.

Unfortunately there were few animals less than 55 mm and none between 35 and 44 mm. The first evidence of gamete development in our samples was in one urchin 32 mm in length. The remainder of nine animals sampled in the 24–46 mm size range had a tiny transparent gonad visible under the microscope but when sectioned showed only connective tissue. The next smallest urchins with gametes were mature animals in the 50–54 mm size class where 2 (100%) of the urchins were mature. In the absence of many animals in this size range it is difficult to estimate but we expect 50% maturity is achieved somewhere between the 45 and 54 mm. For urchins of 60 mm and larger our sample of 164 urchins included 84 males and 80 females. This represents no departure from a 1:1 sex ratio (chi square analysis $P = 0.754$). Above 85 mm our sample of 8 included 7 males and 1 female (chi square analysis $P = 0.034$) suggesting a preponderance of males among larger urchins, but this is a small sample size.

3.4 Spawning behaviour and gonad condition

Of 30 urchins injected with KCl, 5 males and 4 females had spawned after 90 minutes. In addition a small number of males and females spawned spontaneously during measurement (Fig. 4), probably from the stress of desiccation. Table 4 shows the relationship between spawning condition and gonad size. Ripe females have larger gonads than ripe males. Only 1 male (1.3%) had a gonad size of 4 or above, compared with 29 (37%) females. Few gonads (2 or 1.3%) less than gonad size 3 in either males or females were ripe. More females (46 or 58%) had ripe gonads compared to males

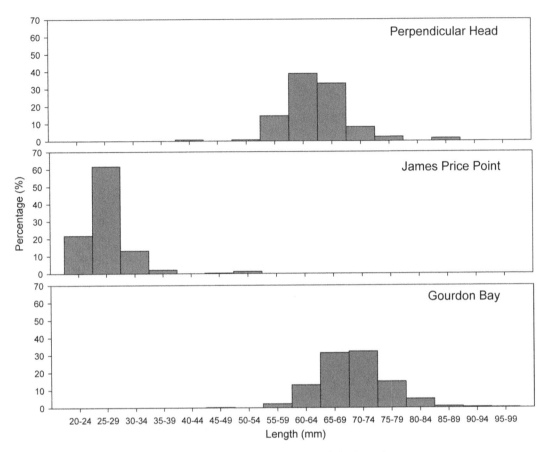

Figure 2.   Size frequency of *Breynia desorii* at three locations in the Kimberley region.

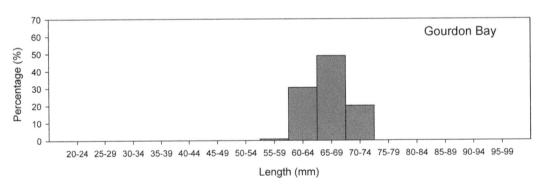

Figure 3.   Size frequency of *Lovenia elongata* at Gourdon Bay in the Kimberley region.

(15 or 18%). Sixty-three (82%) of males were partly or fully spawned compared to 42% of females. Table 5 shows the relationship between gonad colour and spawning condition. Gonad colour was not a good indicator of sex. Urchins with whitish-yellow or orange gonads were predomi-

nantly male, but 3 (17%) of urchins with orange gonads were female. Ninety-six or 62% of all urchins had light brown gonads. A higher proportion of urchins with brown or dark brown gonads were fully spawned (21 or 55%) than urchins with light brown, whitish yellow or orange gonads

Table 3. Summary data for *Breynia desorii* collected for reproductive analyses.

Size class (mm)	Sample size	Site	Gonad size range	Males	Females	Indeterminate	Unknown*
20–24	1	JP	0			1	
25–29	4	JP	0			4	
30–34	4	JP	0		1	3	
35–39	0						
40–44	0						
45–50	1	GB	0				1
50–54	2	JP	1–2	2			
55–59	13	JP, GB	0–4	4	6	3	
60–64	36	GB	1–5	20	16		
65–69	36	GB	1–4	19	17		
70–74	32	GB	1–5	11	21		
75–79	32	GB	1–5	16	16		
80–84	20	GB	1–5	11	9		
85–89	5	GB	1–4	3	1		1
90–94	3	GB	2–3	3			
95–99	1	PH	3	1			
Total	190			90	87	11	2

*sample lost.

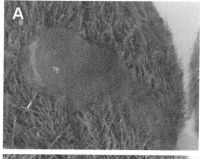

Figure 4. Spawning *Breynia desorii:* A. female; B. male.

Table 4. Relationship between gonad size and spawning condition in *Breynia desorii* (60–84 mm only).

Sex/gonad condition	Gonad size						Total
	0	1	2	3	4	5	
Males							
Fully spawned		4	10	1			15
Partly spawned		3	24	21			48
Ripe			1	12	1		14
Females							
Fully spawned		5	8	2			15
Partly spawned		2	6	9	1		18
Ripe			1	17	21	7	46
Total	0	14	50	62	23	7	**156**

(9 or 7.6%). Figure 5 shows the various stages of gonad condition revealed by the macroscopic dissection and histological section.

3.5 Biomass

Wet weight, dry weight and ash free dry weight of *B. desorii* all increased with length (Fig. 6); the relationships were best described by power curves (all $r^2 \geq 0.85$). *Breynia desorii* of average size at Gourdon Bay (70 mm) had a wet weight of 82.6 g, a dry weight of 47.3 g and an ash free dry weight of 1.89 g or only 4% of dry weight. In contrast, all biomass parameters of *L. elongata* were highly variable in relation to length (Fig. 7). As a result, estimates of weight of an average size urchin (66 mm) cannot be accurately made, but may be between 32–49 g wet weight, 19–23 g dry weight and 0.9–1.3 g ash free dry weight. *Breynia desorii* were heavier than *L. elongata* of the same size.

The relationships for length to weights were not strong for *L. elongata*; this is likely the result of a narrower size range of individuals with no

Table 5.   Relationship between gonad colour and spawning condition in *Breynia desorii* (60–84 mm only).

Sex/gonad condition	Gonad colour					Total
	Dark brown	Brown	Light brown	Orange	Whitish yellow	
Males						
Fully spawned	2	8	3	2		15
Partly spawned	1	10	26	10	1	48
Ripe			8	3	3	14
Females						
Fully spawned	2	9	4			15
Partly spawned	2	3	11	2		18
Ripe		1	44	1		46
Total	7	31	96	18	4	**156**

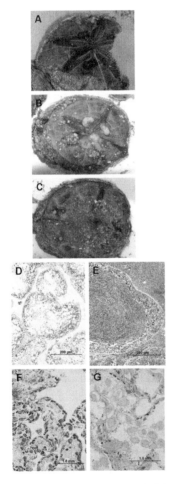

Figure 5.   Examples of gonad stages- A. dark brown, fully spawned; B. whitish yellow, ripe male; C. light brown, ripe female; D. fully spawned male; E. ripe/partly spawned male; F. fully spawned female; G. ripe/partly spawned female.

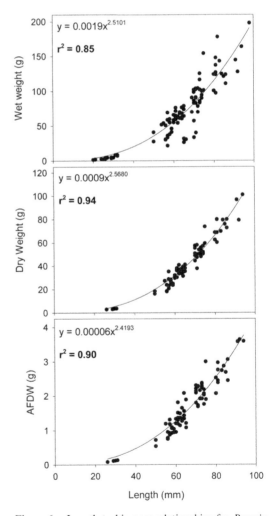

Figure 6.   Length to biomass relationships for *Breynia desorii*.

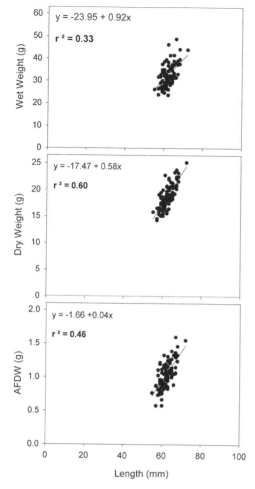

Figure 7.   Length to biomass relationships for *Lovenia elongata*.

Table 6.   Component analysis of heart urchins (DW: dry weight, AFDW: ash free dry weight) (mean ± S.E.).

	Breynia	*Lovenia*	*P*-value
Test			
% organic content	5.04	7.33	
% DW of total DW	37.1 ± 0.9	36.8 ± 1.1	0.8508
% AFDW of total AFDW	43.1 ± 2.5	52.6 ± 2.4	0.0221
Gonad			
% organic content	81.52	59.55	
% DW of total DW	0.1 ± 0.01	0.1 ± 0.02	0.6361
% AFDW of total AFDW	2.7 ± 0.3	1.4 ± 0.3	0.0128
Sand			
% organic content	3.07	3.50	
% DW of total DW	62.8 ± 0.9	63.1 ± 1.1	0.8458
% AFDW of total AFDW	54.3 ± 2.5	46.0 ± 2.2	0.0441

Figure 8.   Dissected *Breynia desorii* showing intestine full of sand within test.

animals below 50 mm or above 75 mm available to be analysed.

As the weight of sand formed such a significant component of total weight, the dry weight and AFDW of different body components of the two species was assessed (Table 6). The organic content (AFDW relative to dry weight) of the test was higher in *L. elongata* (7.3%) compared to 5.0% in *B. desorii*. The organic content of gonad in *B. desorii* was higher (81.5% compared with *L. elongata* 59.55%) probably due to the greater level of ripeness in the former species. All of the *L. elongata* dissected had very small gonads. The organic content of the sand component was 3.1 to 3.5% for both species. There was no significant difference between *B. desorii* and *L. elongata* in

the proportion that different body components contributed to total dry weight (Table 6), with the dry weights of the test approximately 37%, gonad 0.1% and sand approximately 63%. However, there was a significant difference between the proportion each component contributed to total AFDW (Table 6), with the test of *L. elongata* contributing 53% of total body AFDW compared with 43% in *B. desorii* and the sand contributing 54% of total AFDW in *B. desorii* and only 46% in *L. elongata*. These differences can be explained by greater volume of sand and higher volume of gonad in *B. desorii* and the higher AFDW content of the test in *L. elongata* compared with *B. desorii*.

If we consider the maximum densities of *B. desorii* (9766 per ha) and *L. elongata* (3183 per ha) found at Gourdon Bay then these equate to a dry weight (ash free dry weight) per hectare of

462 kg (18.5 kg) for *B. desorii* and 60–73 kg (2.8–4.1 kg) for *L. elongata*.

3.6 Algal content of food source

Newly ingested sediment from the foregut of *B. desorii* (Fig. 8) had mean chlorophyll *a* concentrations of 3.53 ± 0.42 µg per g sediment dry weight and phaeopigments 6.01 ± 0.62 µg per g sediment dry weight. The organic matter content of sand is 3.07% (Table 6), of which 0.01% is chlorophyll *a*.

4 DISCUSSION

The observation of ripe, spawning and post spawning urchins in the samples analysed suggest a winter or dry-season spawning period for *B. desorii* in northwestern Australia. However some other spatangoids that have been studied have protracted spawning seasons (e.g. Chesher 1969, Pearce 1969, Ferber & Lawrence 1976, Pearce & McClintock 1990) and ours is too short an observation period to be conclusive. Our limited histological examination of *B. desorii* seems to accord closely with the detailed study of *Echinocardium cordatum* by Nunes & Jangoux (2004) with many of the same stages of spawning and post spawning gonads observed. A longer time series of observations will be required to characterise reproductive periodicity. The methods used by Nunes & Jangoux (2004) would be well suited to this species although the remote location of the population we studied makes regular sampling problematic.

Although gonadal development was observed in one urchin 32 mm in length, maturity in *B. desorii* is thought to occur between 45 and 54 mm in length, about half the maximum size observed in this study. This is possibly in the second year as discussed below. Chesher (1969) made similar observations noting small (39–52 mm) *Meoma ventricosa* developed genital pores while gonads did not mature until about 90 mm (during the second year). Chesher (1969) found the maximum size of *Meoma ventricosa* varied with location between 124 and 169 mm. We found a 1:1 sex ratio in *B. desorii*, although at the larger sizes (more than 85 mm) there were more males than females suggesting differential growth rates or survivorship. We found no other reference in the literature to this in spatangoid urchins. Nunes & Jangoux (2004) found a 1:1 sex ratio *in Echinocardium cordatum*. Although we did not measure a time series of gonad indices we found that ripe females had much larger gonads than males of the same size. Pearce & McClintock (1990) found no difference in gonad indices between sexes for the spatangoids

Abatus shackletoni and *A. nimrodi*. Chesher (1969) and Nunes & Jangoux (2004) make no mention of differential gonad indices between the sexes in *Meoma ventricosa* and *Echinocardium cordatum* respectively.

Size frequency distributions were found to vary between locations. At two (Gourdon Bay and Perpendicular Head) of the three locations where large size samples were collected the populations comprised a unimodal distribution of animals predominantly larger than 60 mm suggesting an accumulation of age classes. At James Price Point a very high density population of 20–34 mm urchins was sampled, perhaps representing a cohort from a strong recruitment event and suggesting that recruitment may vary significant in a spatial context. At all three sites very small numbers of an intermediate size class (40–54 mm) were recorded. Although we know nothing of the growth rates of *B. desorii*, given the sampling period coincided with the spawning period it is possible that the small animals at James Price Point represent a cohort of 1 year old animals with a mode of 25–29 mm. In addition, that the accumulated cohort of animals with modes of 60–69 mm represent 3+ year old animals with an intermediate age class of 2 year olds present in very small numbers. The only published studies which analysed size frequency distributions of spatangoids to infer growth were that of Beukema (1985) for *Echinocardium cordatum* and Ferber & Lawrence (1976) for *Lovenia elongata*. The observation by Beukema (1985) of irregular recruitment, rapid growth and an accumulation of age classes may mirror what is occurring in the populations of *B. desorii* in this study. The population of *L. elongata* sampled by us had a unimodal population size structure of 60–74 mm. *L. elongata* studied by Pearce (1969) in the Gulf of Suez were of similar size ranging 40–76 mm. These are much larger than the population of the same species studied in the Gulf of Elat by Ferber & Lawrence (1976), which had a maximum size of 40–45 mm. These authors recorded maturity size of 29 mm and measured growth rates of 8–12 mm per year and estimated longevity of at least 7 years.

We did not learn anything about the burrowing and emergence behaviour of *B. desorii* in this study. We only sampled during the day and video footage taken during the sled tows revealed large numbers of urchins on the surface of the seabed. Some species are known to have a pattern of diurnal behaviour (e.g. *Meoma ventricosa* Chesher 1969) emerging at night and remaining buried during the day. Nichols et al. (1989) recorded periodic emergence behaviour (hours to days) of *Brisaster latifrons* that was associated with changes in local oceanography and hypothesised these emergence patterns could be associated with food availability (phytoplankton deposition).

While high densities of spatangoids represent high biomass in wet and dry weight terms, much of this is ingested sediment. We found that organic matter content as estimated from ash free dry weight in *B. desorii* was very low, just 4% of dry weight. Organic matter content of sediment in the gut made up 2% of this. The proportion of AFDW biomass can vary with the seasonal cycle as gonad tissue has a high organic content (89% for *Echinocardium cordatum* Nakamura 2001). However, even with such low proportion of organic biomass, spatangoids can represent a very high proportion of the total biomass in soft sediment habitats when present in high densities (e.g. Nakamura 2001).

McNamara (1982) reported on the taxonomy, post-settlement ontogeny and range of *Breynia desorii* noting it was particularly common on the Dampier Peninsula, the location of our study. Spatangoid urchins can be very abundant with adult populations up to 60 individuals per m² (*Brissopsis lyrifera* Ursin 1960, *Echinocardium spp.* Lohrer et al. 2005) and up to 200 per m² immediately following successful recruitment (*Lovenia elongata* Ferber & Lawrence 1976). The highest densities in our study were approximately 22 per m². At these densities *B. desorii* may have the capacity to have a significant influence, perhaps the most significant influence, on consumption of primary production at the water sediment interface and on reworking of sediments in these extensive soft sediment habitats. A number of studies have pointed to the importance of spatangoids in the biogeochemistry and ecology of soft sediment habitats. As urchins burrow they oxygenate sediments, thus influencing carbon and nitrogen cycling, microbial food webs and they resuspend nutrients making them available for water column productivity (Osinga et al. 1997, Lohrer et al. 2004, 2005, Vopel et al. 2007).

Sediment ingestion and reworking rates for spatangoids have been provided by Chesher (1969), Lohrer et al. (2005), Hollertz & Duchene (2001) and Nakamura (2001). It is difficult to compare these studies across the different species used and the different method calculation and data presentation given. Several of these authors concluded the ecological consequences of burrowing behaviour and sediment reworking is likely to be greater than the actual impact of sediment ingestion. For example, Hollertz and Duchene (2001) estimated that the volume of sediment reworked was 60–150 times greater than the amount of sediment ingested. We have no measurements of the ingestion rate or volume of sediment reworking by *B. desorii* but expect it could be high based on our observations of video footage taken during our study showing high urchin densities, grazing and reworking sediments with a rich cover of microphytobenthos. We found ingested sand in *Breynia desorii* comprised 0.0004% chlorophyll *a*. Estimates of reworking rates of *Meoma ventricosa* are that 1 urchin of an average size of 112 mm would rework 1 m² sediment to 6 cm every 7.6 days (recalculated from Chesher 1969) and for *Echinocardium cordatum* that 1 urchin of an average size of 25 mm would rework 1 m² sediment to 5 cm every 13.3 days (recalculated from Lohrer et al. 2005). At our site where the density of *B. desorii* of average size 27 mm was 22 per m², applying the calculations from Lohrer et al. (2005) suggests complete sediment reworking in the area every 0.6 days or 14.4 hours. Bioturbation of surface sediments by spatangoids and other animals has been shown to be important in maintaining high levels of biodiversity (Widdicombe & Austen 1998, Widdicombe et al. 2000). Lohrer et al. (2004) identified potentially major consequences for biodiversity from anthropogenic induced losses of spatangoid urchins and thus it remains an important task to assess the ecological significance of *B. desorii* in soft sediment habitats in the Kimberley region of Western Australia. This region has some of the worlds largest reserves of natural gas (Government of Western Australia 2006) and been identified for major industrial development.

ACKNOWLEDGEMENTS

We thank Gary Fry, Ted Wassenberg, Bob Pendrey, Greg Smith from CSIRO and Jamie Colquhoun from the Australian institute of Marine Science for their assistance during the vessel operations and Mat Vanderklift and Damian Thomson who collected and measured some of the urchins for us. We also thank Ray Masini and Cameron Sim from the Western Australian Department of Environment and Conservation for helping to arrange funding for the field work and for providing copies of video footage. Ryan Crossing and Jenelle Ritchie assisted with sample preparation in the laboratory. Jeremy Allen, Kim Elliot and Tai Le from the Western Australian Department of Agriculture prepared the histological mounts Sharon Yeo prepared the photographs of the histological specimens. Ashley Miskelly from Sydney and Loisette Marsh from the Western Australian Museum (WAM) identified the echinoids and material from this study is lodged at the WAM. Bill de la Mare (CSIRO) and an anonymous reviewer provided helpful comments on the manuscript.

REFERENCES

Beukema, J.J. 1985. Growth and dynamics in populations of *Echinocardium cordatum* living in the North Sea off the Dutch north coast. *Netherlands Journal of Sea Research* 19(2), 129–134.

Chesher, R.H. 1969. Contributions to biology of Meoma ventricosa (Echinoidea—Spatangoida). *Bulletin of Marine Science* 19(1), 72-&.

Ferber, I. & Lawrence, J.M. 1976. Distribution, substratum preference and burrowing behaviour of *Lovenia elongata* (Gray) (Echinoidea—Spatangoida) in Gulf of Elat (Aqaba), Red Sea. *Journal of Experimental Marine Biology and Ecology* 22(3), 207–225.

Fry, G., Heyward, A., Wassenberg, T., Ellis, N., Taranto, T., Keesing, J., Irvine, T., Stieglitz, T. & Colquhoun, J. 2008. *Benthic habitat surveys of potential LNG hub locations in the Kimberley region.* Unpublished report. 131 pp.

Government of Western Australia. 2006. *Kimberley Economic Perspective. An update on the economy of Western Australia's Kimberley Region.* Perth: Kimberley Development Commission and Department of Local Government and Regional Development.

Haese, R.R. 2004. Macrobenthic activity and its effects on biogeochemical reactions and fluxes. In G Wefer, D. Billet, D. Hebbeln, B.B. Jørgensen, M. Schlüter & T.C.E. van Weering (eds.) *Ocean Margin Systems.* Heidelberg-Berlin: Springer. p 219–234.

Hollertz, K. & Duchene, J.C. 2001. Burrowing behaviour and sediment reworking in the heart urchin Brissopsis lyrifera Forbes (Spatangoida). *Marine Biology* 139(5), 951–957.

Knap, A., Michaels, A.. Close, A.. Ducklow, H., & Dickson, A., (eds.). 1996. *Protocols for the Joint Global Ocean Flux Study (JGOFS) Core Measurements.* JGOFS Report Nr 19. Reprint of the IOC Manuals and Guides No. 29, UNESCO 1994. 145p.

Lohrer, A.M., Thrush, S.F. & Gibbs, M.M. 2004. Bioturbators enhance ecosystem function through complex biogeochemical interactions. *Nature* 431(7012), 1092–1095.

Lohrer, A.M., Thrush, S., F., Hunt, L., Hancock, N. & Lundquist, C. 2005. Rapid reworking of subtidal sediments by burrowing spatangoid urchins. *Journal of Experimental Marine Biology and Ecology* 321(2), 155–169.

McNamara, K.J. 1982. Taxonomy and evolution of living species of *Breynia* (Echinoidea: Spatangoida) from Australia. *Records of the Western Australian Musuem* 10(2): 167–197.

Nakamura, Y. 2001. Autoecology of the heart urchin, *Echinocardium cordatum,* in the muddy sediment of the Seto Inland Sea, Japan. *Journal of Marine Biology Association U.K.* 81: 289–297.

Nichols, F.H., Cacchione, D.A., Drake, D.E. & Thompson, J.K. 1989. Emergence of burrowing urchins from California continental shelf sediments- A response to alongshore current reversals? *Estuarine, Coastal and Shelf Science* 29: 171–182.

Nunes, C. & Jangoux, M. 2004. Reproductive cycle of the spatangoid echinoid *Echinocardium cordatum* (Echinodermata) in the southwestern North Sea. *Invertebrate Reproduction & Development* 45(1), 41–57.

Osinga, R., Kop, A.J., Malschaert, J.F.P. & Van Duyl, F.C 1997. Effects of the sea urchin *Echinocardium cordatum* on bacterial production and carbon flow in experimental benthic systems under increased organic loading. *Journal of Sea Research* 37:109–121.

Pearce, J.S. 1969. Reproductive periodicities of Indo-Pacific invertebrates in the Gulf of Sues. I. The echinoids *Prionocidaris baculosa* (Lamarck) and *Lovenia elongata* (Gray). *Bulletin of Marine Science* 19: 323–350.

Pearse, J.S. & McClintock, J.B. 1990. A comparison of reproduction by the brooding spatangoid echinoids *Abatus shackletoni* and *Abatus nimrodi* in McMurdo Sound, Antarctica. *Invertebrate Reproduction & Development* 17(3), 181–191.

Ursin, E. 1960. A quantitative investigation of the echinoderm fauna of the central North Sea. *Meddr Danm Fisk Havunders* 2: 1–204.

Vopel, K., Vopel, A., Thistle, D. & Hancock, N. 2007. Effects of spatangoid heart urchins on O-2 supply into coastal sediment. *Marine Ecology-Progress Series* 333, 161–171.

Widdicombe, S. & Austen, M.C. 1998. Experimental evidence for the role of *Brissopsis lyrifera* (Forbes, 1841) as a critical species in the maintenance of benthic diversity and the modification of sediment chemistry. *Journal of Experimental Marine Biology and Ecology* 228: 241–255.

Widdicombe, S., Austen, M.C., Kendall, M.A., Warwick, R.M. and Jones, M.B. 2000. Bioturbation as a mechanism for setting and maintaining levels of diversity in subtidal macrobenthic communities. *Hydrobiologia* 440: 369:377.

Echinoderms in a Changing World – Johnson (ed)
© 2013 Taylor & Francis Group, London, ISBN 978-1-138-00010-0

Ophiopsila pantherina beds on subaqueous dunes off the Great Barrier Reef

E. Woolsey, M. Byrne, J.M. Webster, S. Williams, O. Pizarro, K. Thornborough & P. Davies
University of Sydney, Sydney, Australia

R. Beaman
James Cook University, Cairns, Australia

T. Bridge
James Cook University, Townsville, Australia

ABSTRACT: An autonomous Underwater Vehicle (AUV) was used to generate images of an *Ophiopsila pantherina* population on subaquous dunes at Hydrographers Passage, 200 km off the Australian mainland. High-resolution stereo images captured by the AUV were used to determine population structure of the aggregations, which consisted of adults at a mean density of 418 animals m^{-2} at depths of 65–70 m. *Ophiopsila pantherina* (8–15 mm dd) takes advantage of their elevated position on the lee side of the dunes for suspension feeding. On contact stimulation, the arms emit visible light as a bright green flash that travels down the arm. These aggregated ophiuroid communities in dune fields may be a specialized natural feature for consideration in managing common inter-reefal sandy habitats within the Great Barrier Reef Marine Park.

1 INTRODUCTION

Ophiuroids are well known to form dense aggregations in most of the world's marine zones from intertidal to deep-sea habitats in tropical, temperate, and polar regions (Kingston 1980, Summers and Nybakken 2000, Metaxas and Giffin 2004, Oak and Scheibling 2006, Brooks et al. 2007). These dense brittlestar aggregations show their effectiveness as suspension feeders and are important as conduits between the benthic and pelagic environments (Warner and Woodley 1975, Brooks et al. 2007). Suspension-feeding ophiuroids exhibit a preference for current-dominated habitats where they are supported by high nutrient flow (Broom 1975, Warner 1979). Because cycling and re-suspension of food particles increases the odds of successful capture, suspension feeders are often found on the leeward side of structures, where the eddying and low velocity aids in food capture (Warner 1977).

Here we present data on a recently discovered population of *Ophiopsila pantherina* (Koehler 1898) that form dense aggregations on underwater dunes in the central Great Barrier Reef (GBR) (Byrne 2009). This work was undertaken in association with surveys of the relict drowned or submerged reefs along the GBR margin (Webster et al.

2008, Byrne 2009). The brittlestar genus *Ophiopsila* (Forbes 1843) is widespread throughout temperate and tropical environments and is known to form aggregations on soft sediments (Kingston 1980, Basch 1988). This genus is also known to be bioluminescent (Basch 1988, Grober 1988a,b, Vanderlinden and Mallefet 2004) and this phenomenon is observed here. This species is reported to occur in benthic inshore habitats of the GBR at 0–40 m depth (Clark and Rowe 1971, Kingston 1980, Rowe and Gates 1995). The distribution of *Ophiopsila* in the inter-reefal areas of the GBR Marine Park was obtained from the Seabed Biodiversity Survey (Pitcher et al. 2007).

2 MATERIALS & METHODS

The research was conducted at Hydrographers Passage (20.0° S, 150.4° E), 200 km off the coast of Queensland (Fig. 1) in October 2007 and August 2008 using the *RV Southern Surveyor* (Marine National Facility, voyages SS07/2007 and SS09/2008). An Autonomous Underwater Vehicle (AUV) was used to survey and image the seafloor. The AUV remained at a constant altitude of 2 m and captured time-stamped images every half-second. A Wetlabs EcoPuck Fluorometer mounted

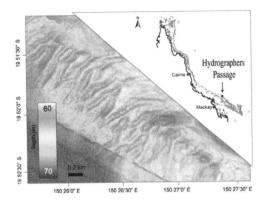

Figure 1. Multibeam bathymetry showing the subaqueous dunes at Hydrographers Passage, a narrow channel in the outer reef matrix of the central GBR, 200 km northeast of Mackay, Queensland.

on the AUV collected data on chlorophyll and turbidity. Bathymetry data of the dunes was documented by multibeam swath mapping using a Simrad EM300 multibeam system.

Samples of ophiuroids and sediments were collected at 65–70 m depths with a Smith-Macintyre grab at five locations along the dune system. In the 2007 survey, one grab (19.8704833° S, 150.4507500° E) was deployed near the aggregation. In the 2008 survey, four grabs were deployed—two on lee dune slopes (19.866500° S, 150.447883° E; 19.867500° S, 150.449483° E) and two on stoss dune slopes (19.867216° S, 150.448033° E; 19.867900° S, 150.450083°E). The Smith-Macintyre grab deployed was a half cylinder with a radius of 18 cm and a width of 30 cm. Although the ophiuroids were likely to have been collected from surface sediments, the total volume of sediment collected was approximately 20 L.

Bioluminescence of live *O. pantherina* collected by grab samples was documented with video on board the *Southern Surveyor*. Images of bioluminescence were collected from 5 specimens. The brittle stars from each grab were preserved in 70% ethanol. Disc diameter and arm length were measured and the incidence of regenerating arms was recorded. Still images taken by the AUV were used to count feeding arms and estimate the population density of the aggregations.

3 RESULTS

The populations of *Ophiopsila pantherina* on dunes at Hydrographers Passage occupy a dynamic dune system on the outer shelf in the central GBR at depths of about 65–70 m. The dune habitat covered approximately 3.4 km² of sea floor in the surveyed area. The dunes were 2–6 m in height and consisted of well-sorted medium-sized carbonate sand. Each AUV image represented approximately 1.5 m² of the seafloor (Fig. 2).

The track compiled from AUV imagery over the dunes showed three separate aggregations of *O. pantherina* (Fig. 3) each of which were 50–70 m long. Examination of the images indicated that the ophiuroids on the seafloor each had four arms extended with the fifth arm buried and out of view. Analysis of randomly selected images indicated that the population of ophiuroids had a mean density of 418 animals m⁻².

The AUV survey indicated that the ophiuroids exhibited a preference for the leeward side of the dunes and this was examined in four grabs, two from a lee slope and two from stoss dune slopes of the dunes. Ophiuroids were only present in the lee slope grabs, confirming the AUV observations.

The AUV images show that the population is dominated by large ophiuroids and this is also

Figure 2. Example image from the AUV. Dimensions are approximately 1.5 × 1.0 m.

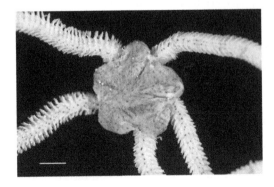

Figure 3. *Ophiopsila pantherina* (3 mm scale bar).

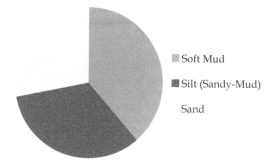

Soft Mud

Silt (Sandy-Mud)

Sand

Figure 4. Sediment types where *O. pantherina* specimens were collected by the Seabed Biodiversity Survey
(Pitcher et al. 2007). Biomass was larger in soft mud
(38%) habitats, followed by silt (33%), and sand (28%)
habitats. Less than 1% of specimens were discovered
in coarse sand and rocky habitats. No specimens were
found in habitats described as sandwaves/dunes, rubble,
stones, or bedrock/reefs.

indicated by the size frequency of the animals
collected. These had disc diameters between 8 and
15 mm (mean 10.17 mm). In the 2007 grabs the mean
diameter was 9.17 mm (n = 6, range = 7–10 mm)
and in the 2008 grabs the diameter was 10.43
(n = 23, range = 5–15 mm). For specimens from
2007 and 2008 that had all arms intact, the longest
arm had a mean arm length of 149.07 mm (n = 27,
range = 80–220 mm). In total, 18 specimens (3 from
2007 and 15 from 2008) were recovered intact in the
grab samples. Ten of these individuals exhibited
regeneration (56%). In respect to the total number
of arms in these specimens 22% exhibited evidence
of regeneration. Juveniles were not present in any
grab samples.

Ophiopsila pantherina is bioluminescent and
exhibited visible light in a bright green flash that
traveled down the arm from proximal to distal.

At 70 m depth, the EcoPuck Fluorometer
mounted on the AUV measured chlorophyll
content as approximately 0.8 µg/L. At the sea surface, the AUV collected a reading of approximately
0.45 µg/L.

The Seabed Biodiversity Survey (Pitcher et al.
2007) indicates the sediment type where *Ophiopsila
pantherina* were found along the GBR (Fig. 4).

4 DISCUSSION

This survey of subaqueous dune systems at
Hydrographers Passage assimilated data from
multibeam, seismic profiles, oceanographic instruments, and AUV imagery to understand the seabed
environment of the sea floor and generate a visual
reconstruction of the *O. pantherina* aggregations.

The *O. pantherina* extended four arms into the
water column to feed and one arm is anchored into
the sand. This activity was recorded during the day
and contrasts with previous observations of cryptic day time behaviour, which was considered to
be an avoidance of predators during the day and
extension of arms to feed with at night (Sides 1987,
Aronson 1998). However the Hydrographers Passage ophiuroids did show evidence of sublethal
predation as indicated by the presence of regenerating arms in 56% of the specimens examined. In
this case the benefits of day and night time feeding by *O. pantherina* in a high nutrient habitat, as
illustrated by the density and extent of the beds,
may outweigh the risks from predators. Dense
aggregations of ophiuroids appear in Ordovician
fossil records and low regeneration rates (<2%) of
these ancient communities suggest that predation
pressures were low (Aronson 1992). The change
in modern ophiuroid lifestyle from dense beds to
cryptic living in previously observed aggregations
has been attributed to the evolution of predators
such as teleosts and decapod crustaceans (Aronson
1989, Aronson 1992).

Ophiuroids are known to aggregate in large
numbers, with hundreds to thousands occurring in
one location (Grange 1991, Jensen and Frederiksen
1992, Pipenburg and Schmid 1996, Summers and
Nybakken 2000, Metaxas and Giffin 2004, Brooks
et al. 2007). The *O. pantherina* beds at Hydrographers Passage showed a mean density of approximately 400 animals m^{-2}. Such dense aggregations
require high food delivery rates that at Hydrographers Passage are likely to come from high tidal
current flow over the dunes and increased nutrient richness at the seafloor (Warner 1979, Uthicke
et al. 2009).

Benthic animals are important for energy cycling
within pelagic environments, as most ophiuroids
feed on suspended organic material (Brooks
et al. 2007, Canuel et al. 2007). The discovery of
O. pantherina on the leeward sides of the dunes
reflects commonly observed behaviour in suspension feeders (Warner 1977). It has been suggested
that leeward positioning on topographic high features among eddying reduces current velocities
and resuspension of fine material provides a clear
feeding advantage (Warner 1977). In addition, by
actively seeking conspecifics in highly dynamic
areas, ophiuroids accomplish both optimum feeding locations and ensure reproductive success
(Broom 1975, Warner 1979).

Upwelling is an important mechanism that
supports marine life in the oligotrophic waters
of the Great Barrier Reef shelf system (Andrews
and Gentien 1982). There is strong upwelling at
Hydrographers Passage, resulting in a nutrient-
rich environment where flood tides flow through

gaps in the relict reef matrix at the shelf break and deliver nutrients to the dune system (Coutis 2000). The chlorophyll level at 70 m depth, 0.8 µg L⁻¹, was higher than at the sea surface, 0.45 µg L⁻¹. The sea surface chlorophyll reading is consistent with the modeled mean surface data over the whole GBR (0.46 µg L⁻¹) as well as off-shore areas off Mackay (0.5 µg L⁻¹) (De'ath and Fabricius 2008).

A feature of *Ophiopsila pantherina*, as documented here, is its luminescence. This is considered to serve as a defensive mechanism by visually stunning any predatory fishes and invertebrates (Basch 1988, Grober 1988a,b, Vanderlinden and Mallefet 2004). Although the feeding biology of *O. pantherina* remains to be documented, it seems likely that this species uses bioluminescence as a defensive response to startle predators, thereby reducing predation on exposed arms. In the northeast Pacific *O. californica* uses the startling effect of bioluminescence for defense, utilizing a luminescent range from dim glows to intense proximal-distal flashes (Basch 1988).

The ability to autotomize and regenerate arms allows recovery from non-lethal predation (Pomoroy and Lawrence 2001). While arm loss in brittlestars has been attributed to environmental stressors, the primary cause of autotomy is likely predation (Woodley 1980, Woodley et al. 1981, Bowmer and Keegan 1983, Sides 1987, Makra and Keegan 1999). Such is likely to be the case in the Hydrographers Passage ophiuroids.

The consistency of disc diameters in the studied populations suggests that the beds were comprised of adults. In the 2007 Seabed Biodiversity Survey report, *O. pantherina* was described as living in soft mud to sandy habitats off the north Queensland coastline, with a particularly high biomass occurring in offshore and mid-shelf areas near Rockhampton (23.28° S) (Pitcher et al. 2007). A high biomass from this study occurs in areas described as soft mud and silt, while our study finds high biomass on the medium-grained san dunes occurring at the limit of the outer shelf.

The dense brittlestar beds found at Hydrographers Passage suggests that similar aggregations of ophiuroids may occur in a broader range of seabed habitats than is currently recorded in the waters of tropical Australia. The *O. pantherina* beds represent a new biotic community associated with active dune systems in the Great Barrier Reef Marine Park. Other dune systems, particularly in areas of upwelling, may support other dense aggregations of suspension-feeding ophiuroids. These aggregated ophiuroid communities in dune systems may be a specialized natural feature for consideration in managing inter-reefal habitats within the Marine Park.

ACKNOWLEDGEMENTS

Thanks to the captain and crew of RV Southern Surveyor (SS07/2007, SS09/2008). Dr. B. Tilbrook (CSIRO) provided support for SS09/2008. Funded by the Marine National Facility Integrated Marine Ocean Observing System and a grant from National Geographic. Thanks to Niel Bruce and the Museum of Tropical Queensland and Tom Savage of The University of Sydney. Thanks to reviewers for helpful comments.

REFERENCES

Andrews, J.C. & Gentien, P. 1982. Upwelling as a source of nutrients for the Great Barrier Reef ecosystem: a solution of Darwin's question? *Marine Ecology Progress Series* 8: 257–269.

Aronson, R.B. 1989. Brittlestar beds: low predation anachronisms in the British Isles. *Ecology* 70: 856–865.

Aronson, R.B. 1992. Biology of scale—independent predator-prey interaction. *Marine Ecology Progress Series* 89: 1–13.

Aronson, R.B. 1998. Decadel-scale persistance of predation potential in coral reef communities. *Marine Ecology Progress Series* 172: 53–60.

Basch, L.C., 1988. Bioluminescent anti-predator defense in subtidal ophiuroid. In: B.e. al (Editor), *Echinoderm Biology*, Rotterdam, Balkema pp. 503–515.

Bowmer, T. & Keegan, B. 1983. Field survey of the occurrence and significance of regeneration in *Amphiura filiformis* (Echinodermata: Ophiuroidea) from Galway Bay, west coast of Ireland. *Marine Biology* 74: 65–71.

Brooks, R.A., Nizinski, M.S., Ross, S.W. & Sulak, K.J. 2007. Frequency of sublethal injury in a deepwater ophiuroid, *Ophiacantha bidenta,* an important component of western Atlantic *Lophelia* reef communities. *Marine Biology* 152: 307–314.

Broom, D.M. 1975. Aggregation behaviour of the brittle-star *Ophiothrix fragilis. J. Mar. Biol. Ass. U.K.* 55: 191–197.

Byrne, M. 2009. Flashing stars light of the Reef's shelf. *ECOS* 150: 28–29.

Canuel, E.A., Spivak, A.C., Waterson, E.J. & Duffy, J.E. 2007. Biodiversity and food web structure influence short-term accumulation of sediment organic matter in an experimental seagrass system. *Limnology & Oceanography* 52: 590–602.

Clark, A. & Rowe, F. 1971. Monograph of shallow-water Indo-west Pacific Echinoderms. Trustees of the British Museum (Natural History): 238 pp.

Coutis, P.F., 2000. Currents, coasts and cays: a study of tidal upwelling and island wakes, University of New South Wales, Sydney.

De'ath, G. & Fabricius, K. 2008. Water quality of the Great Barrier Reef: distribution, effects on reef biota and trigger values for the protection of ecosystem health. Final Report to the Great Barrier Reef Marine Park Authority. Australian Institute of Marine Science (Editor), pp. 104.

Forbes, E. 1843. On the radiata of the Ophiuridae of the eastern Mediterranean. Part I: Ophiuridae. *Trans. Linn. Soc. Lond.* 19: 143–153.

Grange, K. 1991. Mutualism between the antipatharian *Antipathes fiordensis* and the ophiuroid *Astrobrachion constrictum* in New Zealand fjords. *Hydrobiologica* 216/217: 297–303.

Grober, M.S. 1988a. Brittle-star bioluminescence as a aposematic signal to deter crustacean predators. *Animal Behaviour* 36: 493–501.

Grober, M.S. 1988b. Responses of tropical reef fauna to brittle-star luminescence (Echinodermata: Ophiuroidea). *Journal of Experimental Marine Biology and Ecology* 115: 157–168.

Jensen, A. & Frederiksen, R. 1992. The fauna associated with the bank-forming deepwater coral *Lophelia pertusa* (Scleractinaria) on the Faroe Shelf. *Sarsia* 77: 53–69.

Kingston, S. 1980. The Swain Reefs expedition: Ophiuroidea. *Rec. Aust. Mus.* 33: 123–147.

Koehler, R. 1898. Echinodermes, receuillis par *l'Investigator* dans l'Ocean Indien. II Les ophiures littorales. *Bull. Scient. Fr. Belg.* 31: 55–124.

Makra, A. & Keegan, B. 1999. The Population Dynamics of the Brittlestar *Ophioderma brevispinum* in Near and Farshore Seagrass Habitats of Port Saint Joseph Bay, Florida. *Gulf of Mexico Science* 1999 (2): 87–94.

Metaxas, A. & Giffin, B. 2004. Dense beds of the ophiuroid *Ophiacantha abyssicola* on the continental slope off Novia Scotia, Canada. *Deep-Sea Research Part I* 51: 1307–1317.

Oak, T. & Scheibling, R.E. 2006. Tidal activity pattern and feeding behaviour of the ophiuroid *Ophiocoma scolopendrina* on a Kenyan reef flat. *Coral Reefs* 25: 213–222.

Pipenburg, D. & Schmid, M. 1996. Brittle star fauna (Echinodermata: Ophiuroidea) of the Arctic northwestern Barents Sea: composition, abundance, biomass and spatial distribution. *Polar Biology* 16: 383–392.

Pitcher, C.R., Doherty, P., Arnold, P., Hooper, J. & Gribble, N. 2007. Seabed Biodiversity on the Continental Shelf of the Great Barrier Reef World Heritage Area, AIMS/CSIRO/QM/QDPI CRC Reef Research Task Final Report. 320 pp.

Pomoroy, C. & Lawrence, J. 2001. Arm regeneration in the field in *Ophiocoma echinata* (Echinodermata: Ophiuroidea): effects on body composition and its potential role in a reef food web. *Marine Biology* 135: 57–63.

Rowe, F. & Gates, J. 1995. Echinodermata. In: A. Wells (Editor), Zoological Catalogue of Australia. CSIRO Australia, Melbourne.

Sides, E. 1987. An experimental study of the use of arm regeneration in estimating rates of sublethal injury on brittle-stars. *Journal of Experimental Marine Biology and Ecology* 106: 1–16.

Summers, A. & Nybakken, J. 2000. Brittlestar distribution patterns and population densities on the continental slope off central California (Echinodermata: Ophiuroidea). *Deep-Sea Research Part I* 47: 1107–1137.

Uthicke, S., Schaffelke, B. & Byrne, M. 2009. A boom and bust phylum? Ecological and evolutionary consequences of desnity variations in echinoderms. *Ecological Monographs* 79: 3–24.

Vanderlinden, C. & Mallefet, J. 2004. Synergic effects of tryptamine and octopamine on ophiuroid luminescence (Echinodermata). *The Journal of Experimental Biology* 207: 3749–3756.

Warner, G. 1979. Aggregation in Echinoderms. In: G. Larwood and B.R. Rosen (Editors), Biology and Systematics of Colonial Organisms. Academic Press, London.

Warner, G.F. 1977. On the shapes of passive suspension feeders. Proceedings of the eleventh European symposium on marine biology. University College, Galway. Pergamon Press, Oxford, 567–576 pp.

Warner, G.F. & Woodley, J.D. 1975. Suspension-feeding in the brittle-star *Ophiothrix fragilis*. *J. Mar. Biol. Ass. U.K.* 55: 199–210.

Webster, J.M. et al. 2008. From corals to canyons: the Great Barrier Reef Margin. *Eos* 89: 217–218.

Woodley, J.D. 1980. Hurrican Allen destroys Jamaican coral reefs. *Nature* 287: 387.

Woodley, J.D. et al. 1981. Hurrican Allen's impact on Jamaican coral reefs. *Science* 60: 35–45.

Echinoderms in a Changing World – Johnson (ed)
© 2013 Taylor & Francis Group, London, ISBN 978-1-138-00010-0

How to lose a population: The effect of Cyclone Larry on a population of *Cryptasterina pentagona* at Mission Beach, North Queensland

A. Dartnall

Marine Biology and Aquaculture, James Cook University, Townsville, Queensland, Australia

H. Stevens

School of Geography and the Environment, University of Oxford, England

M. Byrne

Schools of Biomedical and Biological Science, University of Sydney, NSW, Australia

ABSTRACT: The asterinid sea star *Cryptasterina pentagona* (Muller and Troschel 1842) occurs along the coast of north Queensland, Australia where numerous populations have been known since the late 1960s with irregular surveys over the following 40 years. These populations occupy boulder and cobble fields on sand and mud. Surveys conducted in 2005 were undertaken 6 months prior to Cyclone Larry which impacted the coast in early 2006. Populations were resurveyed after the cyclone. The search for the population at Mission Beach (Clump Point) failed to find any specimens. It is evident that this population was wiped out by the eye/intense storm surge. The site experienced considerable sand scour with complete loss of the cobble habitat used by *C. pentagona*. In contrast the population south of the influence of the cyclone did not exhibit any apparent change supporting the suggestion of the cyclone caused loss of the Mission Beach population. Although there are many reports of mass mortality of subtidal echinoderm populations due to stranding by storms, disease and anthropogenic influences, this appears to be the first report of a coastal intertidal population lost through natural impact.

1 INTRODUCTION

Asterinid sea stars form extensive populations along the northeast coast of tropical Australia inhabiting the upper intertidal of beaches along the Queensland coast (Dartnall et al. 2003). These species are members of genus *Cryptasterina*, a pan-tropical group of small sea stars that comprise a species complex of difficult to distinguish cryptic species (Dartnall loc cit). The species common along the Queensland coast (north of 20° S), *Cryptasterina pentagona* (Muller and Troschel 1842), as currently understood, has a disjunct distribution ranging from the Philippines to Queensland, Australia (as *Patiriella pseudoexigua* in Rowe and Gates 1995: Dartnall et al. 2003). Throughout this range *C. pentagona* occurs sporadically and little is known of the factors that limit its distribution.

There has been some confusion on the biology of *Cryptasterina* due to the presence of cryptic species (Byrne et al. 2003, Dartnall et al. 2003, Hart et al. 2003). *Cryptasterina pentagona* is a broadcast spawner with a short-lived (4–5 days) lecithotrophic larva while its sister species that occurs south of 20° S, *C. hystera* is a viviparous brooder with intragonadal young (Byrne et al. 2003, Byrne 2005). There may be a further brooding species at the southern end of the known range of *Cryptasterina* (Byrne and Walker 2007, Dartnall and Byrne unpub).

The biology and ecology of Australian populations of *Cryptasterina* along the Queensland coast has been documented in a long series of studies (Dartnall 1971, Ottesen 1976, Gist 1993, Byrne 2005, Lacey 2000, Byrne et al. 2003, Dartnall et al. 2003, Stevens 2005). These studies over 30+ years, involved repeated studies at the same locations, indicating that *Cryptasterina* populations have been a conspicuous component of intertidal fauna for some time. They occupy the mid to high shore 0.7–2.8 m above chart datum (Stevens 2005). The habitats occupied by *C. pentagona* are boulder/cobble beaches where wave exposure and persistence of standing water keeps the habitat wet throughout the tidal cycle (Stevens 2005, Dartnall pers obs).

The biology of populations of *C. pentagona* at Clump Pt., Mission Beach and Rowes Bay, Townsville (Queensland) has been investigated in numerous studies (Ottesen 1976, Gist 1993, Lacey 2000, Byrne et al. 2003, Stevens 2005). Considering the vulnerability of the boulder-cobble habitats to disturbance by storms we revisited these sites following the category 4–5 Cyclone Larry that impacted the region in early 2006 with particular concern for the Mission Beach population which was located in the region that experienced the most damage. Due to coincidence with high tides Cyclone Larry produced significant storm surge with sea level exceeding the predicted tide by 1.75 m at Clump Point (Bureau of Meteorology cyclone report: http://www.bom.gov.au/weather/qld/cyclone_larry).

2 METHODS

2.1 Historic records

Data from surveys of sites in North Queensland, literature records, Queensland Museum, Australian Museum, British Museum of Natural History and our recent surveys (Table 1) were assimilated to provide an overview of the known distribution of C. pentagona in north Queensland starting from studies in the late 1960s (Dartnall 1971).

2.2 Surveys

Populations of C. pentagona at Clump Point, Mission Beach and Rowes Bay, Townsville were surveyed at low tide on 10–18 July 2003. The Mission Beach site was visited on 19 Aug 2006 to check for the presence of C. pentagona on the beach cobble platform and from high intertidal rock pools at the northern end of the beach (northern Clump Point population).

Clump Point and Rowes Bay sites were resurveyed on 6 September 2007. On each survey date five (100 m) transects were placed perpendicular to the shore line within the boulder habitat. At 300 mm vertical intervals 1 m² quadrats were placed with the bottom edge along the transect line and all C. pentagona within the quadrat were counted to obtain an estimate of the population density. The number of exposed animals were counted and then the cobbles were removed to locate individuals concealed among the cobbles. Population counts at Clump Point, Mission Beach were only conducted on the beach cobble platform. Clump Point was also examined on 13 June 2009 and Rowes Bay was examined on 7 August 2010.

3 RESULTS

3.1 Distribution and habitat of Cryptasterina pentagona in North Queensland

Historic and recent records indicated that C. pentagona is known from approximately 14 sites in Queensland, Airlie Beach, Bowen, Townsville and Mission Beach amongst other locations (Table 1). These areas are cobbles and boulders sitting on sand or mud (Fig. 1b,c). With the exception of Low Is, the locations where C. pentagona are found (Table 1) are categorized as reflective beaches by Short (2000) in surveys of the Queensland coast. These beaches are relatively sheltered having a prevailing reflective wave regime from the southeast. This regime maintains a cobble structure that is

Table 1. Known locations of populations of *Cryptasterina pentagona* in North Queensland.

Locality	Latitude °S	Records	Beach reference (in short 2000)
Darnley Is	9°35′	Clark (1946)	
Lizard Is	14°40′	M. Byrne (pers obs 2005)	
Turtle Is	14°45′	BM (NH) 1968.6.14. 156–160	
Low Is	16°23′	Stephenson (1931)	
Mourilyan	17°35′	Hoedt et al. (2000)	
Bingil bay, mission beach (Fig.1)	17°50′	Lacey (2000); Stevens (2005)	780
Townsville	19°14′	Dartnall (1971); Stevens (2005)	861
Dalrymple pt, bowen	20°01′	Lacey (2000)	
Queens beach, bowen	20°02′	Dartnall (1971); Stevens (2005)	950
Dingo beach	20°05′	Dartnall (pers obs 2005)	984
Cannonvale	20°16′	Stevens (2005)	1014/1015
Airlie beach	20°16′	Dartnall (1971)	1016
Midge point	20°40′	Dartnall (pers obs 2003)	1061
Halliday bay	20°40′	Dartnall (pers obs 2003)	1091

BM (NH), British Museum of Natural History.

Figure 1. a. *Cryptasterina pentagona* on the underside of a boulder. b. Airlie Beach site boulder habitat. c. Mission Beach site before Cyclone Larry, boulders on sand.

irrigated and kept free of particulates that would otherwise preclude sea stars from the habitat.

At Rowes Bay the habitat was composed of ca. 300 m of cobbles and larger boulders. Waves strike the shore at about 45 degrees and the site faces NW. The *C. pentagona* habitat at the Mission Beach platform (Fig. 1c) was composed of about 150 m of cobbles often overlain by larger boulders. This site faces NW and waves strike the shore at about 45 degrees.

3.2 Survey

In 2003 the mean density of *C. pentagona* at Mission Beach was 8.2 m^{-2} (SE = 1.33, n = 5, range 3.8–10.8 m^{-2}). The highest density was at 1.64–2.54 m above chart datum. At Townsville the mean density was 2.3 m^{-2} (SE = 0.62, n = 5, range 0.25–3.8 m^{-2}). The highest density at this site was at 1.14–2.04 m above chart datum.

The absence of *C. pentagona* from Clump Point and the northern Mission Beach sites was first noted on 19 Aug 2006. No specimens were detected in the survey a year later in September 2007 or in June 2009. In September 2007 the Rowes Bay population had a mean density 3.2 m^{-2} (SE = 0.66, n = 8, range 2.3–4 m^{-2}) relatively unchanged from previous counts. In August 2010 there was no perceptible change in the Rowes Bay population.

4 DISCUSSION

The locations where *C. pentagona* is found are sheltered cobble-boulder shores most of which (Table 1, Fig. 1b,c) are categorized as reflective beaches (Short 2000). Reflective beaches make up only 3.3% or 6.6 km of the Queensland coast and some of these are sandy (Short, 2000). Thus, although *C. pentagona* is locally abundant, its presence along the coast is patchy and constrained with respect to the distribution of suitable habitat.

The six coastal locations where *C. pentagona* populations occur have been investigated sporadically since the late 1960s (e.g. Dartnall 1971, Byrne et al. 2003) indicating the long term presence of this sea star at these locations. The exception is the Mission Beach location where, as determined here, the population of this sea star disappeared following Cyclone Larry. This region was located within the eye/intense storm surge of the cyclone. Examination of the site 6 months after the cyclone revealed that both the sea stars and its boulder habitat were not present. The two population of *C. pentagona* at this site were wiped out when their cobble habitat was scoured by wave-born sand. This species was not found in subsequent searches at these sites in 2007 and 2009. In contrast, the population south of the influence of the cyclone where *C. pentagona* has been known to occur for at least 30 years did not exhibit any perceptible change. This supports the suggestion that the disappearance of the *C. pentagona* population at Mission Beach was due to the cyclone and not due to another cause (e.g. disease). Sustained damage on intertidal fauna from waves, sand scouring and moving debris occurs following tropical storms in the Caribbean and tropical Pacific (Moran and Reake-Kudia 1991, Kerr 1992). Waves (>8 m) from a severe cyclone drastically reduced abundance of windward reef flat populations of *Holothuria atra* (Kerr 1992).

In a comprehensive review of mass mortalities in echinoderms due to abiotic causes, Lawrence (1996) details the plethora of cases where factors including temperature, freshwater inflow, storms and anthropogenic factors have caused damage to echinoderm populations. With respect to storms most reports of mortalities are due to stranding of subtidal populations cast ashore by waves. These can be spectacular with thousands of asteroids and echinoids washed ashore following storms. Tropical cyclones are reported to be particularly damaging to *Diadema* populations with many reported cases of mass stranding (Lawrence 1996). Although there are many cases of mass mortality of echinoderms following storms, the disappearance of the *C. pentagona* population at Mission Beach appears to be the first report of complete loss of coastal intertidal populations through natural impact. A previous cyclone that impacted the Rowes Bay region in 2000, Cyclone Tessi (Mabin 2000) did not result in disappearance of the local *C. pentagona*.

The population dynamics of *C. pentagona* is characterized by long-term (decades) stability and a unimodal size distribution dominated by large adults (Ottesen 1976, Gist 1993, Dartnal pers obs). Recruitment appears to be sporadic with a small influx of observable juveniles as reported in September and October 1976 following spawning (Ottesen 1976). It is suggested that juvenile

C. pentagona occupy cryptic microhabitats and so are not detected in general surveys (Ottesen 1976, Gist 1993). It remains to be seen whether recruitment will reverse the population loss of the two Mission Beach populations but the lack of suitable habitat suggests that this is unlikely. The sporadic pattern of successful recruitment in other *C. pentagona* populations Queensland also suggests that this is unlikely (Ottesen 1976, Gist 1993, Dartnall, pers obs). Consequently, renewal of the Mission Beach population may depend on recruitment from a cryptic and as yet unrecognized population.

Like other tropical asterinids, *C. pentagona* is a specialist of the mid intertidal and appears to be robust to the broad temperature range that occurs in these habitats (Byrne and Walker 2007). Thus warming of coastal waters due to climate change may not be so deleterious to this species. In addition sea level rise will probably not be inimical if a steep back beach with reflective wave action and a suitable cobble quarry are maintained. The implications of projected increase in the frequency and intensity of tropical storms in the region due to climate change (Lough 2007, Fabricius et al. 2008) are however serious for *C. pentagona* and other species in intertidal communities along the Queensland coast. The ecological implications of intensifying storms could be severe for populations of organisms such as *C. pentagona* with limited local distributions in intertidal boulder habitat and unconsolidated coral rubble, habitats easily shifted by storms (Moran and Reake-Kudia 1991, Kerr 1992, Byrne and Walker 2007).

ACKNOWLEDGEMENTS

Jean Dartnall helped with the Mission Beach resurvey. Assistance was also provided by Erika Woolsey.

REFERENCES

Byrne, M. 2005. Viviparity in the sea star *Cryptasterina hystera* (Asterinidae)—conserved and modified features in reproduction and development. *Biol. Bull.* 208: 81–91.

Byrne, M.,Walker SJ. 2007. Distribution and reproduction of intertidal species of *Aquilonastra* and *Cryptasterina* (Asterinidae) from One Tree Reef, Southern Great Barrier Reef. *Bull. Mar. Sci.* 81: 209–218.

Byrne, M., Hart, M. Cerra, A., Cisternas, P. 2003. Reproduction and larval morphology of broadcasting and viviparous species in the *Cryptasterina* species complex. Biol Bull. 205: 285–294.

Clark, H.L. 1946. The Echinoderm Fauna of Australia: Its composition and origin. Carnegie Institute of Washington Publication, 214, 1–567.

Dartnall, A.J. 1971. Australian sea stars in the genus *Patiriella* (Asteroidea, Asterinidae). *Proc. Linn. Soc. NSW* 96: 39–51.

Dartnall, A.J., Byrne, M., Collins, J., Hart, M.W. 2003. A new viviparous species of asterinid (Echinodermata, Asteroidea, Asterinidae) and a new genus to accommodate the species of pan-tropical exiguoid sea stars. *Zootaxa* 359:1–14.

Fabricius K.E. et al. 2008. Disturbance gradients on inshore and offshore coral reefs caused by a severe tropical cyclone. *Limnol. Ocean.* 53: 690–704.

Gist, M.J. 1993. The reproductive cycle and population dynamics of *Patiriella pseudoexigua* (Dartnall), a rocky intertidal sea star. MSc Thesis, James Cook University, Townsville. 68 pp.

Hart, M.W., Byrne, M., Johnson, S.L. 2003. Cryptic species and modes of development in *Patiriella psueudoexigua*. *J. Mar. Biol. Assoc. U.K.* 83:1109–1116.

Hoedt, F.E., Choat J.H., Collins, J., Cruz, J.J. 2000. Mourilyan Harbour and Abbot Point Surveys: Port Marine Baseline Surveys and Surveys for Introduced Marine Pests. Ports Corp Qld 49pp.

Kerr, A.M. 1992. Effects of typhoon-generated waves on windward and leeward assemblages of holothuroids. In: 7th Int, Coral Reef Symp., Guam.

Lacey, J.E. 2000. Feeding and digestion of the intertidal sea star *Patiriella pseudoexigua* Dartnall in North Queensland. BSc Hon Thesis, James Cook University, Townsville. 84 pp.

Lawrence, J.M. 1996. Mass mortality of echinoderms from abiotic factors. In: Echinoderm Studies, Vol 5, Jangoux, M., Lawrence, J.M., (eds.), A.A. Balkema, Rotterfam. Pp. 103–137.

Lough, J. 2007. Chapter 2 Climate and Climate Change on the Great Barrier Reef. *In* J. Johnson and P. Marshall [eds.], Climate Change and the Great Barrier Reef: A Vulnerability Assessment Great Barrier Reef Marine Park Authority and Australian Greenhouse Office, Australia.

Mabin M.C.G. 2000. Rowes Bay—Pallarenda Foreshore response to Cyclone Tessi: 3 April 2000 http://www.soe-townsville org/data/from SRI-cyclonetessi.pdf.

Moran D.P., Reaka-Kudla M.L. 1991. Effects of disturbance: disruption and enhancement of coral reef cryptofaunal populations by hurricanes. *Coral Reefs* 9:215–224.

Ottesen, P.O. 1976. The reproductive patterns and population dynamics of two small intertidal starfish, *Patiriella obscura* Dartnall and *Nepanthia belcheri* (Perrier). BSc Hons Thesis, James Cook University, Townsville. 88 pp.

Rowe F.E.W., Gates, J. 1955. Echinodermata. *In* Wells, A (ed) *Zoological Catalogue of Australia.* Vol. 33 Melbourne: CSIRO Australia xiii+510pp.

Short A.S., 2000. Beaches of the Queensland Coast: Cooktown to Coolangatta. Surf Lifesaving Australia Ltd, 360pp.

Stevens, H. 2005. An assessment of the influence of factors influencing the density and distribution of *Cryptasterina pentagona* (an intertidal starfish) inhabiting the Queensland coast, Australia. BSc Hons Thesis, University of Oxford. 82 pp.

Stephenson, T.A. 1931. Review of Low Isles Fauna and Flora. Scientific Reports of the Great Barrier Reef Expedition 3.

Ecology

Echinoderms in a Changing World – Johnson (ed)
© *2013 Taylor & Francis Group, London, ISBN 978-1-138-00010-0*

Patchy and zoned *Diadema* barrens on central Pacific coasts of Honshu, Japan

D. Fujita, R. Ishii & T. Kanyama
Tokyo University of Marine Science & Technology, Tokyo, Japan

M. Abe & M. Hasegawa
Shizuoka Prefectural Research Institute of Fishery, Izu Branch, Shimoda, Japan

ABSTRACT: A tropical sea urchin, *Diadema setosum* has been known for more than a century on the Pacific coasts of Boso and Izu peninsulas in Japan. Around Okinoshima near the northernmost record of the species in Boso Peninsula, *D. setosum* inhabited the ditches and notches of cuesta-like rocks leaving *Eisenia/Ecklonia* forests on the top of ledges and formed patchy barrens on the boulders at 2 to 5 m in depth. When *D. setosum* was removed monthly from a quadrat, *Eisenia* forest was formed, while barren was maintained in a control. In Uchiura Bay on the Izu Peninsula, *D. setosum* formed a zoned barren (2 to 10 m in depth) on boulder beds, leaving shallower *Sargassum* and deeper *Undaria/Eckloniopsis* beds. Monthly removal of *D. setosum* from enclosed boulders and a large rock as well as transplantation of *Sargassum* onto suspended floats resulted in temporally limited success in algal restoration.

1 INTRODUCTION

Urchin barren or urchin-dominated barren ground is one of the most conspicuous habitats in and around seaweed beds around the world (Lawrence 1975, Harold & Pearse 1987). In Japan, urchin barren is a type of 'Isoyake', namely the prolonged decrease of seaweeds; sea urchins most commonly causing urchin barrens are those in the *Strongylocentrotus* (mostly *S. nudus*) in northern Japan, and *Diadema* (mostly *D. setosum*) and *Anthocidaris crassispina* in southern Japan (Fujita et al. 2008). Few studies (Yotsui & Maesako 1993, Dotsu et al. 2002) have reported the southern type of urchin barrens, while hundreds of papers (e.g. Agatsuma 1997, Fujita 1998) have documented the northern type of urchin barrens. Sea urchins in the genus *Diadema* are common in tropical to subtropical areas (Lessois et al. 2001). However, the earliest Japanese echinoderm taxonomist, Shigezo Yoshiwara (1874–1940), reported the occurrence of *Diadema setosum* from the temperate Pacific coasts of Boso and Izu peninsulas in central Honshu, as well as Kyushu, Japan, more than a century ago (Yoshiwara 1898). On the central Pacific coast of Honshu, urchin barrens have never been studied and understanding of the ecology of *Diadema* is limited to grazing behaviour on corals (Atobe and Ueno 2001). Recently, the authors found patchy and zoned *Diadema* barrens on the Boso and Izu peninsulas, respectively, and tried restoration of seaweeds there. In the present paper, we document the ecology of *D. setosum* on these peninsulas around the northern limit of its distribution in the western Pacific Ocean and the results of seaweed restoration trials.

2 STUDY AREA

A map of the study area is provided in Figure 1. On the Boso Peninsula, Kominato (N35°7′, E140°11′)

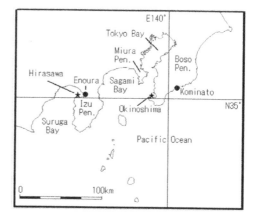

Figure 1. Map indicating the present study sites Okinoshima and Hirasawa (solid asterisks) and two neighboring sites Kominato and Enoura (solid circles) at which Yoshiwara (1898) recorded the occurrence of *Diadema setosum*.

is the northern limit of *D. setosum* on the Pacific coast of Honshu (Yoshiwara 1898, Kawana 1988), but no urchin barrens have been reported there. Okinoshima (N34°59′, E139°49′) is a small tombolo island (1 km in circumstance) located at the southern mouth of Tokyo Bay. The southern coast of the island is composed of cuesta-like bedrock forming ledges (up to 1 m high) and boulders which have been formed when ledges collapsed. Okinoshima is unique because of the influence of the warm Kuroshio Current, so the biota here contains some tropical elements including *Diadema* and corals among temperate elements such as *Eisenia, Ecklonia* and *Undaria* kelps (Chiba Prefecture 1971, Miyata 1995).

On the Izu Peninsula, Yoshiwara (1898) reported the occurrence of *D. setosum* at Enoura (N35°2′, E138°54′) but the present study was conducted at Hirasawa (N35°1′, E138°51′). Both of these sites are in Uchiura Bay, an innermost part of Suruga Bay, influenced by Kuroshio currents. Hirasawa is characterized by boulders, coarse sands and some exposed and submerged rocks. The biota here also contains some tropical elements including corals among temperate elements such as *Eckiloniopsis/Undaria* kelps (Konishi & Hayashida 1992). Interviews with fishermen in 2004 suggested a decline of *Sargassum* beds which had been present at least until the 1970's (Abe et al. 2008).

3 MATERIALS & METHODS

Okinoshima was visited once a month, beginning in April 2004 when some patchy barrens (2 to 5 m in depth) dominated by *D. setosum* were found. Monthly removal of sea urchins was begun in a 5 × 5 m quadrat encircled with a stainless chain (experimental area) leaving the neighbouring control area of the same size in the largest patch (5 × 15 m) in December 2005. Algal cover was monitored monthly on five marked points (50 × 50 cm) in each area for a year. In the treated area, the number of invading sea urchins was also monitored once a month.

Hirasawa was visited once a month beginning in April 2005 when we recognized a zoned barren (2 to 10 m in depth) dominated by *D. setosum*. To clarify the factors determining the lower limit of *Sargassum* beds, algal cover (%) and sea urchin density were monitored along a 50 m long transect from April 2005 to December 2006. For restoring seaweed beds, sea urchins were removed monthly on a huge submerged rock (7 m circumference, surrounded by sandy bottom) from December 2006 to December 2008 and in a 10 × 10 m quadrat on boulders from April 2007 to December 2008. The rock was enclosed with linked cloth bags (60 cm

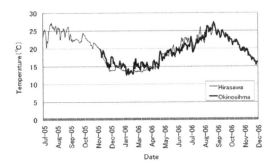

Figure 2. Water temperature monitored at Okinoshima (4 m in depth) and Hirasawa (2 m in depth).

high) to prevent the invasion of sea urchins. The quadrat was only encircled with stainless iron at first but enclosed with linked cloth bags in August 2007 and further enclosed with buoyed plastic sheets and gill nets (60 cm high) in November 2007 and replaced with rolled gill nets (60 cm high) in November 2008. During the study at Hirasawa, we also conducted a cage experiment (Abe et al. 2008), which revealed that intensive grazing by *D. setosum* (but not by herbivorous fish) was the major cause of decrease of transplanted *Sargassum* adults and their offspring.

In both study sites, hourly data of water temperature was monitored using a temperature recorder (MDS-MkV/T, JFE Allec Co Ltd.) and averaged for each day. The ranges of water temperature were 12.5–27.4°C at Okinoshima and 12.8–27.3°C at Hirasawa (Fig. 2). Collected sea urchins and their removed gonads or gut contents were weighed after test diameter was measured to the nearest millimeter.

4 RESULTS & DISCUSSION

4.1 *Patchy barrens at Okinoshima, Boso Peninsula*

At Okinoshima, *D. setosum* inhabited both ledges and boulders (Fig. 3A and B) on the cuesta-like bedrock. Around the ledges, *D. setosum* was always found in the notches and crevices. Their grazing traits were often visible on the encrusting coralline algae covering rocks including vertical walls of the ledges above the notches. *D. setosum* also inhabited interspaces among boulders on the seafloor, forming patchy barrens (usually 5 to 10 m in diameter), among which our experimental patch was the largest, about 15 × 5 m.

In the experimental areas, 179 *D. setosum*, 43 *A. crassispina*, 32 *Pseudocentrotus depressus* and 13 other species of sea urchins were removed

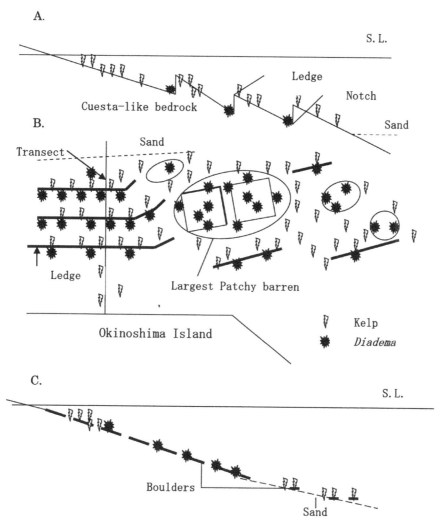

A.

S. L.

Ledge

Notch

Cuesta-like bedrock

Sand

B.

Sand

Transect

Ledge

Largest Patchy barren

Okinoshima Island

ฟ Kelp

✹ *Diadema*

C.

S. L.

Boulders

Sand

Figure 3. Schematic profile of study sites at Okinoshima (A) and Hirasawa (C) and locations of a transect (see A), patchy barrens (ellipses) and experimental areas (squares, right: control, left: removal of sea urchins) at Okinoshima (B). "S.L." = sea level.

at the commencement, among which *D. setosum* was the dominant sea urchin accounting for 67% of individuals, and with a recorded density of ca 7 urchins/m². As the age determination of *D. setosum* was impossible using usual test plate methods (Moor 1935), we estimated ages from the test diameter compositions (Fig. 4) and divided these into five size/age groups; <30 mm (0 year old), 30–39 mm (1 year old), 40–49 mm (2 years old), 50–59 mm (3 years old) and >50 mm (>4 years old) in the monitoring after the urchin removal.

Figure 5 provides a comparison of monthly changes in algal cover between experimental and control areas. In the experimental area, after begin-

ning removal of sea urchins in December 2005, algal cover increased and *Eisenia* forest was formed. After one year, algal cover was 60% in total, among which *Eisenia* and *Sargassum* accounted for 37% and 15%, respectively. Although algal succession has been reported in detail elsewhere (Yamada et al. in prep.), the *Eisenia* forests were maintained for at least three years in the experimental area. In contrast, in the control area, algal coverage was maintained at less than 20% and reached ca 1% after one year. These results show that sea urchins dominated by *D. setosum*, play an important role in maintaining patchy barrens.

Gut contents of invading *D. setosum* collected monthly from the removal area are shown in

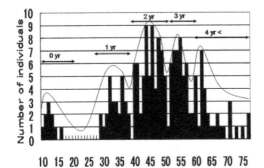

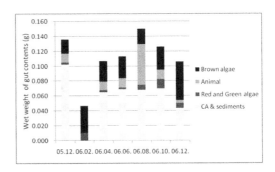

Figure 4. Test diameter composition and estimated age group of *Diadema setosum* removed from the experimental area (5 × 5 m) in the largest patchy barren at Okinoshima.

Figure 6. Bimonthly changes in gut contents of *Diadema setosum* removed from the experimental area at Okinoshima.

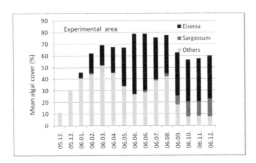

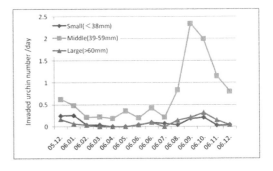

Figure 7. Number/day of *Dimadema setosum* invading the experimental area at Okinoshima.

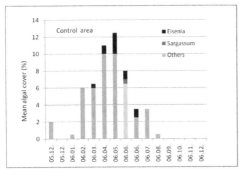

Figure 5. Monthly changes in mean algal cover (%) in the experimental (above) and control (below) areas in a patchy barren at Okinoshima. Note that juveniles of *Eisenia* were included in 'Others' from December 2006 to January 2006 because they were not discriminated from juveniles of *Undaria* and *Ecklonia*.

Figure 6. The contents increased in the warmer season, summer. *Diadema* grazed algal juveniles and deteriorated *Undaria* and barnacles. They tended to consume understory or prostrate algae near the seafloor. During underwater observation

(only in the daytime), neither feeding activity nor feeding marks were found on the canopy composed of *Eisenia* and *Sargassum*.

Figure 7 shows the number/day of *D.setosum* invading the experimental area. *D. setosum* was most active from August to November when the water temperatures were more than 20°C (Fig. 2). The moderate mode of feeding and the limited active period from December to July may be the reasons why urchin barrens remained small and patchy.

4.2 Zoned barren at Hirasawa, Izu Peninsula

At Hirasawa, two major algal assemblages were prominent on the boulders; onshore *Sargassum* beds and offshore kelp beds (Fig. 3C). *Sargassum* species appeared from 0 to 2 m in depth from January to September with a maximum biomass of 800 g/m² in April 2006. *Undaria* and *Eckloniopsis* occurred in the sandy cobble beds from 10 to 18 m in depth from December to July (biomass was not studied). *Diadema* barren was found from 2 to 10 m in depth as a zonation between the two algal assemblages throughout the year. The width of *Diadema* barren was between 50 and 100 m and the length along the coast was at least 1 km. Table 1 shows the seasonal changes in the densities of *D. setosum* and *A. crassispina* in the area shallower than 5 m in depth. The density of *D. setosum* was higher in areas deeper than 2 m, and that of *A. crassispina* was higher in areas shallower than 2 m in depth. Therefore, the distributional pattern of these sea urchins may determine the upper boundary of the *Diadema* barren mentioned above. Our previous report (Abe et al. 2008) showed that transplanted adults of *Sargassum* on the zoned *Diadema* barren could survive whether they were caged or not, but their recruits were heavily grazed by *D. setosum* and survived only when caged and covered (namely, protected also against herbivorous fish). Although no quantitative data were obtained, gut contents of *D. setosum* were comprised of algae and barnacles.

Table 1. Seasonal changes in the densities (individuals/m²) of *Diadema setosum* and *Acanthoicdaris crassispina* along a 50 m transect in urchin barrens on the coast of Hirasawa, Izu Peninsula. Higher densities are represented by more densely shaded cells.

D. setosum	Distance (m) from shore on transect (first row; e.g. 0-5) and mean depth (m, second row; e.g. 1.0)									
	0-5	-10	-15	-20	-25	-30	-35	-40	-45	-50
	1.0	1.2	1.6	1.9	2.2	2.7	3.2	3.7	4.2	4.5
May 05	0.4	1.4	7							
June 05		0.2		1	7.4	5.6	2.6	0.6	0.2	0.6
July 05	1.4	0.4	0.8	5.4	6	8	4	4.6	5.2	5.8
Aug. 05	0.2	0.6	0.4	0.8	3.6	3.2	1.4	2.4	1.4	0.8
Sept.05.	0.8	0.4		3.2	5	5	3.4	6	6.4	6.4
Oct. 05	0.4		0.2	0.2	0.6	1	0.0	0.2	5.4	0.2
Nov.05	0.2	1	3.2	2.8	9.6	3.6	1.8	5.2	3.6	5.8
Dec.12	0.2	1	0.4	0.4	0.2	0.6	0.2	1.4		0.2
Jan. 06		0.2		0.2	1.2	7	0.8			
Feb.06	0.4	0.6	0.8	2	1	2	2.2	1	1.2	1
Mar.06	0.4	2.6	1.2	2.4	5.8	3.4	0.6	0.4	1.8	0.6
Apr.06	0.6	0.2	1.4		2	6			0.6	

A. crassispina	0-5	-10	-15	-20	-25	-30	-35	-40	-45	-50
	1.0	1.2	1.6	1.9	2.2	2.7	3.2	3.7	4.2	4.5
May 05	2.4	1.2	0.2	0.2						
June 05	3.4	1.4		0.8	0.4	0.4			0.2	
July 05	5.6	2.8	1	0.2	0.2	0.2				
Aug. 05	3.6	1.2	0.6							
Sept.05.	3.2	2.4	0.6	0.2	0.2					
Oct. 05	2.2	2	1							
Nov.05	3.8	1.8	0.8	0.4		0.2		0.2		
Dec.12	3.8	2.2	0.8	0.4	0.2			0.2		
Jan. 06	3.8	2.6		0.2		0.2				
Feb.06	3.8	1.2	0.6	0.2	0.2					
Mar.06	3	0.2	0.8		1	0.2				
Apr.06	3.4	1.4	0.6		0.2	0.2				

Figure 8 shows the bimonthly changes in test diameter of *Diadema*. As for Okinoshima, medium sized sea urchins were abundant. Furthermore, occurrence of recruits and their growth could be detected after November.

When an outstanding rock and seafloor were enclosed, turf algae appeared in a month but were removed again through grazing by invading *D. setosum* when the buoyancy of the fence was lost after a few months. Suspended transplantation of *Sargassum horneri* above the seafloor (1 m high) in December 2007 succeeded in the establishment of spring annual forest. The transplanted *S. horneri* grew to 10 m length in March 2008 and matured from February to April. The juveniles derived from the transplanted *S. horneri* formed a dense mat just below the transplantation facility and some of them appeared on rocks and fences (cloth bags, and plastic sheets), but all of them were grazed off by *D. setosum* in April 2008.

4.3 Diadema occurrence on northernmost barrens

Sea urchins of the genus *Diadema*, particularly *D. antillarum* and *D. setosum* are widely distributed in tropical to subtropical areas (Lessios et al. 2001). The importance of *D. antillarum* in limiting algal abundance has been shown repeatedly on Caribbean coral reefs by cage experiments (Ogden et al. 1973, Carpenter 1981, Sammarco 1982, Wanders 1981) and direct observations of algal succession after the mass mortality event in 1982–3 (Levitan 1988, Liddle and Ohlhorst 1986). *D. antillarum* also plays an important role in structuring algal assemblage in the temperate waters of East Atlantic rocky coasts up to a latitude of 32°38′N (Alves et al. 2003, Tuya et al. 2004).

A comparable role of *D. setosum* has been reported for low latitudes in the North Indo-Pacific, (Dart 1972, Gomez et al. 1983, Dotsu et al. 2002) but the present paper is the first to document patchy and zoned *Diadema* barrens around 35°N. These are the northernmost *Diadema* barrens hitherto reported, although *D. setosum* has been found on the Turkey coast (36°08′N), and in the Mediterranean Sea (Yokes and Galil 2006), and a closely related species *D. savignyi* has also been found near the tip of Noto Peninsula (37°12′N), Sea of Japan (Higashide 2002).

At Okinoshima and Hirasawa, the lowest temperature was around 13 °C. A water temperature lower than 12 °C may be critical for the *D. setosum* distribution because mass mortality of this species has been recorded from the Kumano coast on the Kii Peninsula (Japan) in January to February 2002 when water temperatures dropped below 12 °C (Oki et al. 2004). The existence of old records (>100 years old) documenting the occurrence of *D. setosum* on

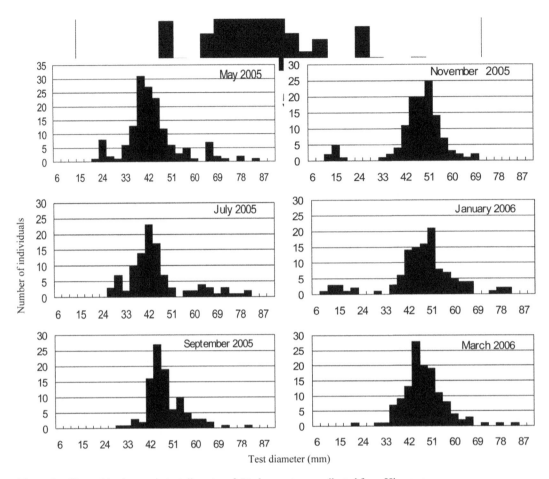

Figure 8. Bimonthly changes in test diameter of *Diadema setosum* collected from Hirasawa.

the Boso and Izu Peninsulas (Yoshiwara 1898) suggests that the distributional range of *D. setosum* has not moved, at least around its northern limit.

However, the abundance of *D. setosum* may have changed over time in these areas. Although there are no reports documenting the timing and process of patchy *Diadema* barren formation in Okinoshima, Yamada et al. (in prep) found that the density of *D. setosum* increased when a top section (2 m long, 50 cm high) of a ledge (ca.30 m long, 1 m high) collapsed and interspaces among broken rocks enlarged, resulting in formation of barrens. As Okinoshima is located near Tateyama City, where the coastline and ports have been frequently modified to be protected from waves, bottom vibration from piling as well as earthquakes may increase rock breakages, and hence the habitat of *D. setosum*. As most of the ledges are still present, of which the tops are good algal refugia, these ledges as well as the moderate mode of feeding and the limited active period (see above) of

D. sestosum may prevent patchy barrens enlarging into zoned barrens.

In contrast, Abe et al. (2008) reported that *Sargassum* beds decreased at Hirasawa after the 1970s. At Hirasawa, *D. setosum* barrens have expanded onshore and replaced the *Sargassum* beds. The reason for this expansion is still unclear. The most prominent coastal changes, namely building of an artificial beach (for swimming) rimmed with stone piles (11 m high) and a breakwater in a fishing port surrounded with stone piles, may have increased habitat availability for *D. setosum*. Moreover these changes cause stagnation of coastal waters, which enhances the activity of movement and feeding of *D. setosum* by decreasing water motion, enhancing stratification (top warm, and bottom cool, as represented in two types of vegetation, *Sargassum* and kelp, respectively) and altering the movement of drift sand. As the topography at Hirasawa is a gentle slope, *D. setosum* could potentially establish zoned barrens on boulders leaving algal refuges on exposed coastline shallower than 2 m and sand-dominant areas deeper than 10 m.

REFERENCES

Abe, M., Ishii, R. & Fujita, D. 2008. Factors limiting growth of *Sargassum yamamotoi* off Hirasawa, Numazu. *Bull. Shizuoka Pref. Res. Inst. Fish.* 43: 13–17.

Alves, F.M.A, Chicharo, L.M., Serrao, E. & Abreu, A.D. 2003. Grazing by *Diadema antillarum* (Philippi) upon algal communities on rocky substrates. *Sci. Mar.* 67: 307–311.

Agatsuma, Y. 1997. Ecological studies on the population dynamics of the sea-urchin *Strongylocentrotus nudus*. *Sci. Rep. Hokkaido Fish. Exp. Sta.* 51: 1–66.

Atobe, T. & Ueno, S. 2001. Grazing effects of *Diadema setosum* on the Scleractiniaon coral, *Acropora tumida* in Suruga Bay, Japan. *Bull. Inst. Oceanic Res. & Devlop., Tokai Univ.* 22: 65–73.

Carpenter, R.C. 1981. Grazing by *Diadema antillarum* (Philippi) and its effects on the benthic algal community. *J. Mar. Res.* 39: 749–765.

Chiba Prefecture 1971. Report of a marine park in Chiba Prefecture. Chiba Prefecture.

Dart, J.K.G. 1972. Echinoids, algal lawn and coral recolonization, *Nature* 239: 50–51.

Dotsu, K., Ohta, M. & Masuhara, H. 2002. Grazing effects of *Diadema* spp. On algal vegetation in the sea around Matsushima Island, Nagasaki pref., western Kyushu. *Rep.mar. Ecol. Res. Inst.* 4: 1–10.

Fujita, D. 1998. Strongylocentrotid sea urchin-dominated barren grounds on the Sea of Japan coast of northern Japan. R. Mooi & M. Telford (eds.) *Echinoderms: San Francisco: Proc. 9th Int. Echinoderm Conf. 5–9 August 1996*, 659–664. Rotterdam. Balkema.

Fujita, D., Machiguchi, Y. & Kuwahara, H. 2008. Recovery from urchin barrens—ecology, fishery and utilization of sea urchins. Seizando-shoten.

Gomez, E.D., Guieb, R.A. & Aro, E. 1983. Studies on the predators of commercially important seaweeds. *Fish. Res. J. Philipp.* 8: 1–17.

Harold, C. & Pearse, J.S. 1987. The ecological role of echinoderms in kelp forests. *Echinoderm Studies* 2: 137–233.

Higashide, M. 2002. Sea urchins: their characteristics and how to identify species. *News Lett. Noto Mar. Ctr,* 17: 2–5.

Kawana, O. 1988. Echinoderms in Chiba Prefecture. *Bull. Biol. Soc. Chiba,* 37: 69–74.

Konishi, Y. & Hayashida, F. 1992. Vegetation of benthic marine algae in Suragua Bay, Central Japan. *J. Sch. Mar.Sc. Tech.Tokai. Univ.* 1: 15–27.

Lawrence, J.M. 1975. On the relationship between marine plants and sea-urchins. *Oceangr. Mar. Biol. Ann. Rev.* 13: 213–286.

Lessios, H.A., Kessing, B.D. & Pearse, J.S. 2001, Population structure and speciation in tropical seas: Global phylogeography of the sea urchin *Diadema*. *Evolution* 55: 955–975.

Levitan, D.R 1988. Algal-urchin biomass responses following mass mortality of *Diadema antillarum* Philippi at Saint John, U.S. Virgin Islands. *J. Exp. Mar. Biol. Ecol.* 119: 167–178.

Liddle, W.D. & Ohlhorst, S.L. 1986. Changes in benthic community composition following the mass mortality of *Diadema* at Jamaica, *J. Exp. Mar. Biol. Ecol.* 95: 271–278.

Miyata, M. 1995. Algal flora of Okinoshima-island in Boso Peninsula, Japan. *J. Nat. Hist. Mus. Inst., Chiba, Special Issue* 2: 113–124.

Moor, H.B. 1935. A comparison of the biology of *Echinus esculentus* in different habitats. Part II. *J. Mar. Biol. Assoc.* 20: 109–128.

Ogden, J.C., Brown, R.A. & Salesky, N. 1973. Grazing by the Echinoid *Diadema antillarum* Philippi: Formation of halos around West Indian Patch Reefs. *Science* 182: 715–717.

Oki, D., Yamamoto, Y. & Okumura, H. 2004. Preliminary observation on the utilization of the sea urchin *Diadema setosum* as food material at the northern coast of Kumano-nada in Mie Prefecture. *Bull. Fish. Res. Div. Mie Pref.* 11: 15–21.

Sammarico, P.W. 1982. Effects of grazing by *Diadema antillarum* Philippi (Echinodermata) on algal diversity and community structure. *J. Exp. Mar. Biol. Ecol.* 65: 83–105.

Tuya, F., Boyra, A., Sanchez-Jerez, P. Barbera, C. & Haroun, R. 2004. Can one species determine the structure of the benthic community on a temperate rocky reef? The case of the long-spined sea-urchin *Diadema antillarum* (Echinodermata: Echinoidea) in the eastern Atlantic, *Hydrobiologia* 519: 211–214.

Wanders, J.B.W. 1977. The role of benthic algae in the shallow reef of Curaçao (Netherlands Antilles) III: The significance of grazing. *Aquatic Botany* 3: 357–390.

Yamada, R., Kanyama, T. & Fujita, D. in prep. Algal succession on new rock surfaces exposed after a collapse of a ledge, Okinoshima, Chiba Prefecture, Japan.

Yokes, N. & Galil, B.S. 2006. The first record of the needle-spined urchin *Diadema setosum* (Leske, 1778) (Echinodermata: Echinoidea: Diadematidae) from the Mediterranean Sea. *Aquatic Invasions* 1: 188–190.

Yoshiwara, S. 1898. Japanese sea urchin (5). *Zool. Sci.* 10: 328–331.

Yotsui, T. & Maesako, N. 1993. Restoration Experiments of *Eisenia bicyclis* beds on barren grounds at Tsushima Islands. *Susanzoshoku* 41: 67–70.

Echinoderms in a Changing World – Johnson (ed)
© 2013 Taylor & Francis Group, London, ISBN 978-1-138-00010-0

Native spider crab causes high incidence of sub-lethal injury to the introduced seastar *Asterias amurensis*

S.D. Ling & C.R. Johnson
Institute for Marine and Antarctic Studies, University of Tasmania, Hobart, Australia

ABSTRACT: The northern Pacific seastar, *Asterias amurensis* (Lütken), is an invasive species established throughout the Derwent Estuary, southeast Tasmania. Here we report on field observations of predation on the seastar within its new environment. The spider crab *Leptomithrax gaimardii* (family Majidae), which characteristically aggregates in shallow water in winter, used its chelae to tear the body wall of *A. amurensis* and completely consume the seastar. Typically, the predatory interaction resulted in sub-lethal injury (arm damage) to the seastar. Sampling in the area of an aggregation of spider crabs revealed that ~70% (ranging from 50–87%) of the *A. amurensis* population incurred sub-lethal arm damage consistent with spider crab attack. Spider crabs were observed at other sites in the estuary, but were transient and rates of arm injury at these sites were much lower, ranging from 7–33%. The most common form of injury sustained by *A. amurensis* in the presence of spider crabs was damage to arm tips (62% of all injuries), followed by full arm loss (34%) and half-arm loss (5%). Near the peak of the ~2 month spider crab aggregation, very few injured seastars displayed evidence of arm regeneration, however rates of arm regeneration increased rapidly as the localised spider crab aggregation dispersed. Smaller *A. amurensis* were disproportionately represented among the injured seastars, while observations *in situ* and in aquaria revealed that injury may frequently occur during competition for mussel prey. These observations suggest that native predators within the seastar's new range may be capable of inflicting localised seasonal impacts on this important introduced pest.

1 INTRODUCTION

Establishment of introduced species in new environments is dependent on the outcomes of interactions with the recipient community (e.g. Pimm 1989, Vermeij 1991, *reviewed by* Lodge 1993). In the absence of adapted natural predators, introduced species may proliferate unopposed and reach very high population levels, i.e. the "predatory release hypothesis" (e.g. Crawley 1987). The northern Pacific seastar, *Asterias amurensis* (Lütken), was inadvertently introduced to Tasmanian waters sometime in the 1980s where it has undergone rapid population increase particularly in the Derwent Estuary where it has become the dominant benthic predator (Buttermore et al. 1994, Grannum et al. 1996, Byrne et al. 1997, Ross et al. 2002, 2003). While predators of *A. amurensis* are known from the northern Pacific (i.e. Alaskan king crabs, see Mikulich & Berulina 1972; and other asteroids, see McLoughlin & Bax 1993) little is known about predators within the seastar's introduced range.

Highly disturbed by anthropogenic activity, the benthic community of the Derwent Estuary is greatly impoverished and dominated by exotic species relative to historical baselines established from sediment cores (Edgar & Samson 2004, Edgar et al. 2005). In the heavily urbanised Port of Hobart, *A. amurensis* can be extremely abundant with maximum densities ranging up to 46 m^{-2} (S. Ling *unpub. data*) which is greater than that reported during population outbreaks of the seastar in the northern hemisphere (Nojima et al. 1986). Thus it has been hypothesised that the dominance of *A. amurensis* in the Derwent Estuary may be due to reduced competition for resources or reduced abundances of predatory guilds (e.g. Goggin 1998).

While population expansion of *Asterias amurensis* in Tasmania continues, the occurrence of arm loss among individuals in the Derwent Estuary suggests the presence of predators (Grannum et al. 1996, Lawrence et al. 1999, J. Ross *unpub. data*). Predatory invertebrates such as the large asteroid *Coscinasterias muricata*, and decapod crustaceans including the hermit crab (*Trizopagurus strigimanus*), horse crab (*Cancer novaezealandiae*) and spiny lobster (*Jasus edwardsii*), have been implicated as predators of *A. amurensis* based on observations in aquaria (Parry et al. 2000, L. Turner *pers. comm.*). Observations of predation *in situ* are rare with only another asteroid *C. muricata* witnessed to prey on *A. amurensis*, however this interaction has been observed on only 3 separate occasions during >300 hours of SCUBA diving throughout the Derwent Estuary and surrounding embayments (J. Ross *pers. obs.*).

This paper reports on previously undescribed predation on *Asterias amurensis* in the Derwent Estuary by the spider crab *Leptomithrax gaimardii* (Milne Edwards). This large Majid crab attains a carapace size of ~120 mm, usually occurs in deep water to ~800 m, but is seasonally abundant in shallow inshore embayments and estuaries in Tasmania in winter where it can form large aggregations while moulting and mating (Gardner 1999, Turner 2001). Specifically, this paper describes the predatory interaction between the spider crab and *A. amurensis* and quantifies patterns of sub-lethal injuries inflicted on the seastar during a localised aggregation of the spider crab. This is the first documentation of *in situ* predation on the introduced seastar outside of its native northern hemisphere range.

2 DATA COLLECTION

In mid-July 1999, a large and sustained aggregation of spider crabs (*Leptomithrax gaimardii*) occurred in the Derwent Estuary adjacent to the Bellerive Yacht Club (BYC; Fig. 1). Near the peak of the aggregation, predator-prey interactions and the incidence of sub-lethal damage to *A. amurensis* was assessed *in situ* by divers. Rates of sub-lethal damage were estimated by examining all seastars along 6 belt transects, each 10 × 2 m. Each seastar encountered was assessed for damage which was categorised as either full arm loss (entire arm

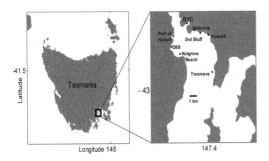

Figure 1. Study sites in the Derwent Estuary (expanded), southeastern Tasmania. All sites were located on soft-sediment, Bellerive Yacht Club (BYC, 6 m depth); Port of Hobart (8–14 m); Derwent Sailing Squadron (DSS, 4 m); Second Bluff (6–10 m); Nutgrove Beach (3–6 m) and Tranmere (6–10 m); Additional sites surveyed by others (Lawrence et al. 1999, J. Ross *unpub. data*) are indicated by open circles, i.e. Bellerive Beach (3–6 m) and Howrah Beach (3–6 m).

missing from base of central disc), half arm loss (~50% arm missing by length), or arm tip damage (observed as a ~2–5 mm excision of the distal region of the arm). Evidence of wound regeneration was also recorded. In addition to the survey at BYC, the frequency of seastar damage was also assessed from fortnightly collections of 30 individuals from BYC and an additional 6 sites (Fig. 1) from late July until ~October depending on site. For each seastar collected, ray length (oral opening to distal end of longest arm) was measured with knife edge callipers to the nearest millimeter. Size distributions of injured and non-injured seastars from BYC were compared using a Kolmogorov-Smirnoff test. Prior to diver collections at each site, the density of *A. amurensis* and *L. gaimardii* was obtained on each sampling occasion by averaging counts across 6 belt transects (each 10 × 2 m at BYC, DSS and Port of Hobart; 30 × 2 m for Nutgrove Beach, Second Bluff and Tranmere). Temporal patterns in seastar density were analysed by linear regression. Statistical tests were performed using the SAS software V9.1.

3 RESULTS

Targeted survey of sub-lethal damage among *Asterias amurensis* at BYC near the peak of the spider crab aggregation revealed that 124 of 211 individual seastars (59%) exhibited damaged arms. The average rate of sub-lethal damage at BYC was much higher than observed elsewhere in the estuary and the average density of *L. gaimardii* was also highest at this site (Table 1). At BYC, *L. gaimardii* (on a single occasion) was also directly observed to completely consume the central disc of *A. amurensis* and thus inflict lethal predation. Lethal predation by *L. gaimardii* was also witnessed on single occasions at both Second Bluff and Nutgrove Beach. More frequently observed were interactions leading to sub-lethal damage to *A. amurensis* as *L. gaimardii* used their chelae to pinch the arm tips of *A. amurensis* and to puncture and tear the body wall. Where wounds remained open, pyloric caecae and gonad material were typically exposed and consumed by *L. gaimardii*. The most common form of sub-lethal injury sustained by *A. amurensis* was severed arm tips, followed by full arm loss and half arm loss (Fig. 2a). Of the injured seastars, most individuals incurred injury to a single arm and the frequency of individuals with multiple injured arms declined as the number of damaged arms increased (Fig. 2b). Partial predation in this way was size-specific since ray length measurements from BYC revealed that sub-lethal injury was biased towards smaller individuals (Fig. 3).

Table 1. Summary of sub-lethal damage incidence for *Asterias amurensis* in the Derwent Estuary. a) Damage incidence and density of *Leptomithrax gaimardii* July–Oct 1999; numbers of seastars examined are shown in parentheses; spider crab densities are mean individuals m^{-2} ± SE averaged across 6 sampling occasions, note that the highly mobile *L. gaimardii* was present on one occasion at the Port of Hobart but not recorded on diver transects. b) Incidence of sub-lethal damage as observed in August 1997 by J. Ross (*unpub. data*). c) Incidence of sub-lethal damage as observed in May 1997 by Lawrence & co-workers (1999).

Site	*A. amurensis* damage incidence	*L. gaimardii* density
(a)		
Bellerive yacht club	65% (481)	0.186 ± 0.074
Nutgrove beach	26% (180)	0.003 ± 0.002
Second bluff (Howrah)	21% (180)	0.005 ± 0.005
Trywork point (Tranmere)	10% (180)	0.004 ± 0.004
Port of Hobart	8% (240)	0.000 ± 0.000
Derwent sailing squadron	7% (240)	0.077 ± 0.072
(b)		
Bellerive	12% (50)	
Howrah	14% (100)	
Tranmere	26% (50)	
(c)		
Bellerive	7% (100)	
Nutgrove beach	7% (100)	

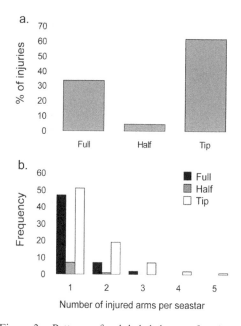

Figure 2. Patterns of sub-lethal damage for *Asterias amurensis* as observed *in situ* near the peak of spider crab aggregation at BYC, Derwent Estuary, August 1999 (N = 211). (a) Distribution of sub-lethal injury types. (b) Distribution of sub-lethal injury types with respect to the number of injured arms per seastar.

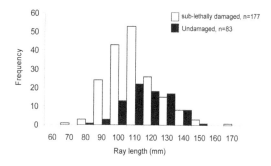

Figure 3. Size-specific patterns of sub-lethal predation on *Asterias amurensis* from BYC. Data are pooled across all collection periods (June–Dec 1999); only individuals with at least one intact, non-regenerating arm were included in this analysis (N = 260). Smaller sizes of *A. amurensis* were disproportionately represented in the "damaged" seastar group (Kolmogorov-Smirnoff comparison yielded a statistically significant difference between size frequency distributions, $P = 0.0021$).

Prior to 16 July, *L. gaimardii* had not been recorded on diver transects at BYC. By 27 July, the density of spider crabs was estimated at 0.62 individuals m^{-2} at this site (Fig. 4a). Interestingly, the spider crab aggregation was sustained at BYC (recorded on transects during 7 of 10 sampling occasions over ~4.5 months, see Fig. 4a) whereas the crab was more transient and only recorded on 3 (or less) of 7 fortnightly samplings at the other 5 sites.

Near the peak of the spider crab aggregation at BYC, <10% of the sub-lethally damaged seastars demonstrated evidence of arm regeneration, i.e. 90% of observed wounds were open. As the density of spider crabs declined post localised aggregation, the incidence of sub-lethal injury did not reveal an obvious trend however the rate of wound regeneration (over-growth of damaged arm tips and entire limb regeneration from central disc) increased dramatically to reach ~70% at final sampling (Fig. 4a). A slight reduction in *A. amurensis* density was observed following the spider crab aggregation at BYC, however the density of seastars appeared to increase and stabilise by the final sampling resulting in a slight, but non-significant,

overall decline in seastar density across the entire sampling period (linear regression: $F_{1,8} = 3.42$, $P = 0.107$), see Figure 4b.

The slight population decrease of *A. amurensis* at BYC was potentially caused by the removal of

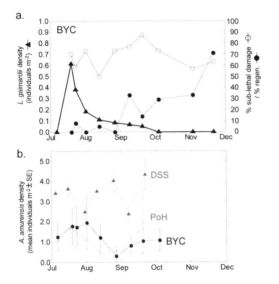

a.

b.

Figure 4. (a) Temporal patterns of sub-lethal injury incidence and rates of injury regeneration for *Asterias amurensis* and *Leptomithrax gaimardii* density at BYC, July–Dec 1999. *L. gaimardii* densities (individuals m⁻²) are displayed as triangles; percentages of *A. amurensis* population showing sub-lethal injuries are displayed as open circles (n = 30), note that data on rates of sub-lethal injury were not assessed prior to the spider crab aggregation; percentages of sub-lethal injuries demonstrating evidence of regeneration are displayed as filled circles. (b) Temporal pattern of *A. amurensis* density in 1999 at BYC (filled circles, solid line) and the high density controls (dashed lines) of Port of Hobart (PoH, open circles) and DSS (triangles).

30 seastars from the population on a fortnightly basis (i.e. at least 210 individuals were removed from the population over the duration of the study). A 'collection effect' likely contributed to population decline of *A. amurensis* at the 'low density' sites (~0.05 individuals m⁻²) of Second Bluff (significant population decline, linear regression: $F_{1,6} = 9.37$, $P = 0.028$; trendline: $y = -0.011x + 0.072$, $R^2 = 0.71$); but non-significant population declines at Tranmere ($F_{1,6} = 3.13$, $P = 0.137$; $y = -0.001x + 0.10$, $R^2 = 0.38$) and Nutgrove Beach ($P = 0.31$; $y = -0.0004x + 0.058$, $R^2 = 0.20$). High density sites, suitable as controls for BYC (i.e. ~2.0 seastars m⁻²), also revealed high variability but no decline in *A. amurensis* density across the sampling period (Fig. 4b), i.e. Port of Hobart ($F_{1,6} = 0.00$, $P = 0.98$, $y = 0.0002x + 1.08$; $R^2 = 0.0001$) and DSS ($F_{1,6} = 0.18$, $P = 0.687$; $y = 0.035x + 3.21$, $R^2 = 0.035$) showed slightly positive population trajectories, indicating that the large population sizes at these sites, and BYC, were likely robust against removal/and or lethal predation on a relatively small number of seastars.

4 DISCUSSION

4.1 *Patterns of spider crab predation & sub-lethal injury*

Direct field observations revealed that the spider crab *Leptomithrax gaimardii* is capable of preying on *Asterias amurensis*. While the predatory interaction was observed to be lethal on several occasions, i.e. central discs of *A. amurensis* were fully consumed, few such events were observed. However, unless observed directly, lethal predation is difficult to detect and therefore invariably underestimated.

Our *in situ* observations were limited to a total of approximately 7 diver hours near the peak of the spider crab aggregation, so the chance of observing lethal predation events was small. In contrast, predatory interactions resulting in sub-lethal injury were readily observed *in situ*. Indeed, the spider crab aggregation at the BYC was associated with a high prevalence of sub-lethal arm damage to *A. amurensis* (50–87%), much higher than at other sites examined concurrently in the Derwent Estuary, and higher than that reported in the estuary during previous years or for *Asterias* globally (Table 1; *reviewed by* Lawrence [2011] this volume). Furthermore, the spider crab aggregation at BYC was coincident with a high prevalence of open wounds among *A. amurensis* while such wounds were rarely observed at the other sites. The clear increase in arm regeneration rates after dispersal of spider crabs at BYC, in combination with direct observations of predation *in situ*, indicates that spider crabs were responsible for the high prevalence of injured seastars at this site. In addition, other candidate predators, such as the asteroid *Coscinasterias muricata* or other crustaceans (i.e. rock lobsters or hermit crabs) were not observed at the BYC site during the study and cannibalism between *A. amurensis*, as noted by J. Ross (*pers. comm.*), was not witnessed.

The types of injury inflicted on *Asterias amurensis* were also consistent with predation by the spider crab. Injuries observed at BYC were similar to those seen in direct encounters *in situ* and in aquaria reflecting use of chelae to either pinch or sever arm tips, or pinch and tear half-way along an arm. Full arm loss (34% of injuries) appeared to be the result of arm autotomy following major trauma to an arm as autotomised arms with approximately half the arm damaged (5% of injuries) were observed on the benthos at BYC. Damage to arm tips (62% of injuries) appeared not to result in arm autotomy, and these injuries were observed to heal, i.e. among the BYC population, the frequency of damaged arm tips that displayed evidence of regeneration increased through time.

The disproportionately high incidence of sub-lethal injuries evident among smaller *Asterias amurensis* at BYC suggests that the interaction between spider crabs and *A. amurensis* is size-specific. Either smaller seastars are more easily manipulated by a spider crab of a given size or, alternatively, smaller seastars are more fragile against the mechanical action of spider crab chelae (e.g. Marrs et al. 2000). Interestingly, the body wall of *A. amurensis* from BYC was notably tougher and more difficult to dissect relative to those of similar sizes from the other sites. The possibility that *Asterias* may increase robustness of its body wall in response to predation risk requires explicit testing.

4.2 Evidence of competitive and trophic interactions

From direct observations at BYC, severed arm tips in *Asterias amurensis* appeared to be the result of competition with *Leptomithrax gaimardii* for mussel prey (*Mytilus* sp.) rather than a predatory interaction *per se*. That is, *L. gaimardii* was witnessed to pinch the arm tips of *A. amurensis* in possession of mussel prey until the seastar relinquished its prey which was then consumed by the spider crab. Supplementary observations in an aquarium suggested that *L. gaimardii* was incapable of preying on mussels (60 mm shell length) unless *A. amurensis* was present, in which case the seastar received damaged arms tips and the mussel was consumed by the crab (for description of similar mechanism between different asteroid species see Gaymer & Himmelman 2008).

While competition for mussel prey appeared to account for arm tip damage, spider crabs at BYC were frequently observed to be in possession of severed *Asterias amurensis* arm segments and to be consuming the body wall and/or pyloric caecae or gonad material, indicating a trophic nature to the crab-seastar interaction. Notably, the pattern of spider crab attack and consumption, i.e. using chelae to pinch and tear the seastar's body wall and consume pyloric caecae, is consistent with that described by Aldrich (1976) for spider crab *Libinia emarginata* (Majidae) predation on the congeneric *Asterias forbesi* in the northern hemisphere. Seastars can clearly be a significant dietary item for crabs, since *A. amurensis* is considered to form an important part of the Alaskan king crab's (*Paralithodes camtschatica*) "lime diet" during pre- and post-moulting periods when the Lithodid crab requires calcium to form new shells (Logvinowich 1945, Kun & Mikulovich 1954 *cited in* Mikulich & Berulina 1972). Because *L. gaimardii* mates and moults in shallow embayments in south eastern Tasmania, it is plausible that *A. amurensis* provides a dietary supplement during the spider crabs annual migration into the Derwent Estuary.

4.3 Seastar behaviour

During the spider crab aggregation at BYC, seastars were observed to seek refuge among physical structure (e.g. bases of wharf pylons, old tyres and sunken wood). Where spatial refuge was absent, *A. amurensis* was observed to aggregate in a manner whereby exposure of arm tips appeared to be minimised. This behaviour may be a defensive response to reduce the likelihood of predation, and has also been noted for *A. amurensis* in the northern hemisphere when the seastar is in the presence of the king crab *Paralithodes camtschatica* in aquaria (Mikulich & Berulina 1972). Notably, foraging behaviour in asteroids is reduced in response to experimental clipping of arm tips and induced arm autotomy (e.g. Harrold & Pearse 1980, Diaz-Guisado et al. 2006, Barrios et al. 2008). Thus, the presence of *L. gaimardii* would appear capable of mediating *A. amurensis* behaviour by restricting foraging rates of the seastar through direct competition and/or predation risk limiting foraging behaviour.

4.4 Potential population effects of sub-lethal predation

Depending on the severity of trauma, injured individuals may be faced with disease, further risk of predation, and compromised reproduction and foraging (*reviewed by* Lawrence [2011], this volume). If the incidence of sub-lethal injury is high, then an overall suppressive effect may occur at the population level (e.g. Harris 1989). The most frequently observed injury in our study, severing of arm tips, involved the removal of the fine apical sensory tube feet and eyespot, the loss of which may influence the efficiency of seastars in sensing both predators and prey (e.g. Harrold & Pearse 1980, Lawrence & Vasquez 1996, Lawrence et al. 1999). However, in contrast to controlled laboratory experiments (e.g. Bingham et al. 2000, Barrios et al. 2008), the high incidence of sub-lethal predation at BYC appeared to have no discernable effect on reproductive investment of *A. amurensis* (i.e. Gonad-Somatic-Index), as average GSIs at this site were high (mean ~30%) and not different to seastar populations at the Port of Hobart and DSS where the spider crab was much reduced in abundance, but where an abundance of mussel prey also maintains high GSIs (S. Ling *unpub. data*). While the high GSIs observed at BYC would have developed before the spider crab aggregation, this observation suggests that high food availability may be capable of offsetting the negative effects of injury such that well provisioned seastars are able to regenerate arms yet maintain large reproductive investment.

In contrast, *A. amurensis* populations existing under conditions of low food availability are likely to be more stressed by sub-lethal injury and are thus more likely to demonstrate population suppression as a result (e.g. Lawrence et al. 1986, Lawrence & Larrain 1994, Diaz-Guisado et al. 2006). Furthermore, *A. amurensis* populations in areas of low food availability in the Derwent Estuary also occur at low density (i.e. Second Bluff, Nutgrove Beach & Tranmere).

5 CONCLUSIONS

Prior to our observations there was a perceived lack of functional predators of *Asterias amurensis* within the seastar's introduced range (Goggin 1998). Our data suggest that large seasonal aggregations of *Leptomithrax gaimardii* in the Derwent Estuary are capable of causing large but localised effects on *A. amurensis*. Clearly, impacts of predation on the seastar will be greatest in areas where aggregations of *L. gaimardii* demonstrate residency over several months and where population size of *A. amurensis* is relatively low. Given these findings, there is a need to consider the resilience of benthic communities against outbreaks of *A. amurensis*, that is, risk of seastar outbreaks may be less likely among benthic communities containing diverse guilds of predators and competitors.

ACKNOWLEDGEMENTS

We would like to thank Craig Mundy, Sean Hooper, Richard Holmes, Simon Talbot, Regina Magierowski and Adam Stephens for field assistance; Jason Beard and Cameron Amos provided assistance with laboratory work.

REFERENCES

Aldrich, J.C. 1976. The Spider Crab *Libinia Emarginata* Leach, 1815 (Decapoda Brachyura), and the starfish, *Asterias forbesi* an unsuitable predator but a cooperative Prey. *Crustaceana* 31:151–156.

Barrios, J.V., Gaymer, C.F., Vasquez, J.A. & Brokordt, K.B. 2008. Effect of the degree of autotomy on feeding, growth, and reproductive capacity in the multi-armed seastar *Heliaster helianthus*. *Journal of Experimental Marine Biology and Ecology* 361: 21–27.

Bingham, B.L., Burr, J. & Head, H.W. 2000. Causes and consequences of arm damage in the sea star *Leptasterias hexactis*. *Canadian Journal of Zoology* 78: 596–605.

Buttermore, R.E., Turner, E. & Morrice, M.G. 1994. The introduced northern Pacific seastar *Asterias amurensis* in Tasmania. *Memoirs of the Queensland Museum* 36: 21–25.

Byrne, M., Morrice, M.G. & Wolf, B. 1997. Introduction of the northern Pacific asteroid Asterias amurensis to Tasmania: reproduction and current distribution. *Marine Biology* 127: 673–685.

Crawley, M.J. 1987. What makes a community invasible? In Gray, A.J., Crawley, M.J. & Edwards, P.J. (eds.), *Colonization, Succession and Stability—the 26th Symposium of the British Ecological Society held Jointly with the Linnaean Society of London*. Blackwell Science, Oxford, pp. 429–453.

Diaz-Guisado, D., Gaymer, C.F., Brokordt, K.B., & Lawrence, J.M. 2006. Autotomy reduces feeding, energy storage and growth of the sea star *Stichaster striatus*. *Journal of Experimental Marine Biology and Ecology* 338: 73–80.

Edgar, G.J., & Samson, C.R. 2004. Catastrophic decline in mollusc diversity in eastern Tasmania and its concurrence with shellfish fisheries. *Conservation Biology* 18: 1579–1588.

Edgar, G.J., Samson, C.R. & Barrett, N.S. 2005. Species extinction in the marine environment: Tasmania as a regional example of overlooked losses in biodiversity. *Conservation Biology* 19: 1294–1300.

Gardner, C. 1999. Spider crab aggregation on Tasmania's northwest coast. *Invertebrata* 14.

Gaymer, C.F. & Himmelman, J.H. 2008. A keystone predatory sea star in the intertidal zone is controlled by a higher-order predatory sea star in the subtidal zone. *Marine Ecology Progress Series* 370: 143–153.

Goggin, L. 1998. Control of the introduced seastar Asterias amurensis in Australian waters. In Carnevali M.D.C.C. & Bonasoro F. (eds.). *Echinoderm Research Proceedings of the Fifth European Conference on Echinoderms Milan, Italy: 477–479*. Rotterdam: Balkema.

Grannum, K.R., Murfet, N.B., Ritz, D.A. & Turner, E. 1996. The distribution and impact of the exotic seastar, *Asterias amurensis* (Lutken) in Tasmania. Final Report to the Australian Nature Conservation Agency, Invasive Species Program No. 80.

Harris, R.N. 1989. Nonlethal injury to organisms as a mechanism of population regulation. *The American Naturalist* 134: 835–845.

Harrold, C. & Pearse, J.S. 1980. Allocation of Pyloric Caecum Reserves in Fed and Starved Sea Stars, *Pisaster Giganteus* (Stimpson): Somatic Maintenance comes before reproduction. *Journal of Experimental Marine Biology and Ecology* 48: 169–183.

Lawrence, J.M., Byrne, M., Harris, L., Keegan, B., Freeman, S., Cowell, B.C. 1999. Sublethal Arm Loss in *Asterias amurensis, A. Rubens, A. Vulgaris*, and *A. Forbesi* (Echinodermata: Asteroidea). *Vie Et Milieu* 49: 69–73.

Lawrence, J.M., Klinger, T.S., McClintock, J.B., Watts, S.A., Chen, C.P., Marsh A., & Smith, L. 1986. Alloction of nutrient resources to body components by regenerating *Luidia clathrata* (Say) (Echinodermata: Asteroidea). *Journal of Experimental Marine Biology and Ecology* 102: 47–53.

Lawrence, J.M. & Larrain, A. 1994. The cost of arm autotomy in the starfish *Stichaster striatus*. *Marine Ecology Progress Series* 109: 311–313.

Lawrence, J.M. & Vasquez, J. 1996. The effect of sublethal predation on the biology of echinoderms. *Oceanologica Acta* 19: 431–440.

Lodge, D.M. 1993. Biological invasions: lessons for ecology. *Trends in Ecology and Evolution* 8: 133–137.

Marrs, J., Wilkie, I.C., Sköld, M. Maclaren, W.J. & McKenzie, J.D. 2000. Size-related aspects of arm damage, tissue mechanics, and autotomy in the starfish *Asterias rubens*. *Marine Biology* 137: 59–70.

McLoughlin, R. & Bax, N. 1993. Scientific discussions in Japan and Russia on the northern pacific seastar. Unpublished Report. CSIRO, Division of Fisheries, 30 pp.

Mikulich, L.V. & Berulina, M.G. 1972. The behaviour of the Amurensis seastar and the Kamchatka crab together in aquarium conditions. *Hydrbiological issues of some areas of the Pacific*: 52–54.

Nojima, S., Soliman, F.A., Kondo, Y., Kuwano, Y., Nasu, K. & Kitajima, C. 1986. Some notes on the outbreak of the seastar *Asterias amurensis versicolor* Sladen, in the Ariake Sea, western Kyshu. *Pub. Amakusa Mar. Biol. Lab.* 8:89–112.

Parry, G.D., Cohen, B.F., McArthur, M.A. & Hickman, N.J. 2000. Victorian Incursion Management Report Number 2. *Asterias amurensis* incursion in Port Phillip Bay: Status at May 1999. Marine and Freshwater Resources Institute Report No. 19, Marine and Freshwater Resources Institute, Queenscliff, Victoria 21pp.

Pimm, S.L. 1989. Theories of predicting success and impact of introduced species. Pages 351–367 In Drake, J.A., DiCastri, F., Groves, R.H., Kruger, F.J., Mooney, H.A., Rejma´nek, M. & Williamson, M.H. (eds). *Biological invasions: a global perspective*. Wiley, New York, New York, USA.

Ross, D.J., Johnson, C.R., & Hewitt, C.L. 2002. Impact of introduced seastars *Asterias amurensis* on survivorship of juvenile commercial bivalves *Fulvia tenuicostata*. *Marine Ecology Progress Series* 241: 99–112.

Ross, D.J., Johnson, C.R. & Hewitt, C.L. 2003. Variability in the impact of an introduced predator (*Asterias amurensis*: Asteroidea) on soft-sediment assemblages. *Journal of Experimental Marine Biology and Ecology* 288: 257–278.

Turner, E. 2001. Recent spider crab aggregations. *Invertebrata* 20.

Vermeij, G.J. 1991. When biotas meet: understanding biotic interchange. *Science* 253: 1099–1104.

Ecotoxicology and heat stress

Echinoderms in a Changing World – Johnson (ed)
© 2013 Taylor & Francis Group, London, ISBN 978-1-138-00010-0

Echinoderm ecotoxicology: Application for assessing and monitoring vulnerabilities in a changing ocean

M. Byrne

Schools of Medical and Biological Science, University of Sydney, Australia

ABSTRACT: Ocean acidification is a major threat to the Echinodermata because it decreases availability of the carbonate ions required for skeletogenesis. Ocean pH will change simultaneously with ocean warming and hypercapnia and all these stressors are likely to impact vulnerable early life history stages. To address questions on future vulnerabilities, data on thermo—and pH/pCO_2 tolerance of fertilisation and development, largely from single stressor physiology and ecotoxicology studies, are placed in a near future climate-relevant setting. Fertilisation exhibited a broad tolerance to warming or acidification beyond stressor values projected for 2100. Early development exhibited susceptibility to warming. For embryos that develop to the larval stage in a warm ocean, hypercapnia and acidification may impair calcification and other physiological processes. Sensitivities will vary between species. Although climate change is potentially dire for the Echinodermata, some species may have a pre-adaptive capacity and this will influence comparative vulnerability to ocean regime change.

1 INTRODUCTION

Global warming and increased atmospheric CO_2 are causing the oceans to become warmer and acidify (IPCC 2007). Uptake of CO_2 is predicted to decrease ocean pH by 0.2 to 0.4 units and increase sea surface temperature by ca. 2°C by 2100 (Feely *et al.* 2004; IPCC 2007, Fabry *et al.* 2008; but see Whooten *et al.* 2008). For the Echinodermata a high pCO_2 ocean has negative impacts on growth, reproduction and development due to direct effects on metabolism (hypercapnia) and decreased bioavailabilty of the carbonate ions needed for skeleton formation (Fabry *et al.* 2008, Pörtner 2008; Widdecome & Spicer 2008). Increased temperature affects everything an organism does through its pervasive physiological impact on all biological functions. Our understanding of the impacts of ocean warming, acidification and hypercapnia is impeded by the scarcity of empirical data and that many studies use stressor levels well beyond values predicted by climate change models (Reviews: Przeslawski *et al.* 2008; Fabry *et al.* 2008; Dupont & Thorndyke 2009).

Echinoderm early life history stages occur in the water column where climate change stressors are likely to have deleterious impacts on development. Understanding the vulnerabilities of these stages is crucial as we endeavor to predict how planktonic stages will fare in the face of climate change. In a long history of research, numerous studies provide data on the effect of temperature or pH on echinoid, asteroid and holothuroid fertilisation, development, reproduction and recruitment (Byrne 2010). A few studies investigate the interactive effects of ocean warming and acidification on echinoid fertilisation and development (Byrne *et al.* 2009, 2010a,b, 2011a; Sheppard Brennand *et al.* 2010).

Due to their sensitivity to environmental perturbation, echinoderm gametes and embryos, particularly those of echinoids, have long been used as a model system for environmental monitoring (Kobayashi 1983; Bay *et al.* 1993; Carr *et al.* 2006; Byrne *et al.* 2008). Ecotoxicological investigations on the response of life history stages to thermal pollution (eg. power plant effluent) and decreased pH (Greenwood & Bennett 1981; Bay *et al.* 1993; Carr *et al.* 2006) provide insights on contemporary vulnerabilities and tolerance limits to climate change stressors. Data on effects of increased pCO_2 on early life history stages of echinoids are available in studies where treatment of seawater with CO_2 gas is used as a standard method to investigate acid pollution (eg. Carr *et al.* 2006). Echinoderm ecotoxicology has a new focus, to assess the impacts of ocean warming, acidification and hypercapnia on gametes, embryos and larvae (Kurihara & Shirayama 2004; Byrne & Davis 2008; Dupont *et al.* 2009; Havenhand *et al.* 2008; Kurihara 2008; Byrne *et al.* 2009; Dupont & Thorndyke 2009).

The extent of ocean warming and acidification will vary among regions due to differences in ocean circulation, coastal processes and the interaction between temperature and pCO_2 (Levitus *et al.* 2005; Poloczanska *et al.* 2007). This is an important consideration in investigating effects on regional biota. In the setting projected for the Australian climate change hot spot (Poloczanska *et al.* 2007) for instance, ocean warming is a more serious near-future climate change teratogen than acidification for the sea urchin *Heliocidaris erythrogramma* (Byrne *et al.* 2009).

Here the thermo—and pH/pCO_2 tolerance of echinoderm fertilisation and early development is reviewed using data from studies of the ecotoxicology and physiology of intertidal and shallow water echinoids. The focus is on species for which data on both temperature and pH are available and where ambient conditions are provided (Tables 1,2). The latter was essential to identify tolerance thresholds with respect to current conditions and to place these within the context of near-future scenarios (2100: ca. SST + 2–4°C; pH – 0.2–0.4 units, IPCC 2007). Some studies also provide pCO_2 conditions.

Determination of the vulnerabilities and thresholds of developmental stages in response to climate change stressors is essential to identify the most relevant biological processes/stages with which to assess risk and monitor change. For instance, if embryogenesis fails due to warming then the question of comprised larval calcification due to acidification is not relevant. Depending on stressor(s), echinoderm development can be perturbed at any stage. Recent studies show different developmental sensitivities to ocean acidification even within closely related species (Kurihara 2008; Dupont & Thorndyke 2009). Echinoderm ecotoxicology provides a model system to determine where vulnerabilities lie, to monitor impacts of climate change on marine biota and to assess risk to marine ecosystems.

Table 1. Echinoid fertilisation: thermo- and pH tolerance.

Species	Sp temp (°C)	Temp range (°C) Fert ≥ 75%	Upper temp (°C) Fert 60–70%	pH range Fert ≥ 75%	Low pH Fert 60–70%
Strongylocentrotus purpuratus[1]	W USA 13	13–25	30	7.3–8.2	7.0
Hemicentrotus pulcherrimus[2,3]	N Japan, 5–19	0–30	ND	7.4–8.0[14]	7.0
Arbacia lixula[4]	W Eur, 14–19	15–32	>32	ND	ND
Arbacia punctulata[5]	E USA ND	ND	ND	6.9–8.6	6.8
Echinometra mathaei[3,6]	TP 28	28–36	ND	7.6–8.1[15]	7.4
Heliocidaris erythrogramma[7,8]	E Aus, 19–24	18–28	>28	7.6–8.2[16]	ND
Heliocidaris tuberculata[9,10]	E Aus, 17–26	17–28	>28	7.6–8.2[16]	ND

Table 2. Echinoid early development: thermo- and pH tolerance.

Species	Sp temp (°C)	Temp range (°C) Dev ≥ 75%	Upper temp (°C) Dev 60–70%	pH range Dev ≥ 75%	Low pH Dev 60–70%
Strongylocentrotus purpuratus[11,12]	W USA 13	13–20	25–30	7.4–8.4	7.3
Hemicentrotus pulcherrimus[3,13]	N Japan, 5–19	5–24	>24	7.6–8.1*[14]	7.4
Arbacia punctulata[5]	E USA ND	ND	ND	7.4–8.6	7.2–7.4
Echinometra mathaei[6]	T Pacific 28	28–34	36	ND	ND
Heliocidaris erythrogramma[7,8]	E Aus, 19–24	18–26	>28	7.6–8.2[16]	<7.6
Heliocidaris tuberculata[9,10]	E Aus, 17–26	17–24	SV	7.6–8.2[16]	<7.6

[1]Farmanfarmaian & Giese 1963; [2]Mita *et al.* 1984; [3]Kurihara & Shirayama 2004; [4]Hagström & Hagström 1959; [5]Carr *et al.* 2006; [6]Rupp 1973; [7]Byrne *et al.* 2009; [8]Nguyen 2007; [9]O'Connor & Mulley 1977; [10]Byrne, unpub; [11]Bay *et al.* 1993; [12]Fujisawa 1993; [13]Fujisawa 1989,1995; pCO_2ppm: [14]365–2360, [15]365–1000, [16]230–700; *cleavage stages scored.

2 ECHINODERM LIFE HISTORIES IN A CHANGING OCEAN

2.1 Fertilisation: influence of warming and acidification—single stressor studies

A broad thermotolerance of fertilisation is reported for echinoids (Table 1, Rupp 1973; Andronikov 1975; Horstadius 1975; Fujisawa 1989,1995; Sewell & Young 1999) and asteroids (Rupp 1973; Lee *et al.* 2004). High fertilisation rates in sea urchins can occur at temperatures well above ambient maxima (ca. + 4–12°C, Table 1). Fertilisation in many echinoderms is robust to increased temperature and, in many cases, is enhanced (Rupp 1973; Mita *et al.* 1984; Fujisawa 1989; O'Connor & Mulley 1977; Sewell & Young 1999). Thermal enhancement of fertilisation is likely due to positive effects on sperm swimming speeds and heightened sperm-egg collisions (Hagström & Hagström 1959; Mita *et al.* 1984). Thermal tolerance of fertilisation is conveyed by protective maternal factors (eg. heat shock proteins) and the thermal independent period post gamete binding (Fujisawa 1989; Sconzo *et al.* 1997; Yamada & Mihashi 1998). The magnitude of thermotolerance varies among species (Table 1) and is likely to be species specific.

With regard to effects of pH/pCO$_2$ on fertilisation, thresholds for deleterious effects varies among species (Table 1). In *Strongylocentrotus purpuratus*, *Hemicentrotus pulcherrimus*, *Arbacia punctulata* and *Echinometra mathaei*, percent fertilisation dropped (<60%) at pH 7.0, 7.0, 6.8 and

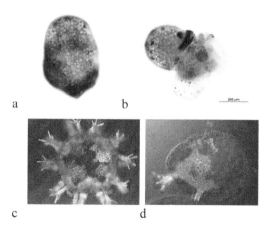

a   b

c   d

Figure 1. Effect of temperature and pH on larval development of the sea urchin Heliocidaris erythrogramma. (a) Normal larva 3 d post fertilisation (dpf). (b) Thermal-effects (+4–6°C SST) larval phenotype 3dpf characteristic of this species (see Kobayashi 1980), (c) Normal juvenile—5dpf. (d) Juvenile (5dpf) reared at +4°C SST and pH 7.6 exhibiting abnormal calcification. Scale bar applies to all images.

7.4, respectively (Table 1). For the holothuroid *Cucumaria frondosa* pH < 7.5 impaired fertilisation (Hamel & Mercier 1996). For the echinoids *Heliocidaris erythrogramma* and *H. tuberculata* high levels of fertilisation were recorded at pH 7.6 (Byrne *et al.* 2009,2010a,b). Low pH retards sperm swimming (Brokaw 1990) and decreased fertilisation in *H. erythrogramma* (pH 8.1: 62%; pH 7.7: 51%) was attributed to impaired sperm motility (Havenhand *et al.* 2008).

Studies on the influence of pH using acid or samples of seawater from field sites known to have low pH also show a broad tolerance of fertilisation in sea urchins (Smith & Clowes 1924; Riveros *et al.* 1996; Kurihara & Shirayama 2004).

The weight of evidence (Table 1) indicates that sea urchin fertilisation is impaired at pH below 7.4 (>1000 ppm CO$_2$), levels well below near future ocean acidification scenarios (IPCC 2007).

2.2 Early development: influence of warming and acidification—single stressor studies

Data on the thermotolerance of echinoderm development is reported for a diverse suite of echinoid, asteroid and holothuroid species (Table 2, Sagara & Ino 1954; Rupp 1973; Andronikov 1975; Fujisawa 1989,1995; Chen & Chen 1992; Sewell & Young 1999; Lee *et al.* 2004; Asha & Muthiah 2005). Comparative data on impacts of temperature or pH/pCO$_2$ tolerance on echinoid development are available for 6 species where pH was adjusted with CO$_2$ gas (Table 2).

Early development in echinoids is tolerant to a ca. + 2–7°C warming and the threshold for <60–70% normal development is ca. + 4–10°C above ambient and this varies among species (Table 2). Thermotolerance range of development is often narrower than that for fertilisation (Table 2). In contrast to the broad thermotolerance of fertilisation, development in *H. erythrogramma* is compromised at +4°C above ambient temperature (ca. 50% normal) and fails at +6°C, indicating vulnerability to near-future ocean warming (Fig. 1a,b). The embryos of some echinoderms (eg. *E. mathaei*, *H. pulcherrimus*, *Asterias amurensis*, *Holothuria spinifera*) however, are highly thermotolerant (Sagara & Ino 1954; Rupp 1973; Kurihara & Shirayama 2004; Lee *et al.* 2004; Asha & Muthiah 2005).

In single stressor studies echinoid embryos are negatively affected by pH < 7.2–7.4 (>1000 ppm CO$_2$) (Table 2). The percentage of normal *S. purpuratus* embryos dropped to 70% at pH 7.3 (Bay *et al.* 2003). These levels of acidification are higher than near future ocean change scenarios. Studies on the influence of pH using acid or environmental samples show a similar tolerance of early development stages to acidification (Smith & Clowes 1924;

Riveros et al. 1996; Kurihara & Shirayama 2004). For the sea cucumber *H. spinifera* pH 7.8 is identified as most suitable for rearing (Asha & Muthiah 2005) and for *C, frondosa*, pH 7.5 is the threshold for deleterious effects (Hamel & Mercier 1996).

2.3 Larval development: influence of warming and acidification—single stressor studies

For those embryos that reach the larval stage, subsequent development may be compromised due to the effect of ocean acidification and hypercapnia on calcification and other physiological processes (Fig. 1c,d). Pluteal skeletons are fragile and may not be produced due to future carbonate under saturation. Echinoplutei of *H. pulcherrimus* reared at pH 7.8 have a shorter arm skeleton, an affect that would negatively impact feeding (Kurihara & Shirayama 2004; Kurihara 2008). Similarly, ophioplutei of *Ophiothrix fragilis* are impacted by acidification (pH 7.7–7.9) with mortality, reduced size and abnormal skeletogenesis (Dupont et al. 2008). However, *Am phiura filiformis* is less sensitive with some larvae (20% less than controls) reaching the juvenile stage (Dupont & Thorndyke 2009). The species-specific nature of the responses to ocean warming and acidification makes it difficult to extrapolate broadly (Dupont & Thorndyke 2009; Byrne 2010; Dupont et al. 2010).

2.4 Interactive effect of warming and acidification on fertilisation and development: multiple stressor studies

While it is important to understand the impact of climate change stressors in isolation to establish based line tolerance thresholds, it is imperative to understand their interactive effects. For echinoderm life histories a series of questions arise with regard to the interactive effect of ocean warming, acidification and hypercapnia: 1) Can fertilisation occur in a warm and high pCO_2/acidic ocean? 2) If fertilisation is successful can the embryos develop? 3) Can embryos reach the larval stage, make their skeleton and reach the juvenile stage? Developmental failure regardless of stage, will cause recruitment failure with flow-on ecosystem effects (Przeslawski et al. 2008). Relatively small perturbations in recruitment can translate to large alterations of adult echinoderm populations (Uthicke et al. 2009), as well demonstrated by the climate change driven range extension of *Centrostephanus rodgersii* (Ling et al. 2008).

Recent multifactorial experiments provide the first data on the interactive effects of ocean warming and acidification on echinoderm life histories (Byrne et al. 2009,2010a,b;2011a). For *H. erythrogramma* increased temperature exerted the major negative effect on development (Fig. 1a,b) with no effect of pCO_2/pH down to pH 7.6. For this species ocean warming is the most serious contemporary and near-future climate change teratogen. Embryos may not reach the skeleton forming stage in a warm ocean, regardless of pH (Byrne et al. 2009,2011).

3 PERSPECTIVES

An understanding of the thermal and pH/pCO2 tolerance of echinoderm life histories is crucial as we endeavour to determine how species may respond to climate change. Tolerance to temperature and decreased pH varies greatly among species. With regard to projected ocean warming (2100), fertilisation and early development of many of the species in this review would be thermotolerant to a +2–4°C change,

Tables 1,2. Temperature and pH tolerance of echinoid fertilization (Table 1) and early development (Table 2) for species where data for both factors are available from single stressor studies. The spawning (Sp) temperatures were obtained from the cited studies or extrapolated from the region and season of the investigation. Experimental pH was adjusted by treatment of seawater with CO_2 gas. Aus, Australia; Dev, early development; Eur, Europe; Fert, fertilisation; ND, no data; SV, seasonally variable; TP, tropical Pacific; Temp, temperature.

But several species showed higher thermotolerance. Our knowledge of tolerance thresholds is largely based on studies of the hardy species used for laboratory research (Tables 1,2). We have a paucity of data on the vast diversity of echinoderms many of which are likely to be vulnerable to ocean change which may already be on the edge due to multiple anthropogenic stressors, particularly pollution and UV (Byrne & Davis 2008; Crain et al. 2008; Przeslawski et al. 2008).

With regard to skeletogenesis predicting the responses to ocean acidification is complicated by the strong influence of temperature on skeleton formation. It has been suggested that increased warming may ameliorate the negative effects of acidification through enhancement and stimulation of the cellular mechanisms underlying calcification (Reynaud et al. 2003; McNeil et al. 2004) and a recent study shows that *A. filiformis* increases calcification at low pH (Wood et al. 2008). Other studies indicated that ocean warming, within limits, can reduce the negative impact of acidification on calcification in developing echinoderms (Sheppard Brennand et al. 2010; Byrne et al. 2011a). More empirical data are needed to place the prospects of calcification of fragile larval and early juvenile skeletons (Fig. 1c,d) in the context of climate change.

The strong influence of adult thermal history, particularly maternal acclimatization and egg imprinting on the thermotolerance of echinoderm fertilisation and development is well established (O'Connor & Mulley 1977; Fujisawa 1989,1995; Johnson & Babcock 1994; Bingham *et al.* 1997). Parental thermal acclimatization in nature dramatically shifts embryonic thermotolerance. This indicates the potential for adaptive phenotypic plasticity with respect to prevailing temperatures. The potential that adaptive phenotypic plasticity may help buffer the negative effects of ocean warming is also suggested for corals (Edmunds & Gates 2008). Widely distributed species may also be able to keep up with a warming world through dispersal of larvae with genotypes adapted to warmer environments as suggested for *Heliodicidaris erythrogramma* (Byrne *et al.* 2011b). In temperate Australia, strong poleward flow of boundary currents may facilitate southerly migration of warm adapted propagules of this species (Byrne *et al.* 2011b).

Although the future appears gloomy for the Echinodermata, the potential for adaptive (genetic) and acclimation (phenotypic plasticity) capacity warrants attention. The in-built tolerance and flexibility (polytopy) of marine invertebrate life histories to environmental stressors is suggested to provide a mechanism to increase the probably of species persistence through geological time (Palmer 1994; Hadfield & Strathmann 1996). Determination of adaptive capacities will provide insights into potential winners and losers in the face of ocean regime change.

ACKNOWLEDGEMENTS

Supported by the Australian Research Council. Assistance provided by N. Soars and E. Woolsey. Thanks to M.Ho, H. Nguyen for photographs.

REFERENCES

Andronikov VB (1975) Heat resistance of gametes of marine invertebrates in relation to temperature conditions under which the species exist. Mar. Biol. 30: 1–11.

Asha PS, Muthiah P (2005) Effects of temperature, salinity and pH on larval growth, survival and development of the sea cucumber *Holothuria spinifera* Theel. Aquaculture 250: 823–829.

Bay S, Burgess R, Nacci D (1993) Status and Applications of Echinoid (Phylum Echinodermata) Toxicity Test Methods. In *Environmental Toxicology and Risk Assessment* (eds W.G Landis, J.S. Hughes and M.A. Lewis), pp. 281–302 Philadelphia: American Society for Testing and Materials.

Bingham BL, Bacigalupi M, Johnson LG (1997) Temperature adaptations of embryos from intertidal and subtidal sand dollars (*Dendraster excentricus*, Wschscholtz). Northwest Sci. 71: 108–114.

Brokaw CJ (1990) The sea urchin spermatozoon. BioEssays 12: 449–452.

Byrne M (2010) Impact of climate change stressors on marine invertebrate life histories with a focus on the Mollusca and Echinodermata. In *Climate alert: Climate Change Monitoring and Strategy*. Y. Yu & A. Henderson-Sellers (eds). Sydney: University of Sydney Press, 142–185.

Byrne M, Davis A (2008) The acid test: responses of benthic invertebrates to climate change. Aust. Antarctic Mag. 15: 8–9.

Byrne M, Oakes DJ, Pollak JK, Laginestra E (2008) Toxicity of landfill leachate to sea urchin development with a focus on ammonia. Cell Biol. Toxicol 24: 503–512.

Byrne M, Ho M, Selvakumaraswamy P, Nguyen HD, Dworjanyn SA, Davis AR (2009) Temperature, but not pH, compromises sea urchin fertilisation and early development under near-future climate change scenarios. Proc Roy Soc B. 276: 1884–1888.

Byrne, M., Soars, N., Selvakumaraswamy, P., Dworjanyn, S.A. & Davis, A.R. 2010a. Sea urchin fertilization in a warm, acidified ocean and high pCO2 ocean across a range of sperm densities. Mar. Env. Res. 69, 234–239.

Byrne M., Soars, N.A., Ho M.A., Wong, E., McElroy D., Selvakumaraswamy P., Dworjanyn, S.A. & Davis, A.R. 2010b. Fertilization in a suite of coastal marine invertebrates from SE Australia is robust to near-future ocean warming and acidification. Mar. Biol. 157: 2061–2069.

Byrne, M., Ho, M.A., Wong, E., Soars, N., Selvakumaraswamy, P., Sheppard Brennand, H., Dworjanyn, S.A. & Davis, A.R. 2011a. Unshelled abalone and corrupted urchins, development of marine calcifiers in a changing ocean. Proc Roy Soc B. (in press).

Byrne, M., Selvakumaraswamy, P., Ho, M.A. & Nguyen, H.D. 2011b. Sea urchin development in a global change hot spot, potential for southerly migration of thermotolerant propagules. Deep Sea Research II 58: 712–719.

Carr RS, Biedenbach JM, Nipper M (2006) Influence of potentially confounding factors on sea urchin porewater toxicity tests. Arch. Environ. Contam. Toxicol. 51: 573–579.

Chen CP, Chen BY (1992) Effects of high-temperature on larval development and metamorphosis of *Arachnoids placenta* (Echinodermata, Echinoidea). Mar. Biol. 112: 445–449.

Crain CM, Kroeker K, Halpern BS (2008) Interactive and cumulative effects of multiple human stressors in marine systems. Ecology Letters 11: 1304–1315.

Dupont S, Havenhand J, Thorndyke W, Peck L, Thorndyke M. (2008) Near-future level of CO_2-driven ocean acidification radically affects larval survival and development in the brittlestar *Ophiothrix fragilis*. Mar. Ecol. Prog. Ser. 373: 285–294.

Dupont S, Thorndyke MC (2009) Ocean acidification and its impact on the early life-history stages of marine animals. CIESM Monograph.

Dupont, S., Ortega-Martínez, O. & Thorndyke, M.C. 2010. Impact of near future ocean acidification on echinoderms. *Ecotoxicology* 19: 440–462.

Edmunds PJ, Gates RD (2008) Acclimatization in tropical coral reefs. Mar. Ecol. Prog. Ser. 361: 307–310.

Fabry VJ, Seibel BA, Feely RA, Orr JC (2008) Impacts of ocean acidification on marine fauna and ecosystem processes. ICES J. Mar. Sci. 65: 414–432.

Farmanfarmaian A, Giese AC (1963) Thermal tolerance and acclimation in the western purple sea urchin, *Strongylocentrotus purpuratus*. Physiol. Zool. 36: 237–343.

Feely R, Sabine, CL, Lee K, Berelson W, Kleypas J, Fabry VJ, Millero FJ (2004) Impact of anthropogenic CO_2 on the $CaCO_3$ systems in the oceans. Science 305: 362–366.

Fujisawa H (1989) Differences in temperature dependence of early development of sea urchins with different growing seasons. Biol. Bull. 176: 96–102.

Fujisawa H (1993) Temperature sensitivity of a hybrid between two species of sea urchin differing in thermotolerance. Develop. Growth Differ. 35: 395–401.

Fujisawa H (1995) Variation in embryonic temperature sensitivity among groups of the sea urchin, *Hemicentrotus pulcherrimus*, which differ in their habitats. Zool. Sci. 12: 583–589.

Greenwood PJ, Bennett T (1981) Some effects of temperature-salinity combinations on the early development of the sea urchin *Parechinus angulosus* (Leske) fertilization. Mar. Biol. Ecol. 51: 119–131.

Hadfield MG, Strathmann MF (1996) Variability, flexibility and plasticity in life histories of marine invertebrates. Oceanol Acta 19: 323–324.

Hagström BE, Hagström B (1959) The effect of decreased and increased temperatures on fertilization. Exp. Cell Res. 16: 174–183.

Hamel JF, Mercier A (1996) Early development, settlement, growth and spatial distribution of the sea cucumber *Cucumaria frondosa* (Echinodermata: Holothuroidea). Can. J. Fish. Aquat. Sci. 53: 253–271.

Havenhand JN, Butler FR, Thorndyke, MC, Williamson JE (2008) Near-future levels of ocean acidification reduce fertilisation success in a sea urchin. Current Biology 18: 651–652.

Hörstadius S (1975). A note on the effect of temperature on sea urchin eggs. J. Exp. Mar. Biol. Ecol. 18: 239–242.

IPCC 2007 Intergovernmental Panel on Climate Change 2007 The fourth assessment report of the IPCC. Cambridge, UK: Cambridge University press.

Johnson LG, Babcock RC (1994) Temperature and the larval ecology of the crown-of-thorns starfish, *Acanthaster planci*. Biol. Bull. 168: 419–43.

Kobayashi N (1980) Comparative sensitivity of various developmental stages of sea urchins to some chemicals. Mar. Biol. 58: 163–171.

Kurihara H, Shirayama Y (2004) Effects of increased atmospheric CO_2 on sea urchin early development. Mar. Ecol. Prog. Ser. 274: 161–169.

Kurihara H (2008) Effects of CO_2-driven ocean acidification on the on the early development stages of invertebrates. Mar. Ecol. Prog. Ser. 275: 275–284.

Lee C-H, Ryu T-K, Choi J-W (2004) Effects of water temperature on embryonic development in the northern Pacific asteroid, *Asterias amurensis*, from the southern coast of Korea. Inv. Rep. Dev. 45: 109–116.

Levitus S, Antonov J, Boyer T (2005) Warming of the world ocean. Geophys. Res. Lett. 32: L02604.

Ling SD, Johnson CR, Frusher SD, King CK (2008) Reproductive potential of a marine ecosystem engineer at the edge of a newly expanded range. Global Change Biology 14: 1–9.

McNeil BI, Matear RJ, Barnes DJ (2004) Coral reef calcification and climate change: The effect of ocean warming. Geophys. Res. Lett. 31: L22309.

Mita M, Hino A, Yasumasu I (1984) Effect of temperature on interaction between eggs and spermatozoa of sea urchin. Biol. Bull. 166: 68–77.

Nguyen HD (2007) Effect of elevated temperature on hsp 70 expression and development in congeneric rocky shore sea urchins. BSc Thesis, University of Sydney.

O'Conner C, Mulley JC (1977) Temperature effects on periodicity and embryology, with observations on the population genetics, of the aquacultural echinoid *Heliocidaris tuberculata*. Aquaculture 12: 99–114.

Palmer AR (1994) Temperature sensitivity, rate of development, and time to maturity: geographic variation in laboratory reared *Nucella* and a cross-phyletic overview. In: Reproduction and Development of Marine Invertebrates (eds Wilson WH Jr., Stricker SA, Shinn GL) pp. 177–194. Johns Hopkins University Press, Baltimore.

Poloczanska ES, Babcock RC, Butler A, Hobday AJ, Hoegh-Guldberg O, Kunz TJ, Matear RJ, Milton DA, Okey TA Richardson AJ (2007) Climate change and Australian marine life. Oceanog. Mar. Biol. Ann. Rev. 45: 407–478.

Pörtner HO (2008) Ecosystem effects of ocean acidification in times of ocean warming: a physiologist's view. Mar. Ecol. Prog. Ser. 373: 203–217.

Przeslawski R, Ahyong S, Byrne M, Wörheide G, Hutchings P. (2008) Beyond corals and fish: the effects of climate change on non-coral benthic invertebrates of tropical reefs. Glob. Change Biol. 14: 2773–2795.

Reynaud S, Leclercq N, Romaine-Lioud S, Ferrier-Pagés C, Jaubert J, Gattuso JP (2003) Interacting effects of CO_2 partial pressure and temperature on photosynthesis and calcification in a scleractinian coral. Glob. Change Biol. 9: 1660–1668.

Riveros A, Zuñiga M, Larrain A, Becerra J (1996) Relationships between fertilization of the Southeastern Pacific sea urchin *Arbacia spatuligera* and environmental variables in polluted coastal waters. Mar. Ecol. Prog. Ser. 134: 159–169.

Rupp JH (1973) Effects of temperature on fertilization and early cleavage of some tropical echinoderms, with emphasis on *Echinometra mathaei*. Mar. Biol. 23: 183–189.

Sagara JI, Ino T (1954) The optimum temperature and specific gravity for the bipinnaria and young of the Japanese starfish Asterias amurensis Lütken. Bull. Jap. Soc. Fish. Sci. 20: 689–693.

Sconzo G, Amore G, Capra G, Giudice G, Cascino D, Ghersi G (1997) Identification and characterization of a constitutive HSP75 in sea urchin embryos. Biochem. Biophys. Res. Commun. 234: 24–29.

Sewell MA, Young CM (1999) Temperature limits to fertilization and early development in the tropical sea urchin *Echinometra lucunter*. J. Exp. Mar. Biol. Ecol. 236: 291–305.

Sheppard Brennand, H., Soars, N., Dworjanyn, S.A., Davis, A.R. & Byrne, M. 2010. Impact of ocean warming and ocean acidification on larval development and calcification in the sea urchin *Tripneustes gratilla*. PlosOne 5: e11372.

Smith HW, Clowes GHA (1924) The influence of hydrogen ion concentration on the fertilization process in *Arbacia*, *Asterias* and *Chaetopterus* eggs. Biol. Bull. 47: 333–334.

Uthicke S, Schaffelke B, Byrne M (2009) A boom and bust phylum? Ecological and evolutionary consequences of large population density variations in echinoderms. Ecol. Monogr. 79: 3–24.

Widdicombe S, Spicer JI (2008) Predicting the impact of ocean acidification in benthic biodiversity: What can animal physiology tell us? J. Exp. Mar. Biol. Ecol. 366: 187–97.

Whooten JT, Pfister CA, Forester JD (2008) Dynamic patterns and ecological impacts of declining ocean pH in a high-resolution multi-year dataset. Proc. Natl. Acad. Sci. USA 105: 18848–18853.

Wood HL, Spicer JI, Widdicombe S (2008) Ocean acidification may increase calcification rates, but at cost. Proc. R. Soc. B. 275: 1767–1773.

Yamada K, Mihashi K (1998) Temperature-independent period immediately after fertilization in sea urchin eggs. Biol. Bull. 195: 107–111.

Echinoderms in a Changing World – Johnson (ed)
© 2013 Taylor & Francis Group, London, ISBN 978-1-138-00010-0

Hsp70 expression in the south-eastern Australian sea urchins *Heliocidaris erythrogramma* and *H. tuberculata*

H.D. Nguyen & M. Byrne
Discipline of Anatomy & Histology, Bosch Institute, University of Sydney, Australia

M. Thomson
School of Biological Sciences, University of Sydney, Australia

ABSTRACT: Heat shock protein (hsp) expression was used as a biomarker to compare the thermal stress response to acute temperature fluctuations in two closely related Australian sea urchins inhabiting different thermal environments, the intertidal *Heliocidaris erythrogramma* and shallow subtidal *H. tuberculata*. Urchins were collected in summer and held in the laboratory at ambient sea surface temperature (22°C) for at least 3 days. Hsp70 expression was examined in *H. erythrogramma* and *H. tuberculata* adults following acute exposure to 18°C, 22°C (control), 25°C and 30°C. Hsp70 expression was similar between species where levels at 30°C were significantly elevated compared to 22°C (control). The increase in hsp70 expression at 30°C was especially marked in *H. erythrogramma* and may be a strategy to buffer against subsequent temperature extremes as often encountered in the intertidal environment.

1 INTRODUCTION

Rocky intertidal habitats are among the most environmentally extreme and variable marine habitats where organisms experience a wide range of thermal conditions due to the tidal cycle (Helmuth 1998, Hofmann 1999, Helmuth 2002). For intertidal marine invertebrates aerial conditions experienced during low tide are physiologically challenging because body temperatures may increase by 10°C–20°C (Hofmann & Somero 1995, Roberts et al. 1997, Buckley et al. 2001, Helmuth et al. 2005, Denny & Harley 2006). In contrast during submersion, body temperatures remain relatively stable and similar to that of sea temperature (Helmuth & Hofmann 2001). Organisms in rock pools also experience temperatures well above or below sea temperature during low tide (Hofmann et al. 2002). At low tide, the intertidal is characterized by a steep thermal stress gradient which increases with shore height (Roberts et al. 1997, Helmuth et al. 2006). Species living higher up the shore are expected to deal with a greater amount of thermal stress than their lower shore counterparts (Roberts et al. 1997, Hofmann 1999). Thus organisms inhabiting different heights of the shore may have developed different physiological strategies to cope with temperature fluctuations.

The heat shock protein (hsp) response has been widely used as a tool to assess organism thermal physiology and response to thermal stress

(Hofmann et al. 2002, Hofmann 2005, Tomanek 2008 for reviews). Hsps are molecular chaperones expressed in higher concentrations when an organism is under physiological stress. Under normal physiological conditions hsps play important roles in protein homeostasis such as transporting proteins across cells, replacing denatured proteins, preventing protein aggregation, and degrading misfolded or aggregated proteins (Parsell & Lindquist 1993, Hofmann & Somero 1995, Feder & Hofmann 1999, Tomanek & Somero 2000). Proteins are thermally sensitive biomolecules and exposure to temperatures outside their thermal stability range corrupts their configuration, rendering them dysfunctional (Somero 1995). Because of their role in stabilising proteins at risk of damage and replacing damaged proteins, hsps contribute to an organism's thermotolerance and so expression levels of hsps provide important insights as to how species may respond to thermal change (Parsell & Lindquist 1993, Feder & Hofmann 1999). There are a number of different heat shock proteins that are named according to their molecular weight (Lewis et al. 1999). The amino acid sequence of hsp70 has been highly conserved throughout evolution such that antibodies raised to hsp70 in one species often display a high cross reactivity to the protein in a different species. This has allowed for the study of hsp70 expression in a wide range of animals (Dahlhoff 2004).

Thus far the use of hsp markers to document thermal ecophysiology in intertidal invertebrates has mainly focused on species in the Northern Hemisphere. Here we investigated hsp70 expression in *Heliocidaris erythrogramma* (Valenciennes) and *H. tuberculata* (Lamarck) sea urchins that are sympatric in south-eastern Australia. Local populations of *H. erythrogramma* are largely intertidal occurring in rock pools or surge gutters at low tide while *H. tuberculata* occurs in intertidal and shallow subtidal habitats (Laegdsgaard et al. 1991, Keesing 2007). These species are important algal grazers in rocky shore habitats (Laegdsgaard et al. 1991, Wright et al. 2005, Keesing 2007). We compared hsp expression between these closely related urchins to provide phylogentically robust insights into thermal tolerance limits in species occupying different thermal habitats. Hsp70 expression was investigated with respect to the thermal conditions these species normally experience in the field. The thermal conditions that they experience were documented through the use of in situ temperature loggers.

2 MATERIALS AND METHODS

2.1 *Animal collection*

Heliocidaris erythrogramma (test diameter 50–65 mm) and *H. tuberculata* (test diameter 50–90 mm) were collected in January 2008 (summer) from intertidal and shallow subtidal (0.5 m) habitats in the Sydney region, at Little Bay (33°58'S, 151°15'E), Chowder Bay (33°50'S, 151°15'E) and Fairlight Beach (33°47'S, 151°16'E). Sea urchins were transported to the laboratory where they were maintained in aquaria at ambient sea surface temperature (SST = 22°C) for 3 days prior to exposure to temperature treatments.

2.2 *Field temperature data*

The thermal environment of *H. erythrogramma* and *H. tuberculata* was recorded using temperature data loggers (iB-Cod submersible logger, Type G) deployed in one shallow subtidal site and a low intertidal rock pool. Temperature measurements were taken every 20 min in summer (December– February). Data for the low intertidal rock pool from 10 Jan–7 Feb is not available as the logger went missing.

2.3 *Heat shock and tissue collection*

The temperature shock treatments were designed to reflect the thermal conditions that the urchins experience with changing tides—a short exposure to decreased or elevated temperatures at low tide, followed by immersion at ambient SST.

In summer ambient SST was 22°C and the experimental temperatures were 18°C, 25°C and 30°C representing –4°C, +3°C and +8°C below or above ambient SST respectively. Low intertidal rock pools during a midday low tide currently reach a maximum of 28°C in summer. The highest temperature used, 30°C, is 2°C above the maximum temperature experienced at low tide in summer (see below) and was used in the experiments to assess if the urchins could upregulate hsp expression at a temperature outside their normal thermal regime. In the control group, urchins were maintained at ambient SST for 90 min while in the experimental treatments, urchins were temperature shocked for 30 min then returned to the control temperature (ambient SST) for 1 h. Ten specimens of each species were used in each treatment.

All urchins were viable after the experimental treatments as indicated by the presence of active tube feet and intact spines. After recovery in the control temperature, sea urchins were dissected on ice and gonad samples were frozen immediately in liquid nitrogen and stored at –80°C until use. The control group urchins were similarly dissected. Gonads were used because of the ease of sampling and to ensure sufficient yield of protein.

2.4 *Heat shock protein analysis*

Hsp70 expression was examined using the western blot technique. Total protein was extracted by homogenising tissue samples with a Teflon micropestle with the addition of 100 µl of homogenisation buffer (50 mM Tris–HCl pH 7.4, 1% Nonidet P40, 0.25% (w/v) sodium deoxycholate, 150 mM NaCl, 1 mM Na_3VO_4, 1 mM PMSF, 1 mg/mL leupeptin and 1 mM NaF). The homogenates were then centrifuged for 15 min at 13200 rpm and supernatants were then collected for protein content estimation and gel electrophoresis. Protein content was estimated using the Pierce BCA Protein Assay according to the manufacturer's protocol.

Protein samples were diluted in electrophoresis sample buffer (50 mMTris–HCl pH6.8, 2% SDS w/v, 10% glycerol and 5% β-mercaptoethanol) to a concentration of 2 mg/ml. Proteins were separated by gel electrophoresis using the Biorad Mini Protean III system. Samples, each containing 10 µl of protein, were loaded in 4–20% Tris Glycine gels and run at 100 V for approximately 2 h. In each gel, one lane was loaded with 7 µl of 1:100 diluted recombinant rat Hsp70 protein (Stressgen) to provide a positive control. Separated proteins were transferred to Polyvinyldine Difluoride (PVDF) membranes using the Bio-Rad Mini Trans-Blot Cell according to the manufacturer's instructions.

The membranes with transferred proteins were blocked by immersion in 5% solution of non-fat milk powder in Phosphate-Buffered Saline (PBS) for 20 min. They were then incubated in a solution of monoclonal anti-mouse hsp70 antibody (Sigma-Aldrich) at a 1:1000 dilution in antibody buffer (PBS containing 1% w/v gelatin) for 1 h. The primary antibody used probed for both the inducible (hsp72) and constitutive forms (hsp73) of hsp70. The membranes were then rinsed twice (ca. 2 sec each) with PBS containing 0.05% (v/v) Tween-20 (TPBS) and further washed four times in TPBS (5 min each). After washing, the membranes were incubated for 1 h with secondary antibody, anti-mouse IgG conjugated with horseradish peroxide (Santa Cruz Biotechnology) at a 1:10000 dilution, and rinsed and washed in TPBS as before. The antibody concentrations producing optimal signal to background intensity were determined in a pilot study.

Finally the membranes were incubated in chemiluminescent substrate (Pierce, west pico) for 1 min and the chemiluminescent signal was detected using Kodak X-Omat LS film. Exposure times of the blots to the film ranged from 10 s to 18 h to obtain optimum signal to background intensity. The films were scanned on a Bio-Rad GS-710 Imaging Densitometer and band intensity was analysed using ImageJ software (NIH). In each western blot, the intensity of the hsp70 bands were normalised against the intensity of the positive control to obtain relative hsp70 level expressed in each urchin. This allowed for multiple western blots to be compared.

2.5 Statistical analysis

A 2-way factorial ANOVA was used to examine the effects of temperature and species on hsp70 expression. Data were $\ln(x+1)$ transformed to meet the assumption of homogeneity of variances. Student Newman-Keuls test was used for *post hoc* comparisons on significant effects. The program used for all statistical tests was SPSS 17.0.

3 RESULTS

3.1 Habitat and temperature

At the collection sites *H. erythrogramma* lives in gutters and shallow rock pools where at low tide, they are often barely covered with water (Fig. 1a). *H. tuberculata* by contrast at these sites is generally covered by at least 1 m of water at low tide (Fig. 1b). Through the summer the temperatures recorded in the low intertidal site occupied by *H. erythrogramma* ranged from 16–28°C (Fig. 2). These were more variable than the temperatures in the shallow subtidal site occupied by *H. tuberculata* which ranged from 17–24°C (Fig. 2).

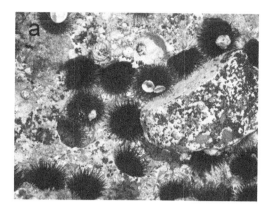

Figure 1. (a) *H. erythrogramma* in a low intertidal rock pool. (b) *H. tuberculata* in the shallow subtidal.

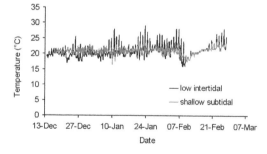

Figure 2. Water temperature monitored in the shallow subtidal and low intertidal rock pool using data loggers in summer (December–February).

3.2 Hsp70 expression

Hsp70 expression was similar between *H. erythrogramma* and *H. tuberculata* with lowest levels at 22°C (control) and a general increase in expression with increasing temperature (Fig. 3). Hsp70 expression peaked at 30°C but the increase was especially marked in *H. erythrogramma*. Both species also

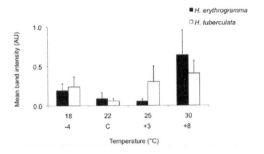

Figure 3. Mean relative hsp70 level (+S.E.) expressed in *H. erythrogramma* (black bars) and *H. tuberculata* (white bars) at different temperature treatments (n = 10).

expressed more hsp70 in the cool (18°C) treatment compared with controls.

ANOVA showed no difference in hsp70 expression between *H. erythrogramma* and *H. tuberculata* ($F_{(3,79)} = 0.6$, $P > 0.05$). However there was a significant effect of temperature on hsp70 expression for both species ($F_{(3,79)} = 3.27$, $P < 0.05$) where levels at 30°C were significantly greater than at 22°C (control; Fig. 3).

4 DISCUSSION

The expression patterns of hsp70 in response to acute temperature shock in southern hemisphere echinoids were documented here for the first time. The low levels of hsp70 in *H. erythrogramma* and *H. tuberculata* across the 18–25°C treatments suggests they are thermotolerant to temperatures within their thermal regime in the low intertidal (16–28°C) and subtidal (17–24°C) respectively.

The significantly higher hsp70 expression at 30°C in *H. erythrogramma* and *H. tuberculata* suggests this is a stressful temperature for the species but also indicates both species are able to upregulate hsp70 at temperatures above their thermal regime. The higher hsp70 levels in *H. erythrogramma* compared to those in *H. tuberculata* at 30°C may reflect the difference in their thermal environments associated with their shore height distribution. As an intertidal species, *H. erythrogramma* has a thermal environment that is more variable throughout the tidal cycle than its subtidal congener. Therefore the higher expression in *H. erythrogramma* at 30°C may be a buffering strategy to cope with temperature extremes that may be encountered over the following tidal cycle.

Here we provided preliminary insights into the heat shock response of *H. erythrogramma* and *H. tuberculata*. Our results show that both species have the physiological capacity to upregulate the hsp protective mechanisms in response to acute

warming under thermal conditions they experience currently. However temperature rises brought about by climate change could exceed their physiological tolerances and consequently have implications for the persistence of their populations. For intertidal marine invertebrates, not only will they be challenged by gradual warming of the oceans, but also by acute aerial extreme events that will affect the thermal conditions they experience during low tide (Tomanek 2008).

Although some characteristics of the heat shock response (e.g. threshold induction temperature) have been demonstrated to shift with thermal history (Tomanek & Somero 1999, Hofmann & Somero 1995, Buckley et al. 2001, Helmuth & Hofmann 2001, Hamdoun et al. 2003, Osovitz & Hofmann 2005), other characteristics of the response are suggested to be genetically fixed (Tomanek & Somero, 1999). Thus the extent to which organisms can modify the heat shock response following acclimation and subsequent heat shock can potentially provide valuable insight into how intertidal marine organisms will fare in near future climate change. *H. erythrogramma* and *H. tuberculata* may be particularly vulnerable to the potential impacts of climate change as their distributions coincide with the South-East Australian climate change hotspot where a 2–3°C warming of coastal waters by 2070 is projected (Poloczanska et al. 2007, CSIRO Mk 3.5 Climate Model).

ACKNOWLEDGEMENTS

Supported by the Australian Government Department of Climate and the Australian Research Council. Many thanks to the members of the Byrne Lab for assistance with animal and sea water collection.

REFERENCES

Buckley, B.A., Owen, M.E. & Hofmann, G.E. 2001. Adjusting the thermostat: The threshold induction temperature for the heat-shock response in intertidal mussels (genus *mytilus*) changes as a function of thermal history. *Journal of Experimental Biology* 204: 3571–3579.

Dahlhoff, E.P. 2004. Biochemical indicators of stress and metabolism: Applications for marine ecological studies. *Annual Review of Physiology* 66: 183–207.

Denny, M.W. & Harley, C.D.G. 2006. Hot limpets: Predicting body temperature in a conductance-mediated thermal system. *Journal of Experimental Biology* 209: 2409–2419.

Feder, M.E. & Hofmann, G.E. 1999. Heat-shock proteins, molecular chaperones, and the stress response: Evolutionary and ecological physiology. *Annual Review of Physiology* 61: 243–282.

Hamdoun, A.M., Cheney, D.P. & Cherr, G.N. 2003. Phenotypic plasticity of hsp70 and hsp70 gene expression in the pacific oyster (crassostrea gigas): Implications for thermal limits and induction of thermal tolerance. *Biological Bulletin* 205: 160–169.

Helmuth, B. 1998. Intertidal mussel microclimates: Predicting the body temperature of a sessile invertebrate. *Ecological Monographs* 68: 51–74.

Helmuth, B. 2002. How do we measure the environment? Linking intertidal thermal physiology and ecology through biophysics. *Integrative and Comparative Biology* 42: 837–845.

Helmuth, B.S.T. & Hofmann, G.E. 2001. Microhabitats, thermal heterogeneity, and patterns of physiological stress in the rocky intertidal zone. *Biological Bulletin* 201: 374–384.

Helmuth, B., Kingsolver, J.G. & Carrington, E. 2005. Biophysics, physiological ecology, and climate change: Does mechanism matter? *Annual Review of Physiology* 67: 177–201.

Helmuth, B., Mieszkowska, N., Moore, P. & Hawkins, S.J. 2006. Living on the edge of two changing worlds: Forecasting the responses of rocky intertidal ecosystems to climate change. *Annual Review of Ecology, Evolution, and Systematics* 37: 373–404.

Hofmann, G.E. 1999. Ecologically relevant variation in induction and function of heat shock proteins in marine organisms. *American Zoologist* 39: 889–900.

Hofmann, G.E. 2005. Patterns of hsp gene expression in ectothermic marine organisms on small to large biogeographic scales. *Integrative and Comparative Biology* 45: 247–255.

Hofmann, G.E. & Somero, G.N. 1995. Evidence for protein damage at environmental temperatures: Seasonal changes in levels of ubiquitin conjugates and hsp70 in the intertidal mussel *mytilus trossulus*. *Journal of Experimental Biology* 198: 1509–1518.

Hofmann, G.E., Buckley, B.A., Place, S.P. & Zippay, M.L. 2002. Molecular chaperones in ectothermic marine animals: Biochemical function and gene expression. *Integrative and Comparative Biology* 42: 808–814.

Keesing, J. 2007. Ecology of *Heliocidaris erythrogramma*. In J.M. Lawrence (ed.), *Edible Sea Urchins: Biology and Ecology* Amsterdam: Elsevier Science.

Laegdsgaard, P., Byrne, M. & Anderson, D.T. 1991. Reproduction of sympatric populations of *Heliocidaris erythrogramma* and *H. tuberculata* (echinoidea) in new south wales. *Marine Biology* 110: 359–374.

Lewis, S., Handy, R.D., Cordi, B., Billinghurst, Z. & Depledge, M.H. 1999. Stress proteins (hsp's): Methods of detection and their use as an environmental biomarker. *Ecotoxicology* 8: 351–368.

Osovitz, C.J. & Hofmann, G.E. 2005. Thermal history-dependent expression of the hsp70 gene in purple sea urchins: Biogeographic patterns and the effect of temperature acclimation. *Journal of Experimental Marine Biology and Ecology* 327: 134–143.

Parsell, D.A. & Lindquist, S. 1993. The function of heat-shock proteins in stress tolerance: Degradation and reactivation of damaged proteins. *Annual Review of Genetics* 27: 437–496.

Poloczanska, E.S., Babcock, R.C., Butler, A., Hobday, A., Hoegh-Guldberg, O., Kunz, T.J., Matear, R., Milton, D.A., Okey, T.A. & Richardson, A.J. (2007). Climate change and australian marine life. *Oceanography and Marine Biology: An Annual Review* 45: 407–478.

Roberts, D.A., Hofmann, G.E. & Somero, G.N. 1997. Heat-shock protein expression in *mytilus californianus*: Acclimatization (seasonal and tidal-height comparisons) and acclimation effects. *Biological Bulletin* 192: 309–320.

Somero, G.N. 1995. Proteins and temperature. *Annual Review of Physiology* 57: 43–68.

Tomanek, L. 2008. The importance of physiological limits in determining biogeographical range shifts due to global climate change: The heat-shock response. *Physiological and Biochemical Zoology* 81: 709–717.

Tomanek, L. & Somero, G.N. 1999. Evolutionary and acclimation-induced variation in the heat-shock responses of congeneric marine snails (genus *tegula*) from different thermal habitats: Implications for limits of thermotolerance and biogeography. *Journal of Experimental Biology* 202: 2925–2936.

Tomanek, L. & Somero, G.N. 2000. Time course and magnitude of synthesis of heat-shock proteins in congeneric marine snails (genus *tegula*) from different tidal heights. *Physiological and Biochemical Zoology* 73: 249–256.

Wright, J.T., Dworjanyn, S.A., Rogers, C.N., Steinberg, P.D., Williamson, J.E. & Poore, A.G.B. 2005. Density-dependent sea urchin grazing: Differential removal of species, changes in community composition and alternative community states. *Marine Ecology-Progress Series* 298: 143–156.

Echinoderms in a Changing World – Johnson (ed)
© 2013 Taylor & Francis Group, London, ISBN 978-1-138-00010-0

Sediment copper bioassay for the brittlestar *Amphiura elandiformis*—technique development and management implications

C.K. Macleod, R.S. Eriksen & L. Meyer
TAFI, University of Tasmania, Marine Research Laboratories, Nubeena Crescent, Taroona, Tasmania, Australia

ABSTRACT: Copper contamination is a concern in many marine and estuarine systems. In the Derwent estuary levels exceed ANZECC sediment quality guidelines in many areas. This study examined copper toxicity in the brittlestar *Amphiura elandiformis* with a view to relating the findings to environmental observations and ecosystem management. Knowing the concentration of copper which adversely affects brittlestars would help managers to interpret the significance of presence/absence data collected during monitoring, and to better understand ecological impacts and the current status of sediments. Species specific sediment bioassay protocols were developed to accommodate the larger than normal amounts of sediment required for brittlestar testing. Acute exposure tests undertaken in spiked and "naturally" contaminated sediments indicated a negative response (e.g. lethargy, surfacing, autotomy) at and above 165 mg Cu/kg. These findings suggest that where sediment copper concentrations exceed 270 mg/kg (ANZECC high guideline value) *A. elandiformis* distribution may be affected.

1 INTRODUCTION

Sediments are a major repository for contaminants that enter the marine environment (Batley and Maher 2001). In urban areas heavy metal levels are often markedly increased as a result of direct industry inputs, stormwater, and as a result of leaching and deposition of antifoulant material associated with shipping (Ranke 2002). The ecological effects of heavy metals are of concern to environmental managers, industry and government stakeholders, and the wider community.

Copper and other contaminants in the marine environment tend to sorb to fine grained particles with a large surface area. Consequently the highest levels of contamination in marine sediments are generally associated with depositional areas where fine-grained material is accumulated in sheltered, low-energy areas (Morrisey et al. 1996, NRC 1997, Matthai and Birch 2001). In the Derwent estuary heavy metals are a significant pollutant issue. The estuary is strongly depositional, with soft sediments throughout most of the estuary. These sediments act as a sink for contaminants to the system and the benthic infauna reflect the cumulative effects of these pollutants.

Despite considerable management improvements in recent years, there are still significant loads of heavy metals in sediments throughout the estuary, with copper, lead, cadmium, mercury, zinc and arsenic all exceeding Interim Sediment Quality Guidelines (ISQG) (ANZECC 2000). Copper levels in the Derwent are primarily a result of industry and stormwater inputs, although there may also be "hotspots", where the input from antifoulants is significant. Levels range from <5 mg/kg to in excess of 2000 mg/kg Dry Matter Basis (DMB); with levels in the mid-estuary well above those previously reported as having a toxic effect (DEP 2007), and well in excess of the ISQG—High value of 270 mg/kg for marine/estuarine sediments, a level above which biological effects are expected (Table 1).

Table 1. Summary of ANZECC guidelines for copper in marine/estuarine sediments with data from other regions included for comparison.

Guideline	Descriptor	Trigger value (mg/kg)
ANZECC (Australia/NZ)	ISQG low	65
	ISQG high	270
NOAA (USA)	ERL	34
	ERM	270
FDEP (USA)	TEL	18.7
	PEL	108

ISQG—Interim Sediment Quality Guideline, ERL—Effects Range Low, ERM—Effects Range Medium, TEL—Threshold Effects Level, PEL—Probable Effects Level.

Benthic infauna plays a critical role in establishing and maintaining sediment health. Numerous studies have shown that elevated levels of heavy metals can have a detrimental effect on the local ecology (e.g. Conradi and Depledge 1999, Marsden 2002). Metals are partitioned between the solid and pore water phases within the sediment, and are available to benthic organisms both from the solid phase (through sediment contact and ingestion) and via exposure to the aqueous phase (through leaching from the porewater or through exposure to overlying water through burrow irrigation etc). Porewater contained in the interstitial spaces between sediment particles may comprise up to 50% of the sediment volume (Batley and Maher 2001). The extent to which any particular organism is exposed to heavy metals will depend on its habitat preferences, feeding strategies and life history (i.e. burrowing vs epibenthic). Porewater exposure is a major pathway for uptake of metals by marine benthic fauna. Solid phase exposure is restricted to animals that either ingest, bioturbate or filter particulate material from the sediments, whilst porewater metals are available to any organisms that burrows within the sediments (Chapman et al. 2002). Brittlestars are suspension feeders that commonly bury within surficial sediment, and may also spend time fully exposed on the surface, and are thus likely to be susceptible to elevated metals by both the solid phase and aqueous pathways.

When sediments accumulate heavy metals the effect on the fauna can be either acute or chronic (Kennish 1996). With acute effects species may be eliminated or their numbers may be greatly reduced, allowing more tolerant species to colonise the impacted area. Such changes in the ecology of an area will affect the natural balance of the system, and may in turn affect the ability of the ecosystem to function.

Until recently Australian research on field effects of heavy metal contamination has largely focused on community impacts rather than effects on individual species (Lindegarth and Underwood 2002). These field surveys provide important information on the changes that have occurred under a particular set of conditions, however, the information is largely retrospective and they provide little predictive information on the mechanisms of effect. Targeted ecotoxicological studies do provide specific information on the direct effects of the contaminants tested to the test species. However, these studies also have limitations; they are generally laboratory based, focused on standard test organisms and constrained by highly defined test conditions. As a result it is often difficult to extrapolate the findings directly to natural communities where there may be other mechanisms influencing the impact. Although there is considerable

information on toxicity of copper to standard test species, there is insufficient information on the effects on the endemic Australian fauna and less on species representative of SE Australian or Derwent benthic communities. Consequently there is a need for locally relevant ecological indicators and ecotox testing under natural conditions.

A number of marine tests have been developed for locally relevant species on mainland Australia. However, there are no such tests for assessment of Tasmanian conditions. Reviews of previous faunal surveys conducted in the Derwent estuary (Macleod and Helidoniotis 2005, Banks 2007) and elsewhere in SE Tasmania (Macleod et al. 2004) in conjunction with discussions with other agencies with direct experience in the development and conduct of whole sediment tests using benthic assays (AAD, CSIRO) identified the brittle star *Amphiura elandiformis* as a suitable candidate for a locally focused bioassay. Brittlestars, including *A. elandiformis*, are important indicators of relatively unimpacted conditions in the Derwent (Macleod and Helidoniotis 2005). In addition this species can be readily identified both in sediment samples and through video footage, and it is currently used as an indicator of "clean" environmental conditions by local regulatory authorities in assessment of the environmental effects of caged fish farming in the adjacent D'Entrecasteaux Channel and Huon Estuary (Fig. 1). Although not found in brackish regions *A. elandiformis* is common in fully marine and lower estuarine conditions.

A. elandiformis is a relatively large species which requires a similarly large amount of sediment in the experimental incubations to ensure an appropriate response. Consequently, this project adapted established protocols for sediment spiking (Simpson et al. 2005) to enable treatment and

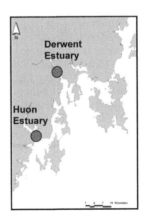

Figure 1.   Location of sampling sites.

testing of larger sediment volumes for copper toxicity. Using these modified protocols standard sediment ecotox experiments were conducted in purpose built benthic mesocosms to evaluate the response of *A. elandiformis* to varying levels of copper contamination. The study then compared the findings from these "spiked" sediment experiments with a further suite of experiments conducted using "natural" sediments with similarly high levels of copper contamination, enabling an examination of the response difference between manipulated "spiked" sediments and that of "natural" contaminated sediments.

A final component of this study examined the effect of standard sediment manipulations on toxicity response. Press sieving is frequently used in sediment toxicity tests to remove organisms that may confound test results, or to homogenise sediments; however, there is debate as to the relative advantages and disadvantages of such sediment manipulations since sieving may significantly alter the chemical and physical nature of the sediments (Simpson et al. 2008).

2 METHODS

Two ecotoxicity assessments (experiments) were conducted:

Experiment 1—Involved spiking of "clean" sediments from the Huon estuary at 3 levels (65, 270 and 1000 mg Cu/kg DMB), and subsequent assessment of sediment toxicity for the brittlestar *Amphiura elandiformis*.

Experiment 2—Involved collection of "contaminated" sediments from the Derwent with copper concentration of 165 mg Cu/kg, and subsequent assessment of sediment toxicity for *A. elandiformis*.

The field collection protocols for each assessment were similar; however, the actual experimental design and laboratory protocols differed.

2.1 Sediment collection

Sediments were collected using a Van-Veen grab from the Huon ("clean") and Derwent ("contaminated") estuaries (Fig. 1). These two estuaries are similar in their biogeochemical, climatic and physical characteristics, except that the Derwent is heavily urbanised and industrialized in the middle estuary (Butler 2005, Macleod and Helidoniotis 2005,). The sites were chosen to be similar in particle size, depth and organic content. Sub-sample cores were collected for particle size characterization, measurement of sediment sulfide and redox potential and for evaluation

of baseline metal levels (techniques described in Eriksen et al. 2009).

2.2 Brittlestar collection and holding

Brittlestars were collected from the Huon estuary in advance of the sediment bioassays and allowed to acclimate for several days to the experimental conditions in holding tanks containing uncontaminated sediments.

2.3 Sediment characterisation

Salinity, temperature and Dissolved Oxygen (DO) were measured in the field. Sediment samples were analysed for pH, redox, sulfide, total organic carbon, % moisture and sediment particle size.

Sediment pH and redox were also measured at the start and end of the experiment, and the pH of the overlying water was monitored in the first experiment. The overlying water, pore water and whole sediment were sampled at the start and end of each test for metal concentrations. Ammonia concentrations were measured every 2 days as an indicator of general water quality. Detailed methods are described in Eriksen et al. (2009). AVS-SEM fractions, were not measured due to project constraints, Whilst recognised as potentially useful for characterising sediment quality, ANZECC (2000) urge caution in the application and interpretation of AVS-SEM model, and its usefulness for copper is suspect.

2.4 Whole-sediment bioassay

Replicate acid washed test containers (4 per treatment) holding approximately 750 g of sediment and 1.5 L of overlying seawater (filtered to 0.2 μm) were allowed to acclimate for approximately 6 days prior to the start of the bioassays. Temperature was consistent between the treatments, varying between 10 and 15°C over the study. Half of the overlying water was exchanged every two days to maintain water quality, and samples were collected at this time for dissolved metals, ammonia, salinity and dissolved oxygen. Six brittlestars were added to each test container at the start of the experiment (Day 0) and there-after daily observations were made of brittlestar behaviour. Upon termination of the test all animals were removed from the sediment, gently washed with filtered seawater and photographed.

2.5 Manipulative experiments—Experiment 1

2.5.1 Sediment spiking

Sediment spiking is an established technique for assessment of sensitivity and response of benthic organisms to contaminants. Clean sediments were collected and dosed with fixed levels of copper. The pH and redox condition of the sediment were monitored throughout this process as the addition of contaminants may change the chemistry of test sediments and this must be controlled for when undertaking sediment spiking (Simpson et al. 2004, Atkinson et al. 2007, Hutchins 2007). After metal spiking sediments were allowed an equilibration period for copper stabilisation of 10–15 days, as recommended by Simpson et al. (2004).

Sediment for the 10-day whole sediment toxicity tests was created by artificially spiking the control (Huon) sediment. The copper concentrations used in the test were 65, 270 and 1000 mg Cu/kg DMB (representing the ANZECC/ARMCANZ—ISQG Low and High trigger levels and an environmentally relevant high concentration respectively). Background concentrations of copper prior to spiking varied between 16.7 and 21.5 mg/kg. Spiked sediments were prepared under nitrogen, with samples rolled frequently throughout the equilibration period to ensure adequate mixing. Sediments were sub-sampled at regular intervals throughout the equilibration period (pore water and total sediment concentrations) to establish equilibration point. Detailed methods are described in Eriksen et al. (2009).

2.6 Manipulative experiments—Experiment 2

Experiment 2 examined (a) the toxicity of natural sediments and (b) the effect of common sediment manipulations (sieving, removal of organisms) on both sediment chemistry and toxicity. Acid washed containers with 750 mL of sediment and 1.5 L of over-lying water were set-up according to each of the three treatment protocols (4 replicates per treatment) outline below:

1. Natural sediments sampled with minimal disturbance
2. Natural sediments sampled as above, press sieved and fauna returned
3. Natural sediments sampled as above, press sieved, fauna excluded.

3 RESULTS AND DISCUSSION

3.1 Manipulative experiments—Experiment 1

A pilot trial (6 kg sediment, spiked to 270 mg/kg) was undertaken in advance of the main sediment spiking experiment to determine (a) whether it would be necessary to buffer for changes in sediment pH during metal addition and (b) whether the desired metal concentration could be achieved. The sediment chemistry was monitored both during and post-spiking, the results indicating that pH

did not change markedly and that the sediments had reached equilibrium after 17 days with a pore water concentration of 1 µg/L. The final copper concentration was slightly higher than 270 mg/kg targeted but all other sediment metals monitored remained stable. On the basis of these results, a pH manipulation step was not recommended in the main sediment spiking experiment.

Sediments spiked with copper at nominal concentrations of 65, 270 and 1000 mg Cu/kg DMB were used in a 10 day acute exposure to the local brittlestar A. elandiformis. Test conditions were broadly consistent with ambient conditions i.e. porewater and overlying water ammonia concentrations were always less than 2 mg/L and frequently below 0.5 mg/L. Overlying water pH ranged between 7.9 and 8.2, and salinity ranged between 34.4 and 36. Dissolved oxygen concentrations always exceeded 10 mg/L in overlying water, and temperature ranged between 10 and 14°C, consistent with ambient environmental conditions at that time of year.

Pore water copper concentrations in the controls varied between 0.35 and 2.8 µg/L (Table 2), which is comparable with previous studies (King et al. 2004, 2006). The overlying water concentrations in the 65 and 270 mg/kg treatment were also within acceptable levels (Table 2). However, the porewater copper concentration in the 1000 mg/kg treatment was extremely high, well in excess of previously reported concentrations (King et al. 2004, 2006). This is likely as a result of incomplete equilibration of the sediments, and suggests that the 17 days indicated by the trial spiking experiment (270 mg/kg) is not sufficient for higher copper concentrations, and contrasts markedly with the equilibration times proposed by Simpson et al. (2005). These results also suggest that at concentrations as high as 1000 mg/kg it will be necessary to adjust the sediment pH after spiking, for porewater concentrations to stabilize in the desired timeframe.

Table 2. Mean dissolved copper concentrations in the pore water and overlying water at the commencement (Day 0) and termination (Day 10) of Experiment 1. (n = 4).

Treatment	Porewater copper (µg/L)		Overlying water copper (µg/L)	
	Day 0	Day 10	Day 0	Day 10
Control 1	1.9	1.8	0.35	0.63
Control 2	2.8	1.25	2.13	0.93
65 mg/kg	2.13	1.95	3.25	1.3
270 mg/kg	17.9	24	22.5	9.1
1000 mg/kg	18490	13034*	266.5	101*

*Note 1000 mg/kg treatment terminated at Day 3.

3.1.1 Ecotoxicological response

Acute toxicity is defined as the point at which mortality occurs or where there is complete lack of response to a standard stimulus (i.e. reburial if exposed, retraction if prodded). However, in this experiment it was difficult to determine a definitive mortality endpoint, consequently sub-lethal behavioural endpoints were employed. Daily observations were made for signs of movement, paralysis, burial ability, response to gentle prodding if exposed and autotomy/mortality. Behavioural responses were scored on a 10 point scale; with natural behaviour (i.e. indistinguishable from controls) scoring 10 and the behavioural response indicating greatest distress (i.e. autotomy) scoring 1. A score of 0 indicates the point at which the test was terminated.

There was a clear effect of elevated copper levels in test sediments. Behavioural response in all animals exposed to the control and 65 mg Cu/kg treatments was normal (Fig. 2). However, in the 270 mg Cu/kg treatment the brittlestars emerged from the sediments within a few days and thereafter remained exposed on the sediment surface. This is not a natural behavioural response, as it would leave them exposed and vulnerable to predation. In the 1000 mg Cu/kg treatment all individuals recorded a severe negative response to the treatment within 24 hours; writhing on the sediment surface and shedding legs and their central disc (autotomy). Consequently, this treatment was terminated after only three days. This excessive behaviour is likely to be in response to the extremely high copper concentrations in both pore and overlying water; however, water only exposures would be required to determine this conclusively (King et al. 2004).

3.2 Manipulative experiments—Experiment 2

3.2.1 Ecotoxicological response
In this test brittlestars were exposed to both "clean" Huon sediments and "contaminated" Derwent sediments under each of the three treatment conditions described earlier (i.e. "with" and "without" the natural fauna to control for any enhancement or mitigating effects associated with the natural fauna). The Derwent sediments had a copper contamination level of approximately 165 mg Cu/kg. The response of the brittlestars was again scored from 1 to 10, with 10 being normal behaviour and 1 indicating extreme distress. The Huon treatments all showed normal behaviour, whilst the Derwent sediment treatments indicated varying levels of negative response, all of which increased with time (Fig. 3).

After four days there was evidence of slowing in response/paralysis in all of the Derwent treatments and after 10 days a large proportion of the replicates contained animals which chose to remain exposed on the sediment surface. This happened most quickly in treatment 3 (press sieved, fauna excluded). However, whether this was as a result of the natural fauna mitigating the sediment impact in some way is not clear. Nonetheless, the results show a definite and measurable toxicity/behavioural effect on brittlestars associated with

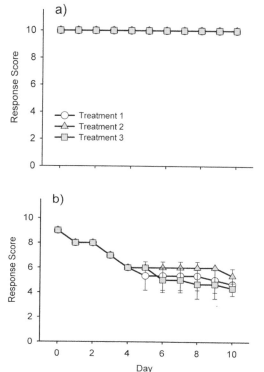

Figure 3. Mean response score (±- standard error) for exposure to (a) Huon and (b) Derwent sediments during 10 day bioassay, where 10 = natural behaviour and 1 = extreme distress.

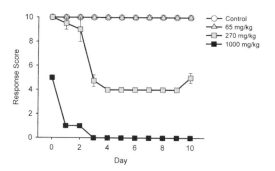

Figure 2. Mean response score (±- standard error) for exposure to copper spiked sediments during 10 day bioassay, where 10 = natural behaviour and 1 = extreme distress.

sediments contamination levels currently recorded from the Derwent (Elwick Bay). Interestingly, although the degree of response varied, the pattern of response in each of the sediment exposures was consistent, regardless of the pre-treatment of the sediments. This suggests that it is not necessary to sieve or remove the fauna for testing with *A. elandiformis* and confirms that exposure to the natural sediments is a relevant test approach.

3.2.2 *Effects of sieving*

There was no marked change in sediment pH, redox, TOC or total copper levels as a result of sieving (Table 3). However, sulphide levels changed markedly and there was a slight increase in the proportion of fine sediments after sieving (Table 3). Changes to sulphide and redox levels are the result of oxygenation of sediments during processing. This is unfortunately unavoidable, unless the whole procedure is carried out under nitrogen, which would be logistically extremely difficult for such large quantities of sediment.

In both the Huon and Derwent samples the treatment pore water concentrations of aluminium, iron, zinc and manganese were markedly different to the field sample levels (Table 4). Iron levels in particular were significantly elevated in the treatment porewaters as compared with the field samples. This suggests that despite attempts to minimise disturbance, there was disruption of sediment/porewater chemistry, which in turn has affected the equilibrium of the sediment metals. For example, hydrolysis of metals, a decrease in pH in conjunction with the oxidation of sediment pore water iron (II) and organic matter may account for the increase in porewater concentrations of zinc and copper (Simpson et al. 2004).

Arsenic increased in porewater samples while lead levels decreased slightly in the sediment manipulations. Treatment 1 showed the least change overall and the magnitude of change was greater in the Derwent sediments than in the Huon sediments.

Table 3. Summary of sediment quality data before and after press-sieving.

Parameter	Before	After
Particle size (% <63 um)	79.2	83.4
TOC (% w/w DMB)	3.8	3.6
pH	7.3	7.37
Redox (mV)	−96	−27
Sulfide (mV)	−355	370
Ammonia (mg-N/l)	0.1	0.194
Total Cu (mg/kg DMB)	22	21.5
Porewater Cu (µg/L)	<1	<1

Table 4. Porewater metal levels for (a) Derwent and (b) Huon samples, at commencement (Day 0) & end (Day 10) of manipulative experiment (µg/L).

Day	Field	Metal (µg/L)						
		Al	As	Cu	Fe	Mn	Pb	Zn
Day	Field	233	24	2.1	706	300	3.7	11
a) Derwent								
0	T1	195	42	1.9	8276	891	2.1	48
	T2	209	50	1.6	2650	602	2.1	61
	T3	220	75	2.1	4932	561	3.1	29
10	T1	284	28	2.5	6458	588	1.9	54
	T2	213	42	4.0	5075	643	2.8	132
	T3	53	34	1.6	5317	676	2.7	105
		Al	As	Cu	Fe	Mn	Pb	Zn
		144	59	1.6	229	70	0.5	7
b) Huon								
0	T1	124	23	1.6	2593	136	0.4	3
	T2	152	13	1.2	785	181	0.3	4
	T3	93	45	9.1	831	193	0.3	4
10	T1	146	39	1.2	552	117	0.5	15
	T2	112	34	1.3	308	141	0.4	8
	T3	146	56	1.4	366	167	0.5	6

Table 5. Total sediment metals for (a) Derwent and (b) Huon samples, at commencement (Day 0) & end (Day 10) of manipulative experiment (µg/L).

Day	Field	Metal (mg/kg)						
		Al	As	Cu	Fe	Mn	Pb	Zn
Day	Field	21766	148	164	23466	286	694	2810
a) Derwent								
0	T1	21800	66	128	23000	183	554	1880
	T2	32500	146	253	36800	283	998	3610
	T3	26700	127	222	29900	255	878	3360
10	T1	21300	71	124	22175	182	520	1940
	T2	22775	81	140	22875	195	578	2160
	T3	21675	75	131	21975	189	541	2038
		Al	As	Cu	Fe	Mn	Pb	Zn
		21766	15	16	24966	62	16	53
b) Huon								
0	T1	25400	15	17	25300	66	17	55
	T2	26200	15	17	25800	66	17	56
	T3	25700	14	17	25600	62	17	56
10	T1	25675	15	17	25775	64	17	56
	T2	25000	14	17	24900	63	17	54
	T3	25775	15	17	25575	65	17	55

Clearly, relative to the Huon the Derwent contained high levels of copper in the sediments (Table 5). This remained the case both in "field" and "treatment" sediments. Levels of As, Mn, Pb and Zn are similarly elevated. Generally, the total sediment metals remained relatively stable, both over time and between treatments, and were consistent with those previously reported from the region (Butler 2005). The sediments from the Huon showed less deviation between field metal concentrations and those of the manipulated treatments than the Derwent sediments. Treatment 1 levels tended to be closer to those of the field samples than the other treatment values. On the whole there was less variation between the Day 0 and Day 10 concentrations in the total sediment metals (Table 5) than in the porewater (Table 4) concentrations, emphasizing the importance of monitoring porewater concentrations and/or surrogate indicators such as pH, redox, and sulphide when determining the impact of sample handling on sediment processes, and in subsequent interpretation of toxicity responses.

4 CONCLUSIONS

Whole sediment toxicity testing is an important tool in assessing the impact of contaminant loadings both on individual species and communities. This study has increased our understanding of how to develop and conduct whole sediment toxicity tests.

The research conducted as part of this project successfully established experimental techniques and protocols for spiking and testing large volumes of sediments which will be extremely useful in future assessments. The advantages of testing larger volumes are that community responses and a wider range of test species can be included in environmental assessments of sediment quality and contamination levels. The study has identified some specific issues that need to be addressed when undertaking large scale sediment spiking. Most notably that sediment spiking at high copper concentrations requires an extended equilibration period if pH manipulation or sediment washing are not undertaken as part of the sediment preparation procedure. In addition a component of the manipulative experiments was to determine the most appropriate exposure approach for testing natural sediments (i.e. whether for comparison with spiked sediments natural sediments needed to be pre-treated in the same way); the results indicate that exposure to undisturbed natural sediments is a relevant test approach.

The toxicity results clearly indicated that the brittlestar *Amphiura elandiformis* is sensitive to elevated copper levels in sediments and is a useful environmental indicator species. At concentrations greater than 270 mg Cu/kg there was a significant effect on the behavioural response of the brittlestars and consequently it is extremely unlikely that this species would occur naturally or thrive under such conditions. Given that this species is mobile, it would remove itself from contaminated areas and therefore where brittlestars are observed it is likely that sediment copper concentrations will be less than 270 mg Cu/kg. This is consistent with observations of the distribution of *A. elandiformis* from the Derwent, where it was not detected in areas where metals levels were elevated (Macleod and Helidoniotis 2005).

The sediment spike test results supports the relevance of the ISQG for Tasmanian conditions, with no sub-lethal effects observed where concentrations were at or below 65 mg/kg, whilst concentrations above 65 mg/kg clearly had the potential to cause significant behavioural effects, autotomy and mortality. The whole sediment tests comparing Derwent and Huon sediments indicated that there were significant behavioural effects at contamination levels lower than the high effect level for copper identified in the current ISQG, with effects evident at a total copper concentration of 165 mg/kg. However, it is likely that this increased response is due to the presence of other metal toxicants contributing to a synergistic effect.

These results provide information that will help managers evaluate change and recovery in urbanised estuaries throughout temperate Australia but specifically in the Derwent. The results of this project will help managers make informed decisions on metal toxicity and develop suitable and sustainable management strategies.

ACKNOWLEDGMENTS

The authors would like to thank the following people for their assistance and/or advice in this project: Tim Alexander, Alan Beech, Quinn Fitzgibbon, Sam Foster, Bob Hodgson, Tom Jackson, Jeff Ross, Camille White & Bill Wilkinson (TAFI); Ashley Townsend (UTas); Cath King (AAD); Graeme Batley, Ian Hamilton, Stuart Simpson, David Spadaro & Jenny Stauber (CSIRO—CERC); Christine Coughanowr & Jason Whitehead (DEP); Lois Koehnken (TAW); Jo Banks (University of Melbourne); Tim O'Hara (Museum of Victoria).

REFERENCES

ANZECC 2000. Australian and New Zealand water guidelines for marine and freshwaters. ANZECC/ARMCANZ.

Atkinson, CA, Jolley, DF & Simpson, SL 2007. Effect of overlying water pH, dissolved oxygen, salinity and sediment disturbances on metal release and sequestration from metal contaminated marine sediments. *Chemosphere* 69: 1428–1437.

Banks, J. 2007. Ecological impacts of pollution on marine soft-sediment assemblages. Honours Thesis, University of Tasmania, Hobart, Tasmania.

Batley, G & Maher, WA 2001. The development and application of ANZECC and ARMCANZ Sediment quality guidelines. *Australasian Journal of Ecotoxicology* 7: 81–92.

Butler, ECV 2005. The Tail of Two Rivers in Tasmania: The Derwent and Huon Estuaries. In *Handbook of Environmental Chemistry* 1–49. Springer-Verlag: Berlin.

Chapman, P.M., Ho, K.T., Munns, W.R., Jr., Solomon, K. & Weinstein, M.P. 2002. Issues in sediment toxicity and ecological risk assessment. *Marine Pollution Bulletin* 44: 271–278.

Conradi, M. & Depledge, M.H. 1999. Effects of zinc on the life-cycle, growth and reproduction of the marine amphipod Corophium volutator. *Marine Ecology Progress Series*. 176: 131–138.

DEP 2007. Water Quality Improvement Plan for the Derwent estuary. Derwent Estuary Program, Hobart, Tasmania.

Eriksen, R., Macleod, C.M. & Meyer, L. 2009. Copper Ecotoxicity Studies—Development of Whole Sediment Toxicity Tests for the Derwent. Tasmanian Aquaculture and Fisheries Institute, DEP/CCI Final report. 42 pp.

Hutchins, C.M., Teasdale, P.R., Lee, J. & Simpson, S.L. 2007. The effect of manipulating sediment pH on the porewater chemistry of copper- and zinc-spiked sediments. *Chemosphere* 69: 1089–1099.

Kennish, M. 1996. Practical Handbook of Estuarine and Marine Pollution. CRC Press—Marine Science Series, Boca Raton, Florida, USA.

King, C.K., Dowse, M.V., Simpson, S.L. & Jolley, D.F. 2004. An assessment of five Australian polychaetes and bivalves for use in Whole sediment toxicity tests: toxicity and accumulation of copper and zinc from water and sediment. *Archives of Environmental Contamination and Toxicology* 47: 314–323.

King, C.K., Gale, S.A., Hyne, R.V., Stauber, J.L., Simpson, S.L. & Hickey, C.W. 2006. Sensitivities of Australian and New Zealand amphipods to copper and zinc in waters and metal -spiked sediments. *Chemosphere* 63: 1466–1476.

Lindegarth, M. & Underwood, A.J. 2002. A manipulative experiment to evaluate predicted changes in intertidal, macro-faunal assemblages after contamination by heavy metals. *Journal of Experimental Marine Biology and Ecology* 274: 41–64.

Macleod, C.K., Crawford, C.M. & Moltschaniwskyj, N. 2004. Assessment of long term changes in sediment condition after organic enrichment: defining recovery. *Marine Pollution Bulletin* 49: 79–88.

Macleod, C.K. & Helidoniotis, F. 2005. Ecological Status of the Derwent and Huon estuaries. Tasmanian Aquaculture and Fisheries Institute, NHT/NAP Final report 46928. 107 pp.

Marsden, I.D. 2002. Life-history traits of a tube dwelling corophioid amphipod, *Paracorophium excavatum*, exposed to sediment copper. *Journal of Experimental Marine Biology and Ecology* 279: 57–72.

Matthai, C. & Birch, G. 2001. Detection of Anthropogenic Cu, Pb and Zn in Continental Shelf Sediments off Sydney, Australia—a New Approach using Normalization with Cobalt. *Marine Pollution Bulletin* 42: 1055–1063.

Morrisey, D. 1996. Effects of copper on the faunas of marine soft-sediments: An experimental field study. *Marine Biology* 125: 199–213.

National Research Council 1997. Contaminated sediments in ports and waterways: clean-up strategies and technologies. National Academy Press.

Ranke, J. 2002. Persistence of antifouling agents in the marine biosphere. *Environmental Science & Technology* 36: 1539–1545.

Simpson, S.L., Angel, B.M. & Jolley, D.F. 2004. Metal equilibration in laboratory-contaminated (spiked) sediments used for the development of whole-sediment toxicity tests. *Chemosphere* 54: 597–609.

Simpson, S.L., Batley, G.E., Chariton, A.A., Stauber, J.L., King, C.K., Chapman, J.C., Hyne, R.V., Gale, S.A., Roach, A.C. & Maher, W.A. 2005. Handbook for sediment quality assessment. CSIRO, Bangor, NSW. Centre for Environmental Contaminants Research, Lucas Heights, NSW. 117pp.

Simpson, S.L., Batley, G. & Chariton, A. 2008. Revision of the ANZECC/ARMCANZ Sediment Quality Guidelines, CSIRO Land and Water Science.

Reproductive biology

Echinoderms in a Changing World – Johnson (ed)
© 2013 Taylor & Francis Group, London, ISBN 978-1-138-00010-0

An ornate fertilisation envelope is characteristic of some *Ophiocoma* species (Ophiuroidea: Ophiocomidae)

P. Cisternas
Bosch Institute, Department Anatomy and Histology, The University of Sydney, NSW, Australia

T.D. O'Hara
Museum Victoria, Melbourne, Victoria, Australia

M. Byrne
Bosch Institute, Department Anatomy and Histology, The University of Sydney, NSW, Australia

ABSTRACT: Ornate fertilization envelopes are unusual among echinoderm eggs, except in the fertilized eggs of several crinoid species and three ophiocomid ophiuroids. We examined the appearance of the fertilization envelope of eggs from other *Ophiocoma* species and described the presence of ridged structures on the eggs of five species. By comparison, the eggs of six other ophiocomids had a smooth fertilization envelope. The presence of ornate fertilisation envelopes in some but not all species of *Ophiocoma* may be of taxonomic significance.

1 INTRODUCTION

Ornate fertilisation egg envelopes appear to be characteristic of some chitons and ascidians (Buckland-Nicks 1993, Cloney 1987). In chitons, elaborate egg projections are involved in directing the sperm to specific entry sites on the egg surface for fertilisation and appear to improve buoyancy of fertilised eggs by reducing sinking rates (Buckland-Nicks 1993).

In echinoderms, the fertilisation envelope is typically smooth, but ornate fertilisation envelopes (with ridges) have been reported for 9 crinoid species (reviewed in Holland 1991, Nakano et al. 2003, Kohtsuka and Nakano 2005). In addition, 3 *Ophiocoma* species in the Ophiuroidea are known to have thorny fertilisation envelopes (Mortensen 1937). Several roles for these ridged structures in crinoid eggs have been suggested; to enhance buoyancy thereby facilitating dispersal of embryonic stages and to provide protection against filter feeding predators (Mortensen 1920, Holland and Jespersen 1973, Mladenov and Chia 1983). Here, we document the presence of ornate fertilisation envelopes in two additional *Ophiocoma* species (Ophiocomidae, Ljungman 1867), and describe the structure of the ridges in *Ophiocoma* eggs using confocal scanning electron microscopy.

2 MATERIALS AND METHODS

We surveyed live *Ophiocoma* specimens that contained gravid ovaries. Fertilised eggs were obtained after spawning was induced by light and temperature shock (Selvakumaraswamy & Byrne 2000). Fertilised eggs were preserved in 2% glutaraldehyde/0.2 um filtered-seawater for examination. Fertilized and unfertilized eggs were initially examined and bright field images captured digitally on a DP70 camera attached to an Olympus BX60 microscope. Serial Z-sections of the fertilization envelope were captured on a Zeiss Meta confocal microscope and 3D reconstructions were generated with the LSM Image software (Zeiss).

In total, we examined gonads from 20 *Ophiocoma* species (Table 1). For six species however,

Table 1. Characteristics of *Ophiocoma* eggs examined.

Smooth fertilization envelope	Ornate fertilization envelope	? (smooth jelly coat present)
O. alexandri	O. aethiops	O. anaglyptica
O. dentata	O. echinata[c,d]	O. brevipes
O. pica[a]	O. erinaceus[a]	O. doederleinii
O. pumila[b]	O. scolopendrina[a]	O. endeani
O. pusilla	O. schoenleinii	O. longsipina
O. wendtii		O. macroplaca
		O. occidentalis

[a]Mortensen 1937, [b]Mladenov 1985, [c]Grave 1898, [d]Mortensen 1920. ? = characteristic of the fertilization envelope not known (live specimens not available). The presence of smooth jelly coat on preserved eggs determined from gravid ovaries.

we could only obtain preserved specimens and of these, *Ophiocoma paucigranulata* and *O. valenciae* lacked mature eggs (Table 1). We supplemented our observations on the appearance of the fertilisation envelope with those from previous studies (Table 1).

3 RESULTS

The unfertilised (spawned) eggs of the majority of *Ophiocoma* species were small (≤100 um in diameter), burgundy-red in colour and smooth in appearance. The exception was the large (≥300 um in diameter), green eggs of *O. endeani*. The appearance of the unspawned eggs of *O. paucigranulata* and *O. valenciae* was not determined, as gonads were not visible in specimens of these species.

The fertilisation envelope of the eggs of five *Ophiocoma* species had a thorny appearance when viewed under the light microscope (Fig. 1a, Table 1). In contrast, a smooth fertilisation envelope was characteristic of another six *Ophiocoma* species (Fig 1b, Table 1). The surface of unspawned eggs and their jelly coat in the ovaries of all *Ophiocoma* species examined had a smooth appearance (Fig. 1c).

Confocal Z-series reconstructions showed that the thorny appearance of the fertilisation envelope

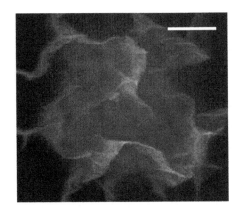

Figure 2. Confocal 3D reconstruction of the fertilization envelope of the egg of *Ophiocoma scolopendrina* showing surface ridges. Scale bar = 20 um.

in these *Ophiocoma* species had a similar morphology and it was characterised by numerous ridged structures (Fig. 2).

4 DISCUSSION

The present study shows that ornate fertilisation envelopes are characteristic of the eggs of two additional *Ophiocoma* species—*O. aethiops* (Lükten 1859) and *O. schoenleinii* (Müller & Troschel 1840). The ornate fertilization envelope in these species was similar to that described for the eggs of *Ophiocoma echinata* (Lyman 1860), *O. erinaceus* (Müller & Troschel 1840) and *O. scolopendrina* (Lamarck 1816) (Grave 1898, Mortensen 1937, Mladenov 1985). The fertilisation envelope of the eggs of all these *Ophiocoma* species had a distinct thorny appearance on visual examination.

Confocal microscopy revealed that these structures were made of ridges that resembled those previously described in fertilized crinoid eggs. The arrangement of the ridges in ophiocomid eggs could not be determined, however, due to breakage during preservation.

In contrast to crinoids, the jelly coat of unspawned eggs of ophiocomids had a smooth appearance. The unspawned eggs of crinoids possess elaborate surface structures, i.e. pits and clumps (Holland 1977). Fine scale examination (e.g. scanning electron microscopy) of the (unspawned) egg surface will be necessary to determine if similar structures are present in these ophiuroids.

Ornate fertilization envelopes have not been reported for other ophiuroids and they are not present on the fertilized eggs of other ophiocomids (Cisternas pers. obs., Hendler 1991, Narasimhamurti 1933). The presence of ornate

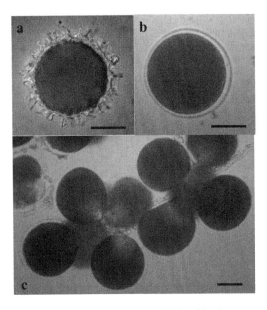

Figure 1. a) Light micrograph of the ridged appearance of the fertilisation envelope in the egg of *Ophiocoma schoenleinii*. b) smooth fertilization envelope of the egg of *Ophiocoma dentata*. c) Unspawned eggs in the ovaries of *O. echinata* showing smooth jelly coat. Scale bar = 50 um.

fertilization envelopes in some but not all *Ophiocoma* species may be of taxonomic importance (Mortensen 1938) and suggests that differences in developmental features may assist with the classification of intra-generic sub groupings proposed by Devaney (1970).

ACKNOWLEDGEMENTS

We would like to thank all those who assisted with collections in the field. Museum Victoria, the Smithsonian Natural History Museum and the Bishop Museum are thanked for access to specimens in their collection. Queensland National Parks and Wildlife Services, the Great Barrier Reef Marine Parks Authority and the crew of the Kerra Lyn. Lizard Island Research Station, One Tree Island RS, Museum Victoria, The University of Sydney Electron Microscope Unit, The Bosch Institute, Naos Marine Labs and Galeta Research Station (Smithsonian Tropical Research Institute, Panama) are thanked for the use of their facilities. This research was supported by funding from the Australian Research Council, The Australian Academy of Science, Museum Victoria, the Raine Island Corporation, the Company of Biologist and the Smithsonian Tropical Research Institute.

REFERENCES

Buckland-Nicks, J. 1993. Hull cupules of chiton eggs: Parachute structures and sperm focusing devices? *Biol. Bull.* 184: 269–276.

Cloney, R.A. 1987. In M. F., Strathmann (ed). Reproduction and development of marine invertebrates of the northern pacific coast: data and methods for the study of eggs, embryos, and larvae: 607–639. Seattle, London: University of Washington Press.

Devaney, D.M. 1970. Studies on ophiocomid brittlestars. 1. A new genus (*Clarkcoma*) of Ophiocominae with a reevaluation of the genus *Ophiocoma*. *Smith. Contrib. Zool.* 51: 1–41.

Grave, C. 1898. Embryology of *Ophiocoma echinata* Agassiz, *Prel. Note. John Hopkins Univ. Circ.* 18: 6–7.

Hendler, G. 1991. Chapter 6: Echinodermata: Ophiuroidea. In A. Giese, J.S. Pearse & V.B. Pearse (eds), *Reproduction of Marine Invertebrates*: 356–511. Pacific Grove, California: The Boxwood Press.

Holland, N.D. 1977. The shaping of the ornamented fertilization membrane of *Comanthus japonica* (Echinodermata: Crinoidea). *Biol. Bull.* 153: 299–311.

Holland, N.D. 1991. Chapter 4: Echinodermata crinoidea. In A. Giese, J.S. Pearse & V.B. Pearse (eds), *Reproduction of Marine Invertebrates*: 247–299. Pacific Grove, California: The Boxwood Press.

Holland, N.D. and Jespernsen A. 1973. The fine structure of the fertilization envelope membrane of the feather star *Comanthus japonica* (Echinodermata: Crinoidea). *Tissue and Cell.* 5(2): 209–214.

Kohtsuka, H. and Nakano, H. 2005. Development and growth of the feather star *Decametra tigrina* (Crinoidea), with emphasis on the morphological differences between adults and juveniles. *J. Mar. Biol. Ass. UK.* 85: 1503–1510.

Mladenov, P.V. 1985. Development and metamorphosis of the brittle star *Ophiocoma pumila*: evolutionary and ecological implications. *Biol. Bul.* 168: 285–295.

Mladenov, P.V. and Chia, F.S. 1983. Development, settling behaviour, metamorphosis and pentacrinoid feeding and growth of the feather star *Florometra serratissima*. *Mar. Biol.* 73: 309–323.

Mortensen, T.H. 1920. *Studies in the development of Crinoids.* Papers from the Tortugas Laboratories (Carnegie Institution) Copenhagen: Gad GEC.

Mortensen, T.H. 1937. Contributions to the study of the development and larval forms of echinoderms III. *Klg. Danske Vidensk. Selsk. Skrifter, Naturv. og Math* 7(1): 1–65.

Mortensen, T.H. 1938. Contributions to the study of the development and larval forms of echinoderms IV. *Klg. Danske Vidensk. Selsk. Skrifter, Naturv. og Math* 7(3): 1–59.

Nakano, H. Hibino, T. Oji, T. Hara, Y. and Amemiya, S. 2003. Larval stages of a living sea lily (stalked crinoid echinoderm). *Nature* 421: 158–160.

Narasimhamurti, N. 1933. The development of *Ophiocomina nigra*. *Q. J. Microscr. Sci.* 76: 63–88.

Selvakumaraswamy, P. and Byrne, M. 2000. Reproduction, spawning and development of 5 ophiuroids from Australia and New Zealand. *Invert. Biol.* 119(4): 394–402.

Behavior

Echinoderms in a Changing World – Johnson (ed)
© *2013 Taylor & Francis Group, London, ISBN 978-1-138-00010-0*

Complexity in the righting behavior of the starfish *Asterina pectinifera*

M. Migita

Shiga University, Otsu, Japan

ABSTRACT: Righting behavior of the starfish *Asterina pectinifera* was investigated to determine whether the starfish can learn to complete the action in a shorter amount of time or to use specific arms as leaders. In this study, *A. pectinifera* individuals were placed on the bottom of an aquarium on the aboral surface, and the righting time and leading arms were recorded. Preliminary results suggested that neither variable showed an apparent adaptive change. Subsequently, a computer model of the righting behavior was constructed from the observations on *A. pectinifera* that described righting in terms of changes in arm states determined in relation to the states of all other arms. Results of the computer simulations suggest that the intrinsic complexity of the process of righting is such that *A. pectinifera* can neither learn to right itself in less time nor use specific arms to do so.

1 INTRODUCTION

Righting behavior is one of the best known behavior patterns of starfish, and it has attracted researcher's interests for more than one hundred years (Jennings, 1907; Ohshima, 1940; Reese, 1966; Polls and Gonor, 1975; Lawrence and Cowell, 1996). Righting in the shortest possible time may reduce the risk of being swept away in surge or attack by predators. Conflicts between arms should be avoided to minimize righting time, which could be accomplished by using specific leading arms. Thus, it could be beneficial for starfish if they could learn to right themselves. Previous studies have explored the possibility that starfish can learn to use specific arms as the leading arms and/or to complete righting in shorter times, but the majority of these have produced negative results (Jennings, 1907; Ohshima, 1940; Reese, 1966). However, recent studies do not focus on the behavioral aspect per se, but deal with the righting behavior as an indication of stress (Diehl et al., 1979; Lawrence & Cowell, 1996; Kashenko, 2003).

Learning in starfish has been observed in several studies (e.g. Reese, 1966; Valentinčič, 1983; Migita et al., 2005). Valentinčič (1983) reports associative learning in *Marthasterias glacialis* to differentiate between harmless tactile stimulus and noxious unconditioned stimulus of electric shock. Reese (1966) reviews an early study on learning by Ven (1922) where *Asterias rubens* could learn to escape from pegs placed between the arms. Migita et al. (2005) observed obstacle avoidance behavior of *Asterina pectinifera*. In this work, some individuals learned effective obstacle avoidance behavior, and individuals that navigated obstacles faster more

commonly used the same leading arm throughout the trials (Migita et al., 2005).

The question addressed in this study is whether starfish are capable of learning to right themselves. More specifically, can the righting process be improved through repetitive trials, if the starfish has the capacity to maintain a certain mode of movement? The behavioral aspect of the righting response of the starfish *A. pectinifera* was investigated, and information gleaned from preliminary trials was then used to construct a simple computer simulation model of the righting behavior designed to assess the extent to which experience contributes to the righting process.

2 RIGHTING BEHAVIOR OF ASTERINA PECTINIFERA

2.1 *Materials and methods*

Adult *A. pectinifera* individuals (4.9 ± 0.3 cm in arm length; n = 5) were collected in Hyogo, Shimane and Okayama, Japan. Starfish were fed boiled clam and crayfish in the laboratory, and their righting behavior was investigated.

Experiments were conducted on *A. pectinifera* in a transparent acrylic tank (30 × 30 × 30 cm) that was filled with seawater to a depth of 7 cm. This depth allowed starfish to right themselves without exposing their arms above the water surface. During each experiment, an *A. pectinifera* starfish was initially placed upside-down (oral side up) on the bottom of the experimental tank. After completing the righting process, it was returned to its original upside-down position. This procedure was repeated and videotaped for a period of one hour. The number of righting events (number of trials), the righting time and the leading arms (Nos. 1–5) were recorded in order to investigate whether certain adaptive changes are involved in the starfish's behavior. Arms were assigned numbers from 1 to 5, counting from the madreporite in a clockwise direction.

2.2 Results of the righting experiments

Righting experiments were conducted on five *A. pectinifera* individuals. Figure 1 shows the righting time of the individual that completed the largest number of trials (n = 27). In the first trial, there was a relatively long latent period (more than 70% of the total righting time) in which the individual showed no apparent response. There was no tendency to speed up the righting time. When the leading arm is plotted in each trial for the same individual (Fig. 2), it is clear that the leading arm(s) alternated throughout the trials except on one occasion when the same arm lead on successive trials.

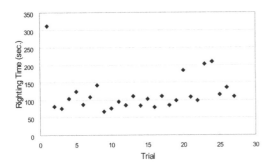

Figure 1. The righting time for each trial during the one-hour righting experiment for the *A. pectinifera* individual that completed the most trials.

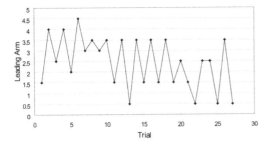

Figure 2. Leading arms in each trial for the *A. pectinifera* individual in Figure 1. Non-integer numbers indicate two leading arms in a trial. For example, if leading arms were arm 1 and 2, they are designated as 1.5 in the graph.

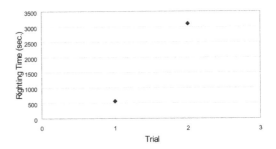

Figure 3. The righting time of an *A. pectinifera* individual that completed only two trials in one hour. The 'folding over' method was used in the second trial, which left the starfish in a deadlock.

Certain righting attempts sometimes resulted in 'deadlock' situations. Following the terminology of Ohshima (1940), *A. pectinifera* usually rights itself using a 'somersault' maneuver, in which arms recovering and contacting the substratum fastest pull the body up to right the individual. However, in this study one *A. pectinifera* individual tried to right itself by folding itself over but appeared to have great difficulty due to its relatively short arms (Fig. 3). Despite appearing to be in good condition, it completed only two trials, suggesting that the righting process is intrinsically complex regardless of the condition of the individuals. These observations reveal that *A. pectinifera* did not improve its righting behavior with repetition.

Stability of leading arms can be interpreted as persistence in orientation, which was referred to as "toy-soldier behavior" by Grabowsky (1994). In toy-soldier behavior, echinoderms persist in locomoting in a particular orientation in spite of disturbance by the experimenter. Since the righting process involves an aspect of locomotion, one can speculate that 'toy-soldier behavior' might emerge through successive righting trials.

3 COMPUTER SIMULATIONS

3.1 Modeling the righting behavior

From these preliminary observations on *A. pectinifera* individuals, it is premature to conclude that a starfish cannot learn to right itself. To further investigate whether the task of righting is intrinsically too complicated to learn, a simplified computer simulation model of righting behavior was constructed based on observations made on righting of *A. pectinifera*.

Starfish do not have a central nervous system controlling all the arms simultaneously (Jennings, 1907; Smith, 1965; Pentreath and Cobb, 1982; Dale, 1999). Neighboring arms are connected via nerve cords around the mouth called a "nerve ring", which is responsible for coordinating arm movements (Suzuki, et al., 1971). This configuration of motor organs makes the task of righting difficult as there must be an inherent delay of information transfer between non-neighboring arms (Yoshida and Otsuki, 1968).

Since the major motor organ of a starfish is the arm, it is natural to model the starfish as a system of arms connected by a nerve net (Suzuki et al., 1971; Dale, 1999). For the computer simulation model, arm states were classified into the following six categories that effectively describe the righting behavior of *A. pectinifera* (Fig. 4):

1. Neutral: the arm is in the neutral state when the individual is placed oral side uppermost on the bottom
2. Anchored: the tip of the arm recovers contact with the substratum
3. Half-righted: half of the arm recovers contact with the substratum
4. Raised: the arm is raised up
5. Float: the arm is floating while other arms are being righted
6. Righted: the arm recovers full contact with the substratum

Righting is achieved by the configuration of the arms in combinations of the above states. The transition between states is restricted to several possible patterns (Fig. 5). Each arm changes its state according to the state of the other arms. Although the simulated starfish was capable of attaining $6^5 = 7776$ states in total, in the real world,

Figure 4. Six categories of arm states used in the computer simulation model in this study.

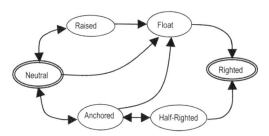

Figure 5. A schematic diagram showing transitions in arm states during the righting process. Arrows represent possible transitions.

considerably fewer states are likely to be possible or significant for righting. I deduced eighty three possible configurations of arms states and constructed state transition rules for every arm. The transition rule for each arm state is summarized in Table 1.

As mentioned above, state transition is essentially dependent on the configuration of multiple arm states. However, there are certain configurations that do not determine changes in arm states, for example, when all arms are in the neutral state. For such cases, probabilistic transition, where an arm changes its state with some probability, is introduced.

3.2 Even condition experiment

In the 'even condition' experiment, the simulated starfish start each righting trial with all arms in the

Table 1. State transition rules for righting.

From	To	Condition
Neutral	Neutral	Probabilistic ($p = 0.3$)
	Anchored	Probabilistic ($p = 0.4$)
	Raised	Probabilistic ($p = 0.3$)
	Float	Remaining neighboring arms will be altered to the Righted state
Anchored	Neutral	Probabilistic ($p = 0.3$)
	Anchored	Probabilistic ($p = 0.3$)
	Half-righted	Neighboring arm(s) will be altered to the Half-righted state, or probabilistic ($p = 0.4$)
	Float	Non-neighboring arms will be alterned to the Righted state
Half-righted	Anchored	Non-neighboring arms will be altered to the Righted state
	Half-righted	Conditions for anchored or righted are not satisfied
	Righted	Neighboring arm will be altered to the Righted state
Righted Raised	Righted	Independent of arms
	Neutral	Neighboring arms will be altered to the Anchored state, or probabilistic ($p = 0.1$)
	Raised	Probabilistic ($p = 0.9$)
	Float	Remaining neighboring arms will be altered to the Righted state
Float	Righted	Remaining neighboring arms will be altered to the Righted state
	Float	The condition above is not satisfied

'neutral' state. At each time step, arm states were updated according to the relevant state transition rule (Table 1). When all arms were determined to be in the righted state, the time steps to complete the righting (interpreted as "righting time" in the computer simulations) and the leading arms were recorded. This procedure was repeated 50 times (the maximum number of trials that *A. pectinifera* could complete in one hour was estimated as 50, based on the shortest righting time observed in actual starfish—see Figure 1. Since the initial configuration of multiple arm states was the same in each trial, the experience gained in the preceding trial did not affect a subsequent trial.

3.3 *The biased condition experiment*

In the 'biased condition' experiment, the simulated starfish started the first righting trial with all arms in the neutral state, just as in the 'even condition' experiment. From the second trial, however, it started with the leading arms in the preceding trial being in the 'anchored' state. This was done in an attempt to simulate "toy soldier behavior" (Grabowsky, 1994), in which echinoderms persist in locomoting in one direction in spite of disturbances. The purpose of this experiment was to investigate whether the propensity for a specific constant leading arm could accelerate the righting process. The number of trials was 50, as for 'even condition' experiment.

3.4 *Results of the simulations*

The righting time was estimated from the steps required for the simulated starfish to complete the righting. The righting times for both the 'even' and the 'biased' condition experiments were 13.2 ± 7.2 and 12.1 ± 6.2 steps respectively. The minimum time steps under each condition were 6 and 4 steps for the 'even' and 'biased' condition experiments respectively, while the maximums were 39 and 32 time steps. The 'biased' starfish righted itself slightly faster than the starfish under 'even' conditions (Fig. 6), however, the difference was not significant (Mann-Whitney's U-test, $P = 0.66$; $Z = -0.42$; n = 50). This suggests that the righting process in *A. pectinifera* is sufficiently intrinsically complex that any advantage provided by the use of leading arms from previous trials to anchor the individual to the substratum faster than would occur in using other arms, is lost during the course of righting.

The significance of the differences between the average righting times in the simulated experiments is dependent on the number of trials. If the number of trials exceeds 90, then significant differences in righting time between the two experiments could

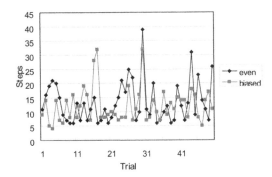

Figure 6. Righting time in computer simulations of *A. pectinifera*.

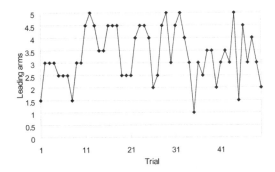

Figure 7. Leading arms in the biased condition simulation experiment. Leading arms alternated frequently and were stable for at most three trials.

be observed. However, this arguably has little biological meaning since this number of righting trials is unlikely to be attainable for a real starfish.

Leading arms often alternated in consecutive trials, even if the simulated starfish started a new trial with leading arms from the preceding trial now in the advantageous 'anchored state' (Fig. 7).

4 DISCUSSION

Based on the results of the computer simulations conducted in this study, it is suggested that the processes of righting is inherently complex in this starfish. Because leading arms easily alternate, it is difficult for starfish to improve righting performance, especially if experience gained in a preceding trial contributes only to the initial state of the following trial. This assumption on the contribution of experience, however, may underestimate the capacity of *A. pectinifera*. In the present study, the computer simulation model was constructed based on the transitions of multiple arm states in each trial. Improvement of the present version of the model is necessary in order to implement the changes in the mode of state transitions during repeated trials.

The model of righting in this study assumes that the state of the arms at a given time step basically determines the state at the next time step except in the case of several probabilistic transitions. However, the righting process of real starfish must be different to this because their arms move according to the history of their movement, a factor that is omitted in the present model.

More importantly, whether learning during the righting process is advantageous or not is still an open question. As learning implies fixed movements of the arms, it may reduce versatility during the righting process (Jennings, 1907). If this is the case, computer simulation models are useful tools for investigating how starfish realize two physical properties vital for righting, namely high efficiency and flexibility of arm movement. Another advantage of simulation models is that one may estimate the behavioral capacity of starfish without the deteriorating effects of fatigue on behavioral performance of a real starfish. Further in-depth analyses of the righting strategy of *A. pectinifera* starfish are necessary to create a more refined and realistic simulation model.

ACKNOWLEDGEMENTS

Thanks are due to K. Furuhashi, A. Hattori, S. Nomura, W. Godo and K. Yoshimura for their support on this study. I am also grateful to C. Lewis, an anonymous reviewer and Prof. C. Johnson for their comments on the manuscript. This study was partly supported by the Ministry of Education, Science, Sports and Culture, Grant-in-Aid for Scientific Research (C) (No. 19500189).

REFERENCES

Dale, J. 1999. Coordination of chemosensory orientation in the starfish *Asterias forbesi. Mar. Fresh. Behav. Physiol.* 32: 52–71.

Diehl, III, W.J., McEdward, L., Proffitt, E., Rosenberg, V. & Lawrence, J.M. 1979. The response of Luidia clathrata (Echinodermata: Asteroidea) to hypoxia. *Comp. Biochem. Physiol.* 62A: 669–671.

Grabowsky, G.L. 1994. Symmetry, locomotion, and the evolution of an anterior end: a lesson from sea urchins. *Evolution* 48(4): 1130–1146.

Jennings, H.S. 1907. Behavior of the starfish, *Asterias forreri* de Loriol. *Univ. Calif. Publ. Zool.* 4: 53–185.

Kashenko, S.D. 2003. The reaction of the starfish *Asterias amurensis* and *Patiria pectinifera* (Asteroidea) from Vostok Bay (Sea of Japan) to a salinity decrease. *Russ. J. Mar. Biol.* 29(2): 110–114.

Lawrence, J.M. & Cowell, B.C. 1996. The righting resposnse as an indication of stress in *Stichaster striatus* (Echinodermata, Asteroidea). *Mar. Fresh. Behav. Physiol.* 27(4): 239–248.

Migita, M., Mizukami, E. & Gunji, Y.-P. 2005. Flexibility in starfish behavior by multi-layered mechanism of self-organization. *BioSystems* 82: 107–115.

Ohshima, H. 1940. The righting movements of the sea-star *Oreaster nodosus* (Linné). *Jap. J. Zool.* 8: 575–589.

Pentreath V.W. & Cobb, J.L.S. 1982. Echinodermata. In G.A.B.Shelton (ed), *Electrical Conduction and Behaviour in 'Simple' Invertebrates*: 440–472. Clarendon Press: Oxford.

Polls, I. & Gonor, J. 1975. Behavioral aspects of righting in two asteroids from the pacific coast of North America. *Biol. Bull.* 148: 68–84.

Reese, E.S. 1966. The complex behavior of echinoderms. In R.A. Boolootian (ed), *Physiology of Echinodermata*: 157–218. Interscience Publishers: New York.

Smith, J.E. 1965. Echinodermata. In T.H. Bullock & G.A. Hoddidge (eds), *Structure and Function in the Nervous Systems of Invertebrates*: 1519–1558. Freeman: San Francisco.

Suzuki, R., Katsuno, I. & Matano, K. 1971. Dynamics of "neuron ring": computer simulation of central nervous system of starfish. *Kybernetik* 8: 39–45.

Valentinčič, T. 1983. Innate and learned responses to external stimuli in asteroids. *Echinoderm Studies*. 1: 111–137.

Ven, C.D. 1922. Sur la formation d'habitudes chez les astéries. *Arch. néerl. Physiol.* 6: 163–178.

Yoshida, M. & Ohtsuki, H. 1968. The phototactic behavior of the starfish, *Asterias amurensis* Lütken. *Biol. Bull.* 134(3): 516–532.

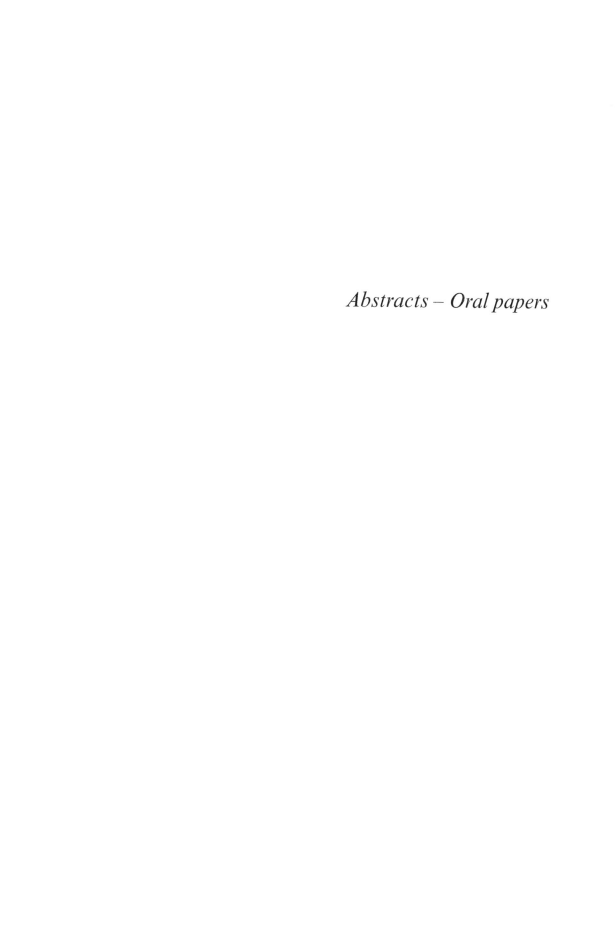

Abstracts – Oral papers

Echinoderms in a Changing World – Johnson (ed)
© 2013 Taylor & Francis Group, London, ISBN 978-1-138-00010-0

Echinoderm biodiversity of the Australian continental margin (100–1500 m depths)

F. Althaus, K. Gowlett-Holmes, F. McEnnulty, A. Williams & P. Dunstan
CSIRO Marine and Atmospheric Research, Hobart, Tasmania, Australia

L. Marsh
Western Australian Museum, Welshpool DC, Western Australia, Australia

T.D. O'Hara & M. O'Laughlin
Museum Victoria, Melbourne, Victoria, Australia

A. Miskelly
Australian Museum, Sydney, New South Wales, Australia

ABSTRACT: Biodiversity surveys of Australia's little sampled western and south-eastern continental margins in 2005 and 2007 found that echinoderms are a major component of deep invertebrate megafaunas. Species-level determinations made by experts from Australian and international institutions showed these collections contained many new species and/or new species records for Australia, particularly of asteroids and crinoids—the deep-sea elements of which are less studied than other echinoderm groups. High resolution video and still images, and multi-beam sonar data, were collected at the same locations and permit faunal distributions and habitat associations to be evaluated.

This talk presents an overview of the diversity and distribution patterns of the echinoderm fauna from these collections, and examines diversity and habitat associations at the multiple scales provided by mapping and imagery—from regional to patch scales (1000s km to 10 s m).

Activity of different populations of *Patiriella regularis* and *Patiriella mortenseni* in lowered salinity

M.F. Barker
Department of Marine Science, University of Otago, Dunedin, New Zealand

ABSTRACT: Unique climatological and hydrographic conditions in Fiordland, southwest New Zealand, produce a near freshwater layer (LSL) on top of fully marine water. The starfish *Patiriella regularis* is often found within this LSL, while *P. mortenseni* occurs immediately below it in full salinity seawater. Laboratory and field experiments have shown that *P. regularis* has extreme tolerances of low salinities, and can even survive in fresh water for several days. *P. mortenseni* is unable to live at salinities much less than 35%.

To examine whether salinity tolerances found in *Patiriella* from fiordland are unusual, the activity of *P. regularis* and *P. mortenseni* was measured in seastars collected from sites on the west and east coast of the South Island and from sites on the east coast of northern New Zealand.

Righting times of seastars bathed in dilutions of 100, 75, 50 and 25% of full salinity seawater immediately after collection show that *P. regularis* from Doubtful Sound have higher activity in low salinity than seastars from all other populations examined. However in experiments in which *P. regularis* from Doubtful Sound and other sites were held at a salinity of 9% for 8 hours before activity was tested, seastars from all sites had similar righting times. All populations of *P. regularis* other than those from Doubtful Sound, have slightly faster righting times in diluted seawater than all populations of *P. mortenseni* tested. The implication of adaptations to survive in low salinities in New Zealand's southern fiords is discussed.

Molecular genetic variation of strongylocentrotid sea urchins in the Eastern Atlantic

T.T. Bizuayehu, N.T. Hagen & T. Moum
Faculty of Biosciences and Aquaculture, Bodø University College, Norway

ABSTRACT: Two common sea urchin congeners, *Strongylocentrotus droebachiensis* and *S. pallidus*, inhabit arctic-boreal waters of the Atlantic and Pacific oceans. The species are closely related, but differ in their ecological impact along the Norwegian coast, as only *S. droebachiensis* is involved in large scale kelp forest destruction. While genetic variation and population history have been assessed in both species throughout most of their distribution range using various molecular markers, eastern Atlantic specimens have been relatively poorly represented in such studies. We collected population samples of more than 20 individuals of each of the two species in Saltenfjord at the Norwegian coast. Based on the level of variation in the complete mitochondrial genome sequences of four individuals from each of the two species, we chose the cytochrome b gene for assessment of organellar genetic variation. The first Internal Transcribed Spacer (ITS-1) of the structural ribosomal RNA region was used to assess nuclear variation. Population diversity indices were low for both species, and consistently lower in *S. droebachiensis* than *S. pallidus*. In *S. droebachiensis*, a shallow and star-shaped haplotype network, excess of rare haplotypes, and negative neutrality statistics of cytochrome b, indicated purifying selection and/or population expansion. For both markers, neutrality statistics and mismatch distributions suggested population expansion and a stationary model was rejected in *S. droebachiensis*. These results are consistent with population fluctuations in the outbreak species *S. droebachiensis* and support previous reports indicating low genetic diversity of the species in the north east Atlantic.

Corrupted urchins: Impact of climate change on development

M. Byrne, M. Ho, H. Nguyen, N. Soars & P. Selvakumaraswamy
Department of Anatomy and Histology, Bosch Institute, University of Sydney, NSW, Australia

A. Davis
Institute for Conservation Biology, School of Biological Sciences, University of Wollongong, NSW, Australia

S. Dworjanyn
National Marine Science Centre, The University of New England and Southern Cross University, Coffs Harbour, NSW, Australia

ABSTRACT: The distribution of the sea urchins *Heliocidaris erythrogramma*, *H. tuberculata* and *Tripneustes gratilla* overlap with the Eastern Australia climate change hot spot where disproportionate ocean warming (CSIRO Climate System Model Mk3.5) and ocean acidification makes regional marine biota particularly vulnerable to climate change. In keeping with near future climate change scenarios, we determined the interactive effects of warming and acidification on fertilisation and development of these echinoids. Experimental treatments (20–26°C, pH:7.6–8.2) were tested in all combinations for IPCC

(2007) 'business as usual' scenario with 20°C/pH 8.2 being ambient. Fertilisation was robust across treatments and elevated temperature significantly impaired early development. For those embryos that were able to develop to the larval stage increased pCO_2 (pH < 8.0) had a negative effect on calcification. In this study of the interactive effects of temperature and pH, we confirm the thermotolerance and pH resilience of sea urchin fertilisation within predicted climate change scenarios, with negative affects at upper limits of ocean warming for early development and impaired calcification for larval development. Although ocean acidification research has focussed on impaired calcification, embryos may not reach the skeletogenic stage in a warm ocean. Our results place previous single stressor studies in context and emphasise the need to design experiments with respect to concurrent ocean warming and acidification.

Larval development in the family Ophiocomidae provides some perspectives on life history trends in the Ophiuroidea

P. Cisternas, P. Selvakumaraswamy & M. Byrne
Department of Anatomy and Histology, Bosch Institute, University of Sydney, NSW, Australia

J. Hodin
Hopkins Marine Station, Stanford University, Pacific Grove, CA, USA

T.D. O'Hara
Museum Victoria, VIC, Australia

ABSTRACT: Comparative data from echinoderms other than echinoids and asteroids with which to assess life history trends are lacking. We documented the developmental diversity in the family Ophiocomidae and generated a molecular phylogeny to examine hypotheses on life history evolution in ophiuroid echinoderms. Specifically we examined the unimodality of egg size distribution, the nature of shifts in developmental mode, independent origins of larval types, and ancestral developmental patterns (Type I vs. Type II) in the Ophiuroidea. The Ophiocomidae is one of the most conspicuous tropical ophiuroid families that exhibit an array of planktotrophic and lecithotrophic life histories (Type I and Type II) and larval types, typical of most ophiuroids. Developmental data from 40 species in the 8 recognised genera revealed that ophiocomids had a distinct bimodal egg size distribution and were characterised by clades with a mixture of planktotrophic and lecithotrophic life histories. The distribution of developmental patterns and distinct larval morphologies among lecithotrophic developers support the idea of multiple independent origins of nonfeeding larvae and the notion that Type II development may be the ancestral pattern in ophiuroids.

Effects of lower seawater pH on sea urchin larvae: Implications for future ocean acidification

D. Clark, M. Lamare & M. Barker
Department of Marine Science, University of Otago, Dunedin, New Zealand

ABSTRACT: Ocean acidification, as a result of increased atmospheric CO_2, is predicted to lower the pH of seawater to between pH 7.5 and pH 7.7 over the next 100 years. The greatest changes are expected in polar waters. Our research aimed to examine how echinoid larvae are affected by lower pH, and if

245

effects are more pronounced in polar species. We examined the effects of lower pH on echinoid larvae from tropical (*Diadema savigyn, Tripneustes gratilla, Echinometra matheii*), temperate (*Pseudechinus huttoni, Evechinus chloroticus*), and a polar species (*Sterechinus neumayeri*) using a series of laboratory experiments. Larvae were reared in a range of lower pH seawater (pH 6.0. 6.5, 7.0, 7.5, 7.7 and 8.1), adjusted by bubbling CO_2 gas. The effect of pH on somatic and skeletal growth, calcification rate, development and survival were quantified, while SEM examination of the larval skeleton provided information on the effects of seawater pH on the fine-scale skeletal morphology (Fig. 1). Lowering pH resulted in a decrease in survival, growth, calcification rate and skeletal structure, although significant effects tended to be restricted to pH levels lower than 7.5. Contrary to expectations, the polar larvae did not demonstrate a higher sensitivity to lower pH when compared with non-polar species.

Response of sea urchin larvae to reduced seawater pH: A comparison among tropical, temperate and polar species

D. Clark, M. Lamare & M. Barker
Department of Marine Science, University of Otago, Dunedin, New Zealand

ABSTRACT: Ocean acidification, due to increasing anthropogenic Carbon Dioxide (CO_2) emissions, is predicted to lower ocean pH to between 7.5 and 7.7 over the next 100 years. Calcifying organisms are expected to be severely impacted because the oceanic uptake of CO_2 indirectly reduces Calcium Carbonate ($CaCO_3$) saturation. Polar regions, where calcification is already difficult due to the cold seawater temperatures and upwelling of CO_2-rich seawater, have been identified as regions where ocean acidification effects may appear earliest. In addition, polar species may be particularly susceptible due to their low calcification rates and slow physiology. Our research aimed to determine how echinoid larvae are affected by lower pH levels and if these effects are more pronounced in polar species. We examined the effects of lower pH seawater on tropical (*Tripneustes gratilla*), temperate New Zealand (*Evechinus chloroticus, Pseduechinus huttoni*) and Antarctic (*Sterechinus neumayeri*) sea urchin larvae using a series of laboratory experiments. Larvae were reared under normal pH (~pH 8.1) or lower pH seawater (pH 6.0, 6.5, 7.0, 7.5, and 7.7), adjusted by bubbling CO_2 gas. Changes in survival, growth, skeletal structure and density, calcification rates, calcite composition, $CaCO_3$ deposition areas, skeletal formation and micro-respiration were examined. Lowering the seawater pH decreased the survival, calcification rates and somatic and skeletal growth of the larvae. In addition scanning electron microscopy showed evidence of pitting and erosion of the skeleton under lower pH conditions. Contrary to expectations, the polar larvae did not demonstrate a higher sensitivity to lower pH when compared with non-polar larvae.

Morphological and molecular systematic data of elasipodid species (Echinodermata: Holothuroidea) from New Zealand's IPY-CAML 2008 NZ survey of the Ross Sea and Scott and Admiralty seamounts, Antarctica

N. Davey

NIWA (National Institute of Water and Atmosphere Research LTD), Port Nelson, New Zealand

ABSTRACT: An extensive biodiversity survey of the Ross Sea and Scott/Admiralty seamounts was carried out by New Zealand in February/March of 2008. Biological sampling using bottom trawls and coarse and fine mesh epibenthic sleds collected 900 holothuroid specimens. These specimens have been identified using traditional morphological systematics. Some molecular systematic analysis using the cyto-Chrome Oxidase-1 (CO 1) gene has been completed by Gustav Paulay at the University of Florida. A total of 20 species of elasipodids were identified from the survey. The biogeographical relationships of these species are discussed in relation to the known ranges of Antarctic elasipodid biogeography. Distribution and depth records have been extended. Some new species, or variations of known species were found. Five of the elasipodid species *Pannychia sp.* cf *moseleyi*, *Benthodytes sanguinolenta*, *Laetmogone wyvillethomsoni*, *Peniagone affinis* and *P. wiltoni* are discussed in relation to circumpolar distribution and occurrence north of the Antarctic Convergence. There is some degree of congruence between morphological systematic data and molecular phylogentics. As well as evidence of sister species north and south of the Antarctic Convergence. Genetic data thus suggest a review of some morphological systematic conclusions.

Evolutionary pathways among shallow and deep-sea echinoids of the genus *Sterechinus* in the Southern Ocean

A. Díaz & E. Poulin

Instituto de Ecología y Biodiversidad, Departamento de Ciencias Ecológicas, Facultad de Ciencias, Universidad de Chile, Santiago, Chile

J.P. Féral

UMR—DIMAR, Centre d'Océanographie de Marseille, Université Aix-Marseille II. Rue de la Batterie des Lions, Marseille, France

B. David & T. Saucède

Laboratoire Biogéosciences, CNRS, Centre des Sciences de la Terre, Université de Bourgogne, Gabriel, Dijon, France

ABSTRACT: Antarctica possesses a narrow and deep continental shelf that sustains a remarkable number of benthic species. The origin of these species and their affinities with the deep sea fauna that borders the continent shelf are not clear. Until now, two main hypotheses have been considered to account for evolutionary connection between both faunas: (1) either shallow taxa moved down to deep waters (submergence) or (2) deep sea taxa colonized the continental shelf (emergence). The regular sea urchin genus *Sterechinus* is a good model to determine the evolutionary relationships between both faunas. It comprises five morphological species with Antarctic or subantarctic distributions and different bathymetric ranges. Phylogenetic relationships and divergence time among species of *Sterechinus* were established according to the COI mitochondrial gene under the hypothesis of the molecular clock. Results show the existence of two genetically distinct groups. The first one corresponds exclusively to the shallow Antarctic species *S. neumayeri*, while the second one includes all the other morphological species, either deep or shallow, Antarctic or subantarctic. Our results suggest that deep-sea species of *Sterechinus* are more closely related to subantarctic ones, which inhabit presently the continental shelf of Argentina and the Kerguelen Plateau, than to the shallow Antarctic species *S. neumayeri*. Divergence time among species suggests an initial separation between Antarctic and Subantarctic shallow species, and a much later colonization to the deep-sea from the Subantarctic region, probably promoted by the geomorphology of the Scotia Arc.

Metal contaminants: A threat to echinoderms in the 21st century?

P. Dubois
Marine Biology, Université Libre de Bruxelles, Bruxelles, Belgium

ABSTRACT: The effects of metals on larval development of echinoderms have been extensively studied and used as embryotoxicology bioassays. Effects on adults, especially in field conditions, were not so intensively investigated. Nevertheless, current knowledge indicates that members of this phylum are effectively contaminated and impacted by metals although some species appear as rather resistant. For the future, if dumping of metal-containing residues in the sea are now seriously reduced, acidification of sea water due to increasing atmospheric CO_2 concentration could modify metal bioavailability and induce the release of metals from secondary sources. Increased metabolism linked due to higher temperature could also result in higher contamination levels of the organisms. In this presentation, we will review the current knowledge on metal contamination and effects in echinoderms, focusing on field situations, and make a tentative assessment of the future risks induced by these contaminants.

How reliable are age estimations of sea urchins?

C.P. Dumont
The Swire Institute of Marine Science & The Division of Ecology & Biodiversity, The School of Biological Sciences, The University of Hong Kong, Pokfulam, Hong Kong, PR China

J.H. Himmelman
Département de Biologie, Université Laval, Québec City, Québec, Canada

M.P. Russell
Biology Department, Villanova University, Villanova, Pennsylvania, USA

ABSTRACT: Knowledge of growth rates and population age structure are needed to manage commercially exploited species. Errors in aging animals can lead to misestimating longevity, an important mechanism that allows populations to persist in their environments, resulting to decisions to overexploit populations. We conducted a mark and recapture study on the circumpolar sea urchin *Strongylocentrotus droebachiensis* over a 3 years period to measure growth rates of juvenile and adult urchins on kelp bed and barrens. We found a much smaller growth rate on barrens than on kelp bed and a slow growth of juveniles which likely remain cryptic for few years before moving on open areas to forage. We compared our growth rates with previous studies on the same species. Also, from a review of 59 urchin growth studies, we identified a variety of methods and models used to estimate growth and age sea urchins. Although there is no a fool-proof method for estimating age of urchins, a number of methods and models are more accurate than others.

The middle Triassic crinoid *Encrinus liliiformis* from SW Germany and its ecophenotypic reaction to habitat diversity

J.F. Dynowski
State Museum of Natural History Stuttgart, Stuttgart, Germany
Institute of Geosciences, University of Tübingen, Tübingen, Germany

J.H. Nebelsick
Institute of Geosciences, University of Tübingen, Tübingen, Germany

ABSTRACT: The investigation of ecophenotypes is important for delimiting the morphological range of species variations. Their investigation is also important for the reconstruction of environmental parameters controlling organism distribution and development. In this study, the well known crinoid *Encrinus liliiformis* from the middle Triassic Muschelkalk of Central Europe was examined in order to identify variations in crown shape. The analysis of two populations reflecting (1) a shallow water habitat and (2) deeper basinal facies, led to distinguishing two ecophenotypes. The shallow water phenotype has shorter arms with a slightly ornamented dorsal surface of secundibrachials; the deeper water phenotype shows longer arms with a higher degree of ornamentation. These differences can be interpreted in terms of water flow regime and predation pressure. Water flow energy and nutrient supply may be reduced in deeper water environments compared to that in shallow water settings. Arm elongation may thus lead to an increase in the catchment area for filtrating particles out of the water. Ornamentation patterns on the secundibrachials may have an influence on the water currents on a microscale, leading to an increased efficiency of filter feeding. In shallow water habitats, predatory intensity on crinoids is higher than in deeper water. Shorter arms provide less contact surface for predatory attacks. The ornamentation may be an adaption against specialized predator attacks. Possible predators of *Encrinus liliiformis* were various durophagous sharks such as *Acrodus* or *Hybodus*, the shell-crushing placodont *Placodus*, as well as cidaroid echinoids, which are known to feed on Recent stalked crinoids.

Problems with life-history comparisons in echinoderms

T.A. Ebert
Department of Zoology, Oregon State University, Corvallis, OR, USA

ABSTRACT: Comparing life histories has taken various forms. Frequently, measures of growth or a combination of growth and survival have been used. If growth follows the Brody-Bertalanffy model where there is an asymptotic size, S∞, and a rate constant, K, then KS∞, or the ln-transformation of this (suggested by Daniel Pauly), have be used for comparisons as has the relationship between K and the mortality constant, M. If growth is not well modeled by Brody-Bertalanffy but is better described by the logistic, Gompertz, or the more general Richards function, then or Pauley's modification cannot be used because the growth curve does not have a constant rate of change. Life history comparisons are even more difficult if the Tanaka function is the appropriate model for growth.

Biomass/production, B/P, is a currency for comparisons but once growth and survival parameters have been determined there still are problems with conversion to production and biomass. Specific problems include estimating ash-free dry weight in organisms that are isotonic with sea water and estimating age-specific spawn mass. A resolution is illustrated using the red sea urchin *Strongylocentrotus franciscanus* that may be useful as a way of comparing different species both within a single region or across many degrees of latitude.

Assessing shape variation of brachials in comatulids

M. Eléaume, L. Hemery & N. Ameziane
DMPA-UMR5178, Muséum National d'Histoire Naturelle, Paris, France

R. Cornette, V. Debat & M. Baylac
S&E-UMR5202, Muséum National d'Histoire Naturelle, Paris, France

ABSTRACT: In crinoids, arms are constituted of a series of articulated ossicles called brachials. When the arm grows up, new brachials add distally which display juvenile characters, whereas more proximal brachials display fully developed articular facet features. Series of brachials therefore represent ontogenetic series. Inter-specific comparison of brachial ontogenetic trajectories can be used to assess the occurrence of heterochronic processes, leading in turn to primary homology hypotheses. This can be investigated by comparing brachials shape variation at different hierarchical levels (i.e. arm, individual, population, species). Landmark based geometric morphometrics was applied to a set of 366 digitized images of distal articular facets of hypozygal brachials along 26 sampled arms representing five specimens and three species of the species complex *Promachocrinus kerguelensis*. The morphometric and statistical analysis were conducted using the TPS software and R-morph library. Preliminary results show a clear discrimination among arms within an individual, as well as between two individuals of the same species and between individuals from different clades. The allometric trajectories can be then visually displayed and compared. These results suggest that taking into account the structuration of morphological variation at these different biological levels should help gain a better understanding of the phenotypic evolution in comatulids.

Using hydrodynamic modelling and genetics of multiple urchin species to infer marine connectivity in Western Australia

P.R. England, D. Alpers & M. Feng
CSIRO Marine & Atmospheric Research, Hobart, Australia

T. Wernberg
Centre for Ecosystem Management, Edith Cowan University, Joondalup, WA, Australia

ABSTRACT: We used hydrodynamic modelling based on particle tracking in three dimensions to predict likely spatiotemporal patterns of marine larval dispersal in south western Australia under the influence of the Leeuwin current system. To test these predictions against observed, realised patterns of dispersal and connectivity we used an invertebrate model system consisting of several sea urchin species with potentially contrasting reproductive strategies. We characterised geographic population structure in these species using DNA sequence variation at mitochondrial and nuclear genes. Patterns of population structure were compared to expectations based on either known or assumed larval biology and modelled connectivity between west and south coast regions. The match of observed population structure with predictions from hydrodynamic models will help us to: 1. refine our ability to predict connectivity in other organisms using models and 2. infer larval strategy in the WA urchin species in which this is unknown.

New asteroids from the North East Pacific: Exceptional diversity and morphological radiation of the family Solasteridae

A. Gale, R. Clark & P. Lambert

University of Portsmouth, Burnaby Building, Burnaby Road, Portsmouth, Hampshire, UK

ABSTRACT: The asteroid fauna of the Pacific northwest is one of the most thoroughly described in the world's oceans on account of a vast and seemingly exhaustive monograph by Walter K. Fisher (1918–1928). However, recent collecting offshore from British Columbia (Canada) and Alaska (Aelutian Islands, Alaska, USA) has discovered many new taxa belonging to a number of families, most notably the *Solasteridae*, which include two new genera and over ten new species. These bring the tally of solasterids known from the Pacific northwest to a total of twenty three species distributed amongst seven genera, by far the greatest diversity of the family known from anywhere; faunas elsewhere in the world typically include a few species belonging to three genera (*Crossaster, Solaster, Lophaster*). The northeast Pacific radiation includes a new genus which homeomorphs the *Paxillosida* in the development of Luidia—like inferomarginals, intermarginal fascioles and a paxilllar abactinal surface, and other forms which superficially resemble ganeriids and asterinids. The reason for this exceptional diversity remains enigmatic, but the unusually restricted diversity of some families in the region (e.g. asterinids—a single species) may have afforded the solasterids an exceptional diversity of niches.

Origin and evolution of the deep sea asteroid fauna

A. Gale

University of Portsmouth, Burnaby Building, Burnaby Road, Portsmouth, Hampshire, UK

ABSTRACT: The present day deep sea asteroid fauna is dominated by a small number of highly specialised families (Porcellanasteridae, Brisingidae, Freyellidae, Benthopectinidae, Pterasteridae, Zoroasteridae) and others, often with more generalised morphologies which are also widespread in shelf seas (e.g. Astropectinidae, Goniasteridae). Cladistic analysis of these families based upon morphological data consistently places taxa which inhabit shallower environments as basal within their respective groups. A new assemblage of well preserved partial individuals and dissociated ossicles from the Early Oxfordian (Late Jurassic) of the Swiss and French Jura includes an undescribed basal pterasterid, an undescribed basal benthopectinid, and a likely basal zoroasterid (*Terminaster cancriformis*). All species are very small with a radius of about 10 mm. The exceptional preservation of ossicles in clay enables fine details of construction to be identified, and demonstrates, for example, that pterasterids had evolved the complex respiratory pump mechanism described in *Pteraster tessellatus* over 160 million years ago. This fauna lived in a water depth of less than 50 metres, and provides evidence that several families which today are characteristic of the deep sea (500 m+) were abundant in Jurassic mid-shelf environments. Although it appears likely that deep sea asteroid faunas originated in shallower settings and subsequently migrated into deeper environments, really deep water Jurassic and Cretaceous asteroid assemblages are presently unknown. Thus, there is a possibility that at least some of the living deep sea families already occupied this habitat in the Mesozoic.

Field and laboratory estimates of growth for the sea urchin *Lytechinus variegates* in Bermuda

V. Garcia & M.P. Russell
Biology Department, Villanova University, Villanova, PA, USA

T.A. Ebert
Department of Zoology, Oregon State University, Corvallis, USA

A. Bodner
Bermuda Institute of Ocean Science, Ferry Reach, St. George's, Bermuda

ABSTRACT: *Lytechinus variegatus* is an abundant sub-tropical and tropical sea urchin found as far north as North Carolina (USA) and extending as far south as Brazil. We conducted a tagging study for 1 year (2005–2006) in Bermuda to estimate growth and longevity. All individuals were collected from two field sites in 2005: Flatts Inlet (n = 245) and Emily's Bay (n = 242). We recorded test diameters, injected them with the fluorchrome calcein, and released them back into the field. Concurrent with the field study we collected a sample to hold in the laboratory (n = 117)—these sea urchins were also tagged with calcein. In the lab sea urchins were held in a concrete tank that was lined with coral rocks and stocked with seagrass on a periodic basis. After one year all samples were collected, test diameters recorded, and skeletal elements cleaned with Sodium Hypochlorite (household bleach). The demipyramids of Aristotle's lanterns were measured and examined under UV illumination for the calcein tag. Field results indicate a high degree of immigration and emigration. We recovered 498 sea urchins with 11 tagged and 22 with zero tagged from Flatts Inlet and Emily's Bay respectively. In the lab only 6 individuals were not tagged. There was a significant difference in the jaw-test allometric relationships between the lab and Flatts Inlet samples indicating the lab sample was food limited. Growth parameters for the Tanaka function will be reported for both the lab and Flatts Inlet samples.

Structure of the species assemblages of conspicuous echinoderms at the Archipelago Espiritu Santo, Baja California Sur, Mexico

I.A. Guzmán-Mendez & M.D. Herrero-Perezru
Centro Interdisciplinario de Ciencias Marinas, Instituto Politécnico Nacional. Av. Instituto Politécnico Nacional S/N Col. Playa Palo de Santa Rita, La Paz B.C.S. México

A.H. Weaver
NIPARAJA A.C., La Paz B.C.S. México

ABSTRACT: Estimates of species richness, abundance, diversity and evenness were calculated from November 2005 to November 2007 at 12 sites in the Archipelago Espíritu Santo in the Gulf of California, México from a total of 336 (25 × 2 m) belt transects. The taxocenosis of echinoderms composed 3 Classes, 9 Orders, 12 Families, 19 Genera and 20 Species. The highest values of relative abundance were observed at La Bonanza (17%), San Gabriel (15%) and El Bajito (12%). The most abundant Class was Asteroidea (12 species), followed by Echinoidea (7) and Holothuroidea (4). The dominant species was the sea urchin *Tripneustes depressus*, followed by the sea star *Phataria unifascialis*. During the study period we observed that species assemblage was different in 2005, which was dominated by the sea star *Pentaceraster occidentalis*. However, co-dominant species were different each year, in 2005 *T. deppressus* was the third in abundance, while in 2006 was *Centrostephanus coronatus* and in 2007 was *Toxopneustes roseus*. Regarding ecological indexes, diversity was statistically different in 2005, but in 2006 and 2007 values were similar. El Bajito was the most diverse, however no statistical differences were observed with the rest of the study sites. We conclude that the taxocenosis of echinoderms is distributed homogenously along the Archipelago, and the community is clearly dominated by *T. depressus* and *P. unifascialis*. Data suggest that there might be a shift in dominance in time, but there is not enough evidence to establish the periodicity of such changes.

Holothurian settlement in two protected reefs at Cozumel, México

M.D. Herrero-Pérezrul & S. Rojero-León
Centro Interdisciplinario de Ciencias Marinas, Instituto Politécnico Nacional, Ave. IPN s/n Col. Playa Palo de Santa Rita, La Paz, Baja California Sur, México

M. Millet-Encalada
Departamento de Monitoreo e Investigación Parque Nacional Arrecifes de Cozumel, Comisión Nacional de Áreas Naturales Protegidas, Cozumel, Quintana Roo, México

ABSTRACT: We analyzed the intensity of settlement of holothurians in two reefs from the Marine Park Arrecifes de Cozumel, using a series of suspended collector systems (Witham type). The collectors were composed of squares of astroturf mat (50×50 cm) attached to plastic frames supported by buoys and weights to keep them straight. Three collectors were deployed on each site in two reefs (Palancar and Tormentos) (total 10 sites) from May 2007 to June 2008. A total of 191 individuals of *Holothuria arenicola* and 114 *Euapta lappa* were counted during the study period. The smallest specimen belonged to *H. arenicola* (0.9 cm) and the largest (10 cm) to *E. lappa* both found in Nov 2007 at Tormentos reef. Size of youngsters was not statistically different between months or between reefs. However, settlement showed differences amongst reefs but not between months, suggesting that reproduction may be continuous throughout the year. Palancar showed the lowest number of recruits. These differences may be explained by the strong currents that influence Palancar reef and by the lack of hard substrate, this reef is located southward bound the Park and is dominated by sea grass and high levels of turbidity, associated to anthropogenic activities. Tormentos reef is located north of the park and is dominated by coral reef and characterized by clear waters. We conclude that both species settled most of the year and that settlement seem to be strongly influenced by currents and water conditions.

Proteomic analysis of sea star coelomocytes

K. Holm
The Royal Swedish Academy of Sciences, Sven Lovén Centre for Marine Sciences, Kristineberg, Fiskebäckskil, Sweden
Department of Zoology, University of Gothenburg, Göteborg, Sweden

B. Hernroth & M. Thorndyke
The Royal Swedish Academy of Sciences, Sven Lovén Centre for Marine Sciences, Kristineberg, Fiskebäckskil, Sweden

ABSTRACT: Many species of echinoderms have a remarkable ability to regenerate lost tissues. The initial step in regeneration is wound healing and injury initiates an immune response, where the circulating cells are activated. In general, invertebrates have a well developed innate immune system that is mediated by circulating blood cells. In the sea star these kinds of cells, the coelomocytes, respond with a rapid and massive accumulation at the wound site. The development of proteomic studies for non model organisms have in the last years got increased attention. We have developed the method two-dimensional gel electrophoresis (2-DE), for analyzing protein contents of coelomocytes of the sea star *Asterias rubens*, comparing the protein expression before and after wounding. The methodology was optimized in terms of sample preparation, pI interval, gel gradient and staining procedure. The analysis of protein spots using MALDI-TOF/TOF mass spectrometry resulted in 9 identified protein homologues out of 18, of which 6 were found significantly up- or down-regulated. Database searches of MS and MS/MS data included searches against the recently sequenced sea urchin *Strongylocentrotus purpuratus*. These results pointed to a distance in homology of proteins between sea star and sea urchin. The development of proteomic methods for sea star coelomocytes has increased the knowledge about the protein expression of these cells, and with 2-DE it is also possible to search for and discover proteins earlier not described. Future proteomic studies on *A. rubens*, may give valuable insights of the underlying mechanisms and molecules involved in wound healing and regeneration.

Is 'barrens' habitat good for sea urchins?

C.R. Johnson, S.D. Ling & E.M.A. Strain
School of Zoology and Tasmanian Aquaculture and Fisheries Institute, University of Tasmania, Hobart, Australia

ABSTRACT: The established view that sea urchin 'barrens' are poor habitats for sea urchins is based on observations that individuals on barrens have lower growth and gonad production, and greater relative jaw size, than their conspecifics in seaweed beds. Here we argue that this notion is '1-dimensional' and that the relationship between sea urchins and barrens habitat is both more subtle and more complex. We review literature and present new data to argue that, in some circumstances, (1) sea urchins demonstrate 'preference' for local barrens patches, and can be disadvantaged by living in seaweed beds, and (2) there can be benefits at a population level to inhabiting barrens habitat. There is strong evidence that, in Tasmania, *Heliocidaris erythrogramma* is negatively affected by competition with *Centrostephanus rodgersii*. However, in experiments designed to separate the effects of *C. rodgersii* from that of the barrens they create, juvenile *H. erythrogramma* recruited at significantly higher densities in barrens patches from which *C. rodgersii* and various guilds of algal regrowth were removed than in adjacent patches in intact seaweed beds. Similarly, adult *H. erythrogramma* occurred at higher densities in barrens patches cleared of *C. rodgersii* than in adjacent areas of intact seaweed communities. Regarding *C. rodgersii*, individuals in seaweed beds have significantly shorter spines than conspecifics living on adjacent barrens because of abrasion by algae. Individuals on barrens are less susceptible to predation, not only because predators are less abundant but also because they have larger spine canopies than their conspecifics living in seaweed.

Identifying management options to minimise risk of development of sea-urchin barrens

C.R. Johnson, J.C. Sanderson, S.D. Ling, C. Gardner, S.D. Frusher, K.S. Redd & H.G. Pederson
School of Zoology and Tasmanian Aquaculture and Fisheries Institute, University of Tasmania, Hobart, Tasmania, Australia

ABSTRACT: Recent establishment of the long-spined sea urchin (*Centrostephanus rodgersii*) in eastern Tasmania as a direct consequence of climate change poses significant challenges for management given its propensity to overgraze seaweeds and benthic invertebrates to form extensive 'barrens' habitat. Formation of sea-urchin barrens results in local collapse of biodiversity, production and key fisheries (abalone, *Haliotis rubra*, and rock lobster, *Jasus edwardsii*). Current research is focussed on identifying options for managers to respond to the threat of *C. rodgersii* barrens in Tasmania. The work has three main elements: (1) using modelling to identify potential management options to increase the biomass of large rock lobsters as the principal predator of C. rodgersii in Tasmania (coastal scale), (2) experiments involving translocation of large numbers of large lobsters (mesoscale), and (3) culling of sea urchins by abalone fishers while fishing (local scale). Results of translocations of thousands of large (~2.5+ kg) rock lobsters are encouraging in that many of the translocated animals are establishing dens at release sites, even on extensive barrens, and at sufficient densities to ensure overlapping home ranges. Moreover, DNA analysis of rock lobster faecal pellets indicates that these large predators are feeding readily on *C. rodgersii* material, while the native sea urchin *Heliocidaris erythrogramma* is a rare component of their diet. Modelling indicates that current fisheries management practice will not result in increased biomass of large predatory lobsters on the east coast of Tasmania over the next decade.

At the root of the problem—phylogeny of post-palaeozoic echinoids

A. Kroh
Natural History Museum Vienna, Department of Geology & Palaeontology, Burgring, Austria

A.B. Smith
Natural History Museum, Palaeontology Department, Cromwell Road, London, UK

ABSTRACT: Previous phylogenetic analyses of echinoids have either examined specific subgroups in detail or have looked at a relatively small number of taxa selected from across the class. Taxon sampling, however, is known to affect the accuracy of cladistic trees, with sparse sampling adversely affecting the accuracy of results. To test ideas about major relationships amongst post-Palaeozoic echinoids we carried out a cladistic analysis in which all named Post-Palaeozoic echinoid families and subfamilies were included. This is a compromise between capturing the diversity of form that exists, and keeping the number of taxa to a level that is practicable for analysis. We scored 170 taxa for 305 skeletal characters (excluding pedicellariae). There was a surprising lack of primary data on plating pattern, lantern, and girdle structure for many supposedly "well known" taxa. Although parts of the resulting trees are only weakly resolved (e.g. the precise sister group of the Irregularia), other parts are unambiguous. The origin of the clypeasteroids, for example, clearly lies within "cassiduloids", with *onoclypus* and oligopygoids rooting as stem-group members in all our analyses. Cassiduloids are paraphyletic, with Echinolampadidae as closest living sister group to the clypeasteroids. Spatangoids and holasteroids, in contrast, appear monophyletic, stemming from a bunch of Jurassic echinoids including *Acrolusia*, *Desorella* and disasterids. Unfortunately, deep branches are often not supported by unique apomorphies. Higher taxa apparently acquire a characteristic set of feature over time only. Diagnoses based on terminal taxa thus fail to adequately encompass fossil stem-group members. Elucidation of the relationships of taxa at the root of large groups is hampered by mosaic type character evolution.

Sea urchins in an acidic and warm high-CO_2 world

H. Kurihara, R. Yin & A. Ishimatsu
Nagasaki University, East China Sea Research Institute, Tairamachi, Nagasaki, Japan

ABSTRACT: The increase of atmospheric CO_2 concentration is changing oceanic carbonate chemistry (elevated seawater pCO_2 and reduced pH), with profound impacts on marine organisms. Echinoderm are one of the most susceptible organisms to these environmental changes since the seawater chemistry change decreases the calcium carbonate saturation state and thereby impairs the calcification process of marine calcifiers. Here we review our studies evaluating the effects of elevated pCO_2/reduced pH conditions on different life stages of sea urchins including egg, larva and adult stages. Morphological, physiological, histological and molecular biological studies were conducted to assess the effects. Our data show that seawater conditions predicted for the end of this century will affect fertilization, larva skeletogenesis, adult physiology and reproduction of sea urchins. Additionally, synergistic effects of ocean acidification and global warming on reproduction and physiology were also observed. Preliminary data evaluating the effects of ocean acidification on the expression of skeletogenesis-related genes will also be presented and potential effects of the future predicted oceanic environment on sea urchin population will be discussed.

Epifauna associated with the sea cucumber *Holothuria mexicana* in Puerto Rico

E.N. Laboy-Nieves
Universidad del Turabo, School of Science and Technology, Gurabo, Puerto Rico

M. Muñiz-Barretto
Universidades do Estadual do Rio de Janeiro. R. São Francisco Xavier, Maracanã, Rio de Janeiro, RJ Brasi, Brazil

ABSTRACT: The epifauna of two populations of the shallow water sea cucumber *Holothuria mexicana*, inhabiting sea grass beds in Guilligan Island y Caribe Keys (Puerto Rico, West Indies) where analyzed. We found that all individuals are covered by a whitish mucilage, probably secreted by them or by microorganisms, to which organic debris (shells, vegetative litter, and coral fragments) and non-organic debris (gravels, sand, and plastic and metallic fragments) are adhered. The epidermis of 122 specimens was gently scrubbed *in situ* with a spatula for ~15 seconds, while washing the animal to collect the epidermic material. Individuals were returned to their habitat. The collected material was preserved in 75% ethanol and samples were filtered with a set of meshes to separate the debris. The debris was separated and organisms were collected and identified. The liquid sample was examined with a compound microscope to identify smaller organisms. 100% of samples contained macro- and micro-invertebrates classified as polychaetes [39.0%], mollusks [29.5%], crustaceans (isopods, copepods, ostracods, stomatopods, amphipods, decapods) [28.0%] and others (zoanthids, ophiuroids, porifers) [3.5%]. The amount of mollusks and crustaceans was significantly different across the two localities. We did not consider whether the existence of this epifauna occurs at random, is a byproduct of commensalisms or symbiosis, or if the mucilage provides feeding or shelter material to them. It is worthy of mention that this sea cucumber species secretes holothurin A and B, a toxin for many potential predators. These preliminary findings constitute baseline data for future projects aimed to assess the ecological niche of *Holothuria mexicana* and its epifauna.

Aspidochirote sea cucumber diversity and status of stocks in the Bunaken National Marine Park (BNMP), North Sulawesi, Indonesia

D.J.W. Lane
University of Brunei, Darussalam Jalan Tungku Link, Bandaer Seri Begawan, Brunei Muara District, Brunei

ABSTRACT: The species richness of aspidochirote sea cucumbers associated with reefs of the BNMP, North Sulawesi is very high, suggesting, along with data for other faunal groups, that this archipelagic region, including other islands neighbouring islands, at the tip of North Sulawesi may constitute a diversity 'hot spot' within the Indo-Malay zone of maximum marine biodiversity. For some forms, abundances are high, but patchily distributed, compared to other areas in the Indo-Pacific, a phenomenon which may reflect localized concentration of settling recruits in areas of eddy currents. Yet in general, abundances of commercial sea cucumber species are low. It is known that, prior to initial designation of this area as a National Marine Park in 1991 and the commencement of protection measures, the reefs and reef slopes around the Bunaken Islands were heavily targeted for commercial sea cucumber species. No reliable data exists for this fishery but recent preliminary findings of sea cucumber surveys indicate that, over the last two decades, recovery of populations has yet to occur.

Sea urchin grazing in seagrass in temperate Western Australia (Luscombe Bay, Cockburn Sound)

M.W. Langdon & M.van Keulen
School of Biological Sciences, Murdoch University, South Street, Murdoch, WA, Australia

E.I. Paling
School of Environmental Science, Murdoch University, South Street, Murdoch, WA, Australia

ABSTRACT: Urchins have been identified as key grazing invertebrates in tropical and subtropical seagrass meadows and at high population densities have been implicated in altering stable states of marine ecosystems i.e. the formation "urchin barrens" in areas that were formerly dominated by macrophytes such as kelps and seagrasses. The seagrasses of Cockburn Sound, temperate Western Australia (32°10'14"S, 115°43'26"E) declined during the 1970s. Although the majority of the seagrass losses were apparently caused by eutrophication, a number of isolated urchin grazing events caused further localised losses. One of these localised events was studied in 1992 and was the site of seagrass transplanting activities in 2004. In this study, I investigated the current status of this localised urchin barren and determined how the extent of seagrass has changed over time. The current urchin population was found to be limited to a small area in the north-east section of the barrens where densities ranged between 2 and 10 individuals m^{-2}. Analyses of historical aerial photographs using GIS showed that the period of rapid seagrass meadow decline coincided with the presence of an unusually large aggregation of the grazing urchin *Heliocidaris erythrogramma* (Valenciennes). The post-1993 seagrass recovery also coincided with the reported collection and removal of the urchins in 1992. Seagrasses have started recolonising around the perimeter of the scar, particularly in narrow sections of the barrens.

Molecular phylogeny of symbiotic pearlfishes

D. Lanterbecq, E. Parmentier, M. Todesco & I. Eeckhaut
University of Mons, Mons, Hainaut, Belgium

ABSTRACT: Carapid fishes are certainly among the most remarkable organisms living in symbiosis with echinoderms. These fishes, known as pearlfishes, belong to the family of Ophidiiform, which is divided in two sub-families, the Pyramodontinae and the Carapinae. Symbiotic pearlfishes belong to the last sub-family. They are found inside bivalves, ascidians, asteroids and holothuroids. Their behavior and morphology are extremely adapted to their symbiotic way of life. Numerous aspects of the biology of these fishes have been recently resolved though not in an evolutionary perspective. We present here the first molecular phylogeny of the symbiotic pearlfishes, including *Echiodon, Onuxodon, Carapus* and *Encheliophis* genera. The phylogenetic relationships of twenty specimens from eight species coming from Mediterranea, the Indian Ocean and the Pacific Ocean have been estimated. Five mitochondrial fragments (3,645 bp) have been sequenced and the phylogenetic trees were obtained via Maximum Parsimony, Maximum Likelihood and Bayesian analyses. The analyses suggest the paraphyly of the *Carapus* group in regard with the monophyletic *Encheliophis*, (ii) *C. boraborensis* is the sister group of all the other sequenced pearlfishes and (iii) *C. bermudensis* and *C. acus* are grouped within a clade. Various characters have been mapped on the trees to estimate the history of these symbiotic fishes.

The geometry of the Allee effect

G.R. Leeworthy
Deakin University, School of Life and Environmental Sciences, Warrnambool Campus, Victoria, Australia

ABSTRACT: Basic geometry can be used to assess how much impact a fishing pattern has on the spawning success of the target species. A hypothesis is presented to suggest why many stochastic stock assessment models, used for the management of fisheries, have traditionally failed. Novel manipulation of fishing patterns may be a way to prevent collapse in fisheries. Preliminary examples from echinoderm fisheries are presented, including reference to stock collapses in the Torres Strait and Red Sea.

Climate-driven range extension of a sea urchin leads to a new and impoverished reef state

S.D. Ling & C.R. Johnson
School of Zoology, and TAFI, University of Tasmania, Hobart, Tasmania, Australia

K. Ridgway & A.J. Hobday
CSIRO Marine & Atmospheric Research, Castray Esplanade, Hobart, Tasmania, Australia

S. Frusher & M. Haddon
TAFI, University of Tasmania, Taroona, Tasmania, Australia

ABSTRACT: Global climate change is predicted to have major impacts on marine biodiversity, particularly when habitat-modifying species undergo range shift. In south east Australia, the barrens-forming sea urchin (*Centrostephanus rodgersii*—Diadematidae) has recently undergone a poleward range extension to Tasmania. The sequential southward appearance of the sea urchin, a pattern of declining age and a general reduction in abundance with increasing latitude along the eastern Tasmanian coastline is consistent with recent warming and dispersal opportunities driven by change in the behaviour of the East Australian Current (EAC). While many species have been documented to undergo range shift to eastern Tasmania, *C. rodgersii* is of major ecological importance as overgrazing by this species results in local losses of ~150 taxa typically associated with Tasmanian macroalgal beds. Such a disproportionate effect by a single range-expanding species demonstrates that climate change may lead to unexpectedly large and non-linear impacts as key habitat-modifying species undergo range modification. Predicted ongoing climate change in the region will favour continued population expansion of *C. rodgersii*, not only as a result of atmospheric forced ocean warming, but also via ongoing intensification of the EAC driving continued poleward supply of larvae and heat. This trend demands explicit effort from managers to build resilience in this important temperate reef ecosystem.

Human-facilitated reproductive hotspots of an introduced seastar

S.D. Ling & C.R. Johnson
School of Zoology and TAFI, University of Tasmania, Hobart, Tasmania, Australia

C.N. Mundy & J.D. Ross
TAFI, University of Tasmania, Taroona, Tasmania, Australia

A. Morris
Department of Primary Industries and Water, Hobart, Tasmania, Australia

ABSTRACT: Opportunities to minimise the risk of continuing spread of an invasive species often rest with limiting its reproductive output. In the Derwent Estuary (SE Tasmania), the introduced northern Pacific seastar *Asterias amurensis* (Lütken) has proliferated to become the chief benthic predator, just as occurs when populations 'outbreak' in its native range. Within the Derwent Estuary, *A. amurensis* is most abundant around wharf and pier structures where it is heavily subsidised by food, namely mussels, that grow on the pylons and fall to accumulate on the seafloor. Sub-populations of *A. amurensis* associated with wharves demonstrate high reproductive capacity in terms of relative and absolute gonad size, and length of spawning season relative to nearby populations away from this novel anthropogenic habitat. Wharf sub-populations also contain larger individuals that remain highly aggregated at extreme densities (equivalent to 'out-break' populations) year around, while elsewhere in the Estuary, *A. amurensis* occurs at much lower densities. Predictions of fertilisation success using a spatially-explicit fertilisation model suggest that wharf sub-populations, while representing <10% of the total population in the Estuary and concentrated in <0.1% of the total area, contribute up to 80–90% of the total zygote production in the Estuary. These human-facilitated 'hotspots' of reproductive output are likely to have been important in the seastars spread beyond the Estuary. Management action targeting 'reproductive hotspots' have the potential to significantly reduce the risk of further spread of invasive species.

Recruitment of echinoderms to artificial habitats in temperate Australia: An assessment of their potential as a surrogate for total biodiversity

R.H. Magierowski & C.R. Johnson
School of Zoology, and TAFI, University of Tasmania, Hobart, Tasmania, Australia

ABSTRACT: Community assembly in macrofauna communities developed in artificial kelp holdfasts was monitored at 1-month intervals over a 13 month period using a sampling design that used systematic patterns of temporal overlap and changes in start and collection dates. The two main aims of this experiment were (1) to study the links between recruitment and community dynamics through the assembly process, and (2) to assess the performance of surrogates of overall community structure through changes in season and community age. Because biodiversity can be defined validly in a number of ways, we examined both univariate and multivariate indices of familial biodiversity. This paper presents results of this experiment with a focus on the phylum Echinodermata. While interactions among species (both positive and negative) were an important determinant of overall community structure in this type of community, these interactions appeared to be less important for echinoderms than for other phyla. This result may explain why familial richness of echinoderms was a good surrogate for total familial richness in communities at an early stage of development, but not for older, more developed communities. In the multivariate consideration of surrogacy of total familial biodiversity, Echinodermata was the worst performing phylum amongst those examined, although none of the phyla examined performed better than random selections of the same number of families. Overall, the diversity of echinoderm families was a poor predictor of the familial diversity of the total assemblage.

Ophiuroidea luminescence: Diversity and distribution, a first analysis

J. Mallefet
Marine Biology Laboratory, Catholic University of Louvain, Louvain-la-Neuve, Belgium

T.D. O'Hara
Museum Victoria, Melbourne, Australia

ABSTRACT: Echinoderms' capability to produce light, i.e. bioluminescence is still poorly understood in terms of phylogenetic and biogeographic distributions. Mainly studied in the Ophiuroidea, it has been well documented but on a rather limited number of species. Thanks to a new sampling effort, comparative study of ophiuroids luminescence has been initiated. Field observations and laboratory experiments were used to determine the luminous status of 171 species. Compiling these data with literature, the total number of luminous ophiuroids species reaches 64. Analysis indicated that these species were mainly collected on soft substratum in deep water where luminescence appeared more intense. Ophiuroid luminescence was less frequently observed in tropical coral reefs than in temperate waters across both shallow and deep sea habitats. A larger sampling effort will be required in order to try to highlight a possible link between luminescence and ophiuroid phylogeny. New field surveys in a variety of marine regions and habitats will be organized to reach this goal. This is also necessary in order to understand why so many brittle stars glow in the dark.

Function of *Ophiopsila aranea* luminescence (Ophiuroidea, Echinodermata)

J. Mallefet, O. Hurbin & A. Jones
Marine Biology Laboratory, Catholic University of Louvain, Louvain-la-Neuve, Belgium

ABSTRACT: In 1995 Herring stated that luminescence in all echinoderms might play an anti-predation role but experimental studies are rather scare. Among the various strategies used to avoid predation it has been suggested that ophiuroids use luminescence as aposematic signal (warning the visual predator to avoid eating the animal) but bright luminescence of arm tips was also interpreted as evidence for sacrificial lure (autotomy of a luminous arms in order to survive). In this work we measured predation rate by two crab species (*Carcinus maenas* and *Pilumnus histellus*) on *Ophiopsila aranea* and on a non-luminous ophiuroid, *Ophiura texturata* in aquarium. Interactions were recorded using one intensified-camera in order to document and quantify prey and predator behaviours. Experiments were also performed using agar blocks stuffed with either fish meat, arms of *O. aranea* or of *O. albida*. Results showed that (i) predation rates were higher on the non-luminous ophiuroid species; (ii) there was more rejection and avoidance behaviours observed when predators were facing *O. aranea*; (iii) the agar blocks containing *O. aranea* arms were eaten less frequently than the other agar block 'treatments' by the crabs. All these data are in favour of the aposematic use of luminescence by *O. Aranea*; the ophiuroid uses light in order to warn predators that they are unpalatable.

Paedomorphosis in brittle stars—postlarval evo-devo

A. Martynov
Zoological Museum of Moscow State University, Moscow, Russia

S. Stöhr
Swedish Museum of Natural History, Department of Invertebrate Zoology, Stockholm, Sweden

ABSTRACT: Heterochronic development has been suggested as a major driving force of evolution. Among echinoderms it has been documented for various taxa, both Recent and extinct (ophiuroids, echinoids, crinoids, edrioasteroids, mesozoic goniasterids). A strongly paedomorphic appearance, defined as the presence of juvenile traits in adult animals, is particularly common in small deep-sea species of the family Ophiuridae. This study is the first systematic attempt to define and describe paedomorphosis in ophiuroids and its phylogenetic implications. We examined external and internal skeletal characters of 20 paedomorphic species of the genera *Anthophiura, Bathylepta, Ophiomastus, Ophiomisidium, Ophiopyrgus, Ophiotypa, Ophiozonella* and *Perlophiura* by SEM. For comparison we examined small juveniles of non-paedomorphic Ophiuridae and other families. Preliminary results suggest that there are indeed parallels between juvenile and paedomorphic morphologies. Typical juvenile characters include the presence of a primary rosette, a limited number of disk scales, elongated arm joints and fused oral papillae. In both juveniles and strongly paedomorphic adults we found elongated jaws, which easily separate into a proximal and a distal plate, a short concave dental plate, bearing only a single tooth, and elongated arm vertebrae that are only partially fused. The degree of paedomorphosis varies between species and even adults with highly differentiated skeleton may retain a small number of juvenile traits. Taxa with similar paedomorphic morphology appear to have evolved separately in distantly related families (i.e. Ophiuridae, Ophiolepididae) and our results suggest systematic changes in several genera, e.g. *Ophiomastus* and *Perlophiura*.

Exploring ecological shifts using qualitative modelling: Formation of sea urchin barrens and alternative states on Tasmanian rocky-reefs

M.P. Marzloff & C.R. Johnson
School of Zoology & TAFI, University of Tasmania, Hobart, Tasmania, Australia

J.M. Dambacher
CSIRO Marine and Atmospheric Research, Castray Esplanade, Hobart, Tasmani, Australia

ABSTRACT: Alternative stable states characterise many natural ecosystems. Subtidal rocky-reefs on the east coast of Tasmania persist in a range of different configurations, including so-called sea urchin 'barrens' and dense seaweed beds with a closed canopy. In creating and maintaining barren habitat on temperate reefs, sea urchins induce major losses of production, biodiversity and physical structure. Two species—the native sea urchin (*Heliocidaris erythrogramma*) and the invasive long-spined sea urchin (*Centrostephanus rodgersii*)—are responsible for the formation of barrens in Tasmania. Formation of urchin barrens on the east coast has been a rising concern in recent decades, in particular because the two most valuable fisheries in the state, for blacklip abalone (*Haliotis rubra*) and southern rock lobster (*Jasus edwardsii*), are not viable on barrens. Thus, identifying triggers of barrens formation is critical in the management of these reefs. Here we explore the dynamics of Tasmanian rocky reef communities using qualitative modelling informed by empirical knowledge of interactions among species. We use loop analysis since it sacrifices precision and maximizes generality and reality in providing a causal understanding of complex systems. The network topology forms emergent feedback patterns that cause meta-stable properties in the system, and the models capture formation of urchin barrens and the mechanisms of the phase shifts. Fishing is identified as a perturbation that can reduce resilience of the highly productive seaweed-dominated state. We show that qualitative loop models can be valuable in identifying the kinds of system dynamics that managers need to consider in ecosystem based management, but they are not designed to act as definitive management models for specific instantiations.

The concerted regulation of acrosome reaction in starfish by intracellular cGMP, cAMP, and Ca2+ levels

M. Matsumoto
Department of Biosciences and Informatics, Keio University, Yokohama, Japan

O. Kawase
Department of Biosciences and Informatics, Keio University, Yokohama, Japan
Obihiro University of Agriculture and Veterinary Medicine, Inada-cho, Obihiro, Hokkaido, Japan

M.S. Islam
Department of Biosciences and Informatics, Keio University, Yokohama, Japan
Department of Biological Science, Graduate School of Science, The University of Tokyo, Tokyo, Japan

M. Naruse
Department of Biosciences and Informatics, Keio University, Yokohama, Japan

M. Hoshi
Department of Biosciences and Informatics, Keio University, Yokohama, Japan
The Open University of Japan, Wakaba, Mihama-ku, Chiba City, Japan

ABSTRACT: The acrosome reaction, namely, the exocytosis of the acrosomal vesicle in the sperm head, is an essential event for fertilization in many species. The acrosome reaction is initiated when a sperm interacts with an egg investment. In many marine invertebrates, including echinoderms, exocytosis is followed by the formation of an acrosomal process that projects from the anterior tip of the sperm head. Acrosome reaction is induced by specific signals present in the extracellular egg coats. In the starfish, *Asterias amurensis*, three components in the jelly coat of eggs, namely Acrosome Reaction-Inducing Substance (ARIS), Co-ARIS and asterosap, act in concert on homologous spermatozoa to induce the Acrosome Reaction (AR). Molecular recognition between the sperm surface molecules and the egg jelly molecules must underlie signal transduction events triggering the AR. Asterosap is a sperm-activating molecule, which stimulates rapid synthesis of intracellular cGMP, pH and Ca2+. This transient elevation of Ca2+ level is caused by a K+-dependent Na+/Ca2+ exchanger, and the increase of intracellular pH is sufficient for ARIS to induce the AR. The concerted action of ARIS and asterosap could induce elevate intracellular cAMP levels of starfish sperm and the sustained increase in [Ca2+], which is essential for AR. The signaling pathway induced by these factors seems to be synergistically regulated to trigger the AR in starfish sperm.

Challenges and breakthroughs in the study of deep-sea echinoderm biology

A. Mercier
Ocean Sciences Centre, Memorial University, St. John's, NL, Canada

ABSTRACT: The general paucity of data available on the life history strategies of deep-sea organisms can be attributed chiefly to irregular sampling opportunities and difficulties associated with keeping live specimens in the laboratory (especially in sufficient numbers and in good enough health for them to exhibit feeding and reproductive cycles consistent with their natural behaviours). In spite of these challenges, important breakthroughs have been made in the past few decades with respect to the feeding and reproductive ecology of deep-sea echinoderms. Studies have initially focused on the analysis of biomarkers, gut contents, body indices and histological sections from preserved samples. Echinoderms were among the first deep-sea animals shown conclusively to display seasonal reproductive periodicity. One of the main hypotheses forwarded to explain such trends include the periodic deposition of organic matter to the deep-sea in regions where seasonal surface production is observed. Variability of food supply on a seasonal or annual basis especially that of critical compounds like carotenoids, may have a considerable effect on the reproduction, especially in deposit-feeding species. Social behaviours (e.g., permanent herding, transient aggregations) are also proposed to be of primary importance in ensuring reproductive success. Experimental studies are now allowing the finer study of prey-predator interactions, reproductive periodicities, larval development and growth patterns. We are slowly beginning to understand how selective pressures such as low population densities, spatially uniform habitats and food limitation may influence life-history traits in deep-sea echinoderms.

Mitochondrial markers reveal many species complexes and non-monophyly in aspidochirotid holothurians

F. Michonneau, K. Netchy, J. Starmer, S. McPherson, C.A. Campbell, S.G. Katz,
L. Kenyon, J. Zill & G. Paulay
Department of Biology, University of Florida and Florida Museum of Natural History, Gainesville, FL, USA

S. Kim & A.M. Kerr
Marine Laboratory, University of Guam, Mangilao, Mangilao GU, USA

M. O'Laughlin
Museum Victoria, Melbourne, Australia

ABSTRACT: Despite their striking presence on coral reefs and the multi-millon dollar industry targeting them, aspidochirotid holothurians have received relatively little taxonomic attention. Heavy reliance on ossicles has lead to a confused taxonomy and masks substantial cryptic diversity. As part of revisionary effort on holothurians, we are sequencing multiple specimens of available species to test for species limits and construct phylogenetic hypotheses. We are using a bottom-up approach, sequencing fast-evolving mitochondrial markers (16S and COI) first, as these are mostly informative at lower taxonomic levels. We have unraveled many species complexes and identified several undescribed species. For instance, our analyses revealed that the circumtropical "species" *Holothuria impatiens* consist of a dozen reciprocally monophyletic, well-defined, Evolutionary Significant Units (ESUs). Broad overlap in the range of some, in combination with recent divergence indicate the rapid evolution of reproductive isolating barriers among these ESUs. Such rapid evolution to sympatric coexistence is also found in several other species complexes we identified, and contrasts with most other marine invertebrates. At a higher taxonomic level, preliminary results show that nonmonophyly of currently recognized taxa is prevalent. Stichopodids emerge from paraphyletic synallactids. While holothuriids appear monophyletic, *Holothuria* is not, as several subgenera in that genus are deeply divergent. We will present the latest phylogenetic hypotheses which further our understanding of speciation and trait evolution in sea cucumbers.

The effect of ocean acidification on early life history stages of the Australian sea urchins *Heliocidaris tuberculata* and *Tripneustes gratilla*

S. Mifsud
Marine Ecology Group, Macquarie University, Sydney, NSW, Australia

J. Havenhand
Tjärnö Marine Biological Laboratory, Göteborg University, SE, Sweden

J. Williamson
Marine Ecology Group, Macquarie University, Sydney, NSW, Australia

ABSTRACT: Rising atmospheric CO_2 is increasing ocean acidity and lowering carbonate ion concentrations of the world's oceans. The IPCC predicts that there will be a three-fold increase in ocean acidity by 2100, thus dramatically compromising the ability of many marine organisms to form calcareous skeletons and shells essential to their survival. As such, ocean acidification is considered one of the most serious long-term threats to marine organisms and ecosystems, yet we have very little information on the underlying mechanisms and impacts. The few existing data pinpoint early life history stages as particularly important for marine invertebrates. This project assessed the impact of ocean acidification on the early life history stages of two common species of sea urchins found in the Sydney region: *Heliocidaris tuberculata* and *Tripneustes gratilla*. Both have larval stages that accrete carbonate skeletons in the early stages of their larval development prior to feeding. Sperm motility, fertilization success, and larval development were assessed for both species in pH conditions predicted to occur by 2050 by the IPCC (2004). Results were compared with those from a similar recent study by Havenhand and Williamson on *Heliocidaris erythrogramma*, a species occurring sympatrically with *H. tuberculata*. Results revealed that the effects of CO_2-induced ocean acidification are variable at different early life history stages and appear to be species-specific. Such data are important in the construction of acidification response scenarios for marine species that will allow managers to identify effective responses in the rapid acidification of the oceans.

Divergent reproductive mode reveals cryptic speciation in the lesser biscuit star *Tosia australis*

K. Naughton
Museum Victoria, Carlton, Victoria, Australia
University of Melbourne, Parkville, Victoria, Australia

ABSTRACT: The evolutionary basis of modern biology relies on an accurate understanding of species relationships. Cryptic species complexes consequently pose a problem for biologists, particularly taxonomists and evolutionary biologists. This study assesses the morphological, molecular and reproductive variation within a possible cryptic species complex, the goniasterid sea-star *Tosia australis*. Two reproductive modes were found to occur within Victorian and Tasmanian *T. australis*, including a brooded, benthically-developing larva and a free-spawned, swimming larva. This is the first documentation of brooded larval development in *T. australis* and in the Goniasteridae. The brooded larvae also lack cilia, a novel feature in asteroid larvae. The molecular study sequenced fragments of two mitochondrial genes and a nuclear gene, revealing the presence of cryptic speciation within the genus.

Heat shock protein expression as a function of temperature and season in *Heliocidaris* sea urchins

H.D. Nguyen
Anatomy and Histology, Bosch Institute, University of Sydney, NSW, Australia

M. Thomson
School of Biological Sciences, Heydon Laurence Building, University of Sydney, NSW, Australia

M. Byrne
Anatomy and Histology, Bosch Institute, University of Sydney, NSW, Australia

ABSTRACT: Eastern Australia is a climate change hot spot where, due to a disproportionate increase in sea temperature, marine life in the region is particularly vulnerable to impacts of climate change. This study investigated the response of two common sea urchins (*Heliocidaris erythrogramma, H. tuberculata*) whose distributions coincide with this hotspot to thermal stress using heat shock protein (hsp) expression as a biomarker to assess their response to environmental change. Endogenous levels of hsps in the 70-kDa class (hsp70) in the gonads were compared between winter and summer in response to acute temperature shock in the laboratory at 18°C, 22°C, 25°C and 30°C. Seasonal variation in hsp70 expression was observed in both species where in summer, hsp70 levels increased in response to cold and heat shock in *H. tuberculata* but only to heat shock at 30°C for *H. erythrogramma*. In winter, hsp70 expression in *H. erythrogramma* was elevated at 25°C but *H. tuberculata* expressed significantly lower hsp70 levels at 25°C and 30°C than at lower temperatures. *H. tuberculata* appears to be less tolerant to wide temperature variability in winter than *H. erythrogramma* and with future ocean warming for Eastern Australia projected to be greatest in winter, *H. tuberculata* may be more vulnerable to climate change.

Echinoderm biogeography in the Southern Hemisphere

T.D. O'Hara
Museum Victoria, Melbourne, VIC, Australia

ABSTRACT: Biogeographers like to divide the earth into discrete faunal regions and assume a common evolutionary history to those species within. However, the oceans are complex 3-dimensional structures and depth is usually the most important geo-positional factor structuring marine assemblages. For example, there are usually greater differences between shelf and slope communities within a region than between regions at the same depth. Isolated seamounts will have many of the same species at similar depths. The marine environment can be thought of as a series of depth strata, with each strata consisting of patches of habitat, connected to varying degrees by dispersal or vicariant events, and consequently exhibiting different regional patterns. Here I present the varying faunal patterns shown by ophiuroids at different depths around Australia and New Zealand. The distributions of 200 species are modelled over large spatial scales using oceanographic and geomorphologic factors. The transition between tropical, temperate and Southern Ocean faunas occurs at different latitudes for different depth strata. Longitudinal connectivity and evolutionary history also varies with depth. Global warming scenarios are analysed and the implications discussed.

A taxonomic review of the family Asteroschematidae (Ophiuroidea) from Japan

M. Okanishi & T. Fujita

National Museum of Nature and Science, Tokyo, Hyakunin-cho, Shinjuku-ku, Tokyo, Japan

ABSTRACT: The family Asteroschematidae, one of the four families of euryaline ophiuroids, is mainly distributed particularly on the hard bottom in deep water habitat, greater than 100 m. It is relatively difficult to collect these poorly known animals and their taxonomic study is behind the other taxa of Ophiuroidea. From Japanese waters, 3 genera and 10 species have been so far recorded. Here the taxonomy of Japanese asteroschematid ophiuroids is reviewed based on the external and internal morphology using the specimens deposited in National Museum of Nature and Science, Tokyo, and the type specimens of Museum of Comparative Zoology at Harvard University, Zoological Museum of Amsterdam and Zoological Museum, University of Copenhagen. The density and shape of dermal ossicles, the layer structure of radial shields and present/absent of conical tubercles on the surface of vertebrate showed clear interspecific differences and they were provided to be useful as taxonomic characters in species level. Observations of these characters revealed following taxonomical findings: two undescribed species of *Asteroschema* and *Astrocharis* were found; *Ophiocreas abyssicolla* is not an asteroschematid but a member of the genus *Astrodia* of the other family, Asteronychidae; *Asteroschema bidwigi*, *A. wrighti* and *A. tubiferum* are junior synonyms of *Ophiocreas sibogae*; *Ophiocreas mortenseni* and *O. glutinosum* are junior synonyms of *O. carnosus*; *Astrocharis gracilis* is a junior synonym of *A. ijimai*. As a result, 3 genera and 17 species including 2 undescribed species and 6 new record species are distributed in Japan.

Observations of reproductive strategies for some dendrochirotid holothuroid species (Echinodermata: Holothuroidea: Dendrochirotida)

M. O'Loughlin & C. Rowley

Marine Biology Section, Museum Victoria, Melbourne, Victoria, Australia

J. Eichler, L. Altoff, A. Falconer, M. Mackenzie & E. Whitfield

Marine Research Group, Field Naturalists Club of Victoria, Australia

ABSTRACT: Some recently observed reproductive strategies by dendrochirotid holothuroid species are reported and illustrated: fissiparity by *Cucuvitrum rowei* O'Loughlin and O'Hara, from SE Australia; evidence of intra-coelomic brood fissiparity in *Staurothyone inconspicua* (Bell), from SE Australia; intra-coelomic brood-protection by an undescribed species of *Parathyonidium* Heding, from Antarctica; evidence of intra-coelomic brood auto-ingestion by *Neoamphicyclus materiae* O'Loughlin, from SE Australia; brood protection in a longitudinal dorsal invaginated marsupium in small specimens of *Cladodactyla crocea* (Lesson), from the Falkland Islands. An analysis is presented of brood-protection in interradial anterior marsupia by species of the *Cucumaria georgiana* (Lampert) group (including *Cucumaria acuta* Massin and *Cucumaria attenuata* Vaney), *Microchoerus splendidus* Gutt, *Psolidiella mollis* (Ludwig and Heding) and *Psolus charcoti* Vaney, all from Antarctica. The report and illustration of anterior interradial marsupial brood-protection by *Psolus koehleri* Vaney, from Antarctica, is rejected.

Some molecular phylogenetic data for holothuroid species from Antarctica, Australia and New Zealand (Echinodermata: Holothuroidea)

M. O'Loughlin
Marine Biology Section, Museum Victoria, Melbourne, Victoria, Australia

ABSTRACT: In collaboration with Gustav Paulay in the University of Florida, DNA phylogenetic data (CO1 gene) are becoming available for holothuroids from Antarctica, New Zealand and Australia. Data for some dendrochirotids and molpadiids and synallactids are reported. There is some good congruence for traditional morphological systematic conclusions and molecular phylogenetic data. These data also indicate a need for review of some morphological systematic conclusions in terms of synonymies and the existence of cryptic species. For example CO1 gene data indicate: *Molpadia musculus* Risso appears to be not cosmopolitan, but comprises at least four discrete species; the Antarctic *Psolus arnaudi* Cherbonnier and *Psolus cherbonnieri* Carriol and Féral appear to be junior synonyms of *Psolus dubiosus* Ludwig and Heding; specimens of *Molpadiodemas morbillus* O'Loughlin and Ahearn from the South Sandwich Trench (5452 m), New Zealand (2500 m) and SE Australia (1993 m) show no genetic variations.

Diversity and distribution of holothuroid species south of the Antarctic Convergence (Echinodermata: Holothuroidea)

M. O'Loughlin
Marine Biology Section, Museum Victoria, Melbourne, Victoria, Australia

ABSTRACT: A morphological systematic overview of current knowledge of the diversity and distribution of holothuroid echinoderm species from all depths south of the Antarctic Convergence is presented. There are 155 species, with 38 undescribed. Species occurrences south of the Convergence at Bouvet Island, Heard and Kerguelen Islands, Prydz Bay, Ross Sea, Bellingshausen Sea, Antarctic Peninsula and Weddell Sea, and north of the Convergence in the Magellanic region, are compared. Based on morphological systematics there is typically a circum-polar distribution south of the Convergence. However, most holothuroid species on the Heard/Kerguelen Plateau do not show a circum-polar distribution. The Convergence is a significant barrier to gene flow north from the Antarctic Ocean.

Integrating electronic technologies in ecological field studies: Assessing movement, habitat use, and behaviour of lobsters as key predators of sea urchins in eastern Tasmania

H.G. Pederson
Myriax Software P/L, Hobart, Tasmania, Australia
School of Zoology, University of Tasmania, Hobart, Tasmania, Australia

C.R. Johnson, S.D. Ling & J.C. Sanderson
School of Zoology, University of Tasmania, Hobart, Tasmania, Australia
TAFI, University of Tasmania, Taroona, Tasmania, Australia

ABSTRACT: Rock lobsters (*Jasus edwardsii*) have been identified as an important predator of the range extending and 'barrens' forming long-spined sea urchin (*Centrostephanus rodgersii*) in eastern Tasmania. To determine the efficacy of rebuilding rock lobster biomass as a means to control of sea urchin populations, a mass translocation of very large (>140 mm carapace length, ~2.5 kg fresh weight) rock lobsters into a barrens habitat and adjacent seaweed bed was undertaken. To understand the behavioural dynamics of translocated individuals we employed an acoustic telemetry array to track the movement, habitat utilisation and interactions between translocated individuals. Detailed habitat maps, derived from acoustic and video surveys, were used to determine the extent of habitat usage by translocated rock lobsters and provide a baseline for on-going monitoring of habitat restoration. We will present results from the experimental translocation, reporting on the movement and behaviour of large predatory lobsters on and around barrens habitat. Utilising the power of a newly developed 4D visualisation and analysis software package (Eonfusion) we will demonstrate a novel technological approach to integrating data on bathymetry, 'landscape' structure, and predator movement and behaviour towards a more holistic approach to ecological field studies.

Molecular detection of echinoderms in the diet of temperate marine predators

K.S. Redd
School of Zoology, University of Tasmania, Hobart, Tasmania, Australia
TAFI, University of Tasmania, Taroona, Tasmania, Australia

S.D. Frusher
TAFI, University of Tasmania, Taroona, Tasmania, Australia

S.N. Jarman
Department of the Environment and Heritage, Australian Antarctic Division, Kingston, Tasmania, Australia

C.R. Johnson
School of Zoology, University of Tasmania, Hobart, Tasmania, Australia
TAFI, University of Tasmania, Taroona, Tasmania, Australia

ABSTRACT: Quantifying predation on echinoderms remains a key challenge for ecologists. Molecular detection techniques provide a robust approach to this problem and other issues in dietary studies of predators. Here we outline a general PCR-based assay for echinoderms and specific quantitative PCR-based assays for the southern Australian sea urchin species, *Centrostephanus rodgersii* and *Heliocidaris erythrogramma*. By analysing faecal pellets collected from southern rock lobsters (*Jasus edwardsii*) for sea urchin DNA, we are able to investigate predation by lobsters on these sea urchin species in eastern Tasmania. We will report on using both controlled feeding experiments and DNA analysis of faecal pellets taken from wild caught rock lobsters to validate the utility of molecular prey detection techniques and to evaluate the quantitative possibilities for these assays. Initial data indicate a higher frequency of *C. rodgersii* detection in faecal pellets of lobsters transplanted to *C. rodgersii* barrens than initially expected. While these preliminary results are encouraging in raising solutions to manage barrens formation, they raise the question of the source of the sea urchin DNA, and whether positive assays largely indicate predation events, or intake by lobsters of *C. rodgersii* from scavenging sea urchin carcasses or urchin faecal pellets. We will assess the utility of the technique to address complex ecological and behavioural issues essential for informed ecosystem based management of rocky reefs.

Variation in reproductive traits of viviparous sea stars (Echinodermata: Asteroidea)

L.M. Roediger & T.F. Bolton
Lincoln Marine Science Centre, Flinders University, Port Lincoln, SA, Australia

ABSTRACT: Levels of maternal energy investment in offspring, and hence offspring size, differ markedly among species. Recently it has been recognised that maternal investment in offspring also differs considerably among individuals within a species, and also within clutches of offspring from individual females in some species. Offspring size in these species is highly plastic influenced by differences in maternal size, maternal nutrition, clutch size and environment. Recent examinations suggest that selection may favour variable offspring size within clutches when environmental conditions are unpredictable. Thus variation in offspring size may represent a bet-hedging strategy that increases offspring fitness on average when environmental conditions fluctuate unpredictably. Conversely, in stable environments, selection should favour a narrow range of offspring sizes that increases offspring fitness under predictable conditions. I examined relationships between levels of variation in clutch size and offspring size in two species of viviparous sea star, *Cryptasterina hystera* and *Parvulastra parvivipara*, that respectively inhabit a stable tropical environment and highly dynamic temperate environment. Coefficients of Variation (CV) in offspring size were 24% and 46% for the tropical and temperate species respectively. Thus, the CV in offspring size of the species inhabiting the variable temperate system was almost twice that of the species inhabiting the stable tropical environment. Additionally, larger adults produced larger offspring. While these results support the theoretical predictions they are confounded by the interspecies comparison. Establishing a causal relationship between levels of variation in offspring size and environmental stability requires comparisons both among and within species.

Changes in protein profiles through the reproductive cycle of the sea urchin *Evechinus chloroticus*

M.A. Sewell, S. Eriksen & K. Ruggiero
School of Biological Sciences, University of Auckland, Auckland, New Zealand

M. Middleditch
Centre for Genomics and Proteomics, School of Biological Sciences, University of Auckland, Auckland, New Zealand

ABSTRACT: The development of export markets for sea urchin roe has prompted experimental research on how artificial diets may enhance the quality and quantity of the roe that is produced. In diet trials a common measure of the effectiveness of a certain diet is a measure of the total protein within the gonad. However, proteomics technology has the potential to allow determination of the particular proteins that are undergoing changes in response to a certain diet. In recent research we have shown using 2-D gels that there were differences in the proteomic profiles of female *Evechinus chloroticus* at different phases of the reproductive cycle. To quantify these differences we used iTRAQ, a protein labelling technology for the simultaneous analysis of the protein profile of four histological stages in female *Evechinus chloroticus* (recovery, growth, premature and mature). Preliminary analysis of the iTRAQ results shows that 105 proteins were recorded from all 4 histological stages, and that the major differences in the protein profile occurred between the growth and premature stages, during the major period of vitellogenesis. In this talk we will highlight particular proteins that show significant changes with the reproductive cycle, and suggest how an iTRAQ approach might be used to assess changes in the protein profiles of gonads from urchins fed different diets.

Fossil clypeasteroids (Echinodermata) of the Indian subcontinent: A review

D.K. Srivastava
Centre of Advanced Study in Geology, University of Lucknow, Lucknow, India

A. Kroh
Natural History Museum Vienna, Burgring, Vienna, Austria

ABSTRACT: Clypeasteroid echinoids are important members of shallow benthic marine communities, both today and during the larger part of the Cenozoic era. They are characterized by their modified corona stabilized by internal support structures and tightly interlocking plate sutures. Historically, research on this group has focused on the Mediterranean and the Caribbean. Here we present a review of the clypeasteroids from the Indian Subcontinent, aiming at improving the knowledge on the evolution of these echinoids in the area of the Indo-Pacific Ocean, where they are most diverse today. While the first clypeasteroids from the Indian Subcontinent were already described in 1837 by Grant, monographic works are largely missing. A notable exception is a series of monographs by Duncan & Sladen. More recently a number of smaller papers on fossil Indian clypeasteroids have been published by Srivastava, Tandon and Jain. In the present project we aim to unite data from published sources and new material collected by one of us (DKS). Old records are revised on base of the type material where possible and new species descriptions, photographs and plating patterns are provided. The stratigraphy of published records is updated based on micropalaeontological data. Up-to-date nine clypeasteroid genera were reported from the study area including: *Clypeaster, Echinocyamus, Cyamidia, Echinodiscus, Amphiope, Tridium, Fibularia, Mortonia* and *Laganum*. The latter ones being relatively rare, while *Clypeaster* and *Echinocyamus* occur widespread. Distributional data reveals surprisingly few ties with East African or Arabian echinoid faunas. In modern echinoid faunas, in contrast, these areas share most clypeasteroid species.

Quantitative analysis of morphological characters in Stichopodidae (Holothuroidea, Aspidochirotida)

J.A. Starmer & G. Paulay
Florida Museum of Natural History, Gainesville, Florida, USA

ABSTRACT: Gross morphology and ossicles from the body wall provide the basic characters used in holothurian taxonomy. However, the plasticity of soft tissue morphology in both living and preserved specimens has focused taxonomic discrimination on the calcareous ossicles. Differentiation between ossicles is typically based on size and qualitative description of form. Variation in ossicle form exists, even within named forms (e.g. button, rosette, tack) within a single individual, and can challenge comparison of study specimens with published descriptions. While ossicles have been recorded from tissues other than the body wall, they have irregularly applied to taxonomy. By evaluating ossicles from tissues other than the body wall within the Stichopodidae and by applying statistical analysis (regression and multivariate analysis) to ossicle size and presence-absence observations, additional quantitative characters are available for differentiating between holothurian taxa.

Negative impacts of the invasive sea urchin (*Centrostephanus rodgersii*) on commercially fished blacklip abalone (*Haliotis rubra*)

E.M.A. Strain & C.R. Johnson
School of Zoology, and TAFI, University of Tasmania, Hobart, Tasmania, Australia

ABSTRACT: Incursion of the long-spined sea urchin (*Centrostephanus rodgersii*) into Tasmanian (Australia) waters in the late 1960s and its subsequent establishment at high densities raises several questions about potential interactions between the sea urchin and another large herbivore on these reefs, namely the black-lipped abalone (*Haliotis rubra*) which supports the State's most valuable fishery (ca. AU$120M p.a.). Surveys show a negative relationship between densities of *C. rodgersii* and *H. rubra* at several spatial scales, suggesting negative interactions. We used manipulative experiments to examine competitive interactions between these species, including effects on behaviour, movement, growth, reproduction, and survival. Experimental removal of *H. rubra* from intact macroalgal beds had no detectable effects on *C. rodgersii* behaviour, movement, growth, reproduction, or survival, so there is no evidence that fishing abalone contributed to the invasion success of the urchin. In contrast, introduction of *C. rodgersii* to intact algal beds causes abalone to flee, seek shelter in cryptic microhabitats, and negatively impacts their growth, reproduction and survival. In other experiments, *H. rubra* would not venture into small 'barren' patches created as a result of overgrazing by *C. rodgersii* irrespective of whether the urchin was present. This suggests that lack of food and/or loss of biogenic habitat structure explains the absence of commercial populations of abalone on extensive urchin barrens. The overall picture from the combined research suggests that management of *C. rodgersii* to optimise the *H. rubra* fishery requires complete removal of the urchin and regrowth of barrens areas.

Isolated lateral arm plates in ophiuroid palaeontology: How diagnostic are they?

B. Thuy
Institute and Museum of Geology and Palaeontology, University of Tübingen, Tübingen, Germany

S. Stöhr
Swedish Museum of Natural History, Department of Invertebrate Zoology, Stockholm, Sweden

ABSTRACT: Disarticulated sclerites of ophiuroids are frequently encountered in clayey and marly sediments and account for the vast majority of the fossil record of the group. Among the isolated skeletal parts of brittle stars lateral arm plates have been regarded as offering the largest set of distinctive characters, making them most suitable for taxonomic assessment. Referring to this assumption, a large number of mainly Mesozoic species have been described based on isolated lateral arm plates only. In order to test the variability and specificity of lateral arm plates, a total of 57 well characterised species from most of the extant ophiuroid families were sampled in this study, macerating arms or arm fragments and extracting lateral arm plates for SEM-imaging. The relevant characters were defined and their changes throughout the arm (distal to proximal, i.e. younger to older segments) were evaluated. For some of the species, intraspecific variability of the characters is assessed, considering specimens of similar size and series of specimens from different ontogenetic stages. Interspecific variability of the characters allows evaluation of the specificity of the lateral arm plates on species and higher taxonomic level. Work is still in progress, but some patterns have already emerged. While most species are recognisable by their lateral arm plates, it appears that in many cases distinction between species solely considering lateral arm plates is hardly possible. The use of isolated lateral arm plates in ophiuroid palaeontology as well as the species that have been described from them are in need of revision.

Life modes of the conspicuously dimorphic brittle star *Ophiodaphne formata* (Koehler, 1905) (Echinodermata: Ophiuroidea)

H. Tominaga
Fukui Prefectural Takefu High School, Fukui, Japan

M. Komatsu
Department of Biology, Graduate School of Engineering and Science, University of Toyama, Toyama, Japan

ABSTRACT: The ophiuroid *Ophiodaphne formata* exhibits conspicuous sexual dimorphism. This ophiuroid has a pairing habit—a larger female and a smaller male couple with each other by clinching their mouths together, and occasionally both sexes attach aborally to the oral surface of irregular sea urchins *Astriclypeus manni* and *Clypeaster japonicas* as 'hosts'. We investigated several themes on the life modes of this ophiuroid: size differences between paired and unpaired individuals of both sexes; the size of mature specimens; and methods of pairing, attaching to the host, and moving around on or toward the host. The disk diameters of paired female and male ophiuroids are larger than those of unpaired ophiuroids. It was estimated that the disk diameter of a mature female paired on the oral plates of the host was more than 3.7 mm, and that of a mature male paired on the plates was more than 0.63 mm, because they developed gonads and then shed gametes. To pair, a male approaches the margin of a female's disk, and then turns over his disk to his oral side, and finally embraces her disk with his mouth clung to hers.

O. formata has three forms of attachment to the host: as paired ophiuroids, as an unpaired female, and as an unpaired male form. This species of ophiuroid crawls under its host, attaching its aboral side to the oral side of the host. After the period of attachment, some ophiuroids leave one host individual to move to another. In this study, the curious life modes of *O. formata* and its ingenious utilization of its host were clarified; only mature individuals paired with the opposite sex, and they spawned in the breeding season after attachment to the host and then exhibited growth and repeated movement behaviours.

Outbreaks and dieoffs: Causes and consequences of large amplitude population density fluctuations in echinoderms

S. Uthicke & B. Schaffelke
Australian Institute of Marine Science, Townsville MC, Australia

M. Byrne
Schools of Biomedical and Biological Science, University of Sydney, NSW, Australia

ABSTRACT: Echinoderms are keystone species in many marine ecosystems and are notorious for large population density fluctuations ('outbreak' or 'dieoff' events). In a review of this phenomenon, we identified 28 species (6 Asteroidea, 8 Echinoidea, 10 Holothuroidea, 4 Ophiuroidea) that exhibit large amplitude population changes. Three generalized patterns were identified: (I) rapid increase and stability at a new population density (e.g. in *Amperima* and *Amphiura*), (II) rapid decreases followed by no or slow (decades) recovery (e.g. in *Diadema* and *Heliaster*), and (III) decadal scale population fluctuations (e.g. in *Strongylocentrotus* and *Acanthaster*). Anthropogenic impact contributed to most population density fluctuations, including species introductions, increased primary productivity through eutrophication or global change, disease and overfishing. Examples of these fluctuations were found from shallow intertidal to the deep sea and from tropical to temperate regions. A distinct similarity amongst the species identified was possession of the (ancestral) planktotrophic larva, which was significantly over-represented in most individual classes and in the combined data set. We conclude that (I) a strong non-linear dependency of larval production on adult densities (Allee effects), (II) a low potential for compensatory feedback mechanisms and, (III) an uncoupling of larval and adult ecology render a life history with planktotrophic larvae a high-risk-high-gain strategy. The alternative (derived) lecithotrophic larval type occurs in 68% of recent echinoderm species and may represent a more buffered life history (e.g. because compensatory feedback between adult densities and larval output is more likely and larvae are less dependent on the vagaries of planktonic food availability).

Molecular diversity and body distribution of the saponins in the sea cucumber *Holothuria forskali*

S. Van Dyck, P. Gerbaux & P. Flammang
University of Mons-Hainaut Pentagone, Mons, Belgium

ABSTRACT: Sea cucumbers contain triterpene glycoside toxins called saponins. We investigated the complex saponin mixture extracted from the common Mediterranean species *Holothuria forskali*. Two different body components were analysed separately: the body wall (which protects the animal and is moreover the most important organ in terms of weight) and the Cuvierian tubules (a defensive organ that can be expelled on predators in response to an attack). Mass spectrometry (MALDI-MS and MALDIMS/MS) was used to detect saponins and describe their molecular structures. As isomers have been found in the Cuvierian tubules, a preliminary chromatographic separation (LC-MS and LC-MS/MS) was performed to identify each saponin separately. A quantitative study was also conducted to compare the amount of toxin in both body components. Twelve saponins have been detected in the body wall and 26 in the Cuvierian tubules. All the saponins from the body wall are also present in the Cuvierian tubules but the latter also contain 14 specific saponins. The presence of isomeric saponins complicated the structure elucidation of the whole set of toxins but 16 saponins have been characterized through their fragmentation pattern. Among these, 3 had already been reported in the literature as holothurinosides A and C, and desholothurin A. Molecular structures have been suggested for the 13 others which, in the present work, have been provisionally named holothurinosides E, F, G, H, I, A_1, C_1, E_1, F_1, G_1, H_1 and I_1 and desholothurin A1.

Daily and seasonal patterns in behaviour of the commercially important sea cucumber, *Holothuria scabra*

S.M. Wolkenhauer

Institute for Biodiversity Research, University of Rostock, Universitätsplatz, Rostock, Germany
CSIRO Marine & Atmospheric Research, Cleveland, QLD, Australia

T. Skewes

CSIRO Marine & Atmospheric Research, Cleveland, QLD, Australia

M. Browne

CSIRO Mathematical and Information Sciences, Cleveland, QLD, Australia

D. Chetwynd

CSIRO Marine & Atmospheric Research, Cleveland, QLD, Australia

ABSTRACT: This study monitored and modelled long-term daily and seasonal patterns in behaviour of adult sea cucumber *Holothuria scabra* in Moreton Bay, Australia. Animals were kept in outdoor tanks for two years and behaviour was recorded each month for a 24-hour period by means of time-lapse video. Behaviour was classified into eight categories and a series of nested conditional, binomial models (Generalized Linear Models) were applied to describe the probabilities of key behaviours occurring. Active behaviours, such as feeding and searching, were negatively correlated to water temperature and were approximately 5-times higher during summer (~16 hr d⁻¹) than during winter (~3 hr d⁻¹). Animals were less likely to bury during summer (December–February), with at least one month where they did not bury at all. There was an 80% probability of animals being inactive during the early hours of the morning (~5:00 hr), irrespective of the time of year; and a 50% probability of animals being fully buried during mid-winter (July/August), irrespective of the time of day. Searching behaviour showed a bimodal pattern, where animals spent more time searching during autumn and spring (~2 hr d⁻¹) than during summer (~1 hr d⁻¹) or winter (~20 min d⁻¹). Describing patterns in holothurian behaviour, especially producing a probability matrix of active behaviour and burying frequencies, is crucial for designing sustainable fisheries management strategies and aquaculture projects. The key findings of this study provide information about optimal timeframes to conduct population surveys, and can be applied to the ecosystem function of tropical holothurians overall.

Evolution of an embryonic gene regulatory network in the sea urchin *Strongylocentrotus purpuratus*

G.A. Wray, D. Garfield & D. Runcie
Department of Biology, Duke University, Durham, NC, USA

ABSTRACT: During embryonic development in echinoderms, genes interact in precise spatial and temporal patterns. This network of interactions is of considerable interest from ecological and evolutionary perspectives, because it influences phenotypic plasticity, stress responses, and trait variation among individuals. In order to better understand how such interactions influence the evolution of development, we are investigating the embryonic gene network of the purple sea urchin, *Strongylocentrotus purpuratus*. We raised embryos from 36 crosses, and analysed the expression of 79 genes with known interactions across eight developmental stages. Natural selection can only operate on heritable traits that vary among individuals. We find that both variability and heritability differ considerably across the network, suggesting that selection can only target specific aspects of network function. Contrary to some predictions, we find that most evolutionary constraint occurs among gene interactions within the late embryo, rather than in the earliest phases of development. Echinoderm embryos are incredibly resilient to environmental perturbations, so we also investigated how network organization influences the robustness of development. We find that most genes are relatively insensitive to the exact expression level of their regulators, which allows the network to dampen rather than propagate perturbations. We also find that some aspects of network organization, such as feedback loops, likely contribute to the robustness of early development. This study provides one of the first glimpses of how a gene network varies within a natural population, and demonstrates how a network approach can inform studies of evolution and ecology.

Evolution and adaptive significance of the gastric caecum in Irregularia (Echinodermata: Echinoidea)

A. Ziegler
Institut für Biologie, Freie Universität Berlin, Königin-Luise-Straße, Berlin, Germany

G. Rolet & C. De Ridder
Laboratoire de Biologie marine, Université Libre de Bruxelles, Bruxelles, Belgium

R.J. Mooi
California Academy of Sciences, Golden Gate Park, San Francisco, California, USA

ABSTRACT: The gastric (or anterior) caecum found in Spatangoida is a large, transparent, non-contractile pouch located atop the anterior part of the stomach and extends backwards towards interambulacrum 5. It is constantly filled with liquid and is connected to the stomach through a non-muscular slit-like opening that prevents sediment from entering. The gastric caecum is a site of microbial activity and presumably plays an important role in the digestive process. However, its true function remains as enigmatic as its evolutionary origin. Here, we use morphological data derived from a set of invasive and non-invasive techniques to compare gastric caeca found in various irregular sea urchin taxa. Gastric caeca of varying size and shape can be found in Holectypoida, "Cassiduloida", Clypeasterina, Holasteroida, and Spatangoida, but are not present in Laganina and Scutellina. Based on a number of characters such as topology, general morphology, histology, mesenterial suspension, and integration into the haemal system we homologize the gastric caecum with the more or less pronounced dilatation (sometimes termed caecum) observed in most "regular" sea urchin taxa. This caecum is largest in the basal euechinoid taxa, such as Aspidodiadematidae and Diadematidae, suggesting that the ancestor of Irregularia and the Acroechinoida already possessed a pronounced caecal structure as part of its anterior stomach. In addition, since the size of the gastric caecum is most pronounced in the predominantly sediment-burrowing irregular taxa, we hypothesize that its evolution is closely linked to the development of more elaborate infaunal lifestyles.

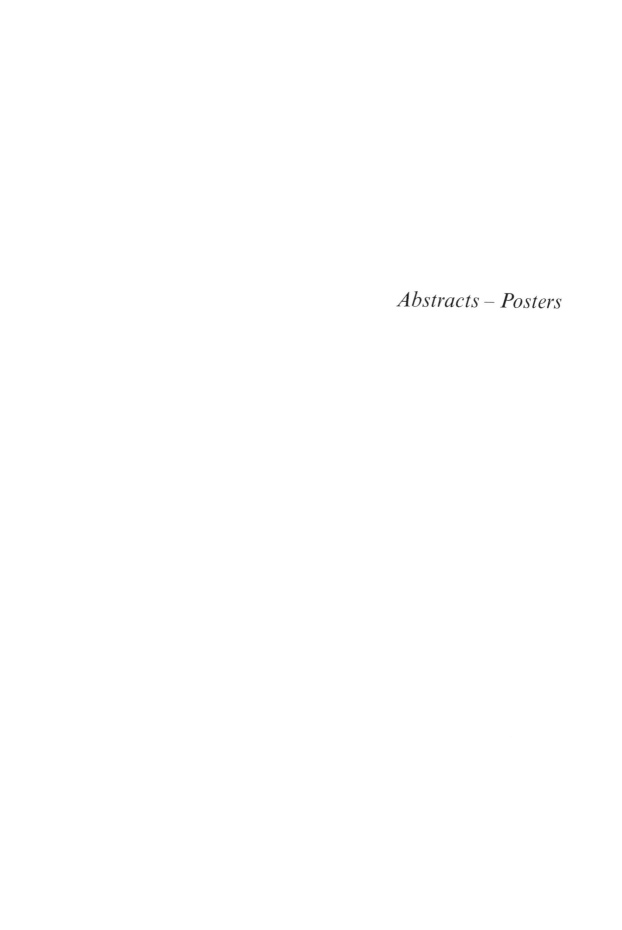

Abstracts – Posters

Concurring evidence of rapid sea-cucumber overfishing in the Sultanate of Oman

K.M. Al-Rashdi, F.I. Al-Ajmi & S.H. Al-Habsi
Ministry of Fisheries, Aquaculture Center, Muscat, Sultanate of Oman

M.R. Claereboudt
Sultan Qaboos University, College of Agriculture and Marine Sciences, Sultanate of Oman

ABSTRACT: A fishery for *Holothuria scabra* recently developed in a small area of the Eastern Coast of the Sultanate of Oman. The area covered by the fisheries is limited to a single shallow embayment only 320 km² in Mahout Bay and involves approximately 400 fishers, and around 50% of them are women. The fishing season (October to May) in 2005 was the first officially on records. However, anecdotal evidence suggests a low level of exploitation since the early 1970s although catch and export data for this period are unavailable. Average size of individuals collected in 2005 varied between 170 and 200 mm in length. The total biomass at the time was estimated at 1500 tons (fresh weight). The following year at least 14.5 tons of processed *H. scabra* were exported towards the United Arab Emirates, corresponding to approximately 145 tons or around 10% of the recorded biomass. Interviews with fishermen and traders, who remain in collection boats following the fishers, revealed that the CPU for sandfish was around 100 individuals h⁻¹ in 2005. The CPU had declined to 10–20 individuals h⁻¹ by 2007 indicating significant pressure on the resource. Over the same timeframe, the value of an average sized *H. scabra*, which was OR 0.1 (US$ 0.25) in 2005, increased to OR 1.5 (US$ 3.75) and still climbs. Concomitantly fishers have commenced targeting the less valuable *H. atra* in large numbers, which command market prices of OR 0.2 (US$ 0.5). Finally, an examination of the processed specimen for sale showed the presence of a significant number of very small individuals (<5 cm processed corresponding to around 10 cm live length). These concurring evidences suggest a rapid decline of the population of *H. scabra* in Mahout Bay while indicating building pressure on *H. atra*.

Echinothurioid diversity in New Zealand seas

O. Anderson
National Institute of Water and Atmospheric Research (NIWA), Kilbirnie, Wellington, New Zealand

ABSTRACT: In New Zealand, the echinoid order Echinothurioida has received relatively little attention from taxonomists, with 7 species recorded from the region, yet the order is represented in NIWAs Invertebrate Collection by over 350 specimen lots, collected between 1961 and 2008. A recent examination of these specimens determined that the three currently recognised families are represented in the New Zealand region by at least thirteen species in eight genera (out of about 50 species worldwide, in 11 genera). The identity of several species is yet to be determined beyond genus, and there are one or more undescribed species. One, a species of *Araeosoma*, is remarkable for its two valved tridentate pedicellariae, a rare form in regular echinoids and one not previously reported in echinothurioids. With between 3 and 108 records available for each species, patterns of distribution could be examined. Some species showed a broad, relatively even distribution within the New Zealand region, while others were strongly restricted to northern regions. The northern Campbell Plateau signifies the southern limit for the order in the region, with few found south of 50° S. Specimens of *Kamptosoma asterias*, the only Antarctic species, were recorded for the first time from deep water to the west of New Zealand. This poster provides an overview of the species of echinothurioids found in New Zealand, including plots of species distributions and depth ranges, comparisons with worldwide distributions of echinothurioids, and outlines some of the identification issues and future research needs.

Gametogenic synchronicity with respect to time, depth and location in three deep-sea echinoderms

S. Baillon
Université de Bretagne Occidentale, Institut Universitaire Européen de la Mer (IUEM), Plouzane, France

J-F. Hamel
Society for the Exploration and Valuing of the Environment (SEVE), St. Philip's, NL, Canada

A. Mercier
Ocean Sciences Centre (OSC), Memorial University, St. John's, NL, Canada

ABSTRACT: This study examined the prevalence of inter-individual synchrony in the gamete synthesis of three deep-sea echinoderms, *Phormosoma placenta* (Echinoidea), *Zygothuria lactea* (Holothuroidea) and *Hippasteria phrygiana* (Asteroidea) collected along the continental slope off the coast of Newfoundland and Labrador (eastern Canada). Analysis of gonad development using histology and gonad indices revealed diverse degrees of asynchrony at the scales examined (within trawls, between trawls over similar or different periods, as well as between depths and locations over the same period). Annual and seasonal patterns were therefore largely masked by heterogeneousness in most samples. These data suggest that determination of so called "continuous" reproductive cycles in many deep-sea species may in fact reflect sampling inadequacies inherent to most deep-sea studies. Assessment of true reproductive patterns and periodicities may require much tighter collection designs as these species are likely to rely on fine-scale cohesion and inter-individual exchanges (i.e. aggregation, chemical communication) to synchronize their breeding activities.

Evolution of maternal provisioning in echinoid echinoderms

M. Byrne, T.A.A. Prowse & I. Falkner
Department of Anatomy and Histology, Bosch Institute, University of Sydney, NSW, Australia

M.A. Sewell
School of Biological Sciences, University of Auckland, New Zealand

ABSTRACT: In the complex life histories of marine invertebrates the egg cell links the adult and larval life stages. Evolution of development is inextricably linked to alterations in maternal provisioning and, in echinoderms, this has involved multiple, independent alterations of oogenesis. Small eggs in species with the ancestral-type planktotrophic larva contain low levels of lipids, while large eggs are stocked with high concentrations of lipids that often result in production of buoyant eggs. Acquisition of large eggs involved modification of the egg lipogenic program. We investigated egg evolution in echinoids to determine how maternal provisioning differs in species with planktotrophic and lecithotrophic development by quantifying total lipid (per lipid class) content on a per egg basis. The small eggs of species with feeding larvae were dominated by structural lipids (sterols and phospholipids) and also contained energetic lipid, which was largely triglyceride. The exception was *Tripneustes gratilla* which has high triglyceride stores in its eggs, a trait that may be important in the success of this species for aquaculture. In contrast the large eggs of species with lecithotrophic larvae were all dominated by energetic lipid in the form of diacylglycerol ethers.

Impact of seawater acidification on sea urchin growth

A.I. Catarino & P. Dubois
Laboratoire de Biologie Marine, Université Libre de Bruxelles, Roosevelt, Bruxelles, Belgium

ABSTRACT: Sea urchins could be particularly vulnerable to seawater acidification due to the nature of their well developed high-magnesium calcite skeleton. This is formed trough an amorphous calcium carbonate precursor form whose solubility exceeds that of aragonite. Being osmo-conforms, they are unable to regulate their coelomic fluid ionic composition, which could result in a poor acid-base regulation. As sea urchins are keystone species in numerous ecosystems, the adverse effects they might experience will have severe ecological repercussions. In this study we assessed the impact of seawater acidification on adults of the temperate sea urchin *Paracentrotus lividus*. Experiments were carried out at controlled temperature and lower pHs were obtained by bubbling CO_2 in seawater. The total alkalinity and pH (seawater scale) were measured and the pCO_2 and total inorganic carbon calculated, as well as the magnesium calcite saturation state. Adult sea urchins were grown at pHs between 7.8–7.3. Spine regeneration was experimentally induced and regenerate sizes were measured using microscopy techniques. Their magnesium content was analysed. Internal pH and pCO_2 was measured. Regenerate elongation did not differ with pH and no change in magnesium concentration was also observed. However, the internal pH of *P. lividus* significantly diminished with a decrease of the seawater pH, indicating a poor acid-base regulation. Available data indicate that regeneration of vital skeleton structures like spines might not be affected by a more acidic environment, but that the sea urchin internal physiology could be altered.

Growth and regeneration of structures involved in ophiuroid disc autotomy

N.A. Charlina & I.Y. Dolmatov
AV Zhirmunsky Institute of Marine Biology FEB RAS, Palchevsky, Vladivostok, Russia, RF

I.C. Wilkie
Glasgow Caledonian University, Glasgow, Scotland, UK

ABSTRACT: Disc autotomy in the ophiuroid *Amphipholis* kochii involves the Genital Bars and their Ligaments (GBLs). The post-autotomy regeneration and origin of these structures were studied. The aboral part of the disc regenerates approximately 7 days after autotomy. Genital bars and GBLs begin to repair only 17 days after autotomy and their regeneration is complete after 25 days. Both the regeneration and origin of the extracellular matrix depend on fibroblasts, which migrate to the future ligament area and synthesize a collagen mass. Collagen fibril organization occurs in special fibroblast vacuoles, whose size varies from 3 to 16 μm depending on the collagen fibril length. Finished fibrils are secreted from the cell by exocytosis. Newly synthesized collagen is a disorganized mass. Collagen fibrils are delimited by numerous microfibrils, which form loose networks. Granules of a few juxtaligamental processes, located in the synthesized collagen, are detached from the cell cytoplasm and enter the extracellular connective tissue matrix. There are connections between granules and collagen fibrils, suggesting that the granules may bring about collagenous tissue stabilization. The mineralized genital bar originates in the connective tissue due to synthesis of the calcite crystal by special cells—sclerocytes, which migrate to the calcification area. In the first stages of calcification there is intracellular calcite crystal formation. Later on it is secreted into the extracellular matrix and the process continues extracellularly. It was found that sclerocytes line up in rows and encircle large cavities. These cells interact with each other and form septate junctions.

Evaluation of tagging techniques for *Australostichopus (Stichopus) mollis* (Hutton, 1872) for potential ranching studies

N. Davey
NIWA (National Institute of Water and Atmosphere Research LTD), Port Nelson, New Zealand

G. Moss & P. James
NIWA (National Institute of Water and Atmosphere Research LTD), Mahanga Bay, Wellington, New Zealand

J. Stenton-Dozey
NIWA (National Institute of Water and Atmosphere Research LTD), Christchurch, New Zealand

ABSTRACT: The sea cucumber *Stichopus mollis* is a common benthic resident in New Zealand coastal waters and is often found in abundance under mussel farms. Since *S. mollis* has potential as a lucrative export commodity for Asian markets, ranching cultured sea cucumbers under farms presents an attractive investment opportunity. The first step in assessing the potential for ranching is to determine whether cultured sea cucumbers placed under a farm remain in residence for later harvesting. To determine residence times under farms we applied a release-recapture design which required the effective tagging of sea cucumbers. We investigated a number of different tagging methods to find the one with the least potential for tag loss. Of the six methods trialled in the laboratory (freeze branding, micro-sand blasting, oxy-tetracycline, pit-tagging, T-bars and Visible Implant Fluorescent Elastomer (VIFE)), T-tags and VIFE had the best results after four weeks in tanks (87%–93% tag retention). These two methods were selected to assess whether tag retention was the same in the field and laboratory. Fifteen tagged sea cucumbers were placed in each of 4 tanks with either sand or mussel shell substrate in the laboratory. Cage experiments in the field with 15 tagged sea cucumbers each ran simultaneously on either sand or mussel shell substrate. Different substrata were tested because we hypothesised that substrate influences sea cucumber movement which in turn influences tag retention. Results showed that sea cucumbers in the field cages lost tags much faster than those in the laboratory tanks. Further, the rougher mussel shell substrate appeared to cause slightly higher tag loss for both tag types. The results have implications for designs to assess residence times of sea cucumbers under mussel farms.

Plasticity of the echinoderm skeleton

P. Dubois, C. Moureaux, C. Borremans, J. Hermans & G. Mannaert
Marine Biology, Université Libre de Bruxelles, Bruxelles, Belgium

L. André
Royal Museum for Central Africa, Section of Mineralogy, Petrography and Geochemistry, Tervuren, Belgium

ABSTRACT: Mineral skeletons are often considered and studied as very constrained structures of precise morphology, composition and mechanical properties. However, environmental conditions may significantly modify the characteristics of mineral skeletons. Some of these modifications are adaptive responses, other are consequences of the induced stresses. In both cases, this plasticity offers an opportunity to investigate the underlying biomineralization processes. In this work, we present the effects of different environmental stresses on the echinoderm skeleton. We investigated skeleton properties of either field specimens living in contrasted environments or animals raised in the laboratory under strictly controlled conditions. Spines of sea urchins living on more exposed shores showed significant lower Young's modulus (E), second moment of inertia (I2) and flexural stiffness. This indicates that both the architecture (I2) and material properties (E) were modulated. Pollution by cadmium and lead significantly affected the architecture (I2) and flexural stiffness of spines from sea urchins collected along a contamination gradient. In this case, material properties (E) were not modified. Magnesium concentration in the echinoderm skeleton is positively linked to temperature. However, laboratory culture showed that this relation showed a plateau in the upper range of temperatures encountered by the studied sea urchin species. This was not due to a growth rate effect but possibly resulted from specific characteristics of the organic matrix of mineralization. Skeletal Mg/Ca and Sr/Ca ratios were positively linked to salinity at which starfish were raised. In conclusion, the echinoderm skeleton clearly showed a significant plasticity in both its mechanical properties and composition.

Impact of CO_2-driven ocean acidification on echinoderm early life-history is species-specific and not easy to predict

S. Dupont
Department of Marine Ecology, Göteborg University, The Sven Lòven Centre for Marine Sciences, Kristineberg, Fiskebäckskil, Sweden

M. Thorndyke
Royal Swedish Academy of Science, The Sven Lòven Centre for Marine Sciences, Kristineberg, Fiskebäckskil, Sweden

ABSTRACT: The world's oceans are slowly becoming more acidic and profound changes in marine ecosystems are certain. Perhaps one of the key marine groups most likely to be impacted by predicted Ocean Acidification (OA) are the echinoderms. Echinoderms are a vital component of the marine environment with representatives in virtually every ecosystem; where they are often keystone ecosystem engineers. In addition, many are indirect developers where both larva and adult have critical (and quite different) episodes of skeletogenic calcification. In contrast others exhibit adult but not larval skeletogenesis. In this way echinoderms offer a valuable and tractable experimental system for exploring the impacts of OA on marine biota. We have used our CO_2-based sea water acidification system to investigate the affects of near future ocean acidification and temperature increase on early developmental success in four brittle stars

(*Amphiura filiformis, Ophiocomina nigra, Ophiothrix fragilis, Ophiura albida*), one seastar (*Asterias rubens*) and five sea urchin species (*Brysopsis lyrifera, Echinus esculentum, Paracentrotus lividus, Strongylocentrotus droebachiensis* and *S. purpuratus*). Our results show that impact of OA on early life-history (1) is not easy to predict and appears to be species-specific, even in closely related taxa (e.g. sea urchins), (2) can be dramatic (e.g. the brittlestar *O. fragilis* showing 100% mortality after 8 days in pH expected in 50 years), (3) can also have positive effects, (4) is not only a calcification issue and, (5) analyses in synergy with other environmental parameters (e.g. temperature) is vital. Our data will be discussed in the light of adaptation potential and future predictions.

Genotype and phenotype in development of the sea star nervous system: The transition from bilateral to radial symmetry

L. Elia, P. Selvakumaraswamy, P. Cisternas & M. Byrne
Department of Anatomy and Histology, Bosch Institute, University of Sydney, NSW, Australia

V. Morris
School of Biological Sciences, University of Sydney, NSW, Australia

ABSTRACT: The nervous system of the larvae and juvenile asterinid sea stars was characterized using immunocytochemistry for neuronal markers and in situ hybridisation for neurogenic homeobox genes. In the feeding brachiolaria larvae of *Patiriella regularis* a conspicuous Immunoreactive (IR) network of nerve cells and fibres were associated with the ciliary bands and in the brachiolar complex. In lecithotrophic brachiolariae (*Parvulastra exigua, Meridastra calcar*), immunoreactive nerve cells were localized to the tip of brachiolar arms. In advanced brachiolaria additional IR appeared in the epithelium of the adhesive disk. In development of the juvenile NS, IR was first seen in primary tube feet, and as the CNS developed, IR increased in the circum oral nerve ring and in the radial nerve cord. Localization of IR reflected the histological organisation of these structures. In situ hybridisation indicated that expression of homeobox genes is largely restricted to juvenile structures. In the developing juvenile CNS homeobox gene expression paralleled that seen with immunocytochemistry. In these asteroids there was no evidence that the larval nervous system persists through metamorphosis. The juvenile CNS formed *de novo*. These data provide insights into potential homologies of the echinoderm body plan with those of other Bilateria.

Preliminary analysis on genetic structure of dense populations of *Ophiura sarsii* on the upper slope around northern Japan

T. Fujita & T. Kuramochi
National Museum of Nature and Science, Hyakunin-cho, Shinjuku-ku, Tokyo, Japan

ABSTRACT: An epibenthic ophiuroid, *Ophiura sarsii*, forms a dense bed (i.e. widespread high-density populations) on the upper bathyal bottom around northern Japan. The distribution, size frequency, foods, predation etc. of the dense bed have been studied ecologically to understand how the dense bed is maintained. In this study, the genetic geographical structure of the dense bed of *O. sarsii* was examined using molecular analyses. A total of 2,426 bps of mitochondrial DNA (cob and cox1 genes were combined) was sequenced and analysed on 32 specimens collected from northern Japan coasts of the western North Pacific, the Sea of Japan and the Sea of Okhotsk. To date, 26 haplotypes have been detected including one shared between individuals from the North Pacific and the Sea of Japan, 14 specific to those from the North Pacific and 11 specific to those from the Sea of Japan. The Fst value was calculated to be 0.1028 ($P = 0.054 \pm 0.024$) when the individuals were tentatively divided into two groups, the North Pacific population and the Sea of Japan-Okhotsk population. The precise nature of the geographic structure and gene flow are still unknown because sample size examined is too small at present. Further examinations will be undertaken using more individuals and analyses among several local populations in the near future.

Stereom differentiation in spines of *Plococidaris verticillata*, *Heterocentrotus mammillatus* and other regular sea urchins

N. Grossmann & J.H. Nebelsick
University of Tübingen, Tübingen, Germany

ABSTRACT: This study is part of a joint project in which sea urchins are being analysed morphologically, mineralogically and mechanically in order to evaluate their biomimetic potential as a model for impact protective systems. Here we give an overview of different stereom structures in the spines of several regular sea urchins. Living as well as freshly preserved specimen of the families Cidaridae and Echinometridae are used. The living sea urchins are kept in aquaria with artificial seawater and other marine organisms. Scanning Electron Microscopy (SEM) is used to study dried specimens. Critical point drying is applied to show organic material within the spines. The medulla (central core of the spine) is similar in all investigated species. It mostly consists of longitudinally oriented, laminar stereom which is arranged around the axis. The medulla of *Plococidaris verticillata* is surrounded by an uninterrupted radiating layer with galleried stereom, and then by the slightly perforated stereom of the cortex, which has a sculptured (crenellated), perforated surface encrusted by numerous epibionts. The radiating layer of *Heterocentrotus mammillatus* shows mostly labyrinthic stereom and is interrupted by several concentric growth lines consisting of slightly perforated stereom. The spine surface is longitudinally striated and covered by an epithelium which lacks encrustation. The spine base consists of dense labyrinthic stereom. Critical point drying reveals small amounts of cells and filaments within the struts of the stereom of the studied species.

Effect of seawater acidification on cidaroid spines

V. Guibourt, A.I. Catarino & P. Dubois
Marine Biology, Université Libre de Bruxelles, Bruxelles, Belgium

ABSTRACT: Due to their well-developed high-magnesium calcite endoskeleton, echinoderms will particularly suffer from seawater acidification. Among them, cidaroid sea urchins could be especially affected since, in this group, fully-grown spines are no more covered by an epidermis, leaving the skeleton in direct contact with seawater. These 'naked' spines are usually heavily colonized by epibionts making these sea urchins 'islands of biodiversity'. This is particularly well evidenced in the Southern Ocean where the presence of cidaroids and their epibionts significantly increase biodiversity of the muddy substrates on which they live. Most models of the ocean-carbon cycle predict that the shallowing of the calcium carbonate saturation horizons due to increasing anthropogenic CO_2 emissions will be particularly important in the Southern Ocean. So, effects of acidification on cidaroid spines are particularly relevant to the ecology of this region. In the present study, we assessed the impact of magnesium-calcite saturation level on cidaroid spines of both experimental and field specimens. In experimental conditions, spines were incubated for up to three weeks in CO_2 enriched seawater of controlled pH (7.2, 7.6 and 8.2) and temperature. Total alkalinity was measured and saturation level calculated. The spine immerged weight was measured, morphological evidence of etching was monitored by scanning electron microscopy and cover of epibionts was evaluated. Dried field specimens from different depths below and above the current saturation horizon were obtained from museum collections. Morphological evidence of etching and magnesium concentration in the skeleton were determined.

Proliferation-inducing activity of mesenchyme cells probably contributes to growth of epithelial cells in starfish embryogenesis

G. Hamanaka
The Center for Biosciences and Informatics, School of Fundamental Science and Technology, Keio University, Hiyoshi, Kohoku, Yokohama, Kanagawa, Japan
DECI Biology Laboratory, Hiyoshi, Kohoku-ku, Yokohama, Kanagawa, Japan

M. Matsumoto
The Center for Biosciences and Informatics, School of Fundamental Science and Technology, Keio University, Hiyoshi, Kohoku, Yokohama, Kanagawa, Japan

H. Kaneko
Department of Biology, Keio University, Hiyoshi, Kohoku, Yokohama, Kanagawa, Japan
DECI Biology Laboratory, Hiyoshi, Kohoku-ku, Yokohama, Kanagawa, Japan

ABSTRACT: The body structure of the starfish embryo is characterized by a monolayer sheet of Epithelial Cells (ECs) and a single type of Mesenchyme Cells (MCs). The ratio of ECs:MCs is approximately 100:1 in the bipinnaria larva that tentatively accomplishes a shape formation. Growth of the embryo is, therefore, mainly ascribable to proliferation of the ECs. In order to examine participation of MCs in ECs proliferation, we designed a novel experiment in which a blastula where ECs are the only cellular constituent is prepared by injecting an aggregate consisting of thirty to forty MCs, and allowed to develop until just before the mid-gastrula stage that corresponds to the beginning of the appearance of MCs in the blastocoel. An aggregate consisting of a similar cell numbers of ECs instead of MCs is used as the experimental control. We performed cytometric analysis of embryos from the MCs-injected condition and compared these with ECs-injected ones. We found that (1) cell numbers in blastulae embryos ranged from ~3000-3500 cells; (2) cell numbers of MCs-injected embryos are significantly greater (1.3 fold) than ECs-injected control embryos; (3) during the experiment, MCs displayed a dispersive distribution beneath the epithelial sheet; and (4) BrdU-uptake was not evident by MCs during the whole period of the experiment. These results suggest that MCs have an ability to induce proliferation of EC(s) without disadvantaging themselves, and therefore we suggest that MCs probably contribute to embryonic growth in the normal development of starfish.

The genus *Antedon* (Crinoidea, Echinodermata): An example of evolution through vicariance

L. Hemery, M. Eléaume & N. Ameziane
DMPA-UMR 5178, Muséum National d'Histoire Naturelle, Paris, France

P. Chevaldonné
2 DIMAR-UMR 6540, Station Marine d'Endoume, Centre d'Océanologie de Marseille, Chemin de la Batterie des lions, Marseille, France

A. Dettaï
DSE-UMR 7138, Muséum National d'Histoire Naturelle, Paris, France

ABSTRACT: The crinoid genus *Antedon* is, at best, paraphyletic and assigned to the polyphyletic family Antedonidae. This genus includes ~16 species separated into two distinct groups. One group is distributed in the north-eastern Atlantic and the Mediterranean Sea, the other in the western Pacific. Species from the western Pacific group are more closely related to other non-*Antedon* species (e.g. *Dorometra*) from this area than to *Antedon* species from the Atlantic-Mediterranean zone. Species from the Atlantic-Mediterranean area show a geographical structure probably linked to the events that followed the Messinian salinity crisis, ~5 Myr ago. To test this hypothesis, a phylogenetic study of the *Antedon* species from the Atlantic-Mediterranean group was conducted using a mitochondrial gene (cytb). The analysis included ~20 specimens representing the seven nominal species described from the region. Phylogenetic hypotheses were inferred using the maximum parsimony and the maximum likelihood criteria with two outgroups sampled within the antedonid subfamily Heliometrinae. Results show that the genus *Antedon* in the Atlantic-Mediterranean zone is composed of five species. *Antedon bifida* is distributed in the Atlantic Ocean and the westernmost part of the Mediterranean (Alboran Sea) while *Antedon mediterranea* is distributed in the rest of the Mediterranean. These results suggest that the divergence between *Antedon bifida* and *Antedon mediterranea* may have occurred by vicariance after the Gibraltar Strait dried up at the onset of the Messinian salinity crisis, and has since been maintained by gyres in the Alboran Sea, acting as an ecological barrier to larval dispersal.

Development of the sea cucumber *Holothuria leucospilota*

Y. Hiratsuka
Graduate School of Engineering and Science, University of the Ryukyus, Senbaru, Nishihara, Okinawa, Japan

T. Uehara
Tropical Biosphere Research Center, University of the Ryukyus, Sesoko, Motobu, Okinawa, Japan

ABSTRACT: Development of the tropical sea cucumber *Holothuria leucospilota* was studied from November to December 2007. Five adult individuals were collected from the northern part of Okinawa Island and maintained in an aerated aquarium. Spawning occurred spontaneously without any stimulation. The fertilized eggs reached the blastula stage 7 h after fertilization at 25°C. The early gastrula stage was reached at 20 h. Hatching occurred at 22 h. Early auricularia were formed after 48 h. The diatom *Chaetoceros gracillis* was added every 2 days during the auricularia stage. By 6 d, the larvae reached the mid-auricularia stage with pronounced pre-oral and post-oral lobes, several pairs of lateral processes, and an ossicle at the posterior end. The auricularia continued to grow during the next two weeks. At this stage, they began to accumulate hyaline spheres at the tip of posterior projection and lateral processes. The late-auricularia stage was reached at 19 to 21 d; it had left and right stomatocoels, as well as a hydrocoel. The transition period from late-auricularia to dolioraria, which was marked by total resorption of all lateral processes, lasted several days. The larvae reached the barrel-shaped dolioraria stage on 23 to 25 d. To induce metamorphosis, the doriolaria were transferred to a Petri-dish covered with a biological film. By 28 d, the doliolaria transformed into early pentactula possessing five primary tentacles and a single podium. Like other congeneric species, *H. leucospilota* has planktotrophic development through the auricularia, doliolaria and pentactula larval stages.

Pattern formation in starfish: Arm stumps, regeneration models, and evolution

F.H.C. Hotchkiss
Marine and Paleobiological Research Institute, Vineyard Haven, MA, USA

ABSTRACT: Starfish arm stumps that healed and did not regenerate the missing arm are analysed using regeneration models. A stump condition having definite structural organization suggests pattern regulation. Wound closure was by symmetrical midline joining like closing a book. Wound closure that precedes arm regeneration is by downward folding of the aboral surface. It is deduced from regeneration models that positional information in the rays of starfish has bilateral symmetry, and that the best fit model for starfish arm regeneration is the *distalization followed by intercalation model*. As there is no discernable difference between regenerating and intact arms, it is proposed that intercalary gap-filling growth applies to both. The similarity of the starfish arm with itself throughout its length suggests that the positional gap between the terminal plate and the last-formed section of arm is never bridged, and growth is indeterminate. It is proposed that distal signalling applies individually to each primary plate series of the arm. Thus abrupt origins/losses of plate series in the arm can be explained by mutations that turn on/off distal signalling behind the terminal plate. Once the organism possesses this signalling mechanism, mutations could cause it to function in more places than just behind the terminal plate. That such mutations should affect the whole organism is predicted by Dawkins' kaleidoscopic embryology concept. Plate series that intercalate in various parts of the body wall could have this origin. Mosaic evolution and recurrent appearance/loss of intercalary plates within and between echinoderm lineages are expected under this proposal.

Comparison of asteroid and ophiuroid trace fossils in the Jurassic of Germany with resting behaviour of extant asteroids and ophiuroids

Y. Ishida
Akabane Commercial Senior High School, Nishigaoka, Kita-ku, Tokyo, Japan

M. Röper
Bürgermeister-Müller-Museum, Solnhofen, Germany

T. Fujita
National Museum of Nature and Science, Shinjuku-ku, Tokyo, Japan

ABSTRACT: Two different size star-shaped trace fossils have been found from the Upper Jurassic Hienheim Formation in Germany. The large and small ones were assigned to *Asteriacites quinquefolius* and *Asteriacites lumbricalis*, respectively, by morphological observations. To identify the producer of

A. quinquefolius and to examine the producing process, resting traces of living asteroids, *Astropecten scoparius* and some deep-sea species, were observed in an aquarium and *in situ*. In an aquarium, *A. scoparius* buries shallowly, keeping its arms in pentamerally symmetric positions. When it moves from the resting position, three front arms bulldoze the substrate in front of these arms. Consequently a star-shape like depression is left behind with radiating four wider and sub-triangular arm furrows tapering toward the tip and one straight or indistinct sub-triangular depression destroyed by the movement of the starfish. Similar shaped resting traces were observed also on the deep-sea floor. *Asteriacites quinquefolius* was very similar to these resting traces of living asteroids, and its producer was estimated to be asteroids. *Asteriacites lumbricalis* was interpreted to be made by ophiuroids based on the previous experimental study of living ophiuroids. The size and shape of trace fossils suggest that the producers may be the ophiuroids, *Sinosura kelheimense* or *Geocoma carinata*, found from Upper Jurassic around Ried.

Photoreception in echinoderms: PCR amplification of opsin-like genes in *Paracentrotus lividus* and *Asterias rubens*

D. Lanterbecq, J. Delroise, P. Flammang & I. Eeckhaut
Marine Biology, University of Mons, Mons, Belgium

ABSTRACT: The evolutionary history of photoreceptors, including eyes *sensu stricto*, has always been an attractive and controversial subject. Whereas anatomical comparisons of photosensory organs in diverse animal lineages suggest their convergent appearance in the evolution, Evo-Devo molecular techniques suggest the presence of multigenic proteins, the opsin family, involved in a phototransduction system in Urbilateria (the common ancestor of organisms with a bilateral symmetry). The knowledge of photoreception mechanisms in Echinodermata is increasing. Besides the recognized photosensory role of the optic cushions in asteroids, data on the other echinoderm classes are rare. The publication of the complete genome of the sea urchin *Strongylocentrotus purpuratus* boosted the investigations on the photosensory system in echinoderms: six opsin proteins have been identified, of which three are involved in photoreception. Here we present preliminary results that concern the identification of primer pairs and development of nested PCR protocols to amplify a fragment of the opsin 4 gene in the five classes of echinoderms. To date, two sequences of opsin-like genes (+/–430 bp) were obtained for *Paracentrotus lividus* and *Asterias rubens*. The ultimate goal of our project is, in an evolutionary context, the identification of photosensory zones in representatives of the five classes of echinoderms by immunoblots and measure of gene expression by QPCR.

Intracellular pathways involved in oocytes maturation induced by dithiothreitol (DTT) and by a new Maturation Inducing Substance (MIS) in sea cucumbers

A. Léonet
Laboratoire de Biologie Marine, University of Mons-Hainaut, Belgium

M. Jangoux
Laboratoire de Biologie Marine, University Libre of Bruxels, Belgium
Aqua-Lab laboratory, Institut Halieutique et des Sciences Marines, University of Toliaria, Madagascar

I. Eeckhaut
Laboratoire de Biologie Marine, University of Mons-Hainaut, Belgium
Aqua-Lab laboratory, Institut Halieutique et des Sciences Marines, University of Toliaria, Madagascar

ABSTRACT: In sea cucumbers, oocyte maturation begins in ovaries but is stopped in prophase I of meiosis. In natural conditions, the blockage is removed during the laying by an unknown mechanism. When oocytes are taken by dissection, the meiotic release, can only be induced by an artificial inductor, the dithiothreitol (DTT). We recently discovered a new Maturation Inducing Substance (MIS) and compared its effects to DTT. The DTT induces the maturation of 91% of sea cucumber oocytes but the fertilization never exceeds 40%. The new MIS induces the maturation of more than 90% of oocytes and all are fertilizable. To identify the intracellular pathway mediated by the DTT and the new MIS, oocytes of sea cucumbers were incubated in presence or absence of DTT or MIS after being treated with various modulators of cAMP, an important regulator of hormone-induced maturation in general. The use of these modulators (forskolin, Isobutylmethylxanthine (IBMX) and hypoxanthine) in presence of DTT shows a decrease in cAMP during meiotic release. On the other hand, results strongly suggest that the way the new MIS acts is cAMP-independent. Whatever the maturation inducer used (either DTT or MIS), the oocyte maturation always requires the synthesis of new proteins and the activation of protein serine/threonine kinase.

Sea cucumbers of Australia

J.S. López
Universidad Autónoma de México, Apdo, México

ABSTRACT: The Australian sea cucumber fauna comprises about 15 families, 69 genera and 211 species. The present research was at the National Museum of Natural History, Smithsonian Institution, where they have about 68 species of Australian holothurians available to study. Each specimen at the Smithsonian was photographed using a digital camera, and the type of ossicles observed under a microscope and photographed. More detailed and precise SEM photographs of ossicles were obtained for 7 species. Other data obtained from each specimen was the region, type of substrate, depth, and other habitat characteristics. This work forms the initial basis of a taxonomic catalogue of sea cucumbers of Australia, with their respective descriptions, photographs of specimens and ossicles, and notes on distribution and habitat. I would like to thank the Smithsonian Institution, Cynthia and John Ahearn, and Dr. David L. Pawson, and Scott D. Whittaker for their kindness, assistance and hospitality, which made this research possible.

Characterisation of *Amphiura filiformis* luciferase (Ophiuroidea, Echinodermata)

J. Mallefet, B. Parmentier & X. Mulliez
Marine biology laboratory, Catholic University of Louvain, Louvain-la-Neuve, Belgium

O. Shimomura
Photoprotein laboratory, Falmouth, MA, USA

P. Morsomme
Physiological biochemistry unit, Catholic University of Louvain, Louvain-la-Neuve, Belgium

ABSTRACT: Despite the number of luminous species, very little is known about the echinoderm luminescence system. The chemical reactions responsible of the emission of visible light by organisms, result from oxidation of a substrate, termed luciferin, by molecular oxygen under catalytic activity of an enzyme, termed luciferase. Many chemically different luciferins and luciferases have been isolated from luminous organisms, and while some luciferins are common to phylogenetically different organisms, luciferases are always specific. The term photoprotein was introduced by Shimomura and Johnson in 1966 to characterize a luminous reaction that does not require oxygen and where the light emitted is proportional to the amount of reacting protein. It is now accepted that photoprotein corresponds to a stable enzyme-substrate intermediate whose light emission is triggered by the presence of a cofactor. The present work is the first attempt to characterize *Amphiura filiformis* luminous system. Crude extracts of the enzyme were tested. Results indicated that (i) coelenterazine is the substrate, (ii) activity is relatively stable (60% remaining after 4 days at −80°C); (iii) temperature does not affect activity in the range encountered by the animals *in vivo* (6–20°C); (iv) optimal pH activity is estimated at 7.4; (v) effects of salt concentrations (NaCl, KCl, MgCl2, CaCl2) have been studied. Purification of this luciferase was initiated by column chromatography (ion exchange followed by hydrophobic interactions); the fraction showing most of the activity was analysed by mass spectrometry with no success. Work is in progress to further characterize and isolate this new luciferase.

Nervous control of *Amphiura arcystata* luminescence (Ophiuroidea, Echinodermata)

J. Mallefet
Marine biology laboratory, Catholic University of Louvain, Louvain-la-Neuve, Belgium

ABSTRACT: For the past 15 years, luminescence control mechanisms have been mainly studied in the class Ophiuroidea (Echinodermata) on a limited number of species. During research documenting the luminescence nervous control of *Ophiopsila californica*, specimens of another brittle star, *Amphiura arcystata* were collected. First experiments revealed that isolated arms of this species are able to produce light in response to potassium chloride depolarisation (KCl 200 mM). Pharmacological results show that acetylcholine (Ach) is the main neurotransmitter involved in light emission. The presence of a cholinesterase activity is suggested since Ach luminescence is highly enhanced by eserine treatment. Ach triggers light emission through the activation of muscarinic receptor's sub-type. KCl and Ach induced luminescence is calcium dependent. Results of this research support the idea of the preponderance of cholinergic control mechanisms in luminescence as suggested by initial research performed on *Amphipholis squamata* and *Amphiura filiformis*. Comparative studies should be undertaken in order to highlight whether cholinergic controls are restricted to Amphiurid families or are general to the Ophiuroidea class.

Taxonomic revision of the genus *Euapta* (Holothuroidea: Synaptidae)

A. Martínez-Melo
Laboratorio de Sistemática y Ecología de Equinodermos, Instituto de Ciencias del Mar y Limnología, Universidad Nacional Autónoma de México. Apdo. México, DF, México

F. Michonneau
Department of Biology, University of Florida and Florida Museum of Natural History, Gainesville, FL, USA

G. Paulay
Florida Museum of Natural History, Gainesville, FL, USA

F.A. Solís-Marín
Laboratorio de de Sistemática y Ecología de Equinodermos, Instituto de Ciencias del Mar y Limnología, Universidad Nacional Autónoma de México. Apdo. México, DF, México

ABSTRACT: Integrative approaches help clarify many previously difficult problems in taxonomy. The sea cucumber family Synaptidae has many such problems, because it is relatively character poor and different workers have often not agreed which characters are informative about species limits. Synaptids lack tube feet, respiratory trees, enlarged haemal system, have thin, simple body walls and only a few types of ossicles. Three large-bodied genera are common on reefs: *Synapta*, *Euapta*, and *Opheodesoma*. Six nominal species of synaptids have been assigned to *Euapta*, three from the West Atlantic: *E. lappa*, *E. polii*, and *E. tobagoensis*, and three from the Indo-west Pacific: *E. godeffroyi*, *E. magna*, and *E. tahitensis*. Heding (1928), a notorious splitter, recognized all (five; *E. tahitensis* was described subsequently) and described two of these species. However most authors (Fisher, Clark, Deichmann, Rowe, Massin, Samyn, Pawson) over the past century have recognized only one in each biogeographic region: *E. lappa* and *E. godeffroyi*. We encountered two color morphs of *Euapta* in the Marshall Islands which raised the issue of whether the current view of a single Pacific species is accurate, and led us to dwell into the differentiation of these and related forms. We show that there are at least two sympatric Indo-west Pacific species of *Eupta*, recognizable by color, ossicle, and DNA sequence characters. We also evaluated the genetic status of East Pacific population and the divergence of *Euapta* across the Isthmus of Panama. Sequence data also support the recognition of *Synapta*, *Euapta*, and *Opheodesoma* as distinct, deeply-divergent, monophyletic genera.

The effects of sunscreen compounds on UV-exposed sea urchin larvae across a latitudinal gradient

A. McCarthy & M. Lamare
Department of Marine Science, University of Otago, Dunedin, New Zealand

ABSTRACT: Sea urchin larvae are particularly vulnerable to enhanced UV-radiation exposure as they are small, highly transparent, undergo rapid cell division and occupy the upper regions of the water column. UV-radiation has the potential to induce a range of deleterious effects on planktonic larvae including increased metabolic costs, DNA damage and mortality. Respiration efficiency is a good measure of metabolic costs. Larvae may mitigate UV induced damage via a number of strategies including the dietary uptake of sunscreen compounds such as Mycosporine-like Amino Acids (MAAs). The link between UV-radiation exposure, MAA protection and respiration was investigated in sea urchin larvae from a range of latitudes, including a tropical (*Tripneustes gratilla*) a temporal (*Pseudechinus huttoni*) and an Antarctic (*Sterechinus neumayeri*) species. Larvae were exposed to varying levels of UV-radiation using *in situ* techniques that were standardised to allow direct comparisons between species. Larvae samples were poured into 125 ml UV transparent bags under two light filters (UV transparent filter, UV opaque filter) and moored at two depths (0.5 m and 2.5 m from water surface). Samples were also fed a range of algae feeds to induce varying MAA levels and left in the field for up to one day. Larval respiration was measured and quantified using a microrespirometer and MAA levels were measured using High Performance Liquid Chromatography (HPLC). Respiration was found to change when exposed to UV-radiation, but changes were species and wavelength specific. Further research will investigate the link between UV exposure, MAA protection and DNA damage within these species.

Arm damage and regeneration of *Tropiometra afra macrodiscus* (Echinodermata: Crinoidea) in Sagami Bay, central Japan

R. Mizui & T. Kikuchi

Graduate School of Environment and Information Sciences, Yokohama National University, Tokiwadai, Hodogaya, Yokohama, Kanagawa, Japan

ABSTRACT: The diurnal comatulid crinoid *Tropiometra afra macrodiscus* is one of the largest and common species in shallow and temperate water of Japan. They occur where current flows swiftly, where they extend their flexible and robust (but not curlable) arms upward. After a strong typhoon which struck Sagami Bay, central Japan, in September 2007, we observed five individuals (one adult and four young) that had been damaged and lost a part of every arm. About a month later, regeneration in some of their arms was observed, however, synchronicity in regeneration of their arms was not observed. By March 2008, two of the young individuals had almost repaired their arms. Average growth rate, on the basis of their longest arm, of the four young individuals of *T. a. macrodiscus* was 8.1 cm/year, while their average regeneration rate was more than twice that of the growth rate (17.6 cm/year). Microscopic observation revealed that the points of mutilation were different in each arm, and only 20% of mutilation had been observed at syzygial articulations. Negative correlation was observed between the length of the stump and the length of regeneration. These results show that *T. a. macrodiscus* has strong powers of recovery from damage and regulates its arm lengths.

Connexions of the larval coeloms in the brachiolaria larva of the sea star *Parvulastra exigua* (Asteroidea, Asterinidae)

V.B. Morris
School of Biological Sciences, University of Sydney, NSW, Australia

P. Selvakumaraswamy & M. Byrne
Department of Anatomy and Histology, University of Sydney, NSW, Australia

R. Whan
Electron Microscope Unit, University of Sydney, NSW, Australia

ABSTRACT: In early development, there is but a single coelom that forms from at the inner end of a short archenteron. Later morphogenetic processes transform this coelom into one with two posteriorly directed lateral processes connected to each other over the inner archenteron by a large anterior coelom.

The anterior coelom projects into the brachia, forming the brachial coeloms. These coeloms are all inter-connected. In later development, this large enterocoele separates into smaller coeloms. We show these separations in confocal images of a lecithotrophic brachiolaria larva, using images in the XY, XZ and YZ planes. The data we present has implications for the trimery theory of the ancestral echinoderm larval body plan.

Variation in stone canal morphology in the brachiolaria larva of the sea star *Parvulastra exigua* (Asteroidea, Asterinidae)

V.B. Morris
School of Biological Sciences, University of Sydney, NSW, Australia

P. Cisternas & M. Byrne
Department of Anatomy and Histology, University of Sydney, NSW, Australia

R. Whan
Electron Microscope Unit, University of Sydney, NSW, Australia

ABSTRACT: The stone canal is a core structure of the echinoderm body plan. It links the hydrocoele to the exterior at the madreporite. Early in its development, one end of the stone canal opens into the hydrocoele where the primary podia form, while the other end opens variously into the anterior coelom, into the pore canal, or opens to the exterior via its own channel. We show confocal images of these variations in a lecithotrophic brachiolaria larvae, using images in the XY, XZ and YZ planes. The connexion of the inner end of the stone canal is quite precise, to podium D of the hydrocoele. The variability of the connexions of the outer end suggests that the channels that connect to the exterior are larval structures, with dependence on the type of larval development.

Maturation stage and extractive components in the gonads from seven Japanese edible species of sea urchins

Y. Murata
National Research Institute of Fisheries Science, Fisheries Research Agency, Fukuura, Yokohama, Japan

T. Unuma
Japan Sea National Fisheries Research Institute, Fisheries Research Agency, Suido-cho, Niigata, Japan

ABSTRACT: In Japan, the seven species of sea urchins, *Anthocidaris crassispina*, *Pseudocentrotus depressus*, *Hemicentrotus pulcherrimus*, *Strongylocentrotus nudus*, *S. Intermedius*, *Tripneustes gratila* and *Diadema setosum* are edible and their fishing is carried out in different seasons and locations. In this study, we investigate the maturation stage and extractive components in the gonads from these species. Urchins of each species were collected on a day in their fishing season. The maturation stage and free amino acids and nucleotide components in the gonads were determined for each individual. All *H. pulcherrimus* were recovering stage (Stage I), while several stages (stage 1, 2, 3 and 4) were observed in the other species. Total free amino acids and nucleotides were different not only among species but also among individuals. The gonads of *D. Setosum* were rich in Tau, while gonads of the other all species were rich in glycine. We are currently examining the relationship between the extractive components and the taste of gonads in each species.

Conserved role of ARIS: Acrosome reaction-inducing substance in the echinodermata

M. Naruse & H. Sakaya
Keio University, Hiyoshi Kohoku-ku, Yokohama-shi Kanagawa-ken, Japan

M. Hoshi
The Open University of Japan, Wakaba Mihama-ku, Chiba-shi Chiba-ken, Japan

M. Matsumoto
Keio University, Hiyoshi Kohoku-ku, Yokohama-shi Kanagawa-ken, Japan

ABSTRACT: In many species the Acrosome Reaction (AR) is a key process in fertilization. Sperm specifically approach homologous eggs and undergo an exocytosis of the acrosomal vesicle in their head. This is followed by morphological change; this exocytosis enables sperm to pass through egg coat by chemical and physical processes. In the starfish *Asterias amurensis*, AR-inducer has been purified from the egg coat as a large proteoglycan-like molecule ARIS (AR-Inducing Substance). Its sugar chain part (named as Fr.1) is responsible for this activity, however its protein components are still unknown. In this study we elucidated that ARIS consists of 3 dependent glycoproteins modified by the Fr.1 sugar chain, named ARIS1, ARIS2 and ARIS3. These 3 ARIS proteins have similar structures and 2 conserved novel domains. The egg coats of other starfishes, *Distolasteras nippon* (Folcipulatida) or *Asterina pectinifera* (Asteroidea) share the ARIS protein structure as *Dn*ARIS1, *Dn*ARIS2, *Dn*ARIS3 and *Ap*ARIS1, *Ap*ARIS2, *Ap*ARIS3 respectively. The BLAST search results of ARISs were highly similar to those obtained with DY635177 and DY625100 in the sea cucumber *Apostichopus japonicas*. Also, in the sea lily *Oxycomanthus japonica* we suggested the presence of ARIS-like glycoproteins in egg coat, although a similar sequence for ARIS couldn't be detected in the sea urchins. Interestingly, in the Cephalochordata *Branchiostoma floridae* BW770556, BW773889 and BW706662 are similar to ARIS1, ARIS2 and ARIS3 respectively. As in ARIS molecules, the protein part is conserved in starfishes, sea cucumbers, sea lilies and *Amphioxus*; ARIS protein part may serve a significant role in the acrosome reaction evolutionarily.

Morphology, taphonomy and distribution of *Jacksonaster depressum* from the Red Sea

J.H. Nebelsick
Institute of Geowissenschaften, Tübingen, Germany

A.A. El-Hamied & A.S. Elattaar
Geology Department, Sohag Faculty of Science, Sohag University, Sohag, Egypt

ABSTRACT: *Jacksonaster depressum* is an Indo-Pacific clypeasteroid sea urchin. The study material originates from the Northern Bay of Safaga, Red Sea, Egypt, where this echinoid occurs commonly in muddy sands in depths of 20–50 m. The sea urchins reach lengths of up to 3 cm and possess flattened tests with internal supports restricted to the outer rim of the skeletons. These sea urchins are poorly studied and probably live infaunally as detritivores. We have investigated the skeletal morphology, taphonomy and distribution of numerous examples from the study area. Both living and dead test collected by various methods were investigated. Morphometric measurements were conducted on specific characters of the skeletons including general size parameters, apical system, petalodium, position and dimensions of the peristome and periproct etc. Additionally, scanning electron microscopy was conducted on spines, pedicellaria, tubercles and ambulacral pores. Stereom differentiation on the oral and aboral side of the test was noted and compared. Variations of size parameters were compared between living and dead test, as well as between different facies types and depths. Taphonomic destruction of the test was studied on both complete as well as fragmented tests. Numerous taphonomic features including disarticulation, fragmentation, encrustation and bioerosion were noted. Predation also plays an important role in the destruction of the test.

The latest fossil record of the genus *Linthia* (Echinoidea: Spatangoida) from the Pleistocene

K. Nemoto
Graduate School of Biological Science, Kanagawa University, Tsuchiya, Hiratsuka, Japan

K. Kanazawa
Department of Biological Sciences, Faculty of Science, Kanagawa University, Tsuchiya, Hiratsuka, Japan

ABSTRACT: Sea urchins of the genus *Linthia* appeared in the late Cretaceous and given the worldwide distribution in Eocene they presumably became extinct around Japan in the Pleistocene. The probable last species, *L. nipponica*, is found in fine grained sand of the Pleistocene Omma Formation. Many of the specimens retain whole tests, indicating that they were buried in their own habitat. The palaeoenvironmental study of foraminifera and molluscs suggests that the habitat was probably neritic (less than 60 m deep) and in a cold water region. From the Tertiary, in addition to *L. nipponica* (which is also found also from the Miocene and the Pliocene), 3 other species of *Linthia* have been reported in Japan: *L. praenipponica* and *L. yessoensis* from the Oligocene and *L. tokunagai* from the Pliocene. *L. tokunagai* occurs from fine sediment similar to that of *L. nipponica*, and biometrical analysis shows these two species are not distinguishable from each other by any morphological character. They would appear to be the same species. Two Oligocene species, *L. praenipponica* and *L. yessoensis* are also found in the fine sediment characteristic of *L. nipponica*. Because of their bad preservation there is no specimen available for precise biometrical analysis, but some morphological characteristics are very similar to those of *L. nipponica*. The similarity of the morphology among the species and of the sediment containing the fossils indicates that *Linthia* would have inhabited the same neritic environment around Japan from the Oligocene until early Pleistocene when the last species of the genus became extinct.

Cloning and gene expression analysis of crinoid *Otx* and *Pax6* orthologs

A. Omori & T.F. Shibata
Misaki Marine Biological Station, The University of Tokyo, Misaki, Miura, Kanagawa, Japan

Y. Nakajima
Department of Biology, Keio University, Hiyoshi, Kohoku-ku, Yokohama, Kanagawa, Japan

D. Kurokawa & K. Akasaka
Misaki Marine Biological Station, The University of Tokyo, Misaki, Miura, Kanagawa, Japan

ABSTRACT: Crinoids (sea lilies and feather stars) are considered as the most basal group of extant echinoderm classes. They are potentially important model organisms for evolutionary developmental biology because they alone retain some primitive morphological features like a well developed aboral nervous system including a ganglion among extant echinoderms. To examine the evolutionary relationship between the mechanism of crinoid nervous systems formation and that of the other animals, we cloned homologs of *Otx* and *Pax6*, which are involved in fundamental processes of anterior neural patterning in vertebrates, from a feather star *Oxycomanthus japonicus*. *In situ* hybridization revealed that *Pax6* is expressed just anterior to the *Otx* expressing area in endomesoderm of the planktonic larvae, which is the same expression pattern in vertebrates' central nervous systems. *Otx* is also expressed in ciliary bands of the planktonic larvae, which are lined with larval nerve tracts. It is also revealed that *Otx* and *Pax6* are expressed in the stomodeum and the podia of the settled larvae and juvenile, suggesting their involvement in the formation of sense organs on crinoids podia. These results suggest that the involvement of *Otx* and *Pax6* in anteroposterior patterning and in nervous system formation is acquired separately in basal echinoderms and more basal animals.

Characterisation and quantification of the intradigestive bacterial microflora in *Holothuria scabra* (Jaeger, 1833) (Holothuroidea), major macrodeposit feeder of recifal intertidal zones

T. Plotieau & I. Eeckhaut

Laboratoire de Biologie Marine, Université de Mons-Hainaut, Belgium

ABSTRACT: For a long time, sea cucumbers were considered as non-selective deposit feeders but recent research suggests that they absorb nutrients of some part of the sediment. The sediment is a complex environment, a mix of mineral material and organic fragments that are both dead and alive (*e.g.* micro-organisms like bacteria, fungi, protists and meiofauna). The aims of the present study are to characterise and quantify the intestinal microbial community in *H. scabra* and that of the substratum on which it lives. To investigate the intradigestive microflora three methods were used: *(i)* gut bacterial cultures followed by 16S rDNA sequencing; *(ii)* Denaturing Gradient Gel Electrophoresis (DGGE) and *(iii)* cloning. The bacterial count was achieved by DAPI coloration (4',6-diamino-2-phenylindole) and the different sizes of particles of the substratum were analyzed by sieving. The results suggest that *H. scabra* does not select a particular size of particles and that the number of bacteria is high in the foregut of the animal indicating that bacteria are either selected from the substrate or cultured in the foregut. It is suggested that *H. scabra* is a nonselective deposit feeder (*i.e.*, it ingests all particles of the sediment without selection) but that the nutrients absorbed by the sea cucumber come mostly from the living fraction of the sediment.

A nearly articulated aspidochirote holothurian from the Late Cretaceous (Santonian) of England

M. Reich

Geowissenschaftliches Zentrum der Universität Göttingen, Museum, Sammlungen & Geopark, Göttingen, Germany

ABSTRACT: With the exception of the Maastrichtian, only scarce data exist on fossil holothurians from the Upper Cretaceous, especially the Santonian (84 myr ago). Up to now, only two records have been mentioned: (1) from the Münsterland area, Germany, and (2) from the Montsec area, Spain; both referred to the Apodida and Dendrochirotida. Here I report the discovery of a nearly articulated holothuroid specimen from the "Upper Chalk" (Santonian; BMNH E 48764) of England (exact locality unknown). The unique specimen shows two different types of ossicles: (A) tables and (B) buttons, demonstrating aspidochirote affinities, very likely to the family Holothuriidae. The 'buttons' share characteristics with tube feet and papillae ossicles of *Holothuria (Mertensiothuria)*, *H. (Microthele)*, and *H. (Semperothuria)* and *H. (Theelothuria)*. Only a part of the former body skeleton is preserved; the anterior part of the skeleton (calcareous ring) is missing. Due to the well-preserved body-wall ossicles and the formation of the ossicles of the new find, I can exclude a digestion and excretion of the sea cucumber by any predator. After comparison with modern members of the Holothuriidae, we can speculate that the body of our new Santonianspecies was also cylindrical, elongate and scattered with numerous podia over the entire body, dorsally as papillae, ventrally as cylindrical tube feet. A wider distribution of the new genus/species can be assumed, because comparable long button-shaped sclerites were also known from Early and Late Maastrichtian chalk sediments of the islands of Møn (Denmark) and Rügen (NE Germany) as well as the Baltic Sea. Considering how important aspidochirote sea cucumbers are today (nearly 1/4 of all known modern species), their fossil record is meagre and poorly understood.

Unusual holothurians (Echinodermata) from the Late Ordovician of Sweden

M. Reich
Geowissenschaftliches Zentrum der Universität Göttingen, Museum, Sammlungen & Geopark, Göttingen, Germany

ABSTRACT: Compared to other modern echinoderms, the early evolutionary history of sea cucumbers is poorly understood. In part, this is due to their disjunct endoskeleton with ossicles and calcareous ring elements, which are released following decomposition of the surrounding tissue. Newly sampled öjlemyr flints from the northwestern part of Gotland, Sweden, yield well-preserved echinozoan echinoderms, including holothurian ossicles. The studied material is Ashgill (upper Pirgu stage; 447 myr) in age and reveals the presence of several new or poorly known taxa of elasipodid, aspidochirote, and apodid holothurians or stem group representatives. The minute ossicles, around 70–400 µm in length/diameter, exhibit an impressive morphology with arms, spires, perforations, teeth etc. The new material differs from previously described Palaeozoic, Mesozoic and most of the modern material by unusual symmetry. This study shows that the Holothuroidea diversified significantly through Late Ordovician times.

Phenotypic plasticity and the developmental regulatory network in the purple sea urchin

D.E. Runcie, D.A. Garfield, C.C. Babbitt, A. Pfefferle, W.J. Nielsen & G.A. Wray
Duke University, Durham, NC, USA

ABSTRACT: Plasticity is a ubiquitous, but often neglected characteristic of the development of nearly every organism. Larvae of many echinoderm species have an adaptive plastic response to nutritive stress. However, despite extensive investigations into various aspects of sea urchin morphogenesis, almost nothing is known about the genetic basis of phenotypic plasticity in any sea urchin species. Here, we are taking a systems approach to investigate the developmental basis for phenotypic plasticity in larvae of the purple sea urchin, *Strongylocentrotus purpuratus*. Plasticity in sea urchins is characterized by changes in the growth rate of the larval skeleton and stomach. The endomesodermal gene regulatory network characterized in this species is a set of genes known to contribute to the specification and formation of the larval skeleton, and to the endodermal tissue of the larval gut. We use a high-throughput gene expression assay to ask if this set of genes is involved in the control of developmental plasticity in *S. purpuratus* larvae. Our results show that the expression of several of the genes in the endomesodermal network is altered when the larvae are cultured under reduced amounts of food. However, surprisingly, we do not find that the environment-sensitive genes are functionally related within the network, and the expression of key biomineralization genes does not appear highly correlated with the growth rate of the larval skeleton. These data point to the complexity of the relationship between gene expression and organismal phenotypes, especially in the context of whole-organism growth rate differences.

Fission, spawning, and metamorphosis occur continuously in the multi-armed sea-star, *Coscinasterias acutispina* (Stimpson, 1882)

D. Shibata & M. Komatsu
Department of Biology, Graduate School of Science and Engineering, University of Toyama, Toyama, Japan

Y. Hirano
Department of Biology, Graduate School of Science, Chiba University, Chiba, Japan

ABSTRACT: An adult *Coscinasterias acutispina* with 7–10 arms commonly possesses 2–5 madreporites. The species is known as an asexual reproducer by fission, although we have reported that it can also reproduce sexually, having observed the completion of metamorphosis in larvae obtained by artificial fertilization. Its development is indirect: it metamorphoses from a feeding bipinnaria into a brachiolaria, then into a juvenile with 6 arms and 1 hydropore. No spawning has yet been reported in the laboratory or the field. In this study, we fed juveniles on small polychaetes and snails. Twelve months after the completion of metamorphosis, 2 young individuals divided into 2 specimens with 3 arms each. Four months later, a third young individual divided. In regenerating, each of these 6 new specimens formed 4 or 5 new arms and 2–4 madreporites, so that their arm and madreporite counts were similar to those of adults. Moreover, 4 of these 6 specimens underwent a second fission 16 to 17 months after the completion of metamorphosis. In 3 of these 4, the new fission occurred at the same position as the first. Through two episodes of fission, the original 3 specimens reproduced to become 12. Furthermore, 1–3 months after the second fission, some of the specimens thus created exhibited well-developed gonads which shed gametes in an aquarium. The brachiolaria obtained from this spawning event metamorphosed into juveniles. Our observations suggest that each individual of this species has the potential for both asexual and sexual reproduction.

Glacial persistence of North Atlantic populations of *Cucumaria frondosa* (Holothuroidea: Echinodermata)

J. So
Ocean Sciences Centre, Memorial University, St. John's, NL, Canada

S. Uthicke
Australian Institute of Marine Sciences, Townsville MC, Queensland, Australia

J-F. Hamel
Society for the Exploration and Valuing of the Environment, Phils Hill Road, St. Philip's, NL, Canada

A. Mercier
Ocean Sciences Centre, Memorial University, St. John's, NL, Canada

ABSTRACT: In the upper latitudes of the northern hemisphere shallow water invertebrates were vastly impacted by glaciations during the Last Glacial Maximum (LGM) in the Late Pleistocene. Many marine taxa show a genetic signature of extirpation in the Western Atlantic followed by recolonization by Eastern Atlantic populations after glaciers receded. *Cucumaria frondosa*, the most abundant and widely distributed holothuroid in the North Atlantic, was used as a model to test whether its present phylogenetic structure reflects a history of persistence or extirpation and recolonization with respect to Pleistocene glaciations. Mitochondrial DNA (mtDNA) was extracted and sequenced from a total of 334 specimens collected from 20 locations (7 to 5900 km apart) throughout the North Atlantic. Distribution of shared haplotypes indicated groups sampled were part of a large panmictic population with high gene flow between regions. In contrast, exact test for population differentiation was overall significant ($P < 0.001$) and showed differentiation of several pairs of populations, irrespective of their geographic distance signifying genetic patchiness at the local level. High connectivity between regions and estimates that approximate time of expansion took place during the Late Pleistocene suggested persistence through the LGM. Haplotype diversity indicated the western Atlantic was recolonized by populations residing in ice-free areas east of Newfoundland and Labrador. Long larval times (ca. 6 weeks) and ability to tolerate arctic conditions enabled *C. frondosa* to survive on the fringes of glaciated regions and colonize areas as glaciers receded.

Echinopluteus vs. *Echinopluteus transversus*: Evolution of phenotypic plasticity in sea urchin larvae

N.A. Soars, T. Prowse & M. Byrne
Anatomy and Histology, Bosch Institute, University of Sydney, NSW, Australia

ABSTRACT: Echinoid larvae are well known for their phenotypically plastic response to different food regimes. Under laboratory conditions these larvae demonstrate an increase in postoral arm length when food is limited, but a shorter postoral arm length and time to metamorphosis when food is abundant. This increase in postoral arm lenth resutls in an increase in larval feeding capacity through elongation of the ciliated band that runs along the arms, and an increase in surface area: volume ratio. This response is thought to be an indication that there is a priority to allocate energy to feeding structures before development of the adult rudiment, thus reducing the time spent in the plankton exposed to predators. However, these studies have been performed on a limited suite of species, most from the northern hemisphere with a typical 8-armed larva. We tested 2 Australian species occurring in Sydney: one an echinometrid with a typical 8-armed echinopluteus (*Heliocidaris tuberculata*) and one a diademid with an atypical 2-armed *Echinopluteus transversus* type larvae (*Centrostephanus rodgersii*). While the 8-armed larva demonstrated a typical phenotypic response, the 2-armed *Echinopluteus transversus* type larva of *C. rodgersii* did not. This has implications for the evolution of the 2-armed form and indicates that phenotypic plasticity may be influenced by phylogeny.

Resistance of pearlfishes to saponins

M. Todesco & I. Eeckhaut
Laboratoire de Biologie marine, Université de Mons-Hainaut, Belgium

E. Parmentier
Laboratoire de Morphologie fonctionnelle et évolutive, Université de Liège, Belgium

ABSTRACT: Carapid fishes, known as pearlfishes, are anguilliformes fishes that enter and live in some invertebrates, especially in some echinoderms such as holothuroids and seastars. However these echinoderms contain a strong concentration of saponins, a secondary metabolite which acts as a predator deterrent. These saponins, based on a triterpene glycosides structure, show a wide range of biological activities including haemolytic damage and ichtyotoxicity. The present work aims at analyzing the effects of saponins on the gills of fishes including pearlfishes. Saponins were extracted from the sea cucumber *Bohadschia argus* (Tuléar, Madagascar) which is a natural host of some pearlfishes. Five carapid species (*Carapus acus*, *C.boraborensis*, *C. homei*, *C. mourlani* and *Encheliophis gracilis*) and five coral reef fishes species (*Amphiprion akallopisos*, *Dascyllus aruanus*, *D. flavicaudus*, *D. trimaculatus*, *Neoniphon sammara*) were investigated. Animals were kept in tanks during two hours and exposed to various concentrations of saponin. At the end of each test, a sample of gill was taken and fixed in order to realize histological and scanning electron microscope analyses. Coral reef fishes exposed to saponins concentration of 0.2 to 0.5 µl/ml died within two hours and exhibited pathological alterations of gill filaments. The pearlfishes proved to be resistant and had no gill damage.

Analysis of expression pattern of *Hox* cluster genes in feather star *Oxycomanthus japonicas*

T. Tsurugaya, T. Shibata, D. Kurokawa & K. Akasaka
Misaki Marine Biological Station, Graduate school of science, The University of Tokyo Misaki-koajiro, Miura, Kanagawa, Japan

ABSTRACT: To understand the evolution of deuterostomes, study of the body plan of echinoderms is indispensable. So far, the sea urchin has been used as a model echinoderm because of its availability. However, sea urchins also evolved some specific characteristics, such as loss of ganglia as a central nervous system. Hox genes are organized in clusters whose genomic organization reflects their central roles in patterning along the anterior/posterior axis. Recent study of the sea urchin genome revealed that dynamic rearrangement of Hox genes occurred in the cluster during evolution. We assume that the rearrangement caused the specific evolution in sea urchins. Among the extant echinoderm, crinoids is the most basal group retaining ancestral characteristics. In Misaki Marine Biological Station, we have succeeded in cultivating the crinoid *Oxycomanthus japonicus* so that specimens at every developmental stage are available. Taking the advantage of this resource, in the first step, we analyzed the expression pattern of the genes during development. Semi-quantitative RT-PCR reveals that the anterior group (*Hox1* and *Hox2*) are expressed as early as gastrula stage. It should be noted that in sea urchins the anterior group of *Hox* genes are not expressed in the early embryonic and larval stage. The expression of central and posterior group *Hox4*, *Hox7* and *Hox8-Hox9/10* start at the cystidean, and as early as gastrula and hatched embryo stages respectively in this feather star.

Changes in free amino acid composition in the ovary and testis of the sea urchin *Pseudocentrotus depressus*

T. Unuma
Japan Sea National Fisheries Research Institute, Fisheries Research Agency, Suido-cho, Niigata, Japan

Y. Murata
National Research Institute of Fisheries Science, Fisheries Research Agency, Fukuura, Yokohama, Japan

ABSTRACT: The quality of the sea urchin gonad as food products varies with maturation. In *Pseudocentrotus depressus*, immature ovary and testis before gametogenesis have superior quality as food. As the gonad becomes mature, its food quality decreases because of the development of various characteristics, such as the bitterness of the ovary and the oozing of gametes from the gonoduct of the ovary and testis. In the present study, we determined the concentrations of free amino acids, some of which are taste components, in the ovary and testis of *P. depressus* cultured in a tank with kelp *Eisenia bicyclis* as a main food source. Serine, glutamic acid, glutamine, glycine, alanine, lysine, histidine, and arginine were the major free amino acids both in the immature ovary and testis. As gametogenesis proceeded, most of these amino acids decreased, while glycine in the ovary and serine and glycine in the testis increased. Lysine in the ovary and glutamic acid and alanine in the testis kept similar value during gametogenesis. The relationship between free amino acid composition and food quality in the gonad will be discussed.

The major yolk protein in the sea urchin is synthesised mainly in the gut inner epithelium and the gonad nutritive phagocytes before and during gametogenesis

T. Unuma
Japan Sea National Fisheries Research Institute, Fisheries Research Agency, Suido-cho, Niigata, Japan

K. Yamano
National Research Institute of Aquaculture, Fisheries Research Agency, Minami-ise, Mie, Japan

Y. Yokota
Department of Applied Information Science and Technology, Aichi Prefectural University, Nagakute, Aichi, Japan

ABSTRACT: Major Yolk Protein (MYP), a transferrin superfamily protein that forms yolk granules in sea urchin eggs, is also contained in the coelomic fluid and nutritive phagocytes of the gonad in both sexes. MYP is stored in ovarian and testicular nutritive phagocytes prior to gametogenesis, and utilized during gametogenesis as material for synthesizing proteins and other components necessary for eggs and sperm.

To reveal the main production site for MYP, we investigated MYP mRNA expression in immature and maturing *Pseudocentrotus depressus*. Real-time RT-PCR analysis revealed that MYP mRNA was expressed predominantly in the digestive tract (stomach, intestine and rectum) and the gonad of both sexes. The total MYP mRNA amounts in the whole digestive tract and in the whole gonad were at similar levels both in immature and maturing sea urchins. Compared to these organs, the coelomocytes contained only small amount of MYP mRNA. Using *in situ* hybridization analysis, MYP mRNA was detected in the inner epithelium of digestive tract and in nutritive phagocytes of the ovary and testis, but was not detected in the germ cells. We conclude that the adult sea urchin has two predominant production sites for MYP regardless of sex and reproductive stage; the inner epithelium of digestive tract and nutritive phagocytes of the gonad.

Genetic barcoding of commercial bêche-de-mer species

S. Uthicke
Australian Institute of Marine Science, Townsville MC, QLD, Australia

M. Byrne
Department of Anatomy and Histology, Bosch Institute, University of Sydney, NSW, Australia

C. Conand
Universite de La Reunion, Laboratoir Ecologie Marine, Saint-Denis, Réunion, France

ABSTRACT: A recent review conducted by the Food and Agriculture Organisation of the United Nations (FAO) identified 47 species of holothurians used for bêche-de-mer production. With three exceptions from the dendrochirotides all bêche-de-mer species are aspidochirotides. Species identification of many of these is difficult. To improve this, we conducted a review of genetic information available and obtained additional data to test whether genetic barcoding can be used as a tool for bêche-de-mer species identification. Similar to other barcoding projects we focussed on the mitochondrial COI gene. Although some genetic information was available for ca. 50% of all bêche-de-mer species, sufficient information and within-species replication was only available for 6 species. We obtained 96 new COI sequences extending the existing database to cover most common bêche-de-mer species. COI unambiguously identified bêche-de-mer species in most cases and therefore provides excellent genetic barcodes. However, since this marker did not work for two of the most valuable species additional method development will be required. In addition to species identification in adults, COI sequences were useful to identify juveniles, and sequences demonstrated that large (deep) and small (shallow) specimens of *Holothuria atra* belong to the same species. Our study has also demonstrated that further genetic and taxonomic work in this group is essential; it provided evidence for at least three further species which are currently undescribed (e.g. one *Bohadschia* species), or species which are likely to constitute separate species in the Indian and Pacific Ocean (e.g. *H. fuscogilva*).

Ecological similarities between the associated fauna of *Echinometra vanbrunti* and *E. Lucunter* (Echinoidea: Echinometridae) on Colombian Rocky shores

V.A. Vallejo
Laboratorio de Zoología y Ecología Acuática LAZOEA, Universidad de los Andes, Bogotá, Colombia

M. Monroy
Instituto Canario de Ciencias Marinas ICCM, Gran Canaria, España

ABSTRACT: To determine the ecological relationships between the *Echinometra* species of the Colombian shores and the fauna inside their boreholes, two similar research studies were made. The first one, made on a Pacific shore (Málaga Bay), sets *E. vanbrunti* as a possible host in commensal relationships with the porcelain crab *Clastotoechus gorgonensis* and the muricid snail Thais melones. The second one, conducted on Caribbean shores, establishes similar kinds of ecological relationships between *E. lucunter, Clastotoechus vanderhorsti*, the brittlestar *Ophiohtryx synoecina* and the goby fish *Acyrtus rubiginosus*. 27 different accompanying species were found in *E. vanbrunti* boreholes, and 21 in those of *E. lucunter*. The boreholes of the two urchin species may act as a shelter for their commensals; they could avoid predation, cannibalism and desiccation by crawling inside and using urchin spines as protection. A distributional analysis of *Echinometra* species showed that they concentrate on substrata covered with chlorophycean and rodophycean algae, which are their main food sources.

Saponin diversity and body distribution in five tropical species of sea cucumbers from the family Holothuriidae

S. Van Dyck, P. Gerbaux & P. Flammang
University of Mons-Hainaut, Mons, Belgium

ABSTRACT: Sea cucumbers lack structural defences because of their reduced skeleton. To deter predation holothuroids contain feeding deterrent molecules, the saponins, in their body wall and viscera. Saponins are secondary metabolites based structurally on a triterpene glycoside structure. The aim of this study is to analyse and compare the saponins mixtures of 5 sea cucumbers species from the Indian Ocean: *Actinopyga echinites, Bohadschia subrubra, Holothuria atra, H. leucospilota* and *Pearsonothuria graffei*. Mass spectrometry (MALDI-MS and MALDI-MS/MS) was used to detect saponins and describe their molecular structures. LC-MS and LC-MS/MS were also used to separate saponins and, in some case, enable the identification of isomers. Two different body components were analysed separately: the body wall (which protects the animal and is moreover the largest organ) and the Cuvierian tubules (a defensive organ that can be expelled onto predators in response to an attack), except for *H. atra* which lacks Cuvierian tubules. Holothuriid saponins are usually classified into 2 categories, non-sulfated in the genus *Bohadschia* and sulphated in all other genera. Our results do not completely corroborate this distribution. Indeed, although *B. subrubra* encloses only non-sulfated saponins and *A. echinites* and *H. atra* have only sulfated saponins, *H. leucospilota* and *P. graffei* contain both saponin types. The number of saponins in a mixture varies between species but also between the body components of the same species. For each species, some saponins are common to both body components while others are specific to one organ.

Ophiopsila pantherina beds on relict dunes off the Great Barrier Reef

E. Woolsey, M. Byrne, S. Williams, O. Pizarro, K. Thornborough, P. Davies & J. Webster
University of Sydney, Sydney, NSW, Australia

R. Beaman
James Cook University, Cairns QLD, Australia

T. Bridge
James Cook University, Townsville QLD, Australia

ABSTRACT: We used an Autonomous Underwater Vehicle (AUV) to generate the first images of *Ophiopsila pantherina* beds on the relict dunes at Hydrographers Passage near the shelf edge of the Great Barrier Reef (GBR), 200 km off the Australian mainland. High resolution multispectral stereo photographic images captured by the AUV co-registered with bathymetry data from multibeam and seismic were used to map and reconstruct the sea floor. The dunes, 2–6 m in height, project into a dynamic environment characterised by strong tidal currents carrying nutrient rich water upwelled from the continental slope. The AUV tracked along the dunes at an altitude of 2 m above the bottom to capture images of the *O. pantherina* beds. *O. pantherina* takes advantage of their elevated position on the lee side of the dunes for suspension feeding. The reliable high current environment of the dunes is ideal for these suspension feeders. They live with their central body in a sand burrow from which they extend 4 of their 5 arms into the flow to feed. At a mean density of 380 m^2 *O. pantherina* forms an impressive wall of arms filtering plankton from passing flow. There were little other macrofauna. This ophiuroid is intensely bioluminescent. On contact stimulation, the arms emit visible light in a bright green flash that travels down the arm. Bioluminescence in *Ophiopsila* is a defensive adaptation visually stunning predators that bump into the arm. As a last ditch defence, writhing autotomized arms function as sacrificial flashing lures to distract predators as the brittle star retreats into the burrow. These aggregated ophiuroid communities in dune fields may be a specialised natural feature for consideration in managing inter-reefal sandy habitats the Great Barrier Reef Marine Park.

In which direction do regular sea urchins walk?

K. Yoshimura & T. Motokawa
Department of Biological Sciences, Tokyo Institute of Technology, Meguro-ku, Tokyo, Japan

ABSTRACT: Regular sea urchins have been said to show no preference in the direction they move. Here we report that the regular sea urchin *Hemicentrotus pulcherrimus* walks with a certain part of the test forward. An animal, which has been rested in a square aquarium (resting aquarium), was transferred to a round aquarium of 185 mm diameter (observation aquarium). The animal was placed in the centre of the bottom of the aquarium and the walking behaviour was recorded with a video camera. Sea urchins started walking within 10 s after they touched the bottom. They took a straight path until they reached the aquarium wall. The walking velocity was rather uniform (20–35 mm/min). Sea urchins did not rotate their test during the walk and thus they proceeded keeping a certain part of the test forward. We found that the part of the test facing forward was the part with which sea urchins attached to the wall during the preceding resting period. Sea urchins always stayed at the corner of the resting aquarium with their tube feet attaching to both sides of the corner. A line that was drawn from the animal centre bisecting the area of the test with attached tube feet coincided with the line that was drawn from the animal centre to the leading edge of the ambitus in the walk the sea urchin showed when transferred to the experimental aquarium (the V test, $P < 0.05$, $n = 5$).

First report of *Ulophysema* (Crustacea: Ascothoracida) outside northern seas, parasitizing the bathyal Antarctic species *Pourtalesia hispida* (Echinoidea: Holasteroida)

A. Ziegler
Institut für Biologie, Freie Universität Berlin, Berlin, Germany

M.J. Grygier
Lake Biwa Museum, Oroshimo, Kusatsu, Shiga, Japan

ABSTRACT: The genus *Ulophysema* Brattström, 1936 belongs to the family Dendrogastridae, an entirely endoparasitic taxon of crustaceans. Parasites of this family are characterized by a large to giant female with spacious, highly modified carapace lobes that serve as brood sacs. Currently, two species are ascribed to this genus. *Ulophysema oeresundense* Brattström, 1936 is found usually in the coelomic cavities, but occasionally in the gonads or ampullae, of three species of spatangoid sea urchins from southern Scandinavian waters. *Ulophysema pourtalesiae* Brattström, 1937 was found in the coelomic cavities of *Pourtalesia jeffreysi* Wyville Thomson, 1873, a deep-sea holasteroid species from the North Atlantic. Here, we report the first finding of *Ulophysema sp.* in *Pourtalesia hispida* Agassiz, 1879, a species solely occurring in Antarctic waters, based on a single infected specimen dredged from a depth of 1353 m in the Weddell Sea. We thereby considerably extend the known distributional range of the genus *Ulophysema*. The female endoparasite has two broad carapace lobes extending deep into the perivisceral coelom of the host and is firmly attached to the host's calcitic endoskeleton near the genital plates in close vicinity to the gonads. Size, location, and form of the parasite within the host, as well as its effect on the gross morphology of the sea urchin's internal organs, are described based on dissection and non-invasive tomographic data.

Interdependence of internal organ systems and evolution of the axial complex in sea urchins (Echinodermata: Echinoidea)

A. Ziegler
Institut für Biologie, Freie Universität Berlin, Königin-Luise-Straße, Berlin, Germany

C. Faber
Institut für Klinische Radiologie, Universitätsklinikum Münster, Waldeyerstraße, Münster, Germany

T. Bartolomaeus
Institut für Evolutionsbiologie und Zooökologie, Universität Bonn, An der Immenburg, Bonn, Germany

ABSTRACT: The echinoid axial complex is composed of various primary and secondary body cavities. Structural differences of the axial complex in 'regular' and irregular species have been observed, but the reasons underlying these differences are not understood. Furthermore, a better knowledge of axial complex diversity could not only be useful for phylogenetic inferences, but improve also an understanding of the function of this enigmatic structure. We analyzed numerous species of almost all sea urchin orders employing magnetic resonance imaging, dissection, histology, and transmission electron microscopy.

A parallel comparison of our data with results from published studies spanning almost two centuries was performed. These combined analyses demonstrate that the axial complex is present in all sea urchin orders. It exhibits considerable morphological variation within the Irregularia, where gradual changes in topography, size, and internal architecture of the axial complex can be observed. We postulate that these are probably a consequence of the growing size of the gastric caecum as well as the rearrangement of the digestive tract morphology. The structurally most divergent axial complex was observed in the highly derived Atelostomata. Our findings demonstrate a structural interdependence of various internal organs, including digestive tract, mesenteries, and the axial complex.

Author index

Conferees

Khalfan Al-Rashdi
Oman Aquaculture Center
OMAN
omanaba@yahoo.com

Franziska Althaus
CSIRO Marine & Atmospheric Research
Tasmania
AUSTRALIA
franzis.althaus@csiro.au

Owen Anderson
NIWA
NEW ZEALAND
o.anderson@niwa.co.nz

Akwasi Apeanti
Taishan Medical University
Shandong Province
CHINA
francissarpong@yahoo.com

Mike Barker
Dept of Marine Science
University Of Otago
Otago
NEW ZEALAND
mike.barker@otago.ac.nz

Todd Beaumont
University Of Otago
Otago
NEW ZEALAND
toddy203@hotmail.com

Colin Buxton
University of Tasmania
Tasmania
AUSTRALIA
colin.buxton@utas.edu.au

Maria Byrne
Bosch Institute
University Of Sydney
NSW
AUSTRALIA
mbyrne@anatomy.usyd.edu.au

Ana Catarino
Université Libre De Bruxelles – ULB
BELGIUM
ana.catarino@ulb.ac.be

Natalia Charlina
Institute of Marine Biology
Vladivostok
RUSSIA
ncharlina@yandex.ru

Ana Christensen
Lamar University
Texas
USA
christenab@my.lamar.edu

Paula Cisternas
University of Sydney
NSW
AUSTRALIA
paula@anatomy.usyd.edu.au

Dana Clark
University Of Otago
Otago
NEW ZEALAND
dclark@ps.gen.nz

Niki Davey
NIWA
NEW ZEALAND
n.davey@niwa.co.nz

Matt Dell
Myriax Software P/L
Tasmania
AUSTRALIA
fionae@myriax.com

Angie Diaz Lorca
Universidad De Chile
CHILE
angie.ddl@gmail.com

Philippe Dubois
UniversitŽ Libre de Bruxelles
Bruxelles
BELGIUM
phdubois@ulb.ac.be

Clement Dumont
The University Of Hong Kong
HONG KONG
cdumont@hku.hk

Janina Friederike Dynowski
University Of Tübingen
Baden-Württemberg
GERMANY
dynowski.smns@naturkundemuseum-bw.de

Thomas Ebert
Oregon State University
Oregon
USA
ebertt@science.oregonstate.edu

Igor Eeckhaut
Université de Mons-Hainaut
Hainaut
BELGIUM
igor.eeckhaut@umh.ac.be

Marc Eleaume
Museum National D'Histoire Naturelle
Paris
FRANCE
eleaume@mnhn.fr

Phillip England
CSIRO
Hobart, Tasmania
AUSTRALIA
phillip.england@csiro.au

Toshihiko Fujita
National Museum of Nature and Science
Tokyo
JAPAN
fujita@kahaku.go.jp

Daisuke Fujita
Tokyo University of Marine Science
    and Technology
Tokyo
JAPAN
d-fujita@kaiyodai.ac.jp

Andy Gale
University Of Portsmouth
Hampshire
UNITED KINGDOM
andy.gale@port.ac.uk

Nils Grossmann
University Of Tübingen
Baden-Württemberg
GERMANY
nils.grossmann@uni-tuebingen.de

Gen Hamanaka
Keio University
Kanagawa
JAPAN
dr064506@hc.cc.keio.ac.jp

Jean-Francois Hamel
Society for the Exploration and Valuing of
    the Environment
Newfoundland and Labrador
CANADA
hamel@seve.cjb.net

Vanessa Hernaman
University Of Queensland
Brisbane, Queensland
AUSTRALIA
v.hernaman@uq.edu.au

Dinorah Herrero-Perezrul
Centro Interdisciplinario
    De Ciencias Marinas-IPN
Baja California Sur
MEXICO
dherrero@ipn.mx

Yuji Hiratsuka
University of the Ryukyus
Okinawa
JAPAN
yujihiratsuka@hotmail.com

Kristina Holm
University Of Gothenburg
Fiskebäckskil
SWEDEN
kristina.holm@zool.gu.se

Frederick Hotchkiss
Marine and Paleobiologcial Research Institute
Massachusetts
USA
hotchkiss@MPRInstitute.org

Seiichi Irimura
Creative Education Center
Tokyo
JAPAN
irimura@f2.dion.ne.jp

Yoshiaki Ishida
Akabane Commercial High School
Tokyo
JAPAN
y-ishida@msi.biglobe.ne.jp

Lindsay Jennings
University of New Brunswick
New Brunswick
CANADA
lindsay.jennings@unb.ca

Craig Johnson
University of Tasmania
Hobart, Tasmania
AUSTRALIA
craig.johnson@utas.edu.au

Ken'ichi Kanazawa
Kanagawa University
Kanagawa
JAPAN
kanazawa@kanagawa-u.ac.jp

John Keesing
CSIRO
Perth, Western Australia
AUSTRALIA
john.keesing@csiro.au

Mieko Komatsu
University of Toyama
Toyama
JAPAN
miekok@sci.u-toyama.ac.jp

Andreas Kroh
Natural History Museum Vienna
Vienna
AUSTRIA
andreas.kroh@nhm-wien.ac.at

Haruko Kurihara
University of Ryukyus
Okinawa
JAPAN
harukoku@e-mail.jp

Miles Lamarc
University Of Otago
Otago
NEW ZEALAND
miles.lamare@otago.ac.nz

David Lane
Universiti Brunei Darussalam
Brunei Muara District
BRUNEI
davelane@fos.ubd.edu.bn

Mark Langdon
Murdoch University
Perth, Western Australia
AUSTRALIA
m.langdon@murdoch.edu.au

John Lawrence
University of South Florida
Florida
USA
lawr@cas.usf.edu

Scott Ling
University Of Tasmania
Hobart, Tasmania
AUSTRALIA
sdling@utas.edu.au

Catriona Macleod
University of Tasmania
Hobart, Tasmania
AUSTRALIA
catriona.macleod@utas.edu.au

Regina Magierowski
University of Tasmania
Hobart, Tasmania
AUSTRALIA
reginam@utas.edu.au

Jérôme Mallefet
University of Louvain
Brabant
BELGIUM
jerome.mallefet@uclouvain.be

Loisette Marsh
Western Australian Museum
Perth, Western Australia
AUSTRALIA
jane.fromont@museum.wa.gov.aa

Martin Marzloff
University of Tasmania
Hobart, Tasmania
AUSTRALIA
martin.marzloff@utas.edu.au

Midori Matsumoto
Keio University
Kanagawa
JAPAN
mmatsumo@bio.keio.ac.jp

Alaric McCarthy
University of Otago
Otago
NEW ZEALAND
alaric_mccarthy@hotmail.com

Annie Mercier
Memorial University
Newfoundland and Labrador
CANADA
amercier@mun.ca

Francois Michonneau
Florida Museum of Natural History
University Of Florida
Florida
USA
francois.michonneau@gmail.com

Stephanie Mifsud
Marine Ecology Group
NSW
AUSTRALIA
stephy_mifsud@hotmail.com

Masao Migita
Shiga University
JAPAN
migita@edu.shiga-u.ac.jp

Ashley Miskelly
Urchinology
NSW
AUSTRALIA
seaurchins1@optusnet.com.au

Ryota Mizui
Yokohama National University
Kanagawa
JAPAN
d05ta009@ynu.ac.jp

Valerie Morris
University of Sydney
NSW
AUSTRALIA
valm@mail.usyd.edu.au

Truls Moum
Bodø University College
Bodø
NORWAY
truls.moum@hibo.no

Yuko Murata
National Research Institute of Fisheries Science
Kanagawa
JAPAN
betty@affrc.go.jp

Masahiro Naruse
Keio University
Kanagawa
JAPAN
masahiro_naruse@yahoo.co.jp

Kate Naughton
Museum Victoria
Melbourne, Victoria
AUSTRALIA
kmnaughton@gmail.com

James Nebelsick
University Of Tübingen
Baden-Württemberg
GERMANY
nebelsick@uni-tuebingen.de

Kazuya Nemoto
Kanagawa University
Kanagawa
JAPAN
r200870189aw@kanagawa-u.ac.jp

Hong Dao Nguyen
University of Sydney
Sydney, NSW
AUSTRALIA
hong@anatomy.usyd.edu.au

Tim O'Hara
Museum Victoria
Melbourne, Victoria
AUSTRALIA
tohara@museum.vic.gov.au

Masanori Okanishi
National Museum of Nature and Science
Tokyo
JAPAN
okanishi@kahaku.go.jp

P. Mark O'Loughlin
Museum Victoria
Melbourne, Victoria
AUSTRALIA
pmo@bigpond.net.au

Akihito Omori Misaki
The University of Tokyo
Kanagawa
JAPAN
omori@mmbs.s.u-tokyo.ac.jp

Hugh Pederson
Myriax Software P/L
Hobart, Tasmania
AUSTRALIA
hugh.pederson@eonfusion.myriax.com

Kevin Redd
University of Tasmania
Hobart, Tasmania
AUSTRALIA
ksredd@utas.edu.au

Mike Reich
University Of Goettingen & Goettingen
    Geoscientific Museum
Niedersachsen
GERMANY
mreich@gwdg.de

Kate Roberts
University of Tasmania
Hobart, Tasmania
AUSTRALIA
kate.roberts@utas.edu.au

Lana Roediger
Flinders University
Adelaide, South Australia
AUSTRALIA
lana.roediger@flinders.edu.au

Daniel Runcie
Duke University
North Carolina
USA
der7@duke.edu

Jo-Anne Ruscoe
Fisheries Research & Development Corporation
Canberra, ACT
AUSTRALIA
jo-anne.ruscoe@frdc.com.au

Michael Russell
Villanova University
Philadelphia, Pennsylvania
USA
michael.russell@villanova.edu

J. Craig Sanderson
University of Tasmania
Hobart, Tasmania
AUSTRALIA
craig.sanderson@utas.edu.au

Mary Sewell
University of Auckland
Auckland
NEW ZEALAND
m.sewell@auckland.ac.nz

Daisuke Shibata
University of Toyama
Toyama
JAPAN
d0771301@ems.u-toyama.ac.jp

Jessica Skarbnik López
Universidad Autónoma De México
Distrito Federal
MEXICO
tambja@hotmail.com

John Starmer
Florida Museum Of Natural History
Florida
USA
jstarmer@yahoo.com

Sabine Stöhr
Swedish Museum of Natural History
Stockholm
SWEDEN
sabine.stohr@nrm.se

Elisabeth Strain
University of Tasmania
Hobart, Tasmania
AUSTRALIA
estrain@utas.edu.au

Jemina Stuart-Smith
University of Tasmania
Hobart, Tasmania
AUSTRALIA
jfduraj@utas.edu.au

Ben Thuy
University of Tübingen
Baden-Württemberg
GERMANY
nebyuht@yahoo.com

Hideyuki Tominaga
Fukui Prefectural Takefu High School
Fukui Prefecture
JAPAN
hide-tom@angel.ocn.ne.jp

Simon Torok
CSIRO
Melbourne, Victoria
simon.torok@csiro.au

Toko Tsurugaya
The University of Tokyo
Kanagawa
JAPAN
toko_nyan@mmbs.s.u-tokyo.ac.jp

Tatsuya Unuma
Japan Sea National Fisheries Research Institute
JAPAN
unuma@fra.affrc.go.jp

Sven Uthicke
Australian Institute of Marine Science
Townsville, Queensland
AUSTRALIA
suthicke@aims.gov.au

Séverine Van Dyck
University of Mons-Hainaut
Mons
BELGIUM
severinevandyck@hotmail.com

Svea-Mara Wolkenhauer
CSIRO
Brisbane, Queensland
AUSTRALIA
swolkenhauer@hotmail.com

Gregory Wray
Duke University
North Carolina
USA
gwray@duke.edu

Sharon Yeo
Murdoch University
Perth, Western Australia
AUSTRALIA
sharon.yeo@sjog.org.au

Kazuya Yoshimura
Tokyo Institute of Technology
Tokyo
JAPAN
yoshimura.k.ad@m.titech.ac.jp

Alexander Ziegler
Freie Universität
Berlin
GERMANY
aziegler@zoosyst-berlin.de

Printed and bound by CPI Group (UK) Ltd, Croydon, CR0 4YY

18/10/2024

01776219-0004